SIXTH EDITION

PHYSICAL SCIENCE

bju press®

Greenville, South Carolina

PHYSICAL SCIENCE Teacher Edition
Sixth Edition

Coordinating Writer
David M. Quigley, MEd

Writers
Christopher D. Coyle
Jeff S. Foster, MS
Elwood Groves, MA

Biblical Worldview
Bryan Smith, PhD
Tyler Trometer, MDiv

Academic Oversight
Jeff Heath, EdD
Rachel Santopietro, MEd
Michael Winningham, MA

Project Editor
Rick Vasso, MDiv

**Cover, Design, and
Interior Concept Design**
Sarah Lompe

Page Layout
Carrie Walker

Illustration
Sarah Lompe

Permissions
Tatiana Bento
Lily Kielmeyer
Hannah Labadorf
Rita Mitchell
Ashleigh Schieber
Elizabeth Walker

Project Coordinator
Chris Daniels

Photo credits appear on pages 581–85.

Teacher Edition Photo Credits: **xii-xiii**, **xix** c DrAfter123/DigitalVision Vectors/Getty Images; **xvi-xvii** Mykyta Dolmatov/iStock/Getty Images; **xviii** Abscent84/iStock/Getty Images; **xix** t,b AlisaRut/iStock/Getty Images

The cover photo shows a close-up of paint in water.

© 2020 BJU Press
Greenville, South Carolina 29609
Fifth Edition © 2014 BJU Press
First Published © 1983 BJU Press

Printed in the United States of America
All rights reserved

ISBN 978-1-62856-508-9

15 14 13 12 11 10 9 8 7 6 5 4 3 2 1

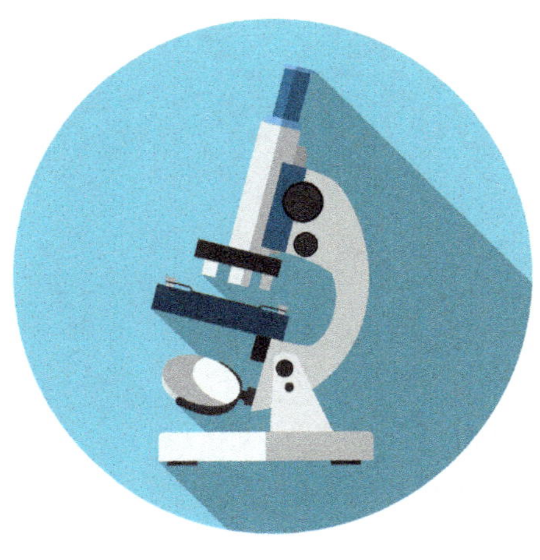

APPENDIXES

TAKING OFF WITH PHYSICAL SCIENCE!

Have you ever thought about how pilots get all the experience they need to safely fly in all conditions? Pilots obviously need actual flight time for training, but there are many things, such as aircraft malfunctions and severe weather, for which a real airplane is not the ideal classroom! That is where *flight simulators* come in. For a simulator to be realistic, the modeling of aircraft movement has to be accurate. Programmers have to model the behavior of the aircraft as well as the aircraft's response to weather conditions and other inputs. By using the computer models in flight simulators, we can prepare pilots to handle both routine and extreme conditions. There's a lot of science and technology that goes into making these simulators realistic enough to prepare pilots for anything they might encounter.

But what exactly is science? You might think that it's discovering truth about the world around us. In reality it is about *creating models* that seek to explain and describe what we observe in the world. You're probably familiar with physical models, perhaps of ships, airplanes, or even atoms. But models can also be conceptual or mathematical—they can even be computer simulations.

PHYSICAL SCIENCE 6th Edition will provide you with the foundation for further study in the fields of chemistry and physics. These two sciences are ones that you do all the time, though you may not realize it. Chemistry impacts your life when you eat, when you use cleaners, and when you cook. Physics is instrumental in playing sports, doing art, and playing an instrument. In your study of physical science, you will learn of the models that scientists have developed and modified throughout history to better explain and describe the world around them. This study of physical science will have you answer the following questions.

1. How do scientists use models in science?

2. What are the limitations to models in science?

3. How accurately have scientific theories predicted results throughout history?

4. What does the periodic table tell me about the physical and chemical properties of elements?

5. How are energy and matter related?

6. How can I use models to describe, explain, and analyze chemical and physical systems?

7. How can I use biblical principles, outcomes, and motivations to decide what is right or wrong in the field of physical science?

So make sure your seat belts are securely fastened and prepare to launch into our study of physical science!

TAKE A PEEK INSIDE!

We've designed this textbook with you in mind. We hope it will help you appreciate the wonders of God's creation even more. Flip through the following pages to see the features that we've included to help you succeed in *Physical Science*. In the back of the book you'll find appendixes, a glossary, an index, and the periodic table of the elements.

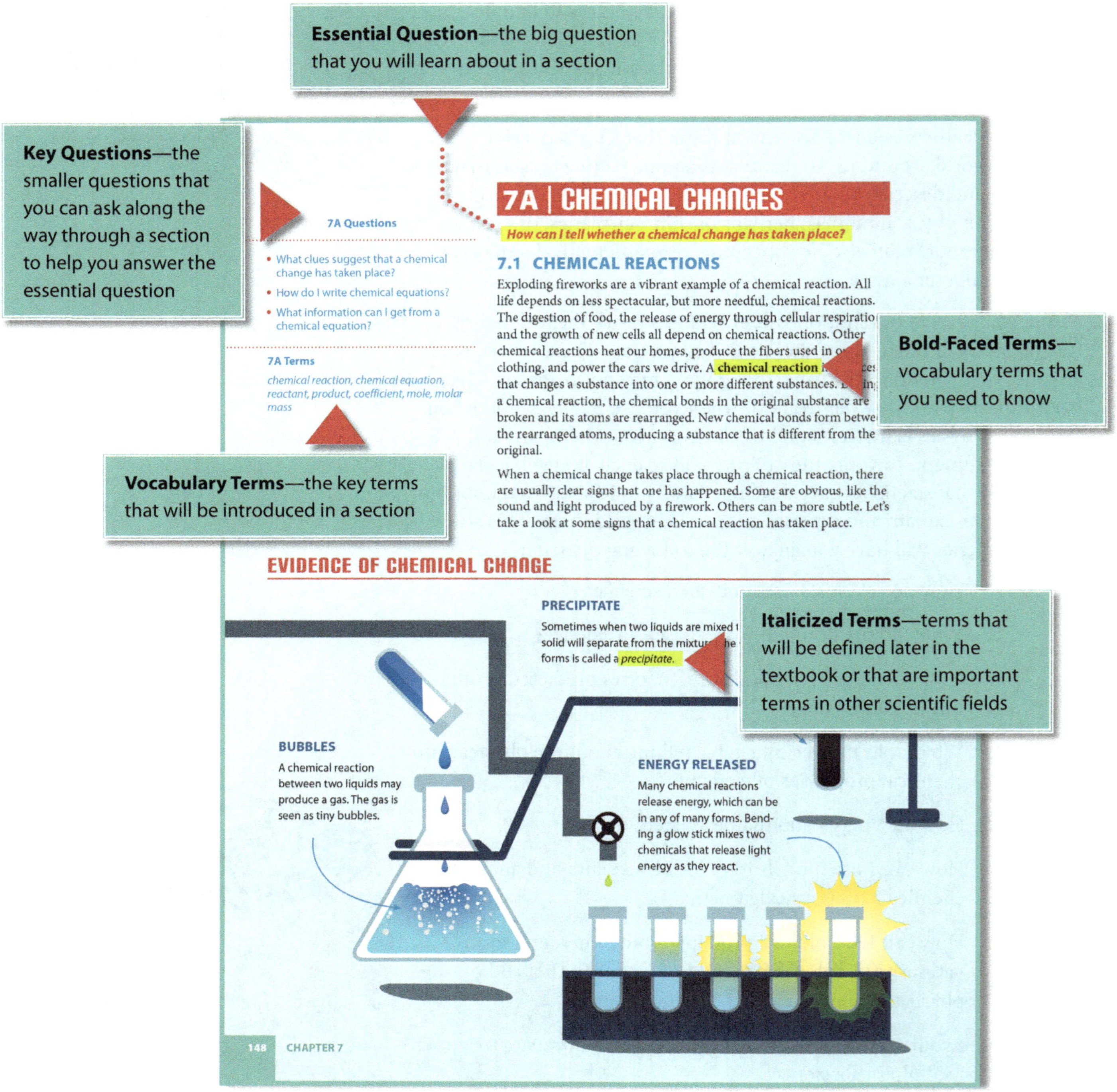

Essential Question—the big question that you will learn about in a section

Key Questions—the smaller questions that you can ask along the way through a section to help you answer the essential question

Bold-Faced Terms—vocabulary terms that you need to know

Vocabulary Terms—the key terms that will be introduced in a section

Italicized Terms—terms that will be defined later in the textbook or that are important terms in other scientific fields

7A | CHEMICAL CHANGES

How can I tell whether a chemical change has taken place?

7A Questions

- What clues suggest that a chemical change has taken place?
- How do I write chemical equations?
- What information can I get from a chemical equation?

7A Terms

chemical reaction, chemical equation, reactant, product, coefficient, mole, molar mass

7.1 CHEMICAL REACTIONS

Exploding fireworks are a vibrant example of a chemical reaction. All life depends on less spectacular, but more needful, chemical reactions. The digestion of food, the release of energy through cellular respiration, and the growth of new cells all depend on chemical reactions. Other chemical reactions heat our homes, produce the fibers used in our clothing, and power the cars we drive. A **chemical reaction** is a process that changes a substance into one or more different substances. During a chemical reaction, the chemical bonds in the original substance are broken and its atoms are rearranged. New chemical bonds form between the rearranged atoms, producing a substance that is different from the original.

When a chemical change takes place through a chemical reaction, there are usually clear signs that one has happened. Some are obvious, like the sound and light produced by a firework. Others can be more subtle. Let's take a look at some signs that a chemical reaction has taken place.

EVIDENCE OF CHEMICAL CHANGE

PRECIPITATE

Sometimes when two liquids are mixed, a solid will separate from the mixture. The solid that forms is called a *precipitate.*

BUBBLES

A chemical reaction between two liquids may produce a gas. The gas is seen as tiny bubbles.

ENERGY RELEASED

Many chemical reactions release energy, which can be in any of many forms. Bending a glow stick mixes two chemicals that release light energy as they react.

148 CHAPTER 7

Worldview Sleuthing Boxes—inquiry-based investigations that help you think through current topics in physical science through the lens of Scripture

Chapter Opener—a short article that highlights issues and developments in physical science that demonstrate how science intersects with your life

CHAPTER 2
Matter

IMITATING GECKOS

Geckos have a superpower: defying gravity. The ability of the gecko on the left is not due to its feet being sticky; it's because they are hairy. The tiny hairs on its feet are about 1/30th the diameter of a human hair, and there are thousands of them on every square millimeter of a gecko's feet! With each step, the gecko spreads his toes, maximizing the contact area of each foot. This action creates a tiny electrostatic attraction with the surface. The combination of these tiny forces creates a substantial attractive force. Happily for the gecko, God also designed geckos so that they can release that attraction!

Materials scientists are studying the gecko to see whether they can copy them for a variety of applications. Some scientists are working on gecko

2A Unders
    Matter
2B Classify
2C States
2D Change

Bulletproof technology is an extension of the ancient art of armor making. Armor proved ineffective against modern weaponry, and metal suits are too heavy for the modern battlefield. Materials scientists started to investigate cloth materials. In the late 1800s, a doctor in Arizona discovered that silk could keep a bullet from penetrating the body, but silk was also extremely expensive. Over time scientists searched for other natural and synthetic (manmade) fibers with similar properties. Today scientists are looking at other applications for these materials.

TASK

You are working as a clothing designer for the nation's Olympic sports team. You have been assigned to research advances in ballistic materials to be used in clothing that will perform well and protect speed skaters from sharp skate blades during accidents. You are to write a one-page proposal for a material to be used for these uniforms.

4. Write your proposal and show it to another person for feedback.

CONCLUSION

We are commanded in Scripture to serve others. We do this by meeting the needs of people around us. Developing materials such as these can protect the lives of athletes. Could they also help protect the lives of military and law enforcement personnel?

CHAPTER 2 REVIEW

2A

UNDERSTANDING MATTER

* Matter is anything that has mass and takes up space. According to the particle model of matter, all matter is made of tiny particles (atoms and molecules) in constant random motion.
* The particle model of matter is very workable as it explains most of our observations about matter.
* Density is calculated by dividing the mass of an object by its volume.
* Mass is the amount of matter in an object, while weight is the force of gravity acting on that object.

2A Terms

matter	26
law of definite proportions	28
particle model of matter	28
atom	29
molecule	29
mass	29
volume	29
density	29
weight	30

CLASSIFYING MATTER

* We classify matter by its physical and chemical properties.
* Matter is classified as either a pure substance or a mixture.
* Pure substances (elements and compounds) contain only one type of substance.
* Mixtures (heterogeneous or homogeneous) are physical combinations of two or more substances in changeable proportions.

2B Terms

pure substance	33
element	33
compound	33
mixture	34
heterogeneous mixture	34
homogeneous mixture	34

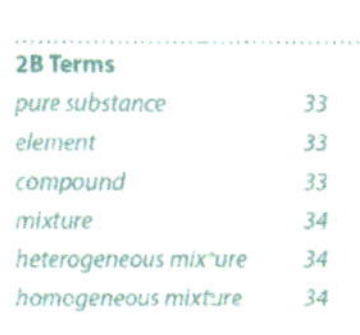

STATES OF MATTER

* Particles in solids have low kinetic energy compared with the forces between particles. Solids have fixed shape and volume, high density, and low compressibility due to their close spacing and low energy.
* Liquid particles have enough kinetic energy to overcome some of the forces between particles. Liquids have fixed volume, high density, low compressibility due to their close spacing and low energy. Their ability to move makes them fluid and gives them a changeable shape.

* Gas particles have sufficiently high energy to overcome all the forces between them. The particles' wide spacing causes gases to have low densities and high compressibility. Their rapid motion allows them to fully occupy their container, no matter its shape or volume.

2C Terms

solid	35
liquid	35
gas	35
plasma	35

2D | REVIEW QUESTIONS

a chemical or a physi-

(III) oxide.

4. Explain what is happening to the particles in a solid as you warm it to its melting point.
5. Define *boiling point*.
6. Compare evaporation and boiling.
7. State the law of conservation of matter.
8. List the changes of state that involve adding energy.

rrent information y by doing keyword allistic materials,"

ar ma- arches ld

ct any required cite your sources.

Chapter Summary—handy statements of the big ideas of the chapter, including vocabulary lists

REVIEW

22. What is the percent by volume of a solution made by mixing 345 mL of ethanol with enough water to form 1325 mL of solution?

23. Create a concept map using the terms *solute, solvent, solution, solubility, concentration, concentrated, dilute, unsaturated solution, saturated solution, supersaturated solution, percent by mass, percent by volume,* and *molarity.*

Critical Thinking

24. Where would saturated, unsaturated, and supersaturated solutions be on the graph on page 220?

25. Calculate the percent by mass of 155 g of a solution that contains 142 g of solvent.

26. How many grams of salt should be added to 100 g of water to make a 14.5% salt solution?

27. Show how you could convert a percent by mass to a percent by volume. (*Hint:* Remember that mass and volume are related by density [$d = m/V$].)

28. Some classmates tell you that if they mix 100 mL of a 3.5 M sugar solution with 100 mL of a 5.0 M sugar solution, they will have 200 mL of an 8.5 M solution. Are they correct? Explain.

Use the Case Study at right to answer Questions 29–32.

29. Which material is the solvent and which is the solute in the sap?

30. What process is being used to separate the water from the sucrose in the sap?

31. What is the percent by mass of sucrose when done?

32. Why does the boiling point rise while the water is boiling off?

Use the Ethics Box below to answer Question 33.

33. Using the strategy presented in Chapter 3, write a one-page essay about how Christians should approach this issue.

ETHiCS — POLLUTION

The Issue: Air Pollution

Air pollution is an issue in many areas around the world. The compounds that cause air pollution come from sources that are both natural (volcanoes, plants and animals, and forest fires) and manmade (fossil fuel power plants, motor vehicles, and aerosols). While most people think of air pollution being outdoors in major urban areas, air pollution can be indoors also. The impact of this pollution is widespread. Air pollution can damage both the natural environment and manmade structures. The health impact is enormous, increasing the number of cases of asthma and allergies as well as lung and heart disease. The World Health Organization estimates that air pollution causes about 7 million deaths each year.

Review Questions—questions that will have you recall facts, demonstrate your understanding of concepts, and cause you to use critical thinking

CASE STUDY: MAPLE SYRUP

Many people love the unique taste of maple syrup, which is a mixture made of primarily sucrose and water. The syrup is made from the sap of the sugar maple tree. The sap is collected in late winter and early spring. All of the collected sap is then boiled, removing much of the water. The process is complete when the boiling point has risen to 104.1 °C. The final syrup contains about 60 g of sucrose for every 100 g of syrup.

Case Studies—opportunities to apply what you have learned in physical science to a real-life example

Career Boxes—information about careers in physical science that can be pursued to wisely use God's world and help people

Ethics Boxes—opportunities to apply a biblical worldview to ethical issues in physical science

SERVING AS A PIPING ENGINEER: KEEPING THE FOOD FLOWING

Natural gas, ketchup, ice cream—each is a fluid product that is transported through a complex system of pipes. Piping engineers are mechanical or materials engineers who specialize in designing pipe systems. Designing a facility with different-sized vats, storage tanks, and a maze of pipes connecting them is not as simple as it may sound. Many factors come into play, such as the temperature and viscosity of the product, pressure changes that occur when the diameter of a pipe changes, and choosing the most efficient route for pipes through a plant. Pipes full of liquid can be heavy, too, so the design must include sufficient support for all the equipment.

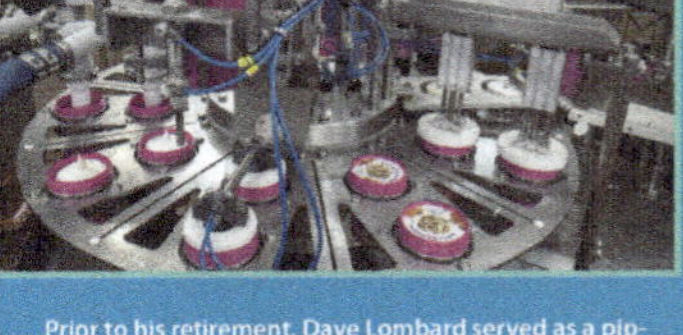

Prior to his retirement, Dave Lombard served as a piping engineer in California's agriculturally rich Central Valley. He helped design many piping systems used in the processing and packaging of foods and beverages. Dave says, "I really enjoy food and beverage work because everyone needs to eat. It's very fulfilling to be a part of that!" If you like challenging work, a career in piping engineering might be a good choice for you!

You might expect that faster-flowing fluids have higher fluid pressures, but in fact they don't. They actually have lower fluid pressures. This relationship between the increasing speed of a fluid and its decreasing pressure is known as **Bernoulli's principle**. A Swiss mathematician, Daniel Bernoulli, described the phenomenon in 1738.

Bernoulli's principle can be put to many practical uses. It partly accounts for the lift generated by aircraft wings. A curved upper wing surface is shaped so that air flows faster over the top of the wing, creating lower pressure on that side. Hose-end sprayers, used for applying fertilizers to lawns and gardens, operate on Bernoulli's principle too. Water passing over a tube inside the sprayer creates low pressure that draws the fertilizer up the tube and into the stream of water. The hourglass shape of a de Laval nozzle creates a region of low pressure in the exhaust gas of a rocket motor. The low pressure increases the speed of the vented gas, providing more thrust for a rocket.

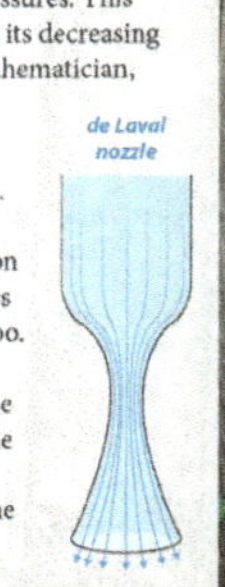

16C | REVIEW QUESTIONS

1. How does Pascal's principle explain the operation of a hydraulic lift?

2. What quantity remains constant within a fluid system regardless of the fluid's velocity or pressure?

3. State Bernoulli's principle.

4. The powerhead shown at right is a type of aquarium pump that circulates water. If a piece of plastic tubing is inserted into the discharge pipe on the powerhead, a stream of air can be drawn into the flowing water. Use Bernoulli's principle to explain how this is possible.

BENDING LIGHT

Rainbows are caused when white light from the sun is refracted and reflected by tiny water droplets in the atmosphere. The different colors within white light don't bend at the same rate, so they separate as they bend. You've probably seen a prism used to produce the same kind of separation. But once the colors are separated, is it possible to recombine them into white light again? Your teacher will use a light ray box and several kinds of lenses to help you think about this question.

Essential Question:
Can a rainbow be undone?

1. Do the colors of light separated by the prism follow the progression described in Subsection 22.1?

2. On the basis of what you observe, which color of light is bent the most and which the least? How can you tell?

3. Do you think it will be possible to recombine the separated colors back into white light using a lens? If so, which lens do you think will work?

4. Describe how each kind of lens affects the separated colors.

CONCLUSION

Different kinds of lenses differ in how they bend light rays. As you read further, you'll learn about these different lenses and see how they can be used to benefit people.

Mini Labs—short hands-on activities to get you thinking and working like a scientist

---

In the example on the right, you can see how light bends as it passes through different media. The ray first bends as it passes from air into water because water's index of refraction is greater than air's. The ray is bent again as it passes into glass because glass has a greater index of refraction than water. Because the boundaries in this example are parallel, the ray returns to its original direction of travel as it passes out of the glass and back into air.

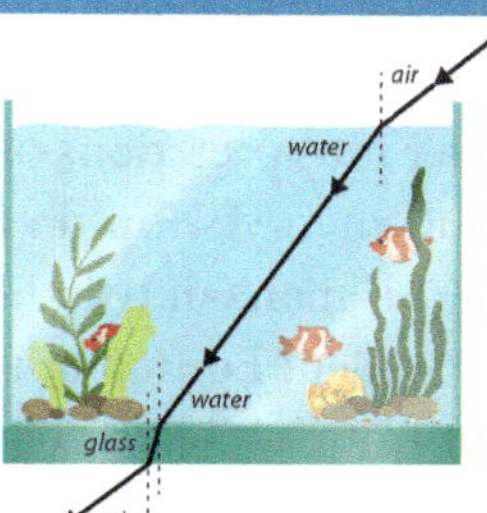

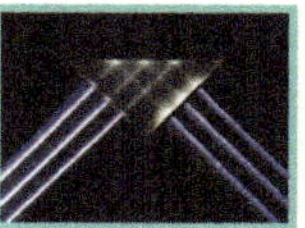

Total Internal Reflection

In some situations, a ray of light *can't* pass from one medium to another. Instead, the ray reflects off the boundary between the two media and remains within the first medium. This phenomenon is called **total internal reflection** (left). It happens when a light ray's angle of incidence exceeds a certain critical value when going into a medium with a lower index of refraction.

18C | USING SOUND WAVES

In what other ways do we use sound waves?

18.7 SOUND TECHNOLOGIES

Acoustic Amplification

Since sound is a form of energy, it makes sense that such energy can be put to good use. And indeed, not only people but animals too use sound energy for many purposes. In this section, we'll explore some of those uses.

Acoustic amplification is the process of making a sound louder. A megaphone is a simple kind of amplifier known as an *acoustic horn*. Normally when a person speaks, the sound waves of his voice spread out over a large area. A megaphone focuses that sound energy into one specific direction, making it sound louder. If you point the large end of a megaphone at the source of a sound and listen at the smaller end, the same thing happens. This is why people sometimes cup their ears to hear better. Amplification also happens in any enclosed space with hard surfaces, where sound reflection occurs easily.

How It Works—examples of physical science applications in everyday items

● HOW IT WORKS

Speakers

Stereo speakers are described with some strange terms. Woofers, midrange, tweeters—just what do these things mean? And how does a speaker work anyway? A speaker contains a cone (1) made of stiff material connected to a voice coil (2) made of wound copper wire. The voice coil is suspended inside a powerful circular magnet (3). An amplified electrical signal enters the voice coil, creating, in essence, an electromagnet (see Chapter 21). The electrical signal causes the voice coil's magnetic field to rapidly change polarity many times per second. As this rapidly changing field interacts with the field of the magnet, the voice coil vibrates back and forth. The vibrating speaker cone pushes against the air, creating sound waves.

The size of the cone relates to the frequency of the tones produced. Large cones are needed to produce deep, low-frequency sounds—the bass end of what humans can hear. These speakers are the "woofers" in a multi-cone speaker. "Tweeters" are the small cones that produce high-pitched, high-frequency sounds. Large speakers may have three or more cones, including mid-range cones. Such speakers are better able to re-create the variety of tones carried by an electrical signal. Single-cone speakers are limited in the range of frequencies that they can produce. They are often unable to produce some low- and high-end frequencies well.

THIS IS FOR YOU!

Welcome to BJU Press *PHYSICAL SCIENCE* 6th Edition! Your students are about to discover and explore the incredible richness of the universe that God has created. During this journey, students will find that people's views of physical science vary drastically, depending on worldview—the overarching narrative that a person uses to see and interpret the world. And everyone has one. It shapes our beliefs and values, and each of us will make choices in life that are based on our worldview.

Since scientists who do physical science all have worldviews, conflicts arise about their interpretations of what they observe. What is the origin of matter and energy? How do we determine the proper use of a new technology? What is man's duty in his treatment of the rest of creation? *PHYSICAL SCIENCE* Student Edition 6th Edition helps students grapple with these questions from a biblical standpoint.

Though a good textbook is important, you the teacher are essential to successful learning. Your job is to guide your students to learn through the lens of the Bible's narrative. Science class should be a time of discovery for both you and them. Your emphasis on glorifying God by obeying the Creation Mandate and helping other people through science will keep science exciting and will impact your students. These are the most important lessons that you can teach.

Make this course fit the needs of your students. *PHYSICAL SCIENCE* Student Edition and *PHYSICAL SCIENCE* Student Lab Manual both include more material than most teachers will be able to cover in one school year. You decide whether you will spend extra time on a particular topic, lab activity, or project.

FEATURES OF THE TEACHER EDITION

You have a lot of great material at your fingertips in this resource! The Teacher Edition features reduced student pages with side and bottom margins packed with educational content. Take a look at what it offers, whether you are a classroom teacher or a home educator.

Lesson Plan Overview

The Lesson Plan Overview beginning on page xx is your year-at-a-glance schedule for teaching physical science. This has been greatly enhanced for the 6th Edition, integrating the accompanying Student Edition, Student Lab Manual, and Teacher Lab Manual to give you an idea of how much time you should spend on each chapter. The table also includes the essential question, objectives, and resources for each section. The schedule for each chapter includes a day of review and a test day.

Teaching the Material

Each section begins with *Teaching the Material* in the bottom margin. This provides an overview of the section. It includes the essential question, objectives, resources, and research-based teaching strategies. There are more teaching strategies than you will have time for, providing you the freedom and flexibility to choose those that best fit your students.

Teacher Edition Margin Notes

Side margin notes provide ideas, background, or support materials for the Student Edition. Many of the margin notes are expanded information for the teaching strategies in the *Teaching the Material* section. These will help you make your teaching come alive. Each chapter begins with an outline of the subsections, which can help you decide whether you want to skip some of the material. Doing so will depend on your schedule and your students' abilities.

Margins also contain the answers to review questions.

Most margin notes are annotated with special icons to alert you to the type of note as described below. Notes that are not accompanied by an icon are additional information that you may find helpful in teaching the material.

Biblical Worldview Shaping—highlights worldview-shaping discussion questions and activities

Demonstration—provides materials and procedures for successful in-class demonstrations

Differentiated Instruction—provides suggestions to help students that need additional support (remediation or enrichment).

Difficult Concept—points out common misconceptions or concepts that students may find difficult

Equipment—highlights notes and warnings about equipment used in different activities

Helpful Tip—provides suggestions for improving students' understanding or lesson management

Lab—flags where a Student Lab Manual activity best fits with the information in the chapter

Math—connects physical science to concepts that are often taught in math classes

Recall—highlights topics from earlier chapters or courses that you may want to review to integrate and reinforce concepts

Resources—points out additional resources that might help in teaching the material, including web resources accessed through BJU Press *PHYSICAL SCIENCE* 6th Edition weblinks at TeacherTools Online.com.

Safety—indicates that a demonstration has special safety considerations that you should consider before you attempt it

Teaching Strategies Icon—indicates notes that suggest ways to apply research-based educational strategies in your classroom, including group activities and discussions, activating prior knowledge, active learning, and formative assessments

Teacher Resources

LAB MANUALS

The Student Lab Manual includes thought-provoking lab activities tied directly to the content of each chapter in the Student Edition. The Teacher Lab Manual provides all the instructions for conducting the activities, including information on the materials and equipment needed; answers to lab activity questions; and example data where applicable. Plan your lab activities on the basis of the equipment that you have and the needs of your students.

New in 6th Edition are inquiry and STEM labs. Inquiry labs provide opportunities for students to direct the investigation of a topic. STEM labs relate to the design process of asking a scientific question, generating ideas, planning a solution, creating a device to solve the problem, testing the solution, and then improving the design as needed. Because inquiry and STEM labs are more open-ended, the Teacher Lab Manual includes a teacher guide for each of these activities.

A companion science equipment kit is available from Logos Science. For more information visit **bjupress.com/category/science-secondary-christian -school-curriculum**.

ONLINE RESOURCES

Resources formerly available on the Teacher Toolkit CD are still available online.

RESOURCES AVAILABLE AT TEACHER TOOLS ONLINE
(PURCHASED SEPARATELY)

- visuals
- video content
- PowerPoint™ presentations
- PDF version of the Teacher Edition
- ExamView™ testing software with editable test database

For more information, visit **TeacherToolsOnline.com**.

PRINTED ASSESSMENTS PACKAGE
(PURCHASED SEPARATELY)

- Quizzes and tests with answer keys

NEW FOR 6TH EDITION

If you have used previous editions of PHYSICAL SCIENCE Student Edition, you will notice some changes in the 6th Edition.

- The writing style, vocabulary, and amount of content have been refined to better facilitate student learning at a ninth-grade level.

- The text focuses on big ideas identified by essential questions.

- Each chapter is identified as foundational, key, or enrichment. Foundational chapters are crucial for students to develop a basic understanding of physical science. Key chapters are very important and may contain material that appears on standardized tests. You may skip enrichment chapters with little adverse effect on students; use them if a significant portion of your class is interested or if you have gifted students who need differentiated instruction.

- Case studies, worldview sleuthing (webquests), How It Works boxes, and career boxes engage student interest.

- Expanded examples model the problem-solving process and provide necessary scaffolding for student mastery of the content.

- Every chapter in the Student Edition includes a mini lab. These are small lab activities that help reinforce the objectives of the section and generally require less time and fewer materials than the lab activities in the Student Lab Manual.

- Ethics boxes found in many chapters present ethical dilemmas in physical science. Students formulate a biblical understanding of the issues and apply it to issues. Students are provided with a good deal of support in the first third of the Student Edition and become more independent as they continue through the remainder of the textbook.

- Each chapter ends with a chapter review that presents in a few brief sentences the primary concepts of each section. This is followed by a set of chapter review questions. The reviews are divided into questions that require students to recall facts, those that compel students to demonstrate a thorough understanding of concepts, and those that require them to apply critical thinking.

- In *PHYSICAL SCIENCE* Teacher Edition 6th Edition, the Lesson Plan Overview has been expanded to include essential questions, objectives, and resources.

- Also in the Teacher Edition, *Teaching the Material* information for each section, located in the bottom margin, offers ideas for research-based teaching strategies.

- No Teacher Toolkit CD is included in the Teacher Edition, but many resources are available online or may be purchased separately in printed form. See Online Resources above.

BIBLICAL WORLDVIEW SHAPING

Developing a Biblical Worldview with BJU Press *PHYSICAL SCIENCE* 6th Edition

Is the study of matter and energy worldview neutral?

As we examine the five biblical worldview themes to answer the questions below (and many more), we will find that a biblical worldview is crucial to a proper understanding of physical science. The answers to these questions not only make a student more confident in his Bible, they also enable him to be a more effective student scientist.

ORDER IN NATURE

Scientists agree that everywhere we look in nature, we see intricate order. While a biblical worldview holds that the universe is a product of God's speech, modern science explains the universe's order without any reference to the Creator.

- ***How should a Christian respond to a secular view of the natural order?***

- ***How should a Christian understand the difference between mere order and creational design?***

Implementation of the biblical worldview themes in their respective chapters

CH 1	CH 4
E	E

See key at right

KEY

R—**Recall** biblical teaching

E—**Explain** biblical teaching

EV—**Evaluate** controversial concepts

F—**Formulate** a Christian understanding of a controversial concept

A—**Apply** a Christian understanding to life

SCIENCE AND DOMINION

Mankind's calling to subdue the earth explains why physical science is vital for human flourishing. Sadly, secularists strive to disconnect this God-given command from the day-to-day work of science.

- *Is there a difference between Christian and secular views of the purpose of science?*

- *How can the mundane work of science declare the glory of God?*

CH 1	CH 8	CH 9	CH 13	CH 14	CH 18
E	EV	EV	F	A	A

MODELING

The work of physical science is the work of creating and using models. Models are simplified representations whose purpose is to produce useful predictions, not absolute truth. In our secular society, however, it is common for people to assume that science is able to produce final answers about our world.

- *What are the limits of science?*

- *How can understanding science as modeling help the student defend the Bible and become a better student scientist?*

CH 1	CH 2	CH 3	CH 11	CH 15	CH 21
R	E	E	E	EV	F

PURSUIT OF SCIENCE

God has built laws into nature. He has also revealed norms that should govern our scientific work and the way that we live our lives. Fallen man, on the other hand, takes these laws and norms and uses them for his own purposes. This practice distorts the pursuit of science.

- *How do secular perspectives distort the work of science?*

- *How can a Christian pursue science according to God's plan?*

CH 1	CH 10	CH 19	CH 22
E	E	F	F

SCIENCE AND ETHICS

Because humans bear God's image, we must learn to apply biblical ethics to our practice of physical science. Scientific naturalism, however, sees things differently. It claims that humans are important because we have been naturally selected to survive and to become the most complex.

- *How should a Christian respond to secular thinking about ethics?*

- *What is the biblical pattern for answering the ethical questions raised in science?*

CH 1	CH 3	CH 5	CH 6	CH 8	CH 9	CH 11	CH 12
E	E	A	E	EV	F	F	F

ACADEMIC RIGOR

Building Academic Rigor
with BJU Press *Physical Science* 6th Edition

How do I ensure that my instruction contains sufficient rigor?

1. DESIRED LEARNING OUTCOMES

What do I want students to learn?

- reading skills in informative texts with discipline-specific vocabulary
- 21st century skills—collaboration, creativity, critical thinking, problem-solving, technology literacy
- perseverance in problem-solving
- the use of the design process
- the connection of science to other disciplines
- the skill to create and test hypotheses
- the use of consistent scientific and mathematical precision and reasoning
- a biblical worldview development regarding the themes of orderliness in nature, physical science for dominion, the role of modeling in physical science, the pursuit of science, and physical science and ethics

2. TEACHING AND LEARNING SUPPORTS

How will I help students learn?

- by asking essential questions
- by modeling skills
- by assigning regular practice of learned skills
- by using integrated studies
- by eliciting and correcting misconceptions
- by using scaffolded Socratic dialogue
- by encouraging collaborative learning
- by assigning STEM learning experiences that require the use of the design process
- by assigning inquiry lab activities

3. EVIDENCE OF UNDERSTANDING

How do I know whether students have learned?

- They can answer high-level questions that require critical thinking.
- They can complete frequent and varied preassessments, formative assessments, and summative assessments.
- They can link representations and phenomena by using and creating models.
- They can create graphic organizers.
- They can make a visual analysis of scientific images and diagrams.
- They can draw conclusions about a scientific question on the basis of data analysis.
- They can design products to solve real-world problems.
- They can formulate biblical positions on ethics and on the nature and work of science.

A Panorama of Academic Rigor

THE TEACHER'S ROLE

The teacher is the person who crafts the proper learning environment. He creates a learning community with high educational expectations, ignites interest and passion to help students persevere in challenging educational tasks, and inspires them by providing models of learning. He instructs, supports, critiques, and praises the efforts of the growing learners. Then he helps students transfer their knowledge, understanding, and skills to life with the purpose of shaping their worldviews.

THE LEARNING ENVIRONMENT

Learning happens in a context. Physical and social environments create this context. For students to be stimulated and engaged and to perform at high levels, they need an environment that connects with and shapes their interests and values. Students need a flexible environment that adapts to remedy their deficits and capitalize on their proficiencies. They need an interactive environment that intrinsically motivates them by giving them structure, freedom, and choices in learning. They need a safe environment that invites questioning, risk taking, and honest discussion. Most importantly, students need positive relationships with mentors and peers and one-on-one time with caring, responsible instructors and parents. This kind of environment does not happen by accident; an educational climate like this must be crafted.

THE LEARNING EXPERIENCE

BJU Press *PHYSICAL SCIENCE* 6th Edition opens doors to the world of the learning experience. It captures student interest and significantly contributes to the educational environment. Educational and technological resources enable the teacher to develop knowledge, understanding, and skills in a scaffolded sequence to build a foundation that students can transfer to life. Teachers use educational tools to craft relevant, authentic learning experiences that develop creativity and problem-solving skills in a variety of contexts. These kinds of learning experiences allow students to take responsibility for their learning and discover the joy of serving God and others with their vocational knowledge and skills.

LESSON PLAN OVERVIEW

LEGEND	SE = Student Edition	TE = Teacher Edition	SLM = Student Lab Manual	TLM = Teacher Lab Manual

Section	SE Pages	TE Pages	Teacher Resources	Essential Questions/Content Objectives
UNIT 1: THE STRUCTURE OF MATTER				
CHAPTER 1: MODELING OUR ORDERLY WORLD (10 DAYS) *FOUNDATIONAL CHAPTER*				
1A Order in Our World	4–11	4–11	Career: Serving as a Forensic Scientist	**EQ:** What is the source of order in nature? **Objectives:** 1A1 Define *physical science*. 1A2 List evidences of order in nature. 1A3 Explain why the world is orderly. 1A4 Explain the role of science in dominion.
Ethics Day	9–10	9–10	Ethics: Christian Ethics	1A5 Outline a biblical framework for science and ethics.
1B Modeling Our World	11–14	10–14		**EQ:** How do scientists do science? **Objectives:** 1B1 Define *model*. 1B2 List different models used in physical science. 1B3 Explain the relationship between hypotheses, theories, and laws. 1B4 Outline the process used to do scientific inquiry. 1B5 Explain why we approach science in an orderly fashion.
Lab Day 1	SLM 1–7	TLM 1–7	Lab 1A: *Based on a True Story*—Thinking Safe in the Laboratory	**EQ:** How can we safely investigate in the laboratory?
1C Using Mathematics for Scientific Inquiry	14–20	14–21	How It Works: Balances and Scales Mini Lab: *Understanding Conversion Factors*	**EQ:** How is math used in scientific inquiry? **Objectives:** 1C1 Compare the US customary and SI systems of measurement. 1C2 Explain the benefits of the SI. 1C3 Use scientific instruments to collect data. 1C4 Use mathematical tools to analyze data. 1C5 Compare accuracy and precision. 1C6 Explain how scientists identify the precision of measurements. 1C7 Convert between SI units.
Lab Day 2	SLM 9–14	TLM 9–14	Lab 1B: *How Do We Measure Up?*—Practicing Measuring	**EQ:** How can we get the most accurate and precise measurements?
Lab Day 3	SLM 15–20	TLM 15–20	Lab 1C: *Visual Data*—Graphing Mass and Volume	**EQ:** How much different information can I find on a graph?
Ethics Day	23	23–23a	Ethics: Reporting Scientific Data	1C8 Explain why it is important for scientists to accurately report data.
Review and Test Days			Chapter 1 Test	

Section	SE Pages	TE Pages	Teacher Resources	Essential Questions/Content Objectives
CHAPTER 2: MATTER (8 DAYS) _FOUNDATIONAL CHAPTER_				
2A Understanding Matter	26–31	26–31	Demonstrations: Matter, Diffusion, Density Mini Lab: _Measuring Volume_	**EQ:** Why is matter so important? **Objectives:** 2A1 Define _matter_. 2A2 Evaluate how well models of matter represent physical matter. 2A3 Calculate mass, volume, and density using the formula for density. 2A4 Explain the difference between mass and weight.
Lab Day 1	SLM 21–25	TLM 21–25	Lab 2A: _Has Mass, Occupies Space_—Modeling Matter	**EQ:** How are physical models created?
2B Classifying Matter	32–34	32–34	Demonstration: Types of Matter	**EQ:** How do scientists classify matter? **Objectives:** 2B1 Compare mixtures and pure substances. 2B2 Classify pure substances as elements or compounds. 2B3 Classify substances as pure substances or mixtures.
2C States of Matter	35–38	35–38	Case Study: How Many States of Matter? Demonstration: Particles in Action Career: Serving as a Materials Scientist	**EQ:** How can particles in a solid be moving? **Objectives:** 2C1 Identify the characteristics of different states of matter. 2C2 Classify different substances as solids, liquids, gases, or plasmas. 2C3 Compare the three common states of matter using the particle model of matter. 2C4 Explain how the particles in a solid can be moving according to the particle model of matter.
2D Changes in Matter	39–44	39–44	Worldview Sleuthing: Bulletproof! Demonstration: Chemical and Physical Changes	**EQ:** How does matter change? **Objectives:** 2D1 Define _physical property_ and _chemical property_. 2D2 Classify changes in matter as physical or chemical. 2D3 Summarize the law of conservation of matter. 2D4 Identify changes of state. 2D5 Relate changes of state to the flow of energy.
Lab Day 2	SLM 27–31	TLM 27–31	Lab 2B: _Something Old, Something New?_—Detecting Physical and Chemical Changes	**EQ:** How can I know whether a new substance forms when I mix two chemicals?
Review and Test Days			Chapter 2 Test	

LESSON PLAN OVERVIEW

Section	SE Pages	TE Pages	Teacher Resources	Essential Questions/Content Objectives
CHAPTER 3: THE ATOM (9 DAYS) _FOUNDATIONAL CHAPTER_				
Lab Day 1	SLM 33–36	TLM 33–36	Lab 3A: _The Fiery Trial_—Using Flame Tests	**EQ:** Why do burning mineral salts produce different colors of light?
3A The Atomic Model	50–55	50–55	Case Study: A World of Models Demonstrations: Electrostatic Charges, Cathode Rays, The Gold Foil Experiment	**EQ:** How have people thought about matter? **Objectives:** 3A1 Identify the key scientist(s) associated with each atomic model. 3A2 Summarize the key discoveries that shaped atomic theory. 3A3 Justify the continued use of the Bohr model of the atom. 3A4 Summarize our current atomic model. 3A5 Explain how workability acted as the driving force in the development of atomic models.
Lab Days 2–4	SLM 37–39	TLM 37–39b	Lab 3B: _Big Time_—Inquiring into Scale Models of Atoms	**EQ:** What would the dimensions of an atom look like if the atom were enlarged to a size that we could easily see?
3B Atomic Structure	56–62, 65	56–62, 65	Case Study: All Models Are Not Equal (p. 65) Mini Lab: _Finding the Atomic Mass of Eggogen_ (p. 63)	**EQ:** What is it like inside an atom? **Objectives:** 3B1 Describe protons, neutrons, and electrons, including their masses, charge, and location. 3B2 Relate the concept of isotopes to variations within the nucleus. 3B3 Determine the number of protons, neutrons, and electrons in an atom using isotope notation. 3B4 Calculate atomic number, mass number, and charge given the number of protons, neutrons, and electrons in an atom. 3B5 Explain how an ion forms. 3B6 Predict the relative abundance of an isotope on the basis of its average atomic mass.
Ethics Day	66–67	66–67	Ethics: Strategies and Protecting People from Radiation	3B7 Explain how we can protect people from exposure to radiation.
Review and Test Days			Chapter 3 Test	

Section	SE Pages	TE Pages	Teacher Resources	Essential Questions/Content Objectives
CHAPTER 4: THE PERIODIC TABLE (7 DAYS) _FOUNDATIONAL CHAPTER_				
4A Organizing the Elements	70–79	70–79	Mini Lab: _Organizing Elements_	**EQ:** How does the periodic table relate to elements in the real world? **Objectives:** 4A1 Identify the contributions of the key scientists associated with the development of the periodic table. 4A2 Explain how workability acted as the driving force in the development of the periodic table. 4A3 Evaluate the predictive power of the periodic table. 4A4 Identify periods, groups, and families on the periodic table. 4A5 Relate the arrangement of the periodic table to our understanding of atomic structure.
Lab Day 1	SLM 41–44	TLM 41–44	Lab 4A: _Bricks and Feathers—Exploring Element Density in a Period_	**EQ:** Are the densities of elements in a row of the periodic table predictable?
4B Classifying the Elements	80–85	80–85	Worldview Sleuthing: Giving Due Credit Demonstration: Sodium's Properties	**EQ:** How is the periodic table useful? **Objectives:** 4B1 Classify elements as metals, nonmetals, and metalloids using a periodic table. 4B2 Compare the properties of metals, nonmetals, and metalloids. 4B3 Identify the families of elements in the periodic table. 4B4 Determine the number of valence electrons in an element according to its family.
Lab Day 2	SLM 45–49	TLM 45–50	Lab 4B: _How Wide Is an Atom?—Exploring Atomic Radii Trends_	**EQ:** Do atomic radii follow a periodic trend?
4C Periodic Trends	86–90, 93	86–90, 93a	Case Study: Allotropes (p. 93)	**EQ:** What can an element's position on the periodic table tell us about the element? **Objectives:** 4C1 Write the electron dot notation for an element. 4C2 Explain why atomic radius changes as it does. 4C3 Arrange elements on the basis of atomic radius. 4C4 Explain the periodic trend in electronegativity. 4C5 Arrange elements on the basis of electronegativity.
Review and Test Days			Chapter 4 Test	

LESSON PLAN OVERVIEW

Section	SE Pages	TE Pages	Teacher Resources	Essential Questions/Content Objectives
CHAPTER 5: BONDING AND COMPOUNDS (8 DAYS)				
FOUNDATIONAL CHAPTER				
5A Principles of Bonding	96–98	96–98	Demonstration: Hydrogen	**EQ:** How do compounds form? **Objectives:** 5A1 Define *chemical bond*. 5A2 Compare elements and compounds. 5A3 Explain how a molecule can be an element or a compound. 5A4 Explain how the octet rule guides chemical bonding. 5A5 Show how the properties of compounds can differ from the elements of which they are made.
5B Types of Bonds	99–105	99–105	Demonstration: Properties of Ionic and Metallic Substances Mini Lab: *Modeling Bonds in Three Dimensions* (pp. 106–7)	**EQ:** Why do atoms bond in different ways? **Objectives:** 5B1 Compare the role of electrons in ionic, covalent, and metallic bonding. 5B2 Interpret the Lewis structures for simple compounds. 5B3 Explain why double and triple bonds form using the octet rule. 5B4 Relate polarity to bonds and molecules.
Lab Day 1	SLM 51–54	TLM 51–54	Lab 5A: *The Solution to a Problem*—Solubility and Chemical Bonds	**EQ:** Does bond type indicate solubility?
Lab Day 2	SLM 55–57	TLM 55–57	Lab 5B: *Electric Lines*—Conductivity and Chemical Bonds	**EQ:** Which type of bond conducts electricity better?
5C Writing Chemical Formulas	108–17, 121	108–17, 121	Case Study: Sticky Situation (p. 121)	**EQ:** How do you know how many atoms are in a chemical formula? **Objectives:** 5C1 Predict the ratios of ions or atoms in ionic and covalent compounds. 5C2 Write the chemical formulas for ionic and covalent compounds. 5C3 Name ionic and covalent compounds.
Ethics Day	118	118	Ethics: Pseudoephedrine	5C4 Justify the use of medications.
Review and Test Days			Chapter 5 Test	

Section	SE Pages	TE Pages	Teacher Resources	Essential Questions/Content Objectives
CHAPTER 6: THE CHEMISTRY OF LIFE (8 DAYS) _ENRICHMENT CHAPTER_				
6A Organic Compounds	124–33	124–33	Demonstration: Straight Chains Mini Lab: _Modeling Hexane Isomers_ (p. 133)	**EQ:** Why is carbon so important? **Objectives:** 6A1 Define _organic compound_ and _hydrocarbon_. 6A2 Explain how isomers are formed. 6A3 Compare saturated and unsaturated hydrocarbons.
Lab Days 1 and 2	SLM 59–61 or SLM 63–66	TLM 59–61b or TLM 63–66	Lab 6A: _Sticky Business—Inquiring into Glues_ or Lab 6B: _Milking Chemistry—Proteins in Food_	**EQ:** How can I improve a casein glue? or **EQ:** Which foods are a source of protein?
6B Substituted Hydrocarbons	134–35	134–35		**EQ:** What other elements can be found in organic compounds? **Objectives:** 6B1 Classify substituted hydrocarbons on the basis of their functional group. 6B2 Give examples of how common substituted hydrocarbons are used.
6C Biochemistry	136–40	136–40	Demonstration: Superabsorbent Polymers Career: Serving as a Food Chemist	**EQ:** What molecules are needed for life? **Objectives:** 6C1 Define _polymer_. 6C2 Compare carbohydrates, lipids, and proteins. 6C3 Explain the role of nucleic acids in cell reproduction.
Ethics Day	143	143a	Ethics: Can Fast Food Be Nutritious?	6C4 Explain the importance of nutrition in a fast-food society.
Review and Test Days			Chapter 6 Test	

LESSON PLAN OVERVIEW

Section	SE Pages	TE Pages	Teacher Resources	Essential Questions/Content Objectives
UNIT 2: CHANGES IN MATTER				
CHAPTER 7: CHEMICAL REACTIONS (10 DAYS) *FOUNDATIONAL CHAPTER*				
7A Chemical Changes	148–54	148–54	Demonstration: Chemical Reaction Indicators Career: Serving as a Toxicologist Mini Lab: *Balanced Diet* (p. 155)	**EQ:** How can I tell whether a chemical change has taken place? **Objectives:** 7A1 List evidences for a chemical change. 7A2 Explain the use of a chemical equation as a model of a chemical reaction. 7A3 Write chemical equations from word equations. 7A4 Balance chemical equations by using the law of conservation of matter. 7A5 Solve simple mole-mass conversion problems.
7B Types of Chemical Reactions	156–58	156–58	Demonstration: Types of Chemical Reactions	**EQ:** Are there different kinds of chemical reactions? **Objectives:** 7B1 Classify chemical reactions. 7B2 Compare oxidation and reduction.
Lab Day 1	SLM 67–71	TLM 67–71	Lab 7A: *Science Fair Revisited—Types of Chemical Reactions*	**EQ:** What happens during a chemical reaction?
7C Energy in Chemical Reactions	159–61	159–61	Demonstration: Endothermic and Exothermic Reactions	**EQ:** Do chemical reactions always give off energy? **Objectives:** 7C1 Explain the flow of energy in exothermic and endothermic reactions. 7C2 Relate exothermic and endothermic reactions to their energy graphs. 7C3 Define *activation energy*.
7D Reaction Rates and Equilibrium	162–68, 172, 173	162–68, 172, 173	Case Studies: Grain Elevators (p. 172), Building Implosion (p. 173) Link: Reaction Simulation	**EQ:** Why do some things burn slowly while others explode? **Objectives:** 7D1 Summarize the collision model. 7D2 Predict how reaction rate is affected by reaction conditions. 7D3 Compare the actions of catalysts, inhibitors, and enzymes. 7D4 Summarize the law of chemical equilibrium. 7D5 Predict adjustments within a reversible chemical reaction in response to specific changes in conditions by using Le Châtelier's principle.
Lab Days 2–4	SLM 73–75	TLM 73–75b	Lab 7B: *It's in the Bag—Inquiring into Chemical Reactions*	**EQ:** How can I tell which reactants cause which effects?
Review and Test Days			Chapter 7 Test	

Section	SE Pages	TE Pages	Teacher Resources	Essential Questions/Content Objectives
CHAPTER 8: NUCLEAR CHANGES (7 DAYS) _ENRICHMENT CHAPTER_				
8A Radioactive Decay	176–84	176–84	Case Study: Vikings Demonstration: Half-Life How It Works: Smoke Detectors	**EQ:** Why do only some isotopes decay? **Objectives:** 8A1 Compare physical, chemical, and nuclear changes. 8A2 Explain why some isotopes decay and others do not. 8A3 Define _radioactive decay_. 8A4 Classify nuclear decay by type. 8A5 Write balanced nuclear decay equations. 8A6 Define _half-life_. 8A7 Predict how much of a sample remains after a given amount of time on the basis of its half-life.
Lab Day	SLM 77–83 or SLM 85–88	TLM 77–83 or TLM 85–88	Lab 8A: _Flipping Out_—Modeling Radioactive Decay or Lab 8B: _Radioactive!_—Exploring Radiation Dose	**EQ:** How can probability predict how long a sample remains radioactive? or **EQ:** What is my annual radiation exposure?
8B Fission and Fusion	185–91	185–91	Case Study: Tsar Bomba Mini Lab: _Modeling Chain Reactions_	**EQ:** Why is the sun so hot? **Objectives:** 8B1 Compare nuclear fission and fusion. 8B2 Relate chain reaction and critical mass to nuclear fission. 8B3 Explain how nuclear reactions don't violate the law of conservation of matter.
8C Nuclear Changes: Benefits and Risks	191–96	190–96	Worldview Sleuthing: Nuclear Waste	**EQ:** What are the benefits and risks of nuclear changes? **Objectives:** 8C1 Explain how radioactivity is used in medical technology. 8C2 Compare genetic and somatic damage. 8C3 Explain how we detect radiation in order to protect people from harmful radiation. 8C4 Justify applications of nuclear changes.
Ethics Day	199	199	Ethics: Nuclear Power Generation	8C5 Evaluate the positions for and against generating energy from nuclear sources.
Review and Test Days			Chapter 8 Test	

LESSON PLAN OVERVIEW

Section	SE Pages	TE Pages	Teacher Resources	Essential Questions/Content Objectives
CHAPTER 9: SOLUTIONS (8 DAYS) _FOUNDATIONAL CHAPTER_				
9A Mixtures and Solutions	202–11	202–11	Worldview Sleuthing: Sports Drinks How It Works: Hot and Cold Packs Demonstrations: Mentos® and Diet Soda, The Tyndall Effect, Mixtures, Energy in Solution, Rates of Dissolving, Boiling Links: Mentos and Diet Soda, Dissolving Rate Applets	**EQ:** How do things dissolve? **Objectives:** 9A1 Compare suspensions, colloids, and solutions. 9A2 Demonstrate the process by which solutions are made. 9A3 Explain energy changes during dissolving. 9A4 Compare different methods for separating mixtures.
Lab Days 1 and 2	SLM 89–91	TLM 89–92	Lab 9A: _All Mixed Up_—Inquiring into Separating Mixtures	**EQ:** How can I separate the components of a mixture?
9B Solution Concentration	212–18, 221	212–19, 221	Case Studies: Road Salt, Maple Syrup (p. 221) Mini Lab: _Mass and Volume in Solutions_	**EQ:** How can we describe the amount of solute in a solution? **Objectives:** 9B1 Compare unsaturated, saturated, and supersaturated solutions. 9B2 Explain measures of concentration, including percent by mass, percent by volume, and molarity. 9B3 Relate colligative properties to measures of concentration. 9B4 Assess the hidden costs of the usage of salt or a brine solution to prevent roads from freezing.
Lab Day 3	SLM 93–95	TLM 93–95	Lab 9D: _That's Cold_—Investigating Freezing Point Depression	**EQ:** How can I cool a solution below its solvent's normal freezing point?
Ethics Day	221	221	Ethics: Pollution	9B5 Formulate a position on pollution mitigation from a Christian perspective.
Review and Test Days			Chapter 9 Test	

Section	SE Pages	TE Pages	Teacher Resources	Essential Questions/Content Objectives
CHAPTER 10: ACIDS, BASES, AND SALTS (7 DAYS) _KEY CHAPTER_				
10A Acids and Bases	224–28	224–28	Case Study: The King of Chemicals Demonstrations: An Acid Reacting with a Metal, Conductivity, Indicators	**EQ:** What's the difference between an acid and a base? **Objectives:** 10A1 Define _acid_ and _base_ and give common characteristics of each. 10A2 Identify a substance as an acid or base on the basis of its characteristics.
10B Acidity and Alkalinity	228–32	228–32	Demonstrations: Concentration and pH, A Buffered System	**EQ:** Why are strong acids and bases dangerous? **Objectives:** 10B1 Explain the relationship between acid and base strengths. 10B2 Relate the terms _concentration_ and _strength_. 10B3 Explain the pH scale. 10B4 Relate pH values with acid or base strength and concentration.
Lab Day	SLM 97–101 or SLM 103–6	TLM 97–101 or TLM 103–6	Lab 10A: _pH pHun—_ Determining pH or Lab 10B: _Feeling the Burn_—Comparing the Concentrations of Basic Solutions	**EQ:** Do pH meters and pH paper provide similar results? or **EQ:** Do all basic solutions neutralize acid equally well?
10C Salts	232–38	232–38	Demonstration: Neutralization Mini Lab: _Basic Problem_	**EQ:** What happens when acids and bases mix? **Objectives:** 10C1 Explain how a neutralization reaction occurs. 10C2 Identify the cation of a base and the anion of an acid. 10C3 Predict the salt compound that will be formed in a neutralization reaction.
Ethics Day	238	238	Ethics: Antacids	10C4 Relate the properties of buffers with how they can benefit people.
Review and Test Days			Chapter 10 Test	

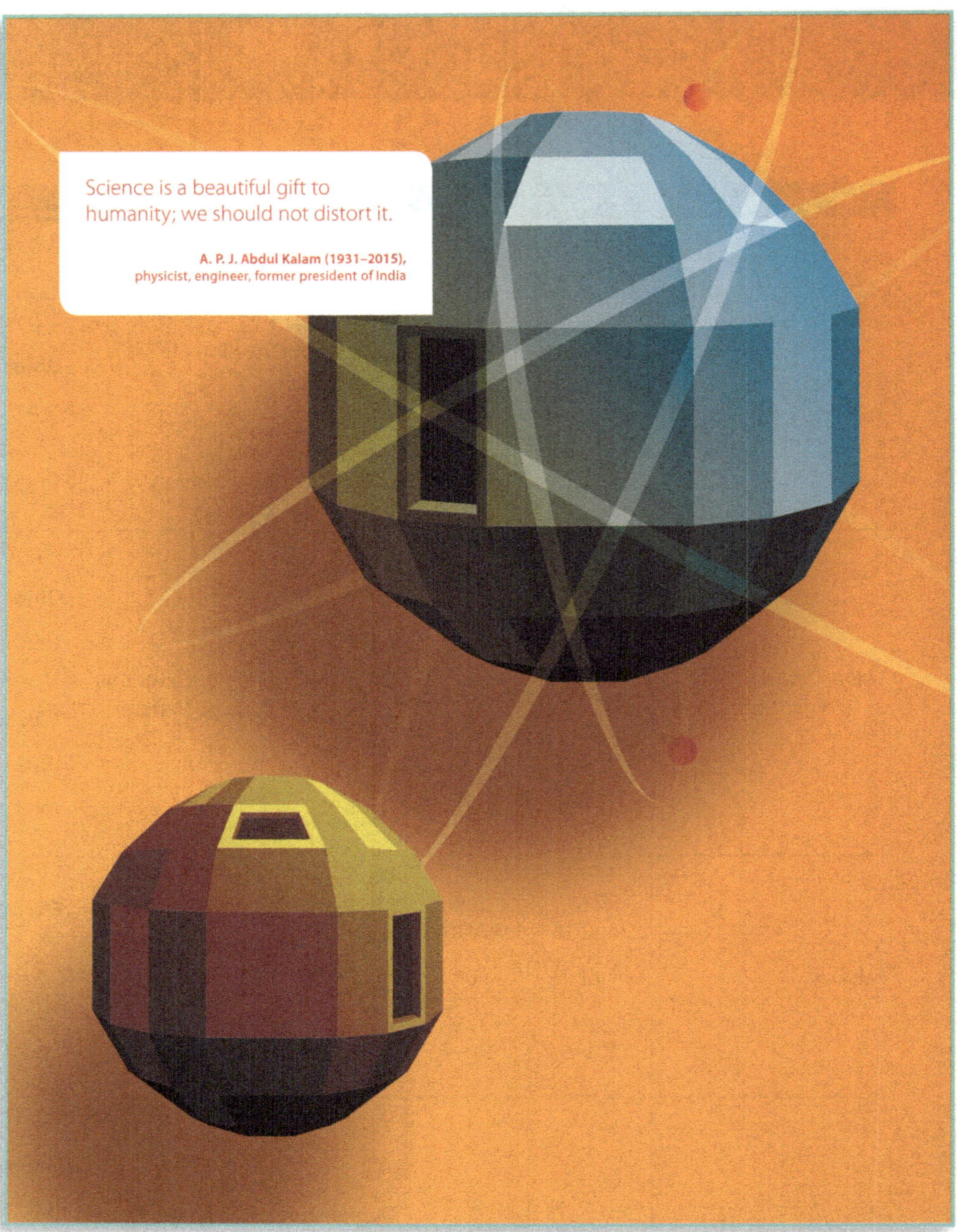

Unit 1 Quote

We are commanded to have dominion over the world that God created. Scientific study is part of having dominion. We must view science through the lens of the Bible and not through the distorting view of naturalism.

First Day of School Activities

To get students thinking like scientists, place an object in a box and ask them to try to determine what is in the box without looking inside.

Another activity involves the use of three wooden blocks (one set for each group of students) with information on all sides.

Block 1: numbers 1–6, like a die (1 opposite 6, 2 opposite 5, and 3 opposite 4)

Block 2: first letter of the numbers 1 through 6 (O, T, T, F, F, and S), like a die (O opposite S, T opposite F, and T opposite F)

Block 3: names and numbers on the sides.

The names consist of similar male and female names on opposite sides. Before each name write a superscript number indicating the number of letters in common with the opposite name. After each name have a number indicating the total number of letters in the name. For example, ^{2}Alma^4 would be opposite ^{2}Albert^6, ^{3}Rob^3 opposite ^{3}Roberta^7, and ^{4}Frank^5 opposite ^{4}Francine^8.

Place Block 1 on the table with any number downward but each group's block oriented the same way. Students can do anything except look at the bottom. Have them hypothesize about what is on the bottom. Repeat with Block 2 (O or S downward, but each group's block oriented the same way) and then Block 3 (any name downward, but each group's block oriented the same way).

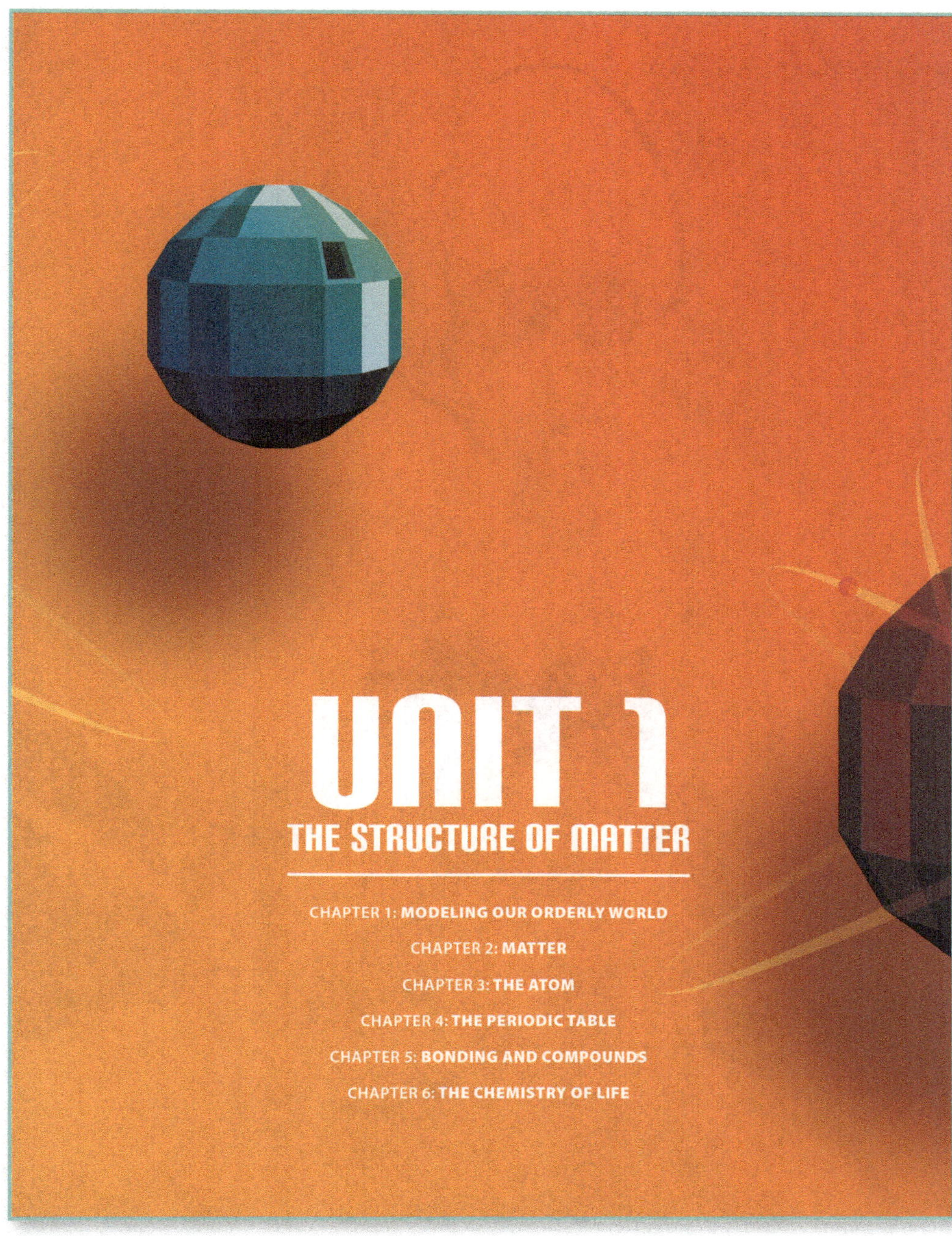

UNIT 1
THE STRUCTURE OF MATTER

Forensic scientists collect samples as evidence that could be used in a court of law. They use this evidence to try to reconstruct events that were not observed or that are in question to ensure that justice is carried out. Forensic scientists take samples and often must wear protective clothing to protect themselves from pathogens. They must also be extremely careful with the evidence so that it doesn't become contaminated.

Visual Analysis: *Forensic Scientist*

Ask students one or more of the following questions to get them thinking about the Chapter opener image.

1. Why is this person wearing so much protective clothing? *(She could be trying to prevent herself from coming in contact with hazardous chemicals or biological agents. She could also be trying to prevent contamination of the collected sample.)*

2. What do you think she is collecting? *(She appears to be collecting something that is biological in nature. It may be a sample of a possible pathogen or body fluids that may carry pathogens.)*

3. Why do you think she is collecting it? *(She may be testing to determine what the sample is. It could be evidence from a crime scene and thus needed for evidence.)*

4. Why is it important to collect the sample in a special bag? *(The bag protects others from coming in contact with the hazardous material. It also protects the integrity of the sample for evidence in a court case.)*

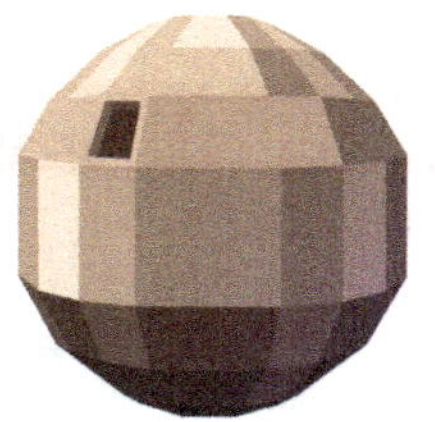

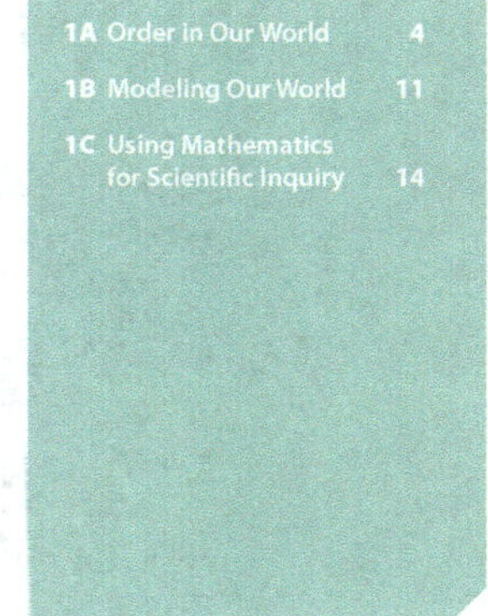

CHAPTER 1

Modeling our Orderly World

SUPER SLEUTHING SCIENTISTS

Drops of blood under the table, beige powder dusting the countertop, and a footprint in the front hall. Each of these raises multiple questions for the forensic scientist. Whose footprint is that? What is that beige powder? Whose blood is that? Is this evidence related to the crime? What other trace evidence can I find?

Forensic scientists collect evidence and conduct investigations to determine what occurred during a crime. Their task is most difficult when there are no witnesses. Forensic scientists may gather evidence at the crime scene or analyze data in a lab or on computers. Whatever their task, they must be observant so that they don't overlook evidence. They must be meticulous in handling the evidence. They conduct tests and experiments to determine the evidence's significance. Finally, they come to conclusions about evidence, helping the police arrest the perpetrator of a crime.

The culmination of the forensic scientists' work occurs in the courtroom. They present their findings and provide expert opinions about what the evidence means. They have to be both accurate and precise in their testimony. They must be well spoken and have the ability to present their findings. Every day forensic scientists use scientific inquiry to see that justice is served.

Variety in Forensic Science

Often students will think of a TV portrayal of forensic science. But there are many other aspects of forensic science that people don't consider. While much of forensics involves DNA and fingerprints, there are also forensic botanists, entomologists, dentists, artists, anthropologists, engineers, and physicists, to name a few. People from many science fields can enjoy the challenges of forensic science.

 Research Topic

If a student has an interest in forensic science, you could have him do some research and report back to the class. Many students don't realize that there are many specialized fields within forensic science. If more than one student is interested in this research topic, challenge those who are to look at different fields.

Overview

Chapter 1 is foundational—it lays the biblical and philosophical groundwork for physical science. Students need an understanding of why studying science is a worthwhile endeavor for a Christian. Be sure to cover this chapter!

1A Order in Our World
- **1.1** Physical Science
- **1.2** Why Is Order Important?
- **1.3** Why We Do Science
- **1.4** Science and Ethics

1B Modeling Our World
- **1.5** Modeling in Science
- **1.6** Scientific Inquiry

1C Using Mathematics for Scientific Inquiry
- **1.7** Measurements
- **1.8** Measuring
- **1.9** Limits of Measurement
- **1.10** Unit Conversions

Lab Activities

Lab 1A: *Based on a True Story*—Students learn about the importance of thinking safe in the laboratory.

Lab 1B: *How Do We Measure Up?*—Students practice measuring with precision and accuracy.

Lab 1C: *Visual Data*—Students practice creating and reading graphs.

Each day have something to engage your students at the very start of class. This can connect to previous content, introduce upcoming content, or just be interesting. The point is to get them thinking.

Collaborative Work:
Opening Question

Write the following questions on the board: What is science? What is physical science? Encourage students to discuss with a neighbor and use resources (e.g., textbook, *internet,* dictionary). Once they have had sufficient time, ask for a few examples of their definitions. Highlight the aspects of each definition that are correct.

If your students are familiar with concept definition maps, you could have them do one for physical science. If not, you may want to use this opportunity to introduce the technique as students share their answers. For more information about concept definition maps, refer to Appendix F.

Reading for Comprehension

Getting Oriented
The beginning of each section in the Student Edition includes an essential question, key questions, and a list of vocabulary terms. These are all great tools for getting students oriented for the section ahead.

Prereading
Prereading—in which students skim through the chapter looking at the titles of the sections, subsections, example problems and the captions of images—gives them an overview or outline of the chapter before they actually read it.

Flash Cards
Some students may benefit from making flash cards of vocabulary and other important terms. There are several apps available that will aid this process.

Class Discussion: *Note Taking*

Note taking is tremendously important and should be encouraged. There are many methods of note taking and each method doesn't necessarily work for every student. Emphasize to students the need to record appropriate notes to help them retain and understand the material. For more information, do an internet search using the keywords "note-taking methods."

1A | ORDER IN OUR WORLD

What is the source of order in nature?

1.1 PHYSICAL SCIENCE

In the Chapter opener, you saw how forensic scientists are always asking and answering questions. This constant questioning is not something that only forensic scientists do. It is a characteristic of all scientists.

Scientists spend their careers asking questions and investigating events or facts called *phenomena* (s. phenomenon). They do this in an attempt to explain or describe the world around them. Some do this because they are naturally curious and want to learn as much as they can about the world around them. Others have a desire to use what they have learned to help other people. No matter their motivation, they all do science. But what *is* science?

Science is the systematic study of the universe to produce observations, inferences, and models. It also includes the products created through this systematic study. Science is initially divided into social science, the study of human societies and relationships, and natural science, the study of the natural world. We further divide natural science into numerous different fields. While we organize and describe these fields of study as if they were completely isolated from each other, they actually are very much interrelated.

4   CHAPTER 1

TEACHING THE MATERIAL

ESSENTIAL QUESTION

What is the source of order in nature?

OBJECTIVES

- 1A1 Define *physical science*.

- 1A2 List evidences of order in nature. **BWS**

- 1A3 Explain why the world is orderly. **BWS**

- 1A4 Explain the role of science in dominion. **BWS**

- 1A5 Outline a biblical framework for science and ethics. **BWS**

RESOURCES

Ethics: Christian Ethics

Serving as a Forensic Scientist

Misconceptions About Science

While many definitions of science focus on the set of data and information that is collected while doing science, always keep students focused on the *process* of doing science. Science is not about discovering truth about the world; it is rather about modeling—explaining and describing—the world.

Fields of Science

Spend time with students investigating the variety of fields of scientific study. Help them to recognize the overlap in many fields. Students often have a range of interests and mistakenly believe that they have to choose between fields.

 ## Outside Research: *Fields of Science*

It may be interesting to have students research some of the fields of science mentioned in the boxes. Have them focus on the overlap of two or more scientific disciplines. Some students may surprise you and research a field not mentioned in the infographic on this page.

STRATEGIES

First Day of School Activities: We all know that the first day of school is often spent dealing with administrative functions. If you have the opportunity, get some scientific thinking going on. Use the First Day of School Activities teacher note on page xxx to get students thinking like a scientist.

Class Opener: Use the Visual Analysis: Forensic Scientist teacher note on page 2 to get students observing and hypothesizing about the scene in the image.

Small Group/Class Discussion: Use the Group Activity: Orderly World teacher note on page 6 and have students brainstorm about order around them.

Biblical Worldview Shaping: Use the Uniformity Versus Uniformitarianism teacher note on page 6 to help students understand the important distinction between these two concepts.

Biblical Worldview Shaping: As you teach Subsection 1.3, focus students on Creation, Fall, and Redemption. See the Creation, Fall, and Redemption teacher note on page 8.

(continued)

 **Group Activity:** *Orderly World*

Have students work in small groups to brainstorm examples of order in the world. Start with how they would define *order*. Then have them discuss order in the classroom, at home, and in their lives. Then have them think of areas that show order in the world.

This is another opportunity for a graphic organizer. A concept definition map would work well here. For more information about graphic organizers, see Appendix F.

Uniformity Versus Uniformitarianism

People often confuse the principle of uniformity with the concept of uniformitarianism. While the principle of uniformity says that nature acts predictably, uniformitarianism is the idea that all geological data can be explained by earlier processes still at work in the present. This view underlies the evolutionary interpretation of geology. It is most problematic for Christians because it denies the reality of supernatural events in Earth's past, including the Creation and Flood accounts in Genesis.

1.2 WHY IS ORDER IMPORTANT?

Why can we study the universe systematically? If phenomena were completely random, we would have to accept just watching events happen, but investigating them would be impossible. However, our world is orderly, and order allows us to study the events that occur.

Evidence of Order

As we wake up each morning, the periodic changes of day into night, days into weeks, weeks into months, and months into years remind us of the order in nature. The cycles that we know as seasons have been guiding farmers in the planting and harvesting of crops since Creation. We see order in the repeated patterns of the chemical and physical properties of elements.

Order in the natural world is so evident that scientists even imitate it. Chemists arrange the elements in a *periodic table* by the repeated patterns in the structure and properties of those elements. Biologists have developed a system for classifying living organisms. This classification system uses the order found in nature to categorize the different species. Order is what allows us to do science successfully.

Result of Order

Watching something random, like the movement of flames in a campfire or raindrops on a pond, may be exciting and even entertaining. But if there is no pattern to the events, then we cannot identify causes, and we certainly cannot make predictions about these events. However, since order is a built-in part of the world in which we live, we can do science. With enough observations, we can even begin to make predictions about future events on the basis of the patterns that we have observed.

One key principle that allows us to study science is the *law of cause and effect*. This law states that every effect has a specific, identifiable cause, and for every cause, there is a definite and predictable effect. We can sum up the law this way: everything happens for a reason; nothing just happens. Therefore, scientists know that they can investigate a particular phenomenon to determine its cause. Similarly, after sufficient investigation, the scientist should be able to predict the effect of a given cause.

Another important aspect of our world is the *principle of uniformity* of nature. This principle declares that nature acts the same today as it did yesterday and that we can fully expect it to act the same way tomorrow. The uniformity in nature is what allows us to make predictions in science. But even the characteristics of uniformity and predictability had to be caused by something or someone.

Ethics: Focus students on the process of ethical decision-making from a Christian perspective described on pages 9–10. Students will use this process in the ethics boxes throughout the Student Edition.

Ticket Out the Door: As a summarizing activity to check for understanding, ask, "What is the source of order in nature?" (See the Class Closer: Ticket Out the Door teacher note on page 10.)

Source of Order

Where does this order come from? Nothing left to itself becomes more orderly than it was before. According to the law of cause and effect, the *effect* of order in the universe had to be the result of a specific, identifiable *cause*.

Genesis 1 outlines the creation of the universe. God created all things out of nothing. The universe that He created is a reflection of His very nature. When He completed His creative work on Day 6, He evaluated everything and declared that everything was good (Gen. 1:31). Some will ask, didn't *God* need a cause? Remember, however, that the law of cause and effect relates to everything *in our world*. As mentioned in Genesis 1:1, God existed before the world He created and is therefore outside of it. And God created the universe with order because He is a God of order.

1.3 WHY WE DO SCIENCE

Order in nature allows us to do science, but what makes it worthwhile? The answer to this question depends on your understanding of the world. We all view the world on the basis of assumptions we have about the world. This is called our *worldview*. As you can imagine, every decision we make is affected by our worldview.

There are many divergent worldviews, some religious and others secular. People who hold to secularism are not necessarily atheistic. Secularists simply believe that we should exclude religious beliefs from the public sphere of discourse.

Secular scientists believe that they can explain the universe and all that is in it by solely naturalistic means. There is no room for the supernatural. According to their worldview, everything began from nothing in the big bang and slowly formed over billions of years. Man is the product of evolutionary changes driven by natural selection over millions of years. Man's purpose is to understand the world around him to help him improve his life.

People with a religious worldview see the world through the lens of the teaching of a particular teacher or text. Christians develop their worldview from the teaching of the Bible. According to a biblical worldview, the world is relatively young and is the product of Creation as outlined in Genesis 1. God created the universe and all that is in it from nothing and for His good pleasure. Everything was created to glorify God. People are a special creation—made in His image.

✝ *A Worldview Is...*

1. a set of basic beliefs, assumptions, and values,

2. that arises from a big story about the world, and

3. that produces individual and group action—human culture.

For more information on a biblical worldview, consider reading BJU Press *Biblical Worldview: Creation, Fall, Redemption*, ©2016.

☑️ *Career Boxes*

Occasionally, the Student Edition will include a box focusing on a career, often by highlighting someone who has left her mark in that field. Use career boxes to show how people serve others in various fields. The individuals highlighted in these boxes are not necessarily Christians, but they are meeting the needs of people around them. People often fulfill aspects of the Creation Mandate and the greatest commandments without even realizing it.

✝️ *Creation, Fall, and Redemption*

Everything in the world was created by God and was originally good. But when Adam and Eve sinned, everything in the physical world was damaged. This includes everything in nature, animate and inanimate, as well as our individual bodies, minds, and relationships with God. But thankfully, God had a plan for the redemption of all that was marred by the Fall. It includes the redemption of nature, our relationships, our body, and even our thinking. The ultimate fulfillment of Redemption will occur with the creation of the new heaven and the new earth. But we can and should be doing all that is possible to work toward redeeming our thinking as we align our thoughts with God's thoughts. We should strive to work toward the redemption of others and the world around us.

SERVING AS A FORENSIC SCIENTIST: SILENT WITNESS

Forensic anthropologist Clea Koff has given a voice to the victims of genocide around the world when entire populations have been murdered with no witnesses. There is often no justice in cases of genocide because it is carried out during civil wars or by the dominant group in a region. Fear of the dominant group often causes these cases to go unreported. Even when reports surface, few witnesses are available or willing to testify. In these situations, teams of forensic scientists from the United Nations get involved.

Clea Koff used the remains in mass graves in Rwanda and Yugoslavia to build cases for genocide in those countries. After concluding her work there, Koff turned her attention to the United States. She used her anthropology skills to help families of missing persons. She also founded the Missing Person Identification Resource Center in Los Angeles.

Crime is a fact of life in a fallen world. Through forensic anthropology, Clea Koff has helped others find closure after loved ones had disappeared.

An artist involved in forensic anthropology works on 3D facial reconstruction.

A Christian's purpose in science is the same as his purpose in life. We were created to love and glorify God (Ps. 86:9; Matt. 22:37) and to serve others (Matt. 22:39). In fulfilling our purpose in life, we can accomplish a crucial aspect of a biblical worldview—redemption. Creation was very good, but man damaged it by introducing sin into the world at the Fall. Each of us needs personal redemption, the forgiveness of our sins. But our thinking also needs to be redeemed (Rom. 12:2), as does the physical world (Rom. 8:22), because every aspect of the Creation was adversely affected by the Fall.

Our purpose in life and science stems from the Creation Mandate. The **Creation Mandate** (Gen. 1:26, 28) directs us to fill the earth and have dominion over it. God's command to humans is to exercise wise and good dominion over His creation for the glory of God and the benefit of their fellow humans. How can we accomplish this as scientists?

We do this by studying and caring for God's creation. We observe and then replicate in our models and understanding of nature the order created by God. We use science to meet the needs of other people, God's image bearers. We imitate God's creative work through inventing and engineering products to benefit people. All that we do in science should honor and glorify God while caring for His creation.

1.4 SCIENCE AND ETHICS

Scientists don't work in isolation; the science they do affects people around the world. As we do science, we must decide what is right and wrong. These decisions can become complicated very quickly. While science can tell us that a person's brain is not working, it cannot tell us whether we should disconnect the person from life support. So how do we decide whether a particular use of science is right or wrong?

What we are talking about is **ethics**—a system of moral values or a theory of proper conduct. Our worldview plays a key role in making ethical decisions. Christian ethics should be based on the three-element foundation of *biblical principles*, *biblical outcomes*, and *biblical motivations*.

Biblical Principles

God created the world and everything in it. He created man as His special creation and gave us His Word to guide our lives. While the Bible doesn't specifically address everything that we might face in life, His Word gives us enough guidance to make right decisions. God tells us what is good and evil. By studying Scripture, we can understand general principles that we can use in all situations. So the first thing that we should ask when we encounter an ethical issue is, "*What does God's Word say?*" The three biblical principles that lead to right thinking are people as God's image bearers, the Creation Mandate, and God's whole truth.

Biblical Outcomes

We also must think about what we want to achieve. God tells us in the Bible the goals that He has for us. Ethical decisions must include striving for right outcomes. A second question we must ask when faced with making an ethical decision is, "*What results are right?*" Clear goals follow from the biblical outcomes of human prospering, a thriving creation, and glorifying God.

Biblical Motivations

When God saves us, He intends for us to be changed to be more like Him. We better reflect His image as we grow. A Christian should do right because it contributes to his growth, not just because the rules tell him to or because he wants good results. A Christian should want to be more like the Lord. So the third and most important question is, "*How can I grow through this decision?*" Three biblical motivations follow from Scripture: faith in God, hope in God's promises, and love for God and others.

So how should a Christian make ethical decisions?

✝ Adaptation of Ethics Triad

Systems of ethical decision-making follow three different methodologies. In the first, ethical decisions are made by adhering to a set of rules or principles, such as a religious book, natural law, or the categorical imperative. This is what ethicists call the *deontological* approach. The second methodology is called the *teleological* approach in which ethical decisions are made on the basis of evaluating potential outcomes; the goal is to achieve beneficial outcomes and avoid negative outcomes. The third methodology is focused on considering internal motivations or emphasizing the living of authentic lives. Ethicists call this the *existential* approach. The best ethical systems use all three methodologies.

As you would expect, biblical ethics includes all three approaches. In BJU Press *Life Science* Student Edition 5th Edition, we introduced two of the three approaches. Students worked through ethical issues by looking at biblical principles and outcomes. To apply biblical principles, students will focus on our unique position as God's image bearers, on the Creation Mandate, and on God's whole truth. When addressing biblical outcomes, students must think about which decision will maximize human prospering, a thriving creation, and glorifying God.

Physical Science Student Edition 6th Edition completes the triad. Students will continue to address biblical principles and outcomes, but they will also have to address the third leg of the ethics triad. They will have to think about their internal motivation: which decision will increase their faith in God, hope in His promises, and love for Him and others.

 ### Class Discussion:
Chunking Information/Reviewing

As you finish this first section, take the opportunity to emphasize to students the benefit of reviewing chunks of material. Suggest that they read through the Section 1A portion of the Chapter Review on page 21. Doing so will allow them to identify any key terms or concepts that they may have missed in the section.

✔ Class Closer: *Ticket Out the Door*

It is a good practice to have some type of summative task at the end of class. This can take many forms; for example, a quick question that students answer on a slip of paper and turn in as they leave—their ticket out the door. A quick survey elicits an indication about their understanding of a key point of the lesson. Use the essential question or key questions to see whether they have grasped the concepts of the section.

Use the information that you gather from class closers to gauge students' level of understanding. Consider reteaching a topic if many in the class have misconceptions.

TEACHING THE MATERIAL

ESSENTIAL QUESTION

How do scientists do science?

OBJECTIVES

- 1B1 Define *model*. **BWS**
- 1B2 List different models used in physical science.
- 1B3 Explain the relationship between hypotheses, theories, and laws.
- 1B4 Outline the process used to do scientific inquiry.
- 1B5 Explain why we approach science in an orderly fashion. **BWS**

RESOURCES

- Lab 1A: *Based on a True Story*—Thinking Safe in the Laboratory

1A | REVIEW QUESTIONS

1. Define *science* in your own words.
2. Graphically display the relationship between social science, natural science, physical science, chemistry, and physics.
3. Give an example of orderliness in nature.
4. Why does order imply a Creator?
5. What is the Creation Mandate?
6. Give two examples to show how we can fulfill the Creation Mandate.
7. Define *ethics*.
8. What are the three questions that guide Christians in making ethical decisions?

1B | MODELING OUR WORLD

How do scientists do science?

1.5 MODELING IN SCIENCE

As we mentioned, scientists seek to explain or describe the world around them. Scientists use models to help them do this. A **model** is a workable explanation or description of a phenomenon. A model may be physical, conceptual, or mathematical.

Two major categories of models in science are theories and laws. A **theory** is a model that *explains* a related set of phenomena. It can be used to predict unobserved aspects of the phenomena. On the other hand, a model, often expressed as a mathematical equation that *describes* phenomena under certain conditions, is called a **law**. It does not attempt to explain the phenomena.

So what is the benefit of using a model? Models allow scientists to focus on a particular portion of the world around them. A model helps them understand and communicate their understanding of the phenomenon being studied.

Types of Models

Because of the complexity of physical systems, scientists use various models in physical science. Physical models are quite common, including atomic models, models of waves, models for the behavior of gases, ball and stick models of molecules, and scale models of vehicles. Conceptual models can take as many forms as there are scientists. Scientists typically use conceptual models for thinking about ideal systems. Mathematical models show up everywhere in physical science. They include equations or computer modeling (virtual models) of weather, atmospheric changes, water flow, or any computer simulation.

1B Questions

- Why are models important in physical science?
- What models are important in physical science?
- How do hypotheses, theories, and laws compare?
- How do we do science?
- Why do we approach science systematically?

1B Terms

model, theory, law, workability, scientific inquiry, hypothesis

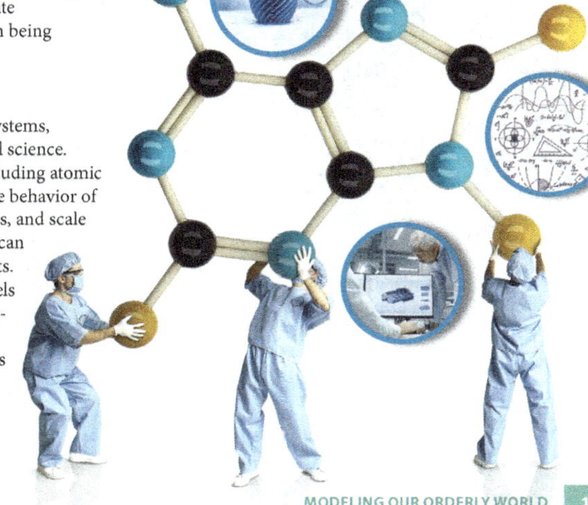

MODELING OUR ORDERLY WORLD 11

Question 2 asks students to graphically represent information. If graphic organizers are new to your students, have them refer to Appendix F. Feel free to have them use one particular graphic organizer, especially if you are teaching them how to use these tools.

1A REVIEW ANSWERS

1. Science is the systematic investigation of the universe that produces observations, inferences, and models. *(p. 4)*

2. Students' answers should be similar to the infographic on page 5.

3. Answers will vary. Students may include cyclical processes or objects and events with obvious design elements. *(p. 6)*

4. Order in nature provides evidence of design. Nothing designs itself and nothing is designed by nothing. Therefore a designer, or Creator, is required whenever there is a creation or a design. *(p. 7)*

5. The command given in Genesis 1 instructed man to fill the earth and have dominion over it. *(p. 8)*

6. Answers will vary. Students may include ways that people can be helped (e.g., improving agriculture, medicine, protective devices) or that nature can be protected (e.g., conservation, pollution mitigation, resource management) *(pp. 8–10)*

7. Ethics is a system of moral values or a theory of proper conduct. *(p. 9)*

8. (1) What does God's Word say? (2) What results are right? (3) How can I grow through this decision? *(p. 10)*

STRATEGIES

Class Opener: Get students thinking about modeling using the Models teacher note on this page.

Formative Assessment: Use the Formative Assessment: Geocentric Model teacher note on page 12 to assess students' understanding of the workability of a model.

Group Activity: Use the Group Activity: Scientifically Cooking a Turkey teacher note on page 13 to help students apply the concepts of scientific inquiry to a common Thanksgiving Day quandary.

Infographic: Use the Scientific Inquiry infographic on page 13 to help students understand what scientists do. Consider selecting a phenomenon and brainstorm how they would work through solving that problem.

Ticket Out the Door: Use either the Formative Assessment: Workability teacher note on page 12 or the Formative Assessment: Scientific Inquiry teacher note on page 13 to check students' understanding.

Models

To get students thinking about modeling, ask them to consider the following questions: What is a model? Why do we use models?

 3D Virtual Models

For access to internet-based 3D modeling software, do an internet search using the keywords "free 3D modeling."

Keep bringing students back to the nature of science as modeling. Scientists attempt to explain or describe the world around them. They "improve" their models (hypotheses, theories, and laws) by improving their workability. Models always vary in their degrees of accuracy, but they are never absolute truth. The key to evaluating models is whether they are workable or not.

Workability is related to how well models explain or describe observations. A more workable model explains/describes more of the observations and explains/describes them better. A workable model is also better at making accurate predictions.

Formative Assessment:
Geocentric Model

Have students work in pairs and explain what was accurate and inaccurate about the geocentric model. Then ask the following questions.

1. How was the geocentric model workable? What did it explain well? *(The geocentric model explained the basic movement of the sun, moon, and stars. It also explained most of the planets' motions.)*

2. In what ways was the geocentric model not workable? *(The model could not explain retrograde motion or why the moon and planets seemed to move closer to and farther from the earth.)*

3. What did scientists do initially to make the geocentric model more workable? *(They modified the model to better explain the evidence.)*

4. What was the result of the changes that they made to the geocentric model? *(The model became too complex and still didn't adequately explain the observations.)*

5. Ultimately what did the scientists have to do? *(They discarded the geocentric model for a more workable model—the heliocentric model.)*

Formative Assessment: *Workability*

Have students answer the following prompt on a slip of paper and turn it in as they leave: Explain what it means for a model to be workable.

Scientific models change because scientists are always learning and improving their understanding of the world around them. People often think about scientific models as being either true or false, right or wrong. But valid models are true though incomplete all the time.

The Goal of Models: Workability

The key to good models is **workability**. What do you think this means? Scientists use two criteria to evaluate models: How well does the model explain or describe what we observe? And how accurate are predictions made with this model? If a model explains or describes the world well and makes accurate predictions, then scientists will retain it. Scientists will modify or discard models that don't do these things well.

To understand this concept of workability, we can think about how scientists have understood the relationship between the sun, earth, and the planets. The geocentric model for this relationship was developed first and stated that the earth was the center of the universe. Everything else revolved in perfect circles around the earth. The model did well at explaining the basic motion of the sun, moon, planets, and stars as seen from the earth. But it couldn't explain how some planets appeared to stop and reverse their direction (now understood as *retrograde motion*). It also couldn't explain the varying distances that the moon and planets were from the earth during their orbits. Scientists tried to adjust the geocentric model, improving its workability, but ending up making it very complicated.

In time, Nicolas Copernicus proposed the heliocentric model, with the sun at the center of our solar systems and the planets (including the earth) orbiting the sun. The heliocentric model explained the observations (retrograde motion and planetary distances) better than the geocentric model did and made better predictions. While it wasn't perfect, it was simpler and more workable than the geocentric model. So the geocentric model was discarded, and the heliocentric model replaced it. As time has passed, the heliocentric model has been modified to make it even more workable.

As we learn more about a particular branch of science, our models get better (more workable). Each successive model explains the phenomena better and makes more accurate predictions. Sometimes models are completely wrong and scientists have to start with a completely new model (a scientific revolution). As long as they remain objective, scientists aren't bothered by this. New or improved models help scientists continue to refine their understanding of the world around them.

1.6 SCIENTIFIC INQUIRY

How do scientists do science? They work through a process called **scientific inquiry**—an ongoing, orderly, cyclical approach used to investigate the world. Inquiry is similar to the scientific method that you have learned previously, but it is a continuous process that is more cyclical than linear.

As their understanding of the solar system changed, scientists initially modified and ultimately discarded the geocentric model and replaced it with the heliocentric model.

Thomas Kuhn

In 1962, the American philosopher of science Thomas Kuhn published *The Structure of Scientific Revolutions*. In this work, Kuhn suggested that science is primarily done in the context of controlling views that he called *paradigms*.

Kuhn's most significant assertion was that the paradigms of science have changed numerous times over the course of history. Thus, if his view is accepted, scientific "truths" are tentative at best and at worst are simply transient ideas to be discarded when more workable models are developed.

Many scientists were uncomfortable with his view of the discipline, but it fits very well with a biblical understanding of the nature of knowledge and the source of real truth.

Incidentally, Kuhn's greatest critic was the Austrian philosopher of science Karl Popper. Popper's view of science was that it was at its heart falsifiable. In other words, if it is impossible to demonstrate that a particular scientific idea is false, then the idea isn't really a scientific idea. Creation scientists have often used Popper's arguments as a critique of evolutionary theory since many aspects of evolutionary theory cannot be falsified.

SCIENTIFIC INQUIRY

Observation. By using his senses, a scientist collects information about the world. Observation is the central activity in science. Scientists must pay great attention to detail and be careful to collect and record data accurately.

Posing Questions. Posing questions is a fundamental skill for the scientist. The questions give him his purpose for doing science. These questions usually stem from something he observed.

Research. Scientists do lots of research. When faced with a new question, a scientist needs to determine what others already know about the topic. This research will allow him to focus his investigation on aspects of the issue that are not well understood.

Forming Hypotheses. As mentioned previously, scientists work with models all the time. One type of model is a **hypothesis**—an initial, testable explanation of a phenomenon that stimulates and guides the scientific investigation.

Investigation. Investigation is the step that most people associate with science. Often this takes the form of an experiment. A scientist designs an experiment to address the hypothesis that he formed on the basis of his question. In a controlled experiment, the scientist changes one variable (the independent variable) and observes the resulting changes in a second (dependent) variable, while holding all other variables constant.

Analysis. Once the scientist has gathered sufficient data, he has to figure out what it means. This analysis may involve graphing, reasoning, and using mathematics. Again the scientist must be careful to analyze his data accurately.

Conclusion. After analyzing his data, the scientist concludes what he has learned. This step connects all the previous activities. The conclusion indicates to what degree the experimental evidence supports or refutes the hypothesis. Hopefully it allows the scientist to answer the question.

Communication. The scientist still has work to do! Science relies on a process of *peer review*. Before publishing his findings, a scientist will have other scientists review and respond to the research. Through this process, models are strengthened, modified, or even discarded. It's part of the reason that scientists must be accurate in collecting, analyzing, and reporting their findings.

 Observing with the Five Senses

Student often think that all observations are done with the sense of sight. But scientists use all their senses to make observations. They also use instruments to make observations for them.

 Group Activity:
Scientifically Cooking a Turkey

You can help students see how scientific inquiry works in a common occurrence by posing the question, "How can you tell when a Thanksgiving turkey has cooked long enough to be safely eaten?"

Have students do research on properly preparing a turkey meal. Then have them work together to form a hypothesis, to suggest means of investigation and analysis, and to explain how they will come to a conclusion about when a turkey is ready to eat.

 Formative Assessment:
Scientific Inquiry

Have students answer the following questions on a slip of paper and turn it in as they leave: What do you think is the most important activity in scientific inquiry? Why?

 Lab 1A: Based on a True Story

Once students have an understanding of the material in Section 1B, they should be ready to work through Lab 1A.

 Active Learning: Scientific Inquiry

Print out the information from each of the boxes in the infographic above on a piece of paper without the bulleted numbers. Give all the boxes to groups of students and have each group arrange the inquiry activities in the order in which they think the activities are done.

Students should notice that some of the activities could occur at different times in the process, while others have very specific positions.

 Peer Teaching: Scientific Inquiry

When teaching the material in this infographic, you could also assign individual boxes to different groups of students. Give each group a few minutes to review their box and then have them explain to the class their activity in the inquiry process.

Scientific Inquiry Versus the Scientific Method

We have specifically chosen not to use the term *scientific method*. Students should understand inquiry as a constant cyclical process of asking and answering questions through a variety of activities. They should not think of science as a set of steps to get through. Science is about wonder.

1. A model is a simplified representation of some aspect of the real world. A model may be physical, conceptual, or mathematical. *(p. 11)*

2. Scientists use models to explain or describe an aspect of the world. Scientists also use models to make predictions. *(p. 11)*

3. A theory is a model that explains a related set of phenomena. *(p. 11)*

4. A law describes the phenomena without attempting to provide any explanation for them, while a theory explains. *(p. 11)*

5. scientific inquiry *(p. 12)*

6. The primary difference is in perception. The scientific method is often understood as a linear series of steps to be followed. Scientific inquiry is a cyclical, multipronged approach to solving the problem. We return and repeat a number of the activities before we are finished. *(p. 12)*

7. Observation is the key to scientific inquiry. Some students may think of investigation, but observation has a wider degree of influence on the process. *(p. 13)*

8. A hypothesis is an initial, testable explanation of a phenomenon. It is the basis for the scientific investigation. *(p. 13)*

9. Since the universe is an orderly place, to study it we need an orderly process. Also, one purpose for studying science is to help others. The most efficient method is one that is orderly. *(pp. 11, 12–13)*

10. Science is validated through a peer review process. Scientists need to be able to clearly and accurately report their findings so that others can review their research. *(p. 13)*

Measurements

Have the following written on the board as the students walk into class: "Without talking, measure the top of your desk and record your measurement on the board." As you begin class, have a discussion about the measurements. Are they all the same? Why are they different? Did everyone measure the same dimension? Do they all include units? Are the units the same? If you can, keep this data to revisit later in class. Taking a photo with a smartphone works well if you can project the image again later.

14 CHAPTER 1

---

1. What is a model?
2. How do scientists use models?
3. What is a theory?
4. How does a law differ from a theory?
5. What do we call the orderly process by which we do science?
6. How does scientific inquiry differ from the scientific method?
7. What is the central activity in the orderly process of science?
8. What is a hypothesis?
9. Why do we need a process to study science?
10. Why is communication important to scientists?

1C | USING MATHEMATICS FOR SCIENTIFIC INQUIRY

How is math used in scientific inquiry?

1C Questions

- Why do scientists use the SI?
- How do we collect data?
- Why do I need math to do science?
- Aren't accuracy and precision the same thing?
- What can I tell about a scientist's instrument from his measurements?
- How can I change units in the SI?
- What happens when scientists are not accurate?

1C Terms

measurement, SI, accuracy, precision

1.7 MEASUREMENTS

Data

As scientists make observations, they record information about what they observed; we call it *data*. Scientists collect two types of data, qualitative and quantitative. Qualitative, or descriptive, data is observations about qualities of the object or event. These observations may be related to color, texture, or relative size, for example. Descriptive data tends to depend on the observer and is therefore less repeatable. Data can also be quantitative, meaning that it is based on numbers or quantities, in other words, **measurements**. Scientists collect quantitative data by using measuring instruments. Measured data is less dependent on the scientist, making the data more repeatable.

We take measurements of different dimensions of an object or event. Dimensions are the measurable aspects of something, such as mass, volume, length, or weight. All measurements have a numerical part and a unit, a standard of measure for comparison.

SI

You are familiar with units such as inches and feet, cups and gallons, and ounces and pounds. Most scientists don't use these units of measurement. Scientists from all around the world use a modern system of standardized metric units. We call this system **SI**, which stands for the Système International d'Unités (International System of Units). Scientists sharing data from their investigations recognized the benefit of having a common set of units. So they developed the metric system with units related to observable phenomena in the world. Over time the standards were modified to easily reproducible standards. For example, scientists defined the meter by the speed of light. This improved standardization is beneficial for all scientists. The scientific community further modified the metric system to base it on just seven fundamental units. The SI that scientists use today is this modified metric system.

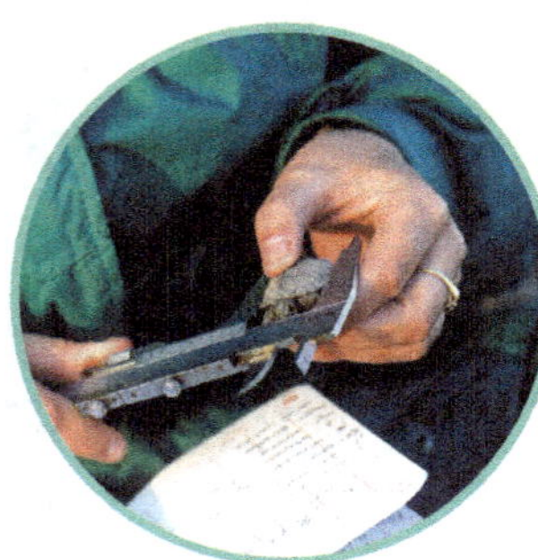

There are as many scientific instruments as there are dimensions to measure.

TEACHING THE MATERIAL

ESSENTIAL QUESTION

How is math used in scientific inquiry?

OBJECTIVES

- 1C1 Compare the US customary and SI systems of measurement.
- 1C2 Explain the benefits of the SI.
- 1C3 Use scientific instruments to collect data.
- 1C4 Use mathematical tools to analyze data.
- 1C5 Compare accuracy and precision.
- 1C6 Explain how scientists identify the precision of measurements.
- 1C7 Convert between SI units.
- 1C8 Explain why it is important for scientists to accurately report data. **BWS**

SI FUNDAMENTAL UNITS

 The fundamental unit of length in the SI is the meter (m).

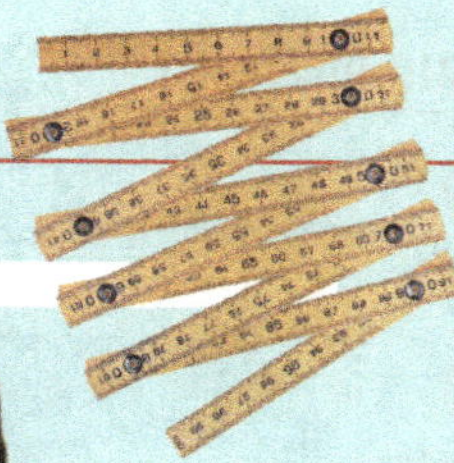

The kilogram (kg) is the SI fundamental unit of mass. 

 The SI uses the second (s) as the fundamental unit for time.

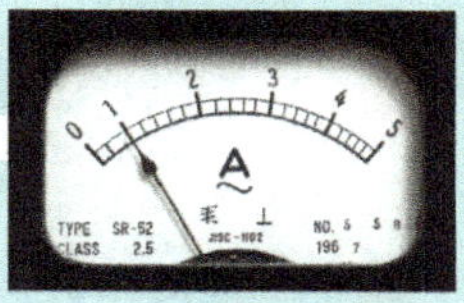

The ampere (A) is used by scientists in the SI to measure electric current. 

 The SI unit for temperature is the kelvin (K).
Notice that we don't use the degree symbol with Kelvin temperatures.

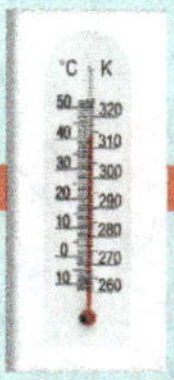

Chemists use the mole (mol) as the SI unit for the amount of a substance. 
A mole is a count that contains 6.022×10^{23} objects.

 Scientists use the candela (cd) in the SI to measure the intensity of a light source.

The SI makes also makes use of *derived units*—mathematical combinations of two or more of the base units. For example, we measure speed in m/s, which scientists derived by dividing the unit for distance (m) by the unit for time (s).

The SI is a decimal system, which means that it is based on powers of ten. Each unit in the metric system can be multiplied or divided by a power of ten. The metric system includes prefixes that correspond to a particular power.

Dimensions

Students often get confused by the two uses of the word *dimension*. They should be familiar with the word when referring to one-, two-, or three-dimensional objects (e.g., a line, a plane, or a prism). But in measuring, *dimension* refers to any measurable quantity (e.g., volume, density, mass).

Measuring

For more information on measuring in general, consider visiting the National Institute of Standards and Technology website of the US Department of Commerce. Do an internet search using the keyword "NIST."

Units

For more information on measuring units, do an internet search using the keywords "dictionary of units."

History of the SI

For more information on the SI, do an internet search using the keyword "history of the SI."

The New and Improved Kilogram

Scientists have been working to redefine the SI fundamental units on the basis of physical constants. One of the last units to be redefined was the kilogram. For more information on the new definition of the kilogram, do an internet search using the keywords "redefining the kilogram."

A Dozen or a Mole?

Make the connection between a dozen being a count that contains 12 of something to a mole being a count that contains 6.022×10^{23} of something.

Dimensions Without Units

Students usually make a connection between certain dimensions and the typical units used to measure them. Remind them that some dimensions we measure *don't* have units. From their study of earth science they may remember that specific gravity—the ratio of an object's density to the density of water—has no units. In physical science, dimensions such as pH, the coefficient of friction, and the index of refraction are also unitless.

RESOURCES

Ethics: Reporting Scientific Data

How It Works: Balances and Scales

Mini Lab: *Understanding Conversion Factors*; Lab 1B: *How Do We Measure Up?*—Practicing Measuring; Lab 1C: *Visual Data*—Graphing Mass and Volume

STRATEGIES

Class Opener: Use the Measurements teacher note on page 14 to help students recognize the uncertainty of measurements.

Class Discussion: Use the Class Discussion: Significant Figures teacher note on page 18 to use the class opener data to begin a discussion about units, accuracy, and precision.

Research/Peer Teaching: Consider having students research the SI. They could focus on the standards, the history, and the prefixes and then teach the other members of their class.

How It Works: Use the How It Works: Balances and Scales box on page 18 to help students understand some of the instruments that they will use in class.

Ticket Out the Door: Have students work in pairs to explain accuracy and precision to each other.

Number Formats

Throughout the Student Edition, you will notice numbers formatted in a particular way. In the SI, every third decimal place is indicated with a thin space. These spaces help you recognize the decimal places in numbers. For instance, the definition of the speed of light is 299 792 458 m/s, and a meter is the distance that light travels in 0.000 000 003 335 640 952 s. Exceptions to this are in numbers with four or fewer places on either side of the decimal point (e.g, 9542 m, 0.2315 L, or 4562.3214 g). Also, numbers used in a nontechnical context will have the standard comma (e.g., There were 12,567 people in attendance).

Benefits of the SI

Have students research the benefits of the SI, not only in science but in daily usage. Research should include information on the number of units for each dimension, the base number for each system, the standards by which each unit is compared, the nations of the world that use particular measurement systems, and a comparison with the US customary system.

Metric Prefixes

Students may wonder how many metric prefixes there are. The prefixes range from 10^{-24} to 10^{24}. An interested student may want to research these other prefixes. Challenge them to determine their origins. Many are from foreign languages.

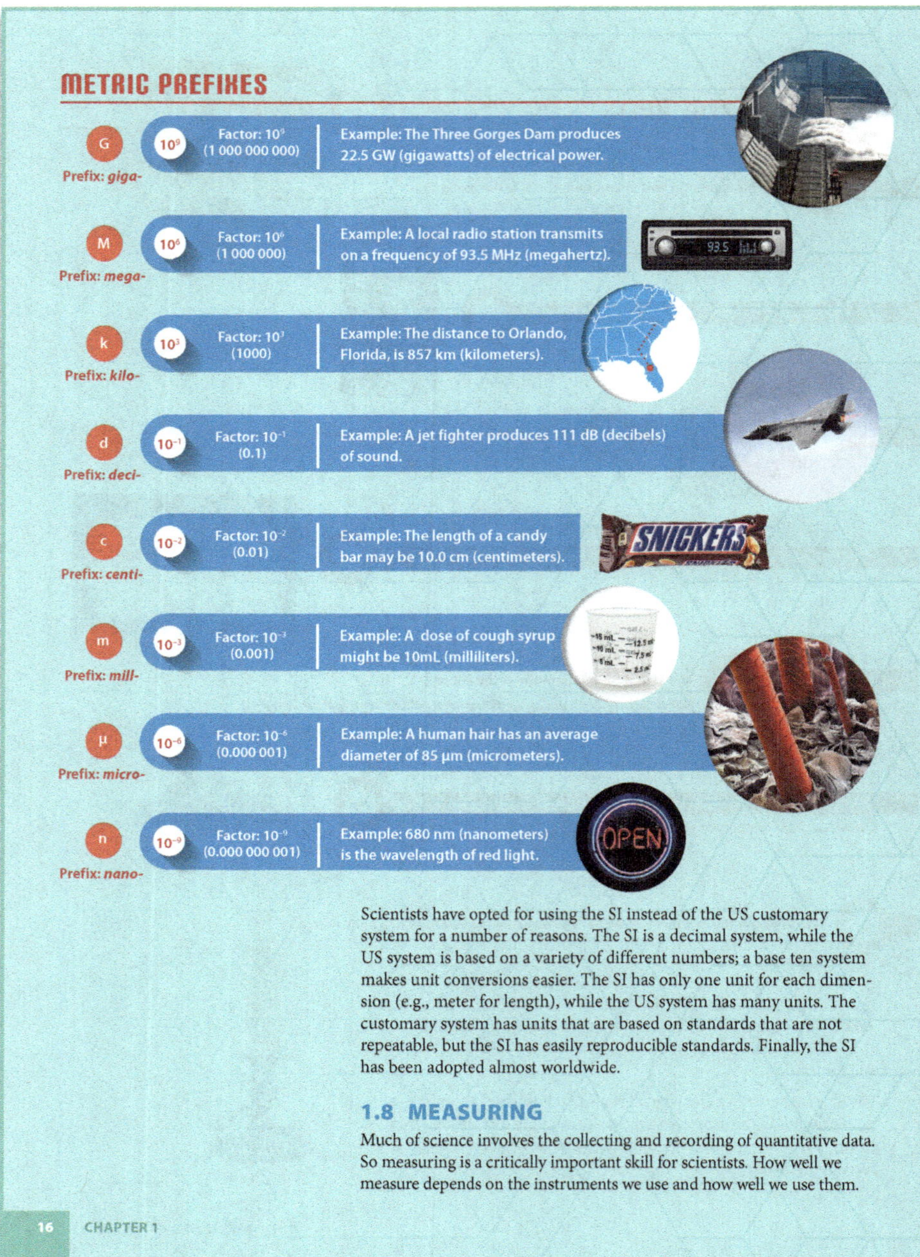

Scientists have opted for using the SI instead of the US customary system for a number of reasons. The SI is a decimal system, while the US system is based on a variety of different numbers; a base ten system makes unit conversions easier. The SI has only one unit for each dimension (e.g., meter for length), while the US system has many units. The customary system has units that are based on standards that are not repeatable, but the SI has easily reproducible standards. Finally, the SI has been adopted almost worldwide.

1.8 MEASURING

Much of science involves the collecting and recording of quantitative data. So measuring is a critically important skill for scientists. How well we measure depends on the instruments we use and how well we use them.

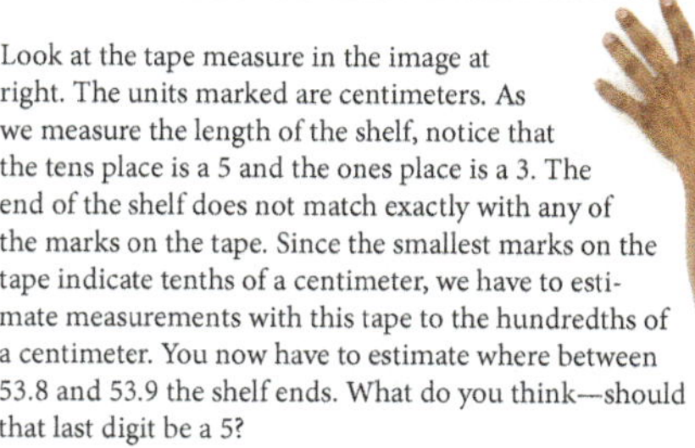

Look at the tape measure in the image at right. The units marked are centimeters. As we measure the length of the shelf, notice that the tens place is a 5 and the ones place is a 3. The end of the shelf does not match exactly with any of the marks on the tape. Since the smallest marks on the tape indicate tenths of a centimeter, we have to estimate measurements with this tape to the hundredths of a centimeter. You now have to estimate where between 53.8 and 53.9 the shelf ends. What do you think—should that last digit be a 5?

1.9 LIMITS OF MEASUREMENT

Uncertainty in Measurements

While quantitative data is less subjective and therefore more repeatable than qualitative data, measurements are never 100% correct. In the example above, you may read the measurement of the shelf as 53.85 cm. Your best friend might tell you that it is 53.84 cm. If you look at the book tonight while doing your homework, you may say that the measurement is 53.86 cm. So which is the correct measurement? As strange as it may seem, all three are perfectly good measurements. Every measurement has a degree of uncertainty to it. The uncertainty comes from both the instrument itself and how the scientist used it.

Accuracy and Precision

Since every measurement has some uncertainty to it, we need a way to tell how good our measurements are. Scientists use two ways to assess measurements—accuracy and precision. Accuracy and precision both compare a measurement to something else.

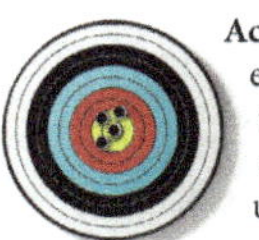

Accuracy compares a measurement to the accepted or expected value of a measurement. When using accuracy, we are looking at how much error is in the measurement. This error is the combination of the instrument uncertainty along with any errors made by the scientist.

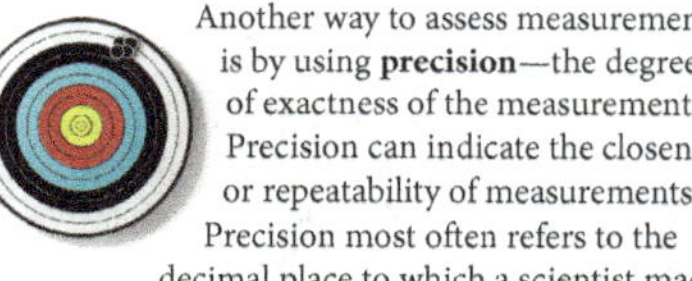

Top: accurate but not precise

Middle: precise but not accurate

Bottom: accurate and precise

Another way to assess measurements is by using **precision**—the degree of exactness of the measurements. Precision can indicate the closeness or repeatability of measurements. Precision most often refers to the decimal place to which a scientist made a measurement. So the instrument has great influence on our precision. An instrument marked to the ones place has a precision to the tenths place because we estimate to one decimal place beyond those marked on the instrument.

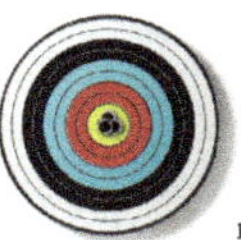

Accuracy

Accuracy is a qualitative description of a measurement. In order to quantify the measurement, we calculate the error.

Error = measured value – accepted value

Error is not always a good indication of accuracy because the accuracy of a measure also depends on how big the measurement is. Making an error of 1 m on the length of a football field is different than making an error of 1 m on the height of a man. For this reason we usually use percent error to quantify accuracy.

$$\%_{error} = \left(\frac{value_{measured} - value_{accepted}}{value_{accepted}} \right) 100\%$$

Sometimes you will see accuracy described as a comparison to the actual value. This description is problematic because, due to the limits of measurements, we almost never know the actual value.

 ### Significant Figures

Students often struggle with the concept of significant figures. At this point it is important to connect significant figures with precision. Students will have ample time and practice with significant figures when they take chemistry and physics.

All calculations in the Student Edition will apply significant figures, but we are not emphasizing it for students. If your students are strong in math, you may consider emphasizing significant figures more. For more information on significant figures, see Appendix B.

Class Discussion: *Significant Figures*

Look again at the data from the class opener. Discuss the precision of the measurements and how many significant figures each measurement recorded. Then write down a new measurement, such as 34.512 g. Ask the following questions.

1. How many significant figures was the measure recorded to? *(5)*

2. To what precision was the measurement made? *(to the thousandth of a gram)*

3. What can you infer about the markings on the instrument? *(If the measurement was taken properly, the instrument is marked to a hundredth of a gram and the thousandths place was estimated.)*

Unit Conversions

Unit conversion utilizes dimensional analysis. Connect the concepts to renaming fractions for fraction addition.

Process

1. Write what you know on the left side of the page.

2. Write the unit to which you are converting on the right side of the page.

3. Write your conversion factors (as many as needed to change to the appropriate unit).

   a. Write a multiplication symbol and the dividing line of a fraction.

   b. Enter the unit that you are trying to convert from into your conversion factor. Position the unit so that it will cancel with the unit that you are trying to convert from.

   c. Enter a unit that you can convert to in the other position of the fraction.

   d. Enter appropriate numbers for each unit into the conversion factor.

4. Cancel any units in the problem.

5. Calculate by multiplying the fractions.

For more practice, do an internet search using the keywords "unit conversion practice."

 ### How It Works Boxes

The Student Edition includes boxes in some chapters that focus on a common device related to the principles being discussed in the chapter. While these contain enrichment information, many students will be interested. Students will gain a greater connection of the content to the world around them.

HOW IT WORKS

Balances and Scales

Pizza! Everyone loves a pizza party. Have you ever been to one and ended up saying, "Wow! I ate way too much pizza!" If so, you may have been reluctant to get on a scale the next day.

Balances and scales basically weigh things. They measure different but related dimensions, and there is much overlap in their use and how we talk about them.

Originally, the balance worked on the principle of, well, balance. It's similar to two kids trying to balance on a seesaw. The double pan balance below was one of the first balances. It works by placing the object that you want to measure in one pan and known masses in the other. When the device is balanced, you know that the unknown mass is equal to the sum of the known masses.

A triple beam balance works on this principle. You place the unknown mass on the platform. You then slide the known masses along the three beams. Once balanced, you record the value of the known mass using the scales on the three beams. This mass is equal to the mass of the unknown object.

Mass and weight are related. On the earth, the relationship between mass and weight is constant. Scales don't measure mass but the related dimension of *weight*—force due to gravity. Within the scale are levers and a spring system. As you push down on the scale platform, the applied force transmits to the spring system. When the spring force is equal to the weight, you can read the weight of the object.

Modern electronic balances act more like scales than balances. Applying a force to the platform causes an electromagnet to create an opposing force. These opposing forces produce an electric signal, which is calibrated to indicate mass.

A measurement made to the thousandths place was made using an instrument marked to the hundredths place. This instrument is much more precise than an instrument marked to the ones place.

Scientists report their precision using *significant figures*. All the known digits of a measurement and the one estimated digit are significant. You will learn more about significant figures in chemistry and physics courses.

1.10 UNIT CONVERSIONS

Your best friend tells you that she did a lot of reading over the summer. She says that she spent 1,260,000 seconds reading. You may have difficulty recognizing whether this is a significant amount of time. If you knew how many hours she read, you could probably better evaluate her claim.

We can convert between any two units of measures (for the same dimension) as long as we know the *conversion factor*. A conversion factor is two quantities with different units that are equivalent to each other and written as a fraction. For example, we know that twelve inches is the same as one foot. So we could write 12 in. = 1 ft. To turn that fact into a conversion factor, divide both sides by 1 ft.

$$\frac{12 \text{ in.}}{1 \text{ ft}} = \frac{1 \text{ ft}}{1 \text{ ft}}$$

$$\frac{12 \text{ in.}}{1 \text{ ft}} = 1$$

We can express this conversion factor in two ways:

$$\frac{12 \text{ in.}}{1 \text{ ft}} \text{ or } \frac{1 \text{ ft}}{12 \text{ in.}}$$

Since the numerator and denominator are equivalent, these conversion factors equal 1, and we write them in a form that allows us to change units. When we change the unit of a measurement using a conversion factor, we really are multiplying by 1. The process is similar to how we rename fractions with different denominators. Let's say we needed to rewrite 1/3 as a fraction with 12 in the denominator. What would we do? Looking at the denominators, we know that we need to multiply the 3 by 4 to get 12. To keep the value the same as our original number, we must multiply by 4/4, which is just another form of 1.

Let's see how much your friend read this summer. This conversion will require two steps.

EXAMPLE 1-1: Unit Conversion

$$1\,260\,000\ \text{s} = ?\ \text{h}$$

$$1\,260\,000\ \text{s}\left(\frac{1\ \text{min}}{60\ \text{s}}\right) = ?\ \text{h}$$

Notice that at this point our units are minutes, but we are looking for hours so we need another conversion factor.

$$1\,260\,000\ \text{s}\left(\frac{1\ \text{min}}{60\ \text{s}}\right)\left(\frac{1\ \text{h}}{60\ \text{min}}\right) = \frac{1\,260\,000\ \text{h}}{3600}$$

$$= 350\ \text{h}$$

Wow! She certainly did do a lot of reading this summer.

EXAMPLE 1-2: Unit Conversion Between Metric Prefixes

How many nanoliters (nL) are in 345ML (megaliters)?

$$345\ \text{ML} = ?\ \text{nL}$$

$$\frac{345\ \text{ML}}{1}\left(\frac{10^6\ \text{L}}{1\ \text{ML}}\right)\left(\frac{1\ \text{nL}}{10^{-9}\ \text{L}}\right) = \frac{345 \times 10^6\ \text{nL}}{1 \times 10^{-9}}$$

$$= 345 \times 10^{15}\ \text{nL}$$

As a final step, convert your answer to *scientific notation*.

$$3.45 \times 10^{17}\ \text{nL}$$

EXAMPLE 1-3: Unit Conversion in Derived Units

What is the speed in km/h of a jet traveling at 153 m/s?

$$\frac{153\ \text{m}}{\text{s}} = ?\ \frac{\text{km}}{\text{h}}$$

$$\frac{153\ \text{m}}{\text{s}}\left(\frac{1\ \text{km}}{10^3\ \text{m}}\right)\left(\frac{60\ \text{s}}{1\ \text{min}}\right)\left(\frac{60\ \text{min}}{1\ \text{h}}\right) = \frac{550\,800\ \text{km}}{1 \times 10^3\ \text{h}}$$

$$= 551\ \frac{\text{km}}{\text{h}}$$

The answer 550.8 kph was rounded to 551 kph to account for the precision of the original measurement. The original measurement had three significant figures, so our answer should have three significant figures.

> **Why Are Unit Conversions Important?**
>
> Scientists often learn from the mistakes that they make. Sometimes those mistakes have huge consequences. There have been some famous cases in which errors have been made in converting units. One costly conversion error resulted in the loss of the $125,000,000 Mars Climate Orbiter. Failure to properly convert between US customary units and SI resulted in the orbiter entering orbit too low, resulting in the destruction of the spacecraft.

 ## Scientific Notation

Students should remember scientific notation from math class. Some will still struggle in knowing what direction to change the exponent or which direction to move the decimal point. A good memory technique is called the *seesaw*. As the decimal part gets smaller, the exponent must get bigger, and as the exponent gets smaller, the decimal number must get bigger.

Practice Problems:

Write in scientific notation.

a. 432 000 000 m *(4.32 × 10⁸ m)*

b. 0.000 000 000 029 74 g *(2.974 × 10⁻¹¹ g)*

c. 0.000 006 58 s *(6.58 × 10⁻⁶ s)*

d. 415.35 × 10⁷ L *(4.153 5 × 10⁹ L)*

Write in standard notation.

a. 3.91 × 10⁶ kg *(3 910 000 kg)*

b. 7.65 × 10⁻⁵ km *(0.000 076 5 km)*

c. 9.772 × 10¹ mL *(97.72 mL)*

d. 5.974 × 10⁻⁸ F *(0.000 000 059 74 F)*

For more practice with scientific notation, do an internet search using the keywords "scientific notation practice."

 ## Converting Between SI and US Customary Units

We don't include any examples of unit conversions between the two systems because we feel it is better to have students use the SI exclusively.

 ## Thinking in SI

Many students never become comfortable with the SI because they never learn to think in SI. They mentally convert between SI and US customary system units. The Student Edition rarely refers to measurements in the US customary system. Using only the SI in the classroom will help students learn the important skill of thinking in SI.

 Mini Labs

Every chapter includes a mini lab. As the name implies, these lab activities are designed to take 15–30 minutes. They are extremely beneficial—they give students hands-on engagement with the content.

Understanding Conversion Factors

Conversion factors can seem as though they were plucked out of thin air. This mini lab emphasizes the equivalence of the two parts of the conversion factor. Students benefit from using their own measurements to determine conversion factors.

Answers

1. A conversion factor is two equivalent quantities with different units arranged as a fraction.

2. Conversion factors are derived by setting two different but equivalent measurements equal to each other (e.g., 12 in. = 1 ft).

3. Answers will vary.

4. $\dfrac{2.54 \text{ cm}}{1 \text{ in.}}$, $\dfrac{1 \text{ in.}}{2.54 \text{ cm}}$

5. The conversion factor came from two measurements of the same dimension expressed in different units. A conversion factor is made of two equal or equivalent measurements and is therefore an equivalence statement.

6. Answers will vary. Students may think of things like defined counts. We don't count out sheets of printer paper at the store—we buy a ream of paper. We don't count out eggs or donuts—we purchase dozens. Others may consider how we pace off a ball field. If we know that a pace is about 3 ft, we would walk thirty paces to measure 90 ft.

Lab 1B: *How Do We Measure Up?*

Lab 1C: *Visual Data*

Once students have an understanding of the material in Section 1C, they should be ready to work through Labs 1B and 1C.

MINI LAB
UNDERSTANDING CONVERSION FACTORS

Essential Question:

Where do conversion factors come from?

Equipment

metric and US customary rulers

rectangular prism

Unit conversions are simple fraction multiplication problems in which you multiply a measurement by a conversion factor to determine the value of that measurement in a different unit. The math is fairly simple, but where does the conversion factor come from?

Answer the questions and follow the lettered steps.

1. What is a conversion factor?
2. How are conversion factors derived?

Procedure

A Using the two rulers, measure the length of the rectangular prism in centimeters and inches.

3. What is the prism's length?

B Use your measurements to create the two possible conversion factors.

$$\frac{\underline{\quad}\,\text{cm}}{\underline{\quad}\,\text{in.}} \qquad \frac{\underline{\quad}\,\text{in.}}{\underline{\quad}\,\text{cm}}$$

4. Determine the conversion factors for cm to in. and in. to cm by reducing each fraction so that the denominators are 1.

C Check your values with a reference source.

Conclusion

5. What does it mean that conversion factors are equivalence statements?

Going Further

6. Give examples of other ways in which we use this equivalence concept.

1C | REVIEW QUESTIONS

1. Compare qualitative and quantitative data.
2. What does SI stand for?
3. Compare fundamental units with derived units.
4. List the seven fundamental SI units, including symbols, with the dimension that each one represents.
5. Give the symbol, factor, and exponential form associated with the metric prefix *micro-*.
6. Explain one way in which the SI is preferable to the US customary system of units.
7. Why can a measurement never be exactly correct?
8. Measure the volume of liquid in the image below.
9. Define *accuracy*.
10. Convert the following:
    a. 37.4 mL into ML
    b. 689 km/hr into m/s
    c. 34.5 m² into mm²

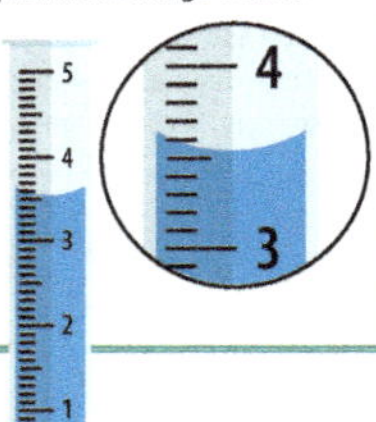

1C REVIEW ANSWERS

1. Both consist of observations about an object or event. Qualitative, or descriptive, data is based on qualities such as color, shape, and texture. Quantitative, or measured, data is based on numbers or quantities, such as length, mass, weight, volume. *(p. 14)*

2. SI stands for *Système international d'unités* (the International System of Units), which is a modified metric system used by scientists. *(p. 14)*

3. Units, both fundamental and derived, are standards of comparison for measurements. Fundamental units are the foundation for all units in the SI and are based on physical constants. Derived units are mathematical combinations of fundamental units. *(pp. 14–15)*

4. meter (m), length; kilogram (kg), mass; second (s), time; ampere (A), electric current; kelvin (K), temperature; mole (mol), amount of a substance; candela (cd), luminosity *(p. 15)*

5. μ; 0.000 001; 10^{-6} *(p. 16)*

6. Accept any of the following: SI is nearly universal (only three countries have not adopted it). SI has one unit for each dimension. SI has easily repeatable standards of comparison. SI is a decimal system, which makes unit conversions easier. *(p. 16)*

CHAPTER 1 REVIEW . . . .

1A ORDER IN OUR WORLD

- Physical science, consisting of chemistry and physics, is a field of natural science that studies matter and energy.
- All evidences of organization and design in nature are examples of order.
- God created all things both good and orderly. Disorder is a consequence of the Fall.
- God commands us to fill the earth and have dominion over it.
- Ethics is the process of deciding between right and wrong.

1A Terms

1B MODELING OUR WORLD

- A model is a physical, conceptual, or mathematical representation of some aspect of the world.
- Chemistry includes models of atoms, chemical reactions, the periodic table of the elements, and the behavior of gases. In physics, we model motion, the behavior of fluids, nuclear reactions, and forces.
- Hypotheses, theories, and laws are all scientific models. Hypotheses and theories explain, while laws describe.
- Scientific inquiry is a multipronged approach to doing science that includes asking questions, doing research, forming hypotheses, conducting investigations, analyzing data, and forming and reporting conclusions.
- The scientific community builds consensus through the peer review process.

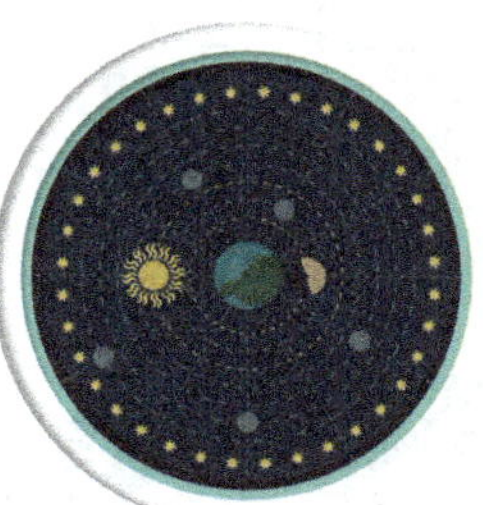

1B Terms

1C USING MATHEMATICS FOR SCIENTIFIC INQUIRY

- The US customary and SI systems of measurement both consist of various units for measuring different dimensions.
- The SI system is the worldwide measurement system for scientific research.
- The SI is beneficial because it is a decimal system, it has few units for each dimension, and it has easily reproduced standards.
- A good measurement is both accurate and precise.
- We do unit conversions through a series of fraction multiplications using conversion factors.

1C Terms

MODELING OUR ORDERLY WORLD   21

---

7. Measurements have some degree of interpretation, and all instruments have some limitation to their precision. *(p. 17)*

8. Accept any number between 3.52 and 3.57. *(pp. 17–18)*

9. Accuracy is a comparison of a measurement to an accepted or expected value. *(p. 17)*

10. a. $37.4 \text{ mL}\left(\dfrac{10^{-3}\text{ L}}{1 \text{ mL}}\right)\left(\dfrac{1 \text{ ML}}{10^{6}\text{ L}}\right) = \dfrac{37.4 \times 10^{-3} \text{ ML}}{1 \times 10^{6}} = 3.74 \times 10^{-8} \text{ ML}$

  b. $689 \dfrac{\text{km}}{\text{h}}\left(\dfrac{10^{3} \text{ m}}{1 \text{ km}}\right)\left(\dfrac{1 \text{ h}}{3600 \text{ s}}\right) = \dfrac{689 \times 10^{3} \text{ m}}{3600 \text{ s}} = 191 \dfrac{\text{m}}{\text{s}}$

  c. $34.5 \text{ m}^{2}\left(\dfrac{1 \text{ mm}}{10^{-3} \text{ m}}\right)\left(\dfrac{1 \text{ mm}}{10^{-3} \text{ m}}\right) = \dfrac{34.5 \text{ mm}^{2}}{10^{-6}} = 3.45 \times 10^{7} \text{ mm}^{2}$ *(all pp. 18–19)*

Recalling Facts

1. Physics is the study of matter and energy and the interactions between them. *(p. 5)*

2. Physical science is the branch of natural science that investigates nonliving matter and energy. *(p. 5)*

3. Answers will vary. Scientists imitate the order in nature when they categorize things and when they design objects. *(p. 6)*

4. the principle of uniformity *(p. 6)*

5. The *principle of cause and effect* states that every effect has an identifiable cause and every cause has a predictable effect. The existence of the world requires a cause that is not a part of the world. God was that cause. The *principle of uniformity* is an aspect of the order that was created into the world. It is reasonable that a God of order would create a world that is predictable. *(pp. 6–7)*

6. Answers will vary. Students should include any elements of design, cyclical processes, or patterns. *(p. 6)*

7. God created order in the universe when He created it. *(p. 7)*

8. We do science to glorify God, to exercise dominion, and to help others. (Accept any two.) *(p. 7)*

9. The Creation Mandate is the command given in Genesis 1:26–28 to fill and subdue the earth. *(p. 8)*

10. What does God's Word say? What results are right? How can I grow through this decision? *(pp. 9–10)*

11. Biblical motivations involve how a particular decision will help us grow in our faith in God, our hope in God's promises, and our love for God and others. *(p. 10)*

12. A scientific law is a model that describes phenomena under certain conditions. *(p. 11)*

13. Answers will vary. Examples include atomic models, the periodic table of the elements, gas laws, wave models, laws of motion, the heliocentric model, and the kinetic-molecular theory. *(pp. 11–12)*

14. observation *(p. 13)*

15. a hypothesis *(p. 13)*

16. peer review *(p. 13)*

17. qualitative data *(p. 14)*

18. SI *(p. 14)*

19. ampere (A) *(p. 15)*

20. kilometer *(pp. 15–16)*

21. precision *(p. 17)*

22. The graduated cylinder on the right is more precise. It is marked to the ones place, which means that measurements can be made to the tenths place. The other cylinder is marked to the tens place, so measurements are precise only to the ones place. *(pp. 17–18)*

CHAPTER 1

CHAPTER REVIEW QUESTIONS

Recalling Facts

1. Define *physics*.
2. Define *physical science*.
3. Describe a way that scientists imitate the order in nature.
4. What principle is the basis for expecting events to happen tomorrow the same as they happened today?
5. How do the universal principles of cause and effect and uniformity support the Christian worldview?
6. List evidences of order in nature.
7. What is the source of order in the universe?
8. Give two reasons for us to do science.
9. What is the Creation Mandate?
10. Write the three questions that lead you through ethical decision-making from a Christian perspective.
11. Summarize the three aspects of biblical motivations.
12. What is a scientific law?
13. Name two models used in physical science.
14. What activity is central to scientific inquiry?
15. What do we call the initial, testable explanation for a phenomenon used by scientists to guide their investigation?
16. Through what process is scientific inquiry assessed?
17. Color, texture, and shape are all examples of what type of data?
18. What system of units do scientists use?
19. What is the SI fundamental unit for electric current? Include its symbol.
20. What unit would be best to measure the distance from New York to San Francisco?
21. What describes the exactness of a measurement?
22. Of the two graduated cylinders on the right, which is more precise? Explain.

Understanding Concepts

23. How does physical science relate to other sciences?
24. Why is order in nature important?
25. Why is studying science a worthwhile activity for a Christian?
26. How are science and ethics related?
27. Explain how Christians should make ethical decisions.
28. Why are models important for studying science?
29. Compare hypotheses, theories, and laws.
30. What does it mean to say that a model is workable?
31. Why do we need a process to study science?

Understanding Concepts

23. Physical science is a branch of natural science and includes chemistry and physics. It overlaps life science in areas such as biochemistry, biophysics, nutrition, and medicine. Physical science overlaps earth and space science in fields such as mineralogy, geophysics, and rocket science. *(p. 5)*

24. Order in nature allows us to do science. It allows us to recognize patterns and make predictions. Order was created by God, and we imitate Him when we create models of the natural world. *(pp. 6–7)*

25. We can fulfill the Creation Mandate through science and can glorify God by studying His Creation. We imitate the

32. You are comparing two measurements: 35.21 cm and 49.6 cm. Which is more precise? Explain.

33. The length of a string is measured to be 22.87 mm. What do you know about the ruler that was used for this measurement?

34. If the accepted value for the length of the string in Question 33 is 24.11 mm, was the string measured accurately? Explain.

35. Convert the following:

  a. 34.5 mA into A

  b. 22.5 m/s into km/h

Critical Thinking

36. At the end of the first day of school, your science teacher shows you two candles (one taller than the other). She asks what would happen if you lit them both and then covered them with a jar. Use the scientific inquiry infographic (p. 13) to outline how you would investigate this problem.

37. Evaluate this statement: I know that the length of the board in my garage is exactly 2.74 m.

38. Why would the concept of significant figures not apply to the fact that there are twenty-three students in your class?

ETHiCS — REPORTING SCIENTIFIC DATA

No scientist can conduct all the research in any one field, so the scientific community depends on the research of others. Scientists publish papers, which are then used by other researchers to conduct future research. Much of the research in biological science is also used to make decisions related to medical treatment. We would like to be able to trust all the academic research we read. But scientists, just like all people, are sometimes tempted to lie.

In 2012, the US Office of Research Integrity concluded a ten-year investigation into a cardiology researcher accused of falsifying data. The investigation detailed forty-five instances of the use of false data in publications. Various government agencies had awarded over $8 million to this researcher on the basis of his distorted data; other researchers could have more legitimately

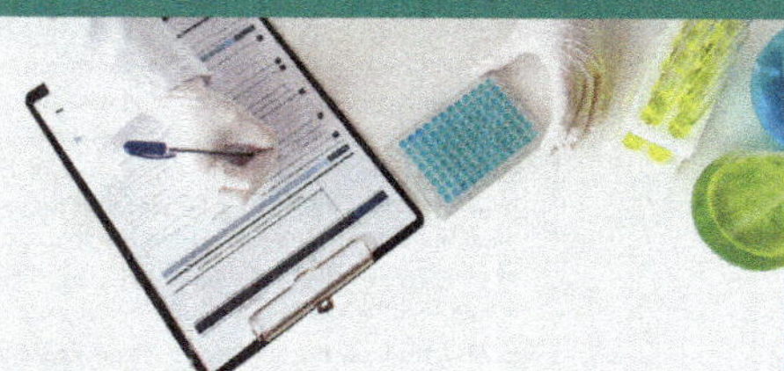

used those funds. Additionally, other researchers cited the papers that were written on the basis of this data over one hundred times. This research may even have caused doctors to make incorrect treatment decisions for cardiac patients.

Use the Ethics box above to answer Questions 39–42.

39. What does God's Word say about how we report data?

40. What are the acceptable results?

41. What benefit does a Christian scientist gain from the acceptable results above?

42. What might motivate scientists to operate in a dishonest way?

---

creative work of God when we make models and invent things. We help others when we use what we learn through science. *(pp. 7–8)*

26. Science cannot answer ethical questions, but there are many areas in which we have to decide ethical issues in the work of science. *(p. 9)*

27. Christians make ethical decisions by applying the biblical principles to achieve biblical outcomes, guided by biblical motives. *(pp. 9–10)*

28. Models allow us to explain and describe aspects of the world. They allow us to study a simplified version of a complex world. *(p. 11)*

29. Hypotheses, theories, and laws are all scientific models. Hypotheses and theories explain, while laws describe. Hypotheses are initial, testable explanations used to guide scientific investigations. *(pp. 11, 13)*

30. A model is workable to the degree that it explains or describes observations and makes accurate predictions. *(p. 12)*

31. The universe is an orderly place; to study it we need an orderly process. Also, one purpose for studying science is to help others, and an orderly method is the most efficient. *(pp. 12–13)*

32. Since 35.21 cm is measured to the hundredths place, it is more precise. *(p. 17)*

33. The ruler was marked in millimeters. Because the measurement is to the hundredth of a millimeter, we know that the smallest marks on the ruler must be tenths of a millimeter. *(pp. 17–18)*

34. The percent error of the measurement is −5.14%.

$$\%_{error} = \left(\frac{value_{measured} - value_{accepted}}{value_{accepted}}\right)100\%$$

$$= \left(\frac{22.87 \text{ mm} - 24.11 \text{ mm}}{24.11 \text{ mm}}\right)100\%$$

$$= \left(\frac{-1.24 \text{ mm}}{24.11 \text{ mm}}\right)100\%$$

$$= -5.14\%$$

Whether a 5.14% error is considered accurate or not is a judgment call. Some students may consider this to be an accurate measurement while others may not. *(pp. 17–18)*

35. **a.** $34.5 \text{ mA}\left(\dfrac{10^{-3}\text{ A}}{1 \text{ mA}}\right) = 3.45 \times 10^{-2}$ A

  b. $22.5 \dfrac{\text{m}}{\text{s}}\left(\dfrac{1 \text{ km}}{10^3 \text{ m}}\right)\left(\dfrac{3600 \text{ s}}{1\text{h}}\right)$

    $= \dfrac{8.1 \times 10^4 \text{ km}}{10^3 \text{ h}} = 81.0$ kph

(all pp. 18–19)

Critical Thinking

36. <u>Observations:</u> relative heights of the candles, the arrangement of candles under the jar

<u>Questions:</u> Why do candles burn? What would make a candle go out? What happens to the wax as the candles burn?

<u>Research:</u> burning in general, candle burning in particular

<u>Hypothesis:</u> Possibilities: (1) Once we light the candles, they will continue to burn until the wax is used up. (2) Both candles will burn until the oxygen is used and will go out at the same time. (3) The candles will both burn, producing carbon dioxide, and the tall candle will go out first because the warm carbon dioxide will accumulate at the top of the jar. (4) The candles will both burn, producing carbon dioxide, and the short candle will go out first because the heavier carbon dioxide will accumulate at the bottom of the jar.

<u>Investigation:</u> Set up two pairs of candles. Light a l four. Cover two with a jar and leave the other two (control) burning uncovered. Observe which candles go out and which candles remain lit.

Observe the temperature of the top and bottom of the jar.

Analysis: Note the order that the candles went out (tall then short, control candles remaining lit). Note that the jar top is much warmer than the bottom. When we extinguished the two control candles, the smoke continued to rise.

Conclusion: The investigation supported the third hypothesis only. *(p. 13)*

37. The measurement may be accurate and is precise to the hundredth of a meter. This statement is false because the word *exactly* was used. *(p. 17)*

38. There can't be partial students in the class. There are either twenty-three or there are twenty-four. We can know counts exactly; therefore there is no need to think about the number of significant figures. *(pp. 17–18)*

 **Ethics Boxes**

The Student Edition includes boxes in some chapters that focus students on ethical issues related to physical science. While it's great for students to gain knowledge about science, they need to be able to make decisions from a biblical framework. These boxes give students an opportunity to think through ethical decisions in an academic setting before they have to make similar decisions in life.

39. Answers will vary. God is clear in the Ten Commandments that we are not to lie. Proverbs 22:1 teaches that a good name is more valuable than great riches. Mark 10:19 teaches us not to defraud others.

40. Answers will vary. We want to get credit for work that we have done, and we should never try to take credit for the work of someone else. We should be willing to share our findings as long as they are true. We will make errors at times and should always correct any as soon as we are aware of them.

41. We are motivated in our work to show ourselves faithful. We can help others best by showing ourselves trustworthy so that our work will be trusted and be useful to others.

---

32. You are comparing two measurements: 35.21 cm and 49.6 cm. Which is more precise? Explain.

33. The length of a string is measured to be 22.87 mm. What do you know about the ruler that was used for this measurement?

34. If the accepted value for the length of the string in Question 33 is 24.11 mm, was the string measured accurately? Explain.

35. Convert the following:

   a. 34.5 mA into A

   b. 22.5 m/s into km/h

Critical Thinking

36. At the end of the first day of school, your science teacher shows you two candles (one taller than the other). She asks what would happen if you lit them both and then covered them with a jar. Use the scientific inquiry infographic (p. 13) to outline how you would investigate this problem.

37. Evaluate this statement: I know that the length of the board in my garage is exactly 2.74 m.

38. Why would the concept of significant figures not apply to the fact that there are twenty-three students in your class?

ETHiCS   REPORTING SCIENTIFIC DATA

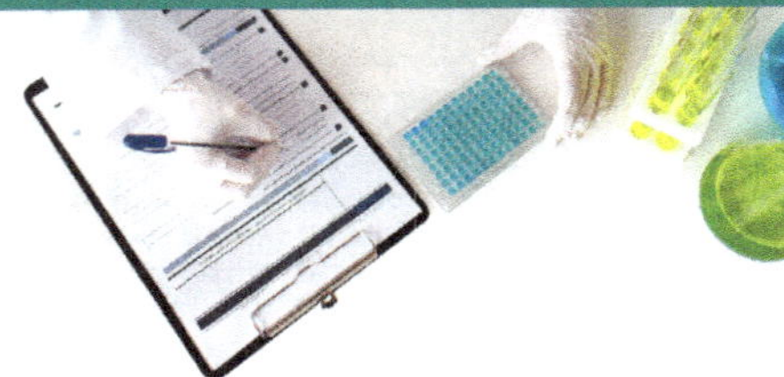

No scientist can conduct all the research in any one field, so the scientific community depends on the research of others. Scientists publish papers, which are then used by other researchers to conduct future research. Much of the research in biological science is also used to make decisions related to medical treatment. We would like to be able to trust all the academic research we read. But scientists, just like all people, are sometimes tempted to lie.

In 2012, the US Office of Research Integrity concluded a ten-year investigation into a cardiology researcher accused of falsifying data. The investigation detailed forty-five instances of the use of false data in publications. Various government agencies had awarded over $8 million to this researcher on the basis of his distorted data; other researchers could have more legitimately used those funds. Additionally, other researchers cited the papers that were written on the basis of this data over one hundred times. This research may even have caused doctors to make incorrect treatment decisions for cardiac patients.

Use the Ethics box above to answer Questions 39–42.

39. What does God's Word say about how we report data?

40. What are the acceptable results?

41. What benefit does a Christian scientist gain from the acceptable results above?

42. What might motivate scientists to operate in a dishonest way?

---

42. In this case, there was a monetary gain for the scientist. Scientists also may want to become well known for their research. Or they may spend decades working on one topic only to discover that they were wrong—and then they may have difficulty accepting what they see as failure. But even when scientists are wrong, the time invested is not wasted as long as the scientist learns from the experience.

Propagation of False Data

While the falsifying of data is a problem in itself, the issue is magnified when the fraudulent research is used as the basis for other research or for making medical decisions. When other researchers cite papers supported by false data, that data has a greater impact: the false data is used to guide new research. The new research may then be again cited and the problem grows larger with each citation.

Main image: *The amazing design of a gecko's feet enables it to climb walls and even walk on the ceiling. The tokay gecko shown in this image effortlessly walks down a wall.*

Inset: *Material scientists at Stanford University have built a stickybot that can climb walls just like a gecko.*

Matter

IMITATING GECKOS

Geckos have a superpower: defying gravity. The ability of the gecko on the left is not due to its feet being sticky; it's because they are hairy. The tiny hairs on its feet are about 1/30th the diameter of a human hair, and there are thousands of them on every square millimeter of a gecko's feet! With each step, the gecko spreads his toes, maximizing the contact area of each foot. This action creates a tiny electrostatic attraction with the surface. The combination of these tiny forces creates a substantial attractive force. Happily for the gecko, God also designed geckos so that they can release that attraction!

Materials scientists are studying the gecko to see whether they can copy them for a variety of applications. Some scientists are working on gecko tape that acts like a gecko's feet. Others are working on robots that can climb walls, enabling them to do jobs that would be dangerous for people to do. Could scientists develop gloves and boots that could allow us to climb walls and ceilings? You'll have to *hang on* and see.

The Future of Material Science

Some students may be interested in researching the future of nanotechnology or another aspect of materials science. Suggest the following keyword searches: "carbon fiber," "carbon nanotubes," "nanobots," and "new construction materials."

CHAPTER 2 OBJECTIVES

- Summarize the particle model of matter.
- Classify matter on the basis of its physical properties.
- Distinguish between different states of matter.
- Explain the relationship between energy and states of matter.

Overview

Chapter 2 is foundational. Chemistry is the study of matter and its interactions and changes. Physics is the study of the interactions of matter and energy. Thus, an understanding of matter is essential to an understanding of physical science. Be sure to teach this chapter.

2A **Understanding Matter**

   2.1 Defining Matter

   2.2 Modeling Matter

   2.3 Measuring Matter

2B **Classifying Matter**

   2.4 Why Classify?

   2.5 How Do We Classify Matter?

2C **States of Matter**

   2.6 States and the Particle Model

2D **Changes in Matter**

   2.7 Physical Properties

   2.8 Chemical Properties

   2.9 Conservation of Matter

Lab Activities

Lab 2A: *Has Mass, Occupies Space*—In this lab activity, students use a smaller model to estimate the mass of a larger object.

Lab 2B: *Something Old, Something New?*—In this lab activity, students observe various physical and chemical changes to determine the type of change that occurred.

Preparation: Use four beakers or other types of containers of the same volume. Fill one with water and one with sand. Leave a third empty. Place a votive candle or birthday candle in the fourth.

Performance: Light the candle and then ask questions to prompt students to think about what constitutes matter.

Discussion:

1. Which of the containers contain matter? *(All of them do, but students may identify certain containers.)*

2. Does the empty container contain any matter? *(Yes. It is full of air.)*

3. Which container contains the most matter? *(The one containing sand contains the most mass, which is the measure of matter. Some students may want to say that they all contain the same amount because they all have the same volume.)*

4. How do you know that the one with sand has the greatest mass? *(It is the densest, meaning that it has the most mass for an equal volume.)*

5. Do any of these contain something that is not matter? *(Yes. They all contain energy too, but the burning candle is obviously releasing thermal energy.)*

2A | UNDERSTANDING MATTER

Why is matter so important?

2.1 DEFINING MATTER

Materials scientists get inspiration from a variety of sources as they try to solve problems. They look for new ways of using existing materials. They change production methods so that the material has different properties. They also work to create new substances, such as gecko tape. But what are these materials made of?

They are all made of *matter*.

In the photo below, what do you see that you think is made of matter? Is there anything that you wouldn't consider a material? What about the lights and sounds that you would see and hear if you were at this place? While you can perceive both, neither of them are matter.

So what is matter? Matter is so fundamental that it's hard to give it a definition that is based on simpler terms. Scientists define **matter** as anything that occupies space and has mass. An easy-to-understand definition is that matter is the stuff things are made of. Notice that most things (metal, wood, even air) fit this definition. But you should also recognize that light, sound, and warmth don't. They are forms of *non-matter*.

TEACHING THE MATERIAL

ESSENTIAL QUESTION

Why is matter so important?

OBJECTIVES

- 2A1 Define *matter*.
- 2A2 Evaluate how well models of matter represent physical matter. **BWS**
- 2A3 Calculate mass, volume, and density using the formula for density.
- 2A4 Explain the difference between mass and weight.

RESOURCES

📷 Demonstrating Matter, Demonstrating Diffusion, Demonstrating Density

🔬 Mini Lab: *Measuring Volume;* Lab 2A: *Has Mass, Occupies Space*—Modeling Matter

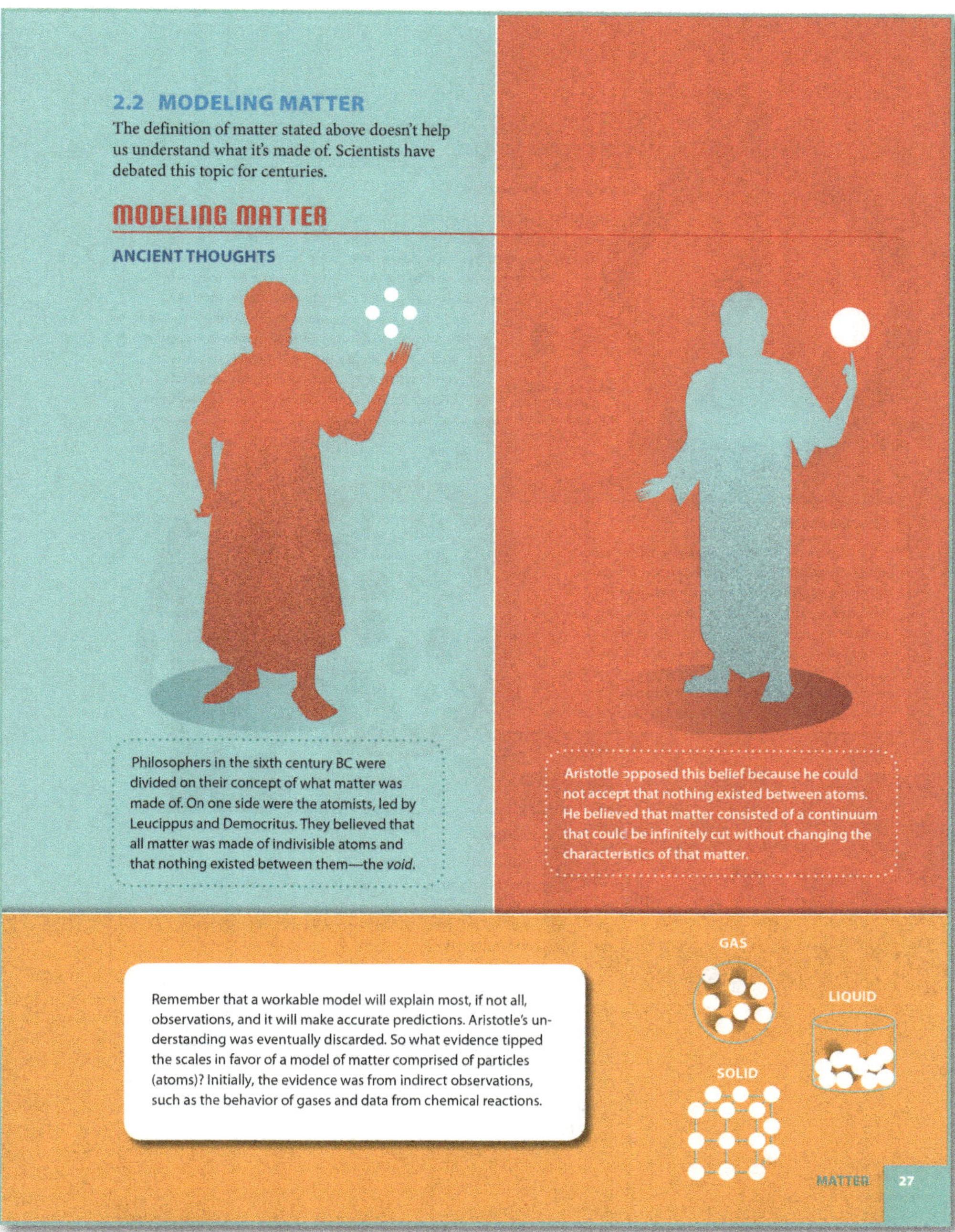

Class Opener: Use the Demonstrating Matter teacher note on page 26 to open a discussion regarding what matter is.

Demonstration: Use the Demonstrating Diffusion teacher note on page 28 to help students understand observations that support the particle model of matter.

Formative Assessment: Use the Formative Assessment: Particle Model of Matter teacher note on page 28 to check students' understanding of this model.

Demonstration: Use the Demonstrating Density teacher note on page 29 to stimulate students' thinking about density.

Formative Assessment: Use the Formative Assessment: Properties of Matter teacher note on page 30 to gauge students' level of knowledge and understanding.

Ticket Out the Door: Explain how air is matter even though we can't see it.

Preparation: Bring a can of air freshener to class. Be aware that some students may have a sensitivity to fragrances.

Performance/Discussion: As you are discussing Brownian motion and diffusion, tell students that you are going to spray the air freshener.

1. Who do you expect will smell the scent first? *(The students should predict that the students who are sitting closest will smell the scent first.)*

2. Will it make a difference if I spray it away from the class? *(Yes and no. It will take longer for the students to detect the scent, but eventually everyone will smell it, and the order in which students notice it should be the same as predicted.)* (Note that some students may assume that the scent will never get to them if you spray it away from them.)

Remind students to raise their hands when they smell the scent, and then spray the air freshener away from the class.

3. How did the scent end up moving toward you when I sprayed it away? *(The gases expanded to fill the container. Once the freshener was out of the can, the room became the container for the gas. As the particles collided, they changed direction and spread throughout the room.)*

(icon) *Formative Assessment:*
Particle Model of Matter

1. What are the building blocks of all matter? *(All matter is made up of atoms. Atoms can combine to make molecules.)*

2. Where did the word *atom* come from? *(The English word* atom *came from the Greek word meaning "indivisible" because the Greeks believed that the atom was the smallest particle and it could not be further divided.)*

3. What was the most common theory that competed with the particle model? *(Aristotle believed that matter was a continuum that could be cut continuously.)*

4. Does the Greek description of atoms as indivisible still hold true? *(No. We now know that atoms are made of protons, electrons, and usually neutrons.)*

The debate between Aristotle's theories and the ideas about particles raged for 2400 years, from the sixth century BC until it was settled in the early 1900s. Finally, sufficient scientific evidence allowed scientists to develop the **particle model of matter**, which states that all physical matter exists in the form of particles (atoms or molecules) in constant motion. This model is also called the *kinetic model* of matter.

5. Is it a problem that the Greek model had to be corrected? *(No. Doing science involves modeling, and scientists are always working to make models more workable.)*

6. What does this understanding tell us about the use of scientific models and truth? *(Models and thus science work in the realm of workability. Scientists need to realize that their models are not producing truth.)*

These particles are known today as atoms and molecules. An **atom** is the building block of all matter. It consists of *protons*, *electrons*, and (usually) *neutrons*. A **molecule** is a distinct group of two or more atoms *covalently bonded* together (see Chapter 5). The particle model is an excellent model because it explains most of the features of matter. Scientists accept it because it is more workable than other models. We will use the particle model as we learn more about matter.

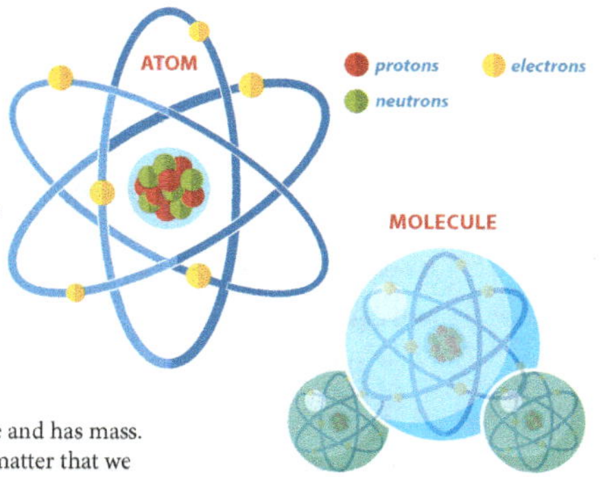

2.3 MEASURING MATTER

As defined above, matter is anything that takes up space and has mass. Mass (m) and space, or volume (V), are two aspects of matter that we can measure. **Mass** is the amount of matter in an object. The SI unit of mass is the gram, and we measure mass with a balance. **Volume** is the space enclosed or occupied by an object. Scientists can measure volume in liters or cubic meters. These are the most basic properties of matter. We can derive a third property—density—from these two. **Density** (d) is the mass of matter contained within a particular volume. The formula for density is shown below.

$$d = \frac{m}{V}$$

You can see how density is derived by dividing mass by volume. Density is an important *physical property* of matter that can help us identify substances. Let's look at some examples.

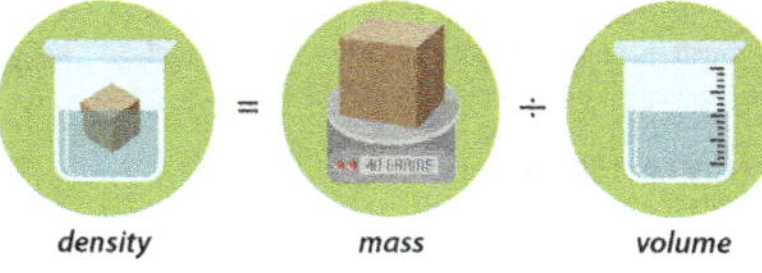

density = *mass* ÷ *volume*

EXAMPLE 2-1: Calculating Density

A sample of aluminum has a volume of 14.8 mL and a mass of 39.96 g. What is the density of aluminum?

Write what you know.

$V = 14.8$ mL
$m = 39.96$ g
$d = ?$

Write the formula and solve for the unknown.

$$d = \frac{m}{V}$$

Plug in known values and evaluate.

$$d = 2.70 \ \frac{g}{mL}$$

Notice that the units don't cancel, so we end up with g/mL, which is the unit for density.

Demonstrating Density

Preparation: You will need a clear container almost completely filled with water. A fish tank will work well. You will also need a can of diet soda and and a can of regular soda. (The demonstration works best if the two cans are the same brand of soda.)

Performance/Discussion: Show students the two cans of soda.

1. Describe the two cans. *(Students should recognize that the cans are the same size and thus have the same volumes. Their relative masses and densities may differ.)*

2. Predict whether the cans will float or sink. *(Predictions will vary.)*

Place both cans in the water, making sure that no air is trapped under the cans. The diet soda should float and the regular soda should sink.

3. Why do you think the cans behaved as they did? *(Both sodas are solutions with material dissolved in water. The regular soda has more mass (sugar) dissolved in the same volume, resulting in a higher density.)*

4. What do you think makes the diet soda less dense than the water? *(The sodas have gas dissolved in them, reducing the density of both, but the diet soda, not having the added sugar, has a total density lower than that of water.)*

Solving for the Unknown

Point out to students that Example 2-1 instructs them to "write the formula and solve for the unknown." In this case no "solving" is necessary, but they should always be thinking about how to solve for the unknown.

Example 2-1 Significant Figures

The answer was rounded to three digits because the volume is measured with a precision of only three significant figures. When multiplying or dividing, the answer is limited by the measurement with the least number of significant figures.

 **Example 2-2 Significant Figures**

In this example the answer was rounded to four digits because the volume, which is the only measured value, has four significant figures.

Example 2-3 Significant Figures

The answer is rounded to four digits because the mass, which is the only measured value, has four significant figures.

EXAMPLE 2-2: Calculating Mass

What will be the mass of an iron bar with a volume of 35.74 mL? Iron's density is 7.874 g/mL.

Write what you know.

$$V = 35.74 \text{ mL}$$
$$d = 7.874 \text{ g/mL}$$
$$m = ?$$

Write the formula and solve for the unknown.

$$d = \frac{m}{V}$$
$$dV = \frac{m}{V}V$$
$$m = dV$$

Plug in known values and evaluate.

$$m = \left(7.874 \frac{g}{mL}\right)(35.74 \text{ mL})$$
$$= 281.4 \text{ g}$$

The multiplication of the units in this problem is like any fraction multiplication. The mL in both the numerator and the denominator cancel, leaving us with grams, the unit of mass.

EXAMPLE 2-3: Calculating Volume

How much space will a sample of gold with a mass of 175.6 g occupy? The density of gold is 19.3 g/mL.

Write what you know.

$$m = 175.6 \text{ g}$$
$$d = 19.3 \text{ g/mL}$$
$$V = ?$$

Write the formula and solve for the unknown.

$$d = \frac{m}{V}$$
$$dV = \frac{m}{V}V$$
$$\frac{dV}{d} = \frac{m}{d}$$
$$V = \frac{m}{d}$$

Plug in known values and evaluate.

$$V = \frac{175.6 \text{ g}}{19.3 \frac{g}{mL}}$$
$$= 9.10 \text{ mL}$$

In this case, the units *grams* cancel out. To understand how we end up with mL, the unit for volume, think about the division of fractions.

$$\frac{g}{\left(\frac{g}{mL}\right)} = g\left(\frac{mL}{g}\right) = mL$$

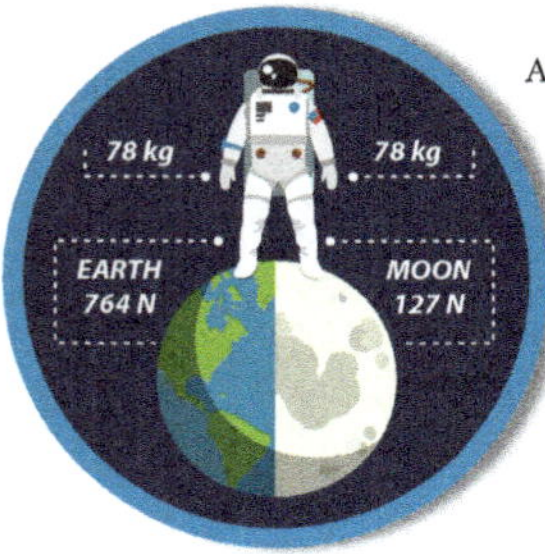

Another property of matter—weight—is closely related to mass but is different. **Weight** is the measure of gravity acting on the matter in an object. The SI unit for weight (a force) is the newton (N). The terms *weight* and *mass* are often used interchangeably because on Earth the relationship between them is constant. Mass doesn't change on the basis of location, but weight can. A 78 kg astronaut on Earth has a weight of 764 N. That same astronaut on the moon still has a mass of 78 kg but weighs only 127 N due to the much lower gravity of the moon.

English System Equivalent

Since American students are not usually familiar with newtons, they will probably ask what 764 N is in pounds. It is approximately 172 lb, found by using the conversion factor 1 lb = 4.45 N.

Varying Weights

The weight of an object can vary slightly on Earth depending on its location. Gravitational force varies slightly due to elevation and variations in the earth's density.

Formative Assessment:
Properties of Matter

1. What are the two properties that define matter? *(mass and volume)*

2. Give three evidences that matter is made of particles. *(the law of definite proportions, Brownian motion, and diffusion)*

3. What property of matter is a derived quantity? *(density)*

2A REVIEW ANSWERS

1. Matter is anything that has mass and takes up space. *(p. 26)*

2. Answers will vary. Air, water, and iron are three examples. *(p. 26)*

3. (1) Matter is made of indivisible particles with empty space between them. (2) Matter is continuous and could be cut infinitely small without changing its properties. *(p. 27)*

4. The atomist theory implied that void or nothingness existed between atoms. Aristotle couldn't accept the concept of the void. *(p. 27)*

5. Brown concluded that the movement of parts of plant spores had to be caused by something. Even though he could not see the particles, he could see the effect of their constant random motion. *(p. 28)*

6. All physical matter exists in the form of particles (either atoms or molecules) in constant motion. *(p. 28)*

7. Atoms and molecules are both particles. An atom is the smallest particle of an element. Molecules are groups of atoms (of the same or different elements) covalently bonded together. *(p. 29)*

MEASURING VOLUME

Fish producers need to know how many trout they can safely transport in a truck. Trout are not spherical or cylindrical, so finding their volume would be difficult. But we can find their volume by water displacement. Fish producers partially fill the truck with water, then add trout until the water rises to the full level. Because of the density of trout, they know that 0.46 kg of trout displaces 0.45 kg of water. This lab activity will help you understand how to determine the volume of an irregularly shaped object.

Procedure

A Measure the mass of the object and record your data on a piece of paper.

B Fill the graduated cylinder about half full with water. Measure and record the volume of water.

  1. What will happen to the water as you lower the object into it?

C While tipping the graduated cylinder, slide the object into the water.

D Stand the graduated cylinder upright. Measure and record the new volume of the cylinder.

E Subtract the initial volume from the final volume and record your answer.

  2. What does this calculated volume represent?

  3. Calculate the density of the irregularly shaped object. Check the value with your teacher.

Conclusion

  4. Why does this method allow you to find the volume of the object?

Going Further

  5. Check online resources for densities of common metals. Use this information to identify the metal you think composes the object.

Essential Question:

How can we find the volume of an irregularly shaped object?

Equipment
laboratory balance
metal object, irregularly shaped
plastic graduated cylinder
water

2A | REVIEW QUESTIONS

1. Define *matter*.
2. Give three examples of matter.
3. Describe the two competing models of matter in ancient Greece.
4. Why was Aristotle opposed to one of the models?
5. How did Robert Brown's observations support the particle model of matter?
6. State the particle model of matter.
7. Compare atoms and molecules.
8. What volume will a sample of copper with a mass of 98.2 g occupy? The density of copper is 8.96 g/mL.
9. Compare mass and weight.

---

8. What we know:
$m = 98.2$ g, $d = 8.96$ g/mL

Unknown: V

Write the formula and solve for the unknown:

$$d = \frac{m}{V}$$

$$dV = \frac{m}{\cancel{V}}\cancel{V}$$

$$\frac{\cancel{d}V}{\cancel{d}} = \frac{m}{d}$$

$$V = \frac{m}{d}$$

Evaluate:

$$V = \frac{98.2 \ \cancel{g}}{8.96 \ \dfrac{\cancel{g}}{mL}}$$

$$= 11.0 \text{ mL } (p.\ 30)$$

9. Mass and weight are two properties of matter. Mass is the amount of matter contained in an object. Weight is the force due to gravity acting on an object. *(pp. 29–30)*

Measuring Volume

Densities

Prior to starting the lab activity, calculate the densities of the objects that you are using. This will allow you to check the answers to Question 3.

Prelab Discussion

1. How can we determine volume of an object like a box? *(On a regularly shaped object, we can determine volume by measuring the dimensions and using a volume formula.)*

2. How could we determine the volume of an irregularly shaped object? *(Irregularly shaped objects are more difficult. In these cases, we can often use a procedure called water displacement.)*

Answers

1. The water level will rise as the object pushes the water out of its way.

2. the volume of the irregularly shaped object

3. Answers will vary.

4. Two objects can't occupy the same space at the same time, so the amount of water displaced is equal to the object's volume.

5. Answers will vary. Students may be able to narrow down the metal to a few choices. Some will look at other properties and may be able to identify the metal.

Archimedes's Principle

For a computer-based activity that applies Archimedes's principle, do an internet search using the keywords "Archimedes's principle interactive simulation" and you will find several. These simulations allow students to manipulate the variables involved with an object's displacement.

Lab 2A: *Has Mass, Occupies Space*

Once students have an understanding of the material in Section 2A, they should be ready to work through Lab 2A.

1. What are some different things that we classify or categorize? *(Students may think of cars, movies, books, animals, or chemicals, to name a few.)*

2. What are some of the different ways that we classify or categorize things? *(size, shape, age, style, genre, behavior)*

3. Why do we classify or categorize things? *(Organization helps us to study, describe, and remember things.)*

Active Learning: *Classifying*

To help students understand the classification concept, you could have them organize themselves. They can get into groups or form a line. They can organize by gender, height, age, or name. You may decide to tell them how to organize, or you may leave that up to them.

Demonstrating Types of Matter

Preparation: You will need a helium-filled balloon, aluminum foil, a bottle of solid iodine, a beaker of water, salt in a salt shaker, a packet of sugar, a bottle of salad dressing, a bag of trail mix, a bottle of iced tea, and a piece of steel. All these items should be labeled. This activity can be done either before or after you have taught the classification system.

Performance/Discussion:

- Ask students to identify the materials as elements, compounds, or mixtures (heterogeneous and homogeneous).

- Elements: a helium-filled balloon, aluminum foil, iodine

- Compounds: water, salt, sugar

- Heterogeneous mixtures: salad dressing, trail mix

- Homogeneous mixtures: iced tea, steel

2B Questions

- How are mixtures different from pure substances?
- How can I recognize pure substances?
- How can I recognize mixtures?

2B Terms

pure substance, element, compound, mixture, heterogeneous mixture, homogeneous mixture

2B | CLASSIFYING MATTER

How do scientists classify matter?

2.4 WHY CLASSIFY?

Man has been classifying things since the beginning of time. Shortly after Creation, Adam named all the animal kinds that God created. Classification provides a structure within which we can conduct a scientific study. Classification systems are scientific models. Today we have the Linnaean classification system, which you learned about in life science class. Biologists classify organisms into domains, kingdoms, phyla, classes, orders, families, genera, and species. Chemists classify elements as metals, nonmetals, and metalloids, as well as into families, such as alkaline earth metals or halogens.

2.5 HOW DO WE CLASSIFY MATTER?

So how can we classify matter? We classify matter according to its properties. These properties can be physical, chemical, or nuclear. Scientists base the broadest classification of matter on whether it is a pure substance or a mixture.

TEACHING THE MATERIAL

ESSENTIAL QUESTION

How do scientists classify matter?

OBJECTIVES

- 2B1 Compare mixtures and pure substances.
- 2B2 Classify pure substances as elements or compounds.
- 2B3 Classify substances as pure substances or mixtures.

RESOURCES

Demonstrating Types of Matter

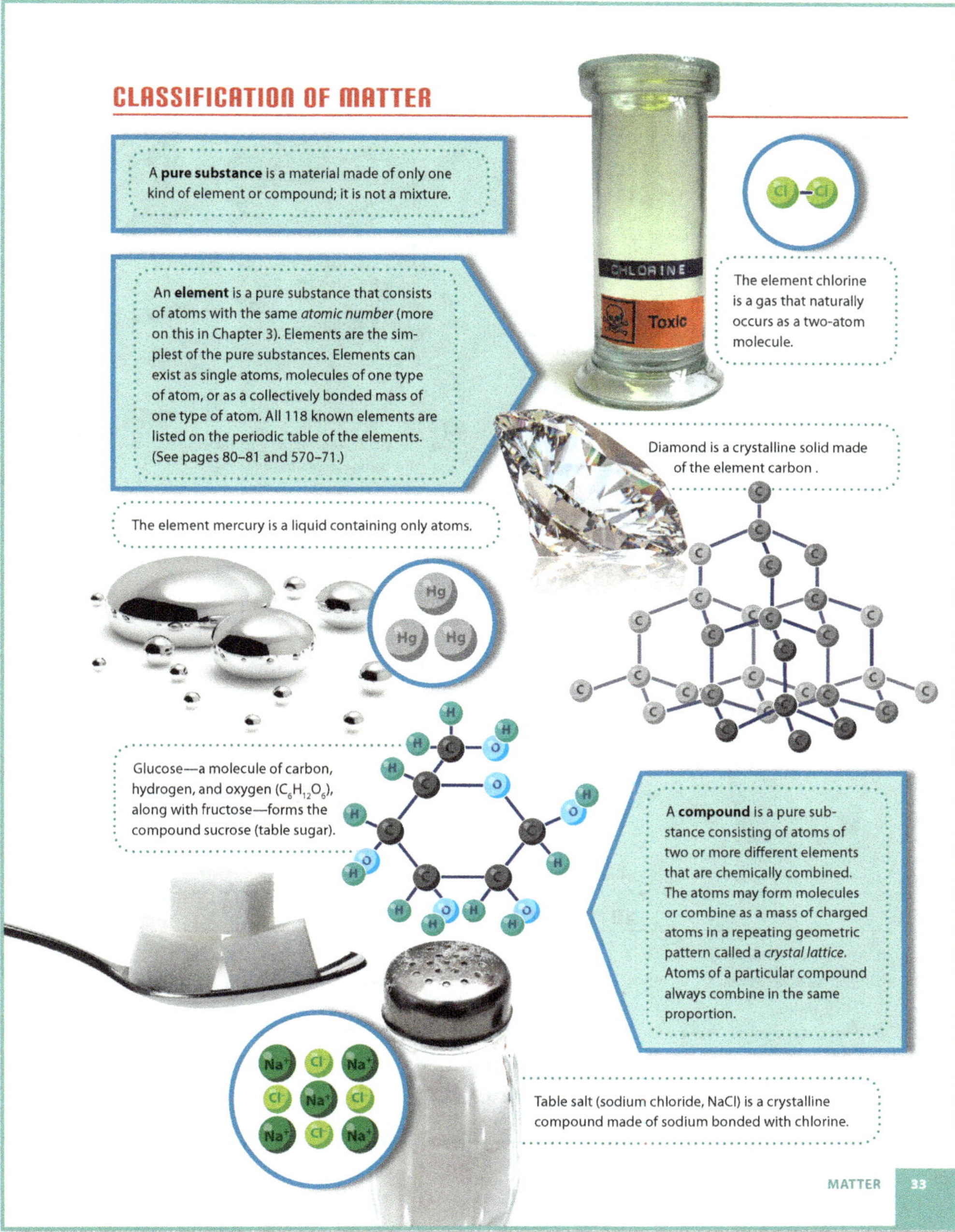

Element, Compound, or Mixture?

Students can determine whether a substance is an element by looking at the periodic table. Compounds can be determined by whether there is a combination of symbols in a formula for the substance. A mixture can be identified by being described as containing more than one pure substance.

Because students at this point may not be familiar with the periodic table or with formulas of compounds, it is best to use more common substances. You could also point out that most materials with which they are familiar are mixtures.

Active Learning/Formative Assessment: *Classifying Matter*

Post labels for the terms *element*, *compound*, *homogeneous mixture*, and *homogeneous mixture* at four locations in the room. Instruct students to classify each of the following examples by moving to the correct label:

1. copper wire *(element)*
2. soil *(heterogeneous mixture)*
3. air *(homogeneous mixture)*
4. carbon dioxide *(compound)*
5. Italian salad dressing *(heterogeneous mixture)*
6. brass *(homogeneous mixture)*
7. aluminum foil *(element)*

STRATEGIES

Class Opener: To stimulate students' thinking, have them make two columns on a piece of paper, one labeled *Mixtures* and the other *Pure Substances*, and then write three examples in each column. Each student should then compare his list with another student's and update as needed. (Let students know that they will hand in this paper at the end of class.)

Alternate Opener: Use the Class Discussion: Classifying teacher note on page 32 to get students to think about why scientists spend so much time classifying and categorizing information.

Active Learning: Use the Active Learning: Classifying teacher note on page 32 to have students put their classifying skills into action.

Demonstration: Use the Demonstrating Types of Matter teacher note on page 32 to help students understand the differences in the classifications of matter.

(continued)

A Mixture of Substances That Don't Mix

Thinking students may ask, "If oil and water don't mix, how can they be called a mixture?" It all depends on how the word *mixture* is defined. The definition we are using stipulates only that substances be "physically combined." If oil and water are in a closed container and then shaken, they will "mix" temporarily, but the heterogeneous nature of the mixture will be obvious. Another similar situation is sand and water, which will mix if they are stirred together, but the sand will eventually settle to the bottom and appear unmixed.

2B REVIEW ANSWERS

1. We classify things to create order in our thinking. Classification allows us to study and compare objects. *(p. 32)*

2. A pure substance is made up of only one type of material; it can be an element or a compound. *(p. 33)*

3. An element is a pure substance that consists of atoms with the same atomic number. It is the simplest of the pure substances. *(p. 33)*

4. A mixture involves two or more substances physically combined in a changeable ratio. *(pp. 33–34)*

5. A *heterogeneous* mixture doesn't have a uniform appearance because its substances clump together. A *homogenous* mixture looks uniform throughout because the substances are evenly mixed. *Both* are physical combinations of two or more substances in a changeable proportion. *(p. 34)*

6. Answers will vary. *Examples of pure substances*: copper (element) and carbon dioxide (compound); *Examples of mixtures*: trail mix (heterogeneous mixture) and sugar solution (homogeneous mixture) *(pp. 33–34)*

7. A compound is two or more elements chemically combined in a fixed ratio. A mixture is two or more substances physically combined in a changeable proportion. *(pp. 33–34)*

8. **a.** compound

   b. element

   c. mixture *(all pp. 33–34)*

CLASSIFICATION OF MATTER

Anything that is not a pure substance is a **mixture**—a physical combination of two or more substances (elements, compounds, or other mixtures) in a changeable ratio. The two key aspects of mixtures are (1) that they are only physically combined, not chemically, and (2) that the proportion of substances within the mixture is variable. Scientists further classify mixtures by their appearance.

A **heterogeneous mixture** does not have a uniform appearance since the substances are unevenly distributed. Sometimes the different substances can be seen with the unaided eye or a low-powered microscope.

Iron and carbon mixed together form the homogeneous mixture steel.

A chocolate chip cookie is a heterogeneous mixture because the chocolate is not evenly distributed throughout the cookie.

When different substances are mixed so that they have a uniform appearance throughout, a **homogeneous mixture** results. Another term for homogeneous mixture is *solution*.

Oil and water is another heterogeneous mixture because the oil and water remain separated with a nonuniform appearance.

Tea is another homogeneous mixture.

2B | REVIEW QUESTIONS

1. Why do we classify things?
2. What is a pure substance?
3. Define *element*.
4. What characteristics make a mixture *not* a pure substance?
5. Compare heterogeneous and homogeneous mixtures.
6. Give two examples of pure substances and two examples of mixtures not included in this section.
7. How does a compound differ from a mixture?
8. Classify each of the following as an element, compound, or mixture.
   a. magnesium chloride ($MgCl_2$)
   b. copper
   c. trail mix

TEACHING THE MATERIAL

Active Learning/Formative Assessment: Use the Active Learning/Formative Assessment: Classifying Matter teacher note on page 33 to test students' understanding of the classification of matter.

Discussion: Once you have introduced the categories of matter, ask, "Which are there more of—pure substances or mixtures? Why?"

Ticket Out the Door: Have students revise their lists from the class opener. If they need to delete or move any of their examples, have them explain why they are doing so.

2C | STATES OF MATTER

How can particles in a solid be moving?

You can see matter in different physical forms, or *states*. The four most common states of matter are **solid**, **liquid**, **gas** (or vapor), and **plasma**. At times we can find a substance existing in more than one state at the same time. Scientists will often refer to these different states as *phases*. Notice the three distinct phases of water (right). Solid ice, liquid water, and water vapor are all present at the same time. The water vapor is in the bubbles within the water and in the beaker between the liquid water and the glass cover. The cloud above the beaker is not water vapor, but rather tiny droplets of liquid water, which is why you can see it.

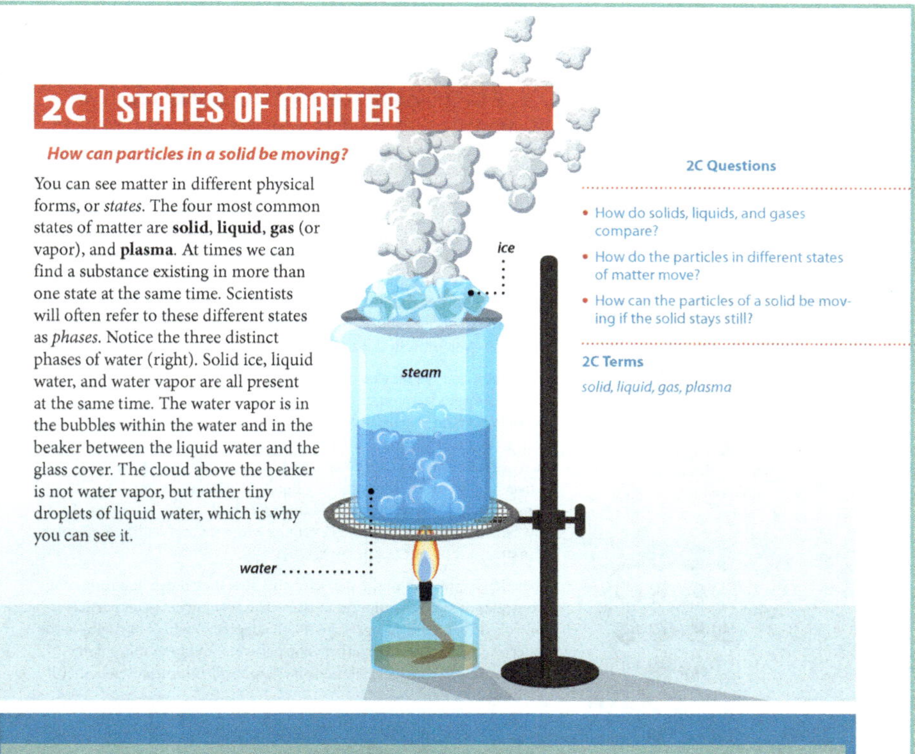

2C Questions

- How do solids, liquids, and gases compare?
- How do the particles in different states of matter move?
- How can the particles of a solid be moving if the solid stays still?

2C Terms

solid, liquid, gas, plasma

SERVING AS A MATERIALS SCIENTIST: MAKING AN INVISIBILITY CLOAK

Imagine a fabric that could bend light around you when you put it on, making you invisible. Magic, right? No, science! Graduate student Joseph Choi and professor John Howell of Rochester University developed the Rochester cloak (right) that bends light around small objects, making them invisible.

Materials science is an interdisciplinary field in which scientists work to develop new materials and improve existing ones. Materials scientists use many approaches to produce an invisibility cloak. Some approaches use retro-reflective projection technology in which an image of the background behind an object is projected onto the surface of that object. Other approaches use carbon nanotubes or other materials to bend light around an object. Some scientists are using optics to achieve invisibility. Materials scientists envision using this technology for aviation, cars, medicine, and many other applications.

The creativity of materials scientists imitates God's creative work. God has given them the skills and understanding to make these advances. The work of these scientists directly affects the lives of those around them.

TEACHING THE MATERIAL

ESSENTIAL QUESTION

How can particles in a solid be moving?

OBJECTIVES

- 2C1 Identify the characteristics of different states of matter.
- 2C2 Classify different substances as solids, liquids, gases, or plasmas.
- 2C3 Compare the three common states of matter using the particle model of matter.
- 2C4 Explain how the particles in a solid can be moving according to the particle model of matter.

RESOURCES

Case Study: How Many States of Matter?

Demonstrating Particles in Action

Serving as a Materials Scientist

⁉️ Misconception About Water Vapor

Students often get confused when thinking of steam. They know that steam is the vapor form of water, but as they watch water boil, they often think that the visible cloud over the pot is steam. This is actually a cloud of very small liquid water droplets caused by the vapor condensing back into liquid water. Vapor consists of individual water molecules, which are too small to be seen.

Demonstrating Particles in Action

Preparation: Start class with two beakers holding equal amounts of water (one almost ice-cold and the other very hot).

Performance/Discussion:

1. Tell me about the two beakers that are in the front of the room. (*From a distance the beakers may look identical. Some students may see "steam" from the hot water depending on conditions.*)

2. Predict what will happen if I add a couple drops of food coloring to each of the beakers. (*Students should predict that the food coloring will mix into the water.*)

Now add a few drops of food coloring to each of the beakers. After the food coloring in the hot water has had time to spread, ask students the following questions.

3. What did you observe? (*The food coloring spread, one faster and to a greater extent than the other.*)

4. Why did it spread? (*Particles of the water were moving. Some students might recall diffusion from Section 2A.*)

5. Why did it spread faster in one container than in the other? (*Students may hypothesize that the water in one of the beakers must be warmer. The warm water particles are moving faster.*)

6. How do you think increasing or decreasing this movement of particles is related to states of matter? (*Answers will vary, but try to get students to see that temperature and particle motion are directly related to each other and that the more the temperature is changed, the greater the potential there is for changes in the states of matter.*)

7. Why do you think that some substances don't have particle motion that is as apparent? (*There are forces that are constraining the motion of the particles, and those forces are different in different materials.*)

(continued)

Throughout history, people have worked with materials to improve them. This is evident by how we name some historical ages, such as the Bronze Age, the Copper Age, and even the Digital Age. The ages are identified by a significant material from that period.

2.6 STATES AND THE PARTICLE MODEL

Recall that the particle model states that all matter consists of tiny particles in constant random motion. The states of matter are determined by the relationship between the *kinetic energy* (energy of motion) of the particles and the attractive forces between them. In a solid substance, for example, the particles vibrate in place, so their movements are very small. Their kinetic energy is insufficient for overcoming the attractive forces within the material. If the substance is warmed, its particles move faster and it becomes a liquid once its kinetic energy is high enough to overcome some of the attractive forces between the substance's particles. If the material continues to be warmed, it eventually gains enough energy to overcome all the attractive forces and become a gas.

Different materials may be in different states of matter even if their kinetic energies (temperatures) are the same. At room temperature, for example, oxygen is a gas, water is a liquid, and aluminum is a solid. This occurs because they have different degrees of attractive forces between their particles. So the oxygen is a gas because it has low attractions between its particles, the water is a liquid because it has medium attractions, but the aluminum is a solid due to its high attractive forces. As you can see, the particle model explains why we observe different states of matter at different temperatures.

The relative position of the particles and how they move determine the properties of solids, liquids, and gases. The spacing between particles controls density and compressibility, the degree to which particles in the material can be pushed together. The particles' ability to move determines how well the material holds its shape and volume and whether it can flow.

TEACHING THE MATERIAL

STRATEGIES

Class Opener: Use the Demonstrating Particles in Action teacher note on page 35 to introduce particle motion.

Video: Do an internet search for a video on "Why does ice float?"

Common Misconception: Use the image on page 35 in the Student Edition and the Misconception About Water Vapor teacher note on the same page to discuss the common misconception about steam over boiling water.

Graphic Organizer: Use the Graphic Organizer: States of Matter teacher note on page 37 to help students organize information about the states of matter.

Formative Assessment: Use the Formative Assessment: States of Matter teacher note on page 37 to check students' progress.

Ticket Out the Door: On a piece of paper, write two things that you learned today and one thing that you still have a question about and hand it in as you leave class.

STATES OF MATTER

SOLID

Particle spacing: *close*
Particle motion: *vibrating in place*
Volume: *fixed*
Shape: *fixed*
Compressibility: *low*
Density: *high*
Fluid? *no*

Crystalline Solids—solids with particles arranged in regular repeating patterns, or lattices

Amorphous Solids—solids that consist of a mass of particles with no discernible pattern

LIQUID

Particle spacing: *close*
Particle motion: *able to slide past each other*
Volume: *fixed*
Shape: *changes to fill a container from the bottom*
Compressibility: *low*
Density: *between that of a solid and that of a gas*
Fluid? *yes*
Viscosity: *The attractive forces between the liquid particles determine the viscosity of a liquid.*

Viscosity *is a measure of a fluid's resistance to flowing.*

GAS (VAPOR)

Particle spacing: *widely spaced*
Particle motion: *high speed*
Volume: *changes to fill the container*
Shape: *changes to fill the container*
Compressibility: *high*
Density: *low*
Fluid? *yes*
Pressure: *due to collisions with container surface*

PLASMA

The most common state of matter in the universe is not one that we encounter often in our daily lives. Plasma is a gas-like state of matter formed at very high temperatures that consists of high-energy ions and free electrons. Plasma is the state of matter of our sun and other stars.

 **Graphic Organizer:** *States of Matter*

Many students will benefit from organizing what they learn about the states of matter. This can be done as a comparison matrix, a Venn diagram, a concept map, or other means. The key is that students organize the information in a way that will help them understand the similarities and differences. For help with graphic organizers, refer to Appendix F.

Formative Assessment:
States of Matter

1. What two opposing properties of matter primarily determine the state of matter? *(kinetic energy or particle motion and the attractive forces between particles)*

2. Why can one substance be a gas and another substance be solid at the same temperature? *(The solid has much greater attraction between its particles than does the gas.)*

3. Which state of matter has a fixed shape and volume? *(solid)*

4. What type of solid has particles that are the most organized? *(crystalline)*

5. Which has the greater viscosity—honey or water? *(honey)*

6. Oil is more viscous than water. How do their densities compare? *(Oil is less dense, even though it is more viscous.)*

Case Studies

Throughout the Student Edition, select chapters will have case studies. These are opportunities for students to apply what they are learning in the chapter to a real-world physical science topic.

Case Study Answers

1. All the states are comprised of particles, whether they are common or uncommon states.

2. high energy and low energy

3. the properties of substances

4. Answers will vary. Scientists may use a particle accelerator like the LHC. Others will use magnetic cooling and laser cooling to supercool materials.

CASE STUDY: HOW MANY STATES OF MATTER?

If someone were to ask you how many states of matter there are, you would probably answer four: solid, liquid, gas, and plasma. But particle physicists today would say that there are thirty-three actual or theoretical states!

What are these twenty-nine other states of matter? They are states of matter that exist only under very precisely controlled conditions. Even when these conditions exist, some of these states are very unstable and decay into some other state quickly. All this matter is made of particles, though the particles aren't always "normal" protons, neutrons, and electrons.

Some of these are high-energy states of matter, such as quark-gluon plasma (QGP) and color-glass condensate (CGC). QGP exists at either extremely high temperatures or very high densities. CGC is a theoretical state of matter that occurs when *nuclei* move near the speed of light (extremely high kinetic energy). Scientists at the Large Hadron Collider have observed behavior in matter consistent with the predicted behavior in a CGC.

Even more of these states of matter are considered low-energy states, such as the Bose-Einstein condensate (BEC) and superfluids. BEC forms in samples with very low densities when scientists cool them to almost *absolute zero* (−273 °C or 0 K). This is the lowest temperature possible. Superfluids are extremely low-temperature fluids that also have extremely low viscosities.

There are other states of matter that don't necessarily fall into the high-energy or low-energy categories. These include states like liquid crystals, quantum spin liquids, and exotic matter. Each different state has properties that make it interesting to study and potentially useful in a variety of applications.

1. How are the four common states of matter similar to the other states?

2. What are two major divisions in the uncommon states of matter?

3. What do scientists use to classify all states of matter, even the unusual states mentioned here?

4. What are some ways that scientists study these other types of matter?

2C | REVIEW QUESTIONS

1. List the four most common states of matter.

2. What factors determine the state of a substance?

3. Describe a solid.

4. Compare crystalline and amorphous solids.

5. What property allows you to describe liquids and gases as fluids?

6. What is viscosity?

7. Compare the degree of motion of the particles in solids, liquids, and gases.

8. What state of matter has changeable shape, low compressibility, and fixed volume?

9. What is a plasma?

2C REVIEW ANSWERS

1. solid, liquid, gas (or vapor), and plasma *(p. 35)*

2. the relationship between the kinetic energy of the substance's particles and the attractive forces between those particles *(p. 36)*

3. The particles of a solid vibrate while being held rigidly in place, so a solid has a fixed shape and volume. The particles are closely spaced, resulting in low compressibility and relatively high density. The particles are arranged in a regularly repeating pattern (crystalline) or without a pattern (amorphous). *(p. 37)*

4. Both are solids, so the closely spaced particles rigidly hold each other in place. In a crystalline solid, the particles are arranged in a regular repeating pattern. In an amorphous solid, the particles have no regular arrangement. *(p. 37)*

5. The word *fluid* means "able to flow." Because liquid and gas particles are not held rigidly in place, they have freedom of movement and can flow. *(p. 37)*

6. Viscosity is a fluid's resistance to flowing. *(p. 37)*

7. Particles in solids are closely spaced, are held rigidly in place, and move only by vibrating. Particles in liquids are closely spaced and move by vibrating and sliding past each other. Particles in gases are widely spaced and move freely and rapidly. *(p. 37)*

8. a liquid *(p. 37)*

9. Plasma is a high-energy state of matter similar to a gas except that it is made up of ions and free electrons. *(p. 37)*

2D | CHANGES IN MATTER

How does matter change?

If you were going to make a pillow, would you make it out of river rocks or duck feathers? You would use feathers, of course! Materials scientists, engineers, and even pillow makers choose materials on the basis of the properties that best suit them for the intended application. These properties depend on the particle nature of the materials. Once again, the particle model of matter demonstrates that it is a good model because it is workable. Scientists categorize the properties of materials as physical, chemical, or nuclear (see Chapter 8 for nuclear properties).

2.7 PHYSICAL PROPERTIES

A **physical property** is anything about a substance that can be observed or measured without altering the substance's chemical composition. Physical properties are helpful in identifying a particular substance and include color, texture, physical state, and so on. Physical properties can be explained according to the particle model of matter.

2D Questions

- What are physical properties?
- What are chemical properties?
- How can I tell whether a change in matter is physical or chemical?
- What does it mean to conserve matter?
- How does energy affect states of matter?

2D Terms

physical property, physical change, chemical change, chemical property, melting, melting point, freezing, vaporization, evaporation, boiling, boiling point, law of conservation of matter, condensation, sublimation, deposition

TEACHING THE MATERIAL

ESSENTIAL QUESTION

How does matter change?

OBJECTIVES

- 2D1 Define *physical property* and *chemical property*.
- 2D2 Classify changes in matter as physical or chemical.
- 2D3 Summarize the law of conservation of matter.
- 2D4 Identify changes of state.
- 2D5 Relate changes of state to the flow of energy.

(continued)

Demonstrating Chemical and Physical Changes

Preparation: Set out a bottle of soda, an empty container to pour the soda into, a package containing effervescent tablets, and a small container of water.

Performance/Discussion:

Open the soda bottle and pour it into the container.

1. What substance is causing the fizzing? *(carbon dioxide gas)*

2. Is the CO_2 being produced or was it already in the soda? *(Because soda is a carbonated beverage, the CO_2 was already in the soda.)*

3. Are any new substances being produced that were not already there? *(No)*

4. If some students think that carbon dioxide is being produced: What other substance is changing to produce it? *(None. If all the CO_2 left the soda, it would still be the same basic flavor but without the fizz.)*

5. What normally keeps the soda from losing its fizz? *(The closed container physically restrains the CO_2 from escaping.)*

6. How can you know this? *(by the sound made when the container is opened and by the fact that it goes flat if left open)*

Open the effervescent tablet package and drop a tablet in water to illustrate a chemical change. Then ask these questions.

7. What is causing the fizzing? *(CO_2 gas)*

8. Was there anything physically restraining the CO_2? *(No)*

9. How do you know? *(There was no sound when the package was opened.)*

10. So is the CO_2 being produced or just released? *(It is being produced.)*

TEACHING THE MATERIAL

RESOURCES

- Worldview Sleuthing: Bulletproof!
- Demonstrating Chemical and Physical Changes
- Lab 2B: *Something Old, Something New?*—Detecting Physical and Chemical Changes

STRATEGIES

Class Opener: To introduce physical and chemical properties and changes, ask students to give examples of properties that can be used to identify different types of matter. Write down those that are given and ask which would involve changing the composition of each substance.

Demonstration: Use the Demonstrating Chemical and Physical Changes teacher note on page 39 to stimulate a discussion regarding physical and chemical changes.

Any change in matter that does not alter the composition of a substance is called a **physical change**. Physical changes will change the appearance of a substance. We could change the shape, texture, color, dimensions, state, and so on. The key to recognizing that a change is physical is knowing that the substance is chemically the same after the change has occurred. Some examples include melting butter, making ice cubes, and crumpling paper.

2.8 CHEMICAL PROPERTIES

Some changes alter the chemical composition of substances. Scientists call these **chemical changes**, or *chemical reactions*. You will learn more about chemical changes in Chapter 7.

Chemical changes occur according to the chemical properties of a substance. A **chemical property** describes how a substance changes in the presence of another substance or under certain conditions. There are many chemical properties, but two are particularly important for this course.

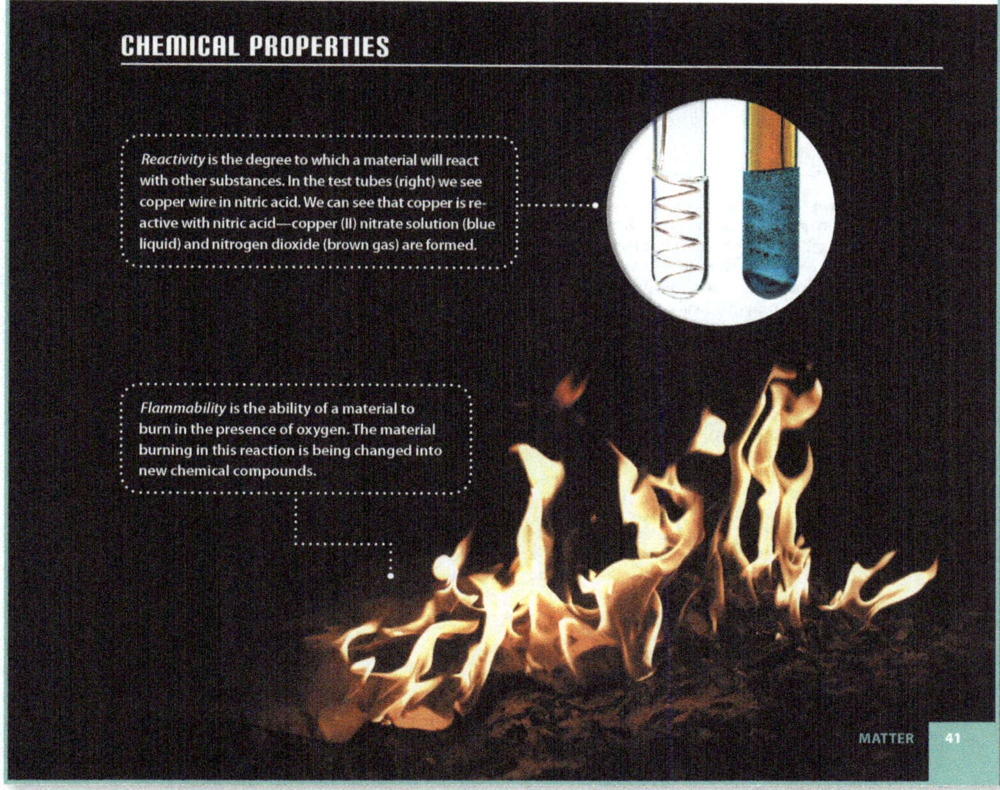

Active Learning: Use the Active Learning/Formative Assessment: Changes of State teacher note on page 42 to help students learn the correct terms for changes of state.

Formative Assessments: Use the Formative Assessment: Changes in Matter teacher note on page 43 to gauge students' level of understanding.

Biblical Worldview Shaping: Use the Worldview Sleuthing: Bulletproof! activity on page 44 to help students see materials research as a means of serving others.

Ticket Out the Door: Explain what the law of conservation of matter means with regard to physical and chemical changes.

Do an internet search using the keywords "states of matter interactive simulation." You will find several simulations that students can use to manipulate the conditions affecting the states of matter of a substance.

🧭 **Active Learning/Formative Assessment:** *Changes of State*

Post labels around the room for the terms *deposition*, *freezing*, *condensation*, *melting*, *sublimation*, *vaporization*, *evaporation*, and *boiling*. Describe changes of state and have students move to which word best fits the description. Tell them to be ready to justify their choices. Be sure to include questions regarding the movement of energy relative to these changes of state.

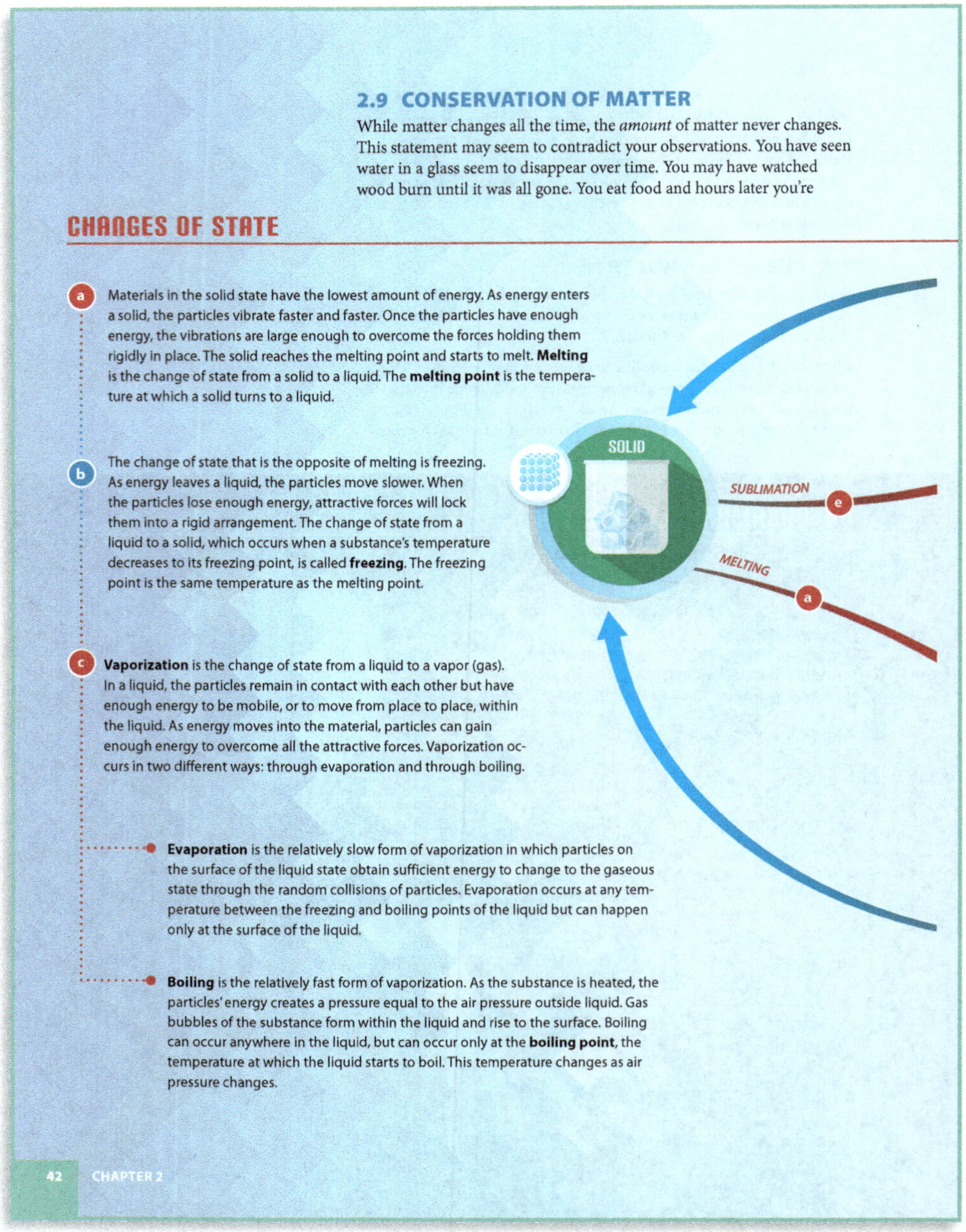

2.9 CONSERVATION OF MATTER

While matter changes all the time, the *amount* of matter never changes. This statement may seem to contradict your observations. You have seen water in a glass seem to disappear over time. You may have watched wood burn until it was all gone. You eat food and hours later you're

CHANGES OF STATE

a Materials in the solid state have the lowest amount of energy. As energy enters a solid, the particles vibrate faster and faster. Once the particles have enough energy, the vibrations are large enough to overcome the forces holding them rigidly in place. The solid reaches the melting point and starts to melt. **Melting** is the change of state from a solid to a liquid. The **melting point** is the temperature at which a solid turns to a liquid.

b The change of state that is the opposite of melting is freezing. As energy leaves a liquid, the particles move slower. When the particles lose enough energy, attractive forces will lock them into a rigid arrangement. The change of state from a liquid to a solid, which occurs when a substance's temperature decreases to its freezing point, is called **freezing**. The freezing point is the same temperature as the melting point.

c **Vaporization** is the change of state from a liquid to a vapor (gas). In a liquid, the particles remain in contact with each other but have enough energy to be mobile, or to move from place to place, within the liquid. As energy moves into the material, particles can gain enough energy to overcome all the attractive forces. Vaporization occurs in two different ways: through evaporation and through boiling.

Evaporation is the relatively slow form of vaporization in which particles on the surface of the liquid state obtain sufficient energy to change to the gaseous state through the random collisions of particles. Evaporation occurs at any temperature between the freezing and boiling points of the liquid but can happen only at the surface of the liquid.

Boiling is the relatively fast form of vaporization. As the substance is heated, the particles' energy creates a pressure equal to the air pressure outside liquid. Gas bubbles of the substance form within the liquid and rise to the surface. Boiling can occur anywhere in the liquid, but can occur only at the **boiling point**, the temperature at which the liquid starts to boil. This temperature changes as air pressure changes.

hungry again. It seems as if matter is disappearing in all of these instances. But what is happening is that the matter is changing into a different state or substance. Matter can't disappear—it's against the law!—the **law of conservation of matter**. It is a fundamental natural law that states that matter can neither be created nor destroyed, but can only change forms. This happens when a substance changes state.

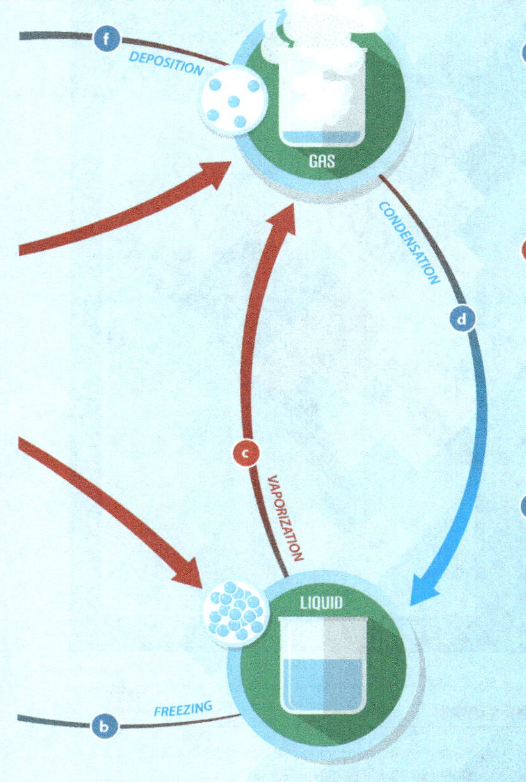

d **Condensation** is the change of state from a vapor (gas) to a liquid and is the opposite of vaporization. As particles in a gas lose energy to their surroundings, the particles slow down. When they are moving slowly enough, the particles can get trapped in the liquid state. Dew forming on a cool summer night is an example of condensation.

e While not as common, it is possible for a solid to transition directly to a vapor without melting first. This change is called **sublimation** and occurs when solid particles get enough energy to change to vapor. It occurs only at the surface of the solid and is the opposite of deposition. If you have ever noticed that ice cubes seem to shrink in the freezer, then you have noticed sublimation. Dry ice—solid carbon dioxide—and moth crystals are two examples of substances that sublime at room temperature.

f On a cold winter morning, as you scrape frost from the windows of the family car, you are not scraping frozen water droplets. The solid frost formed directly from a vapor (gas), bypassing the liquid state, in a process called **deposition**. This process is the opposite of sublimation.

When scientists are working with changes of state, they will often use a *phase diagram*. The diagram at right shows the pressures and temperatures at which each state, or phase, exists. The lines between the phases each represent changes in state. The red circle on the graph is a unique point called the *triple point*—the temperature and pressure at which solid, liquid, and gas exist in equilibrium.

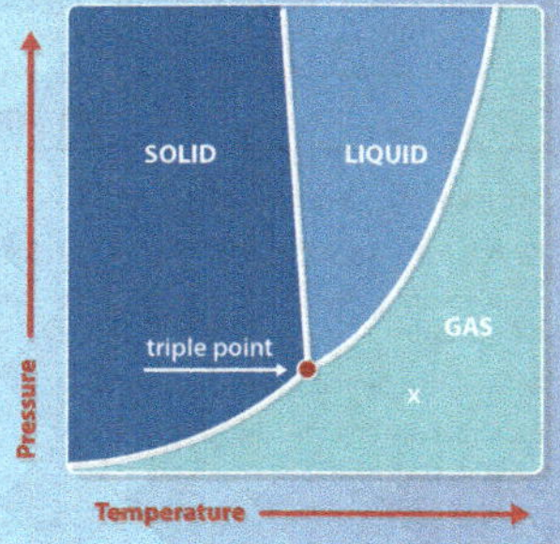

Triple Point of Water

The triple point for pure water is 273.16 K (0.01 °C) and 0.0060 atm. These are the conditions at which liquid water, ice, and steam exist in equilibrium.

 Triple Point Visualized

Do an internet search using the keywords "triple point of water" to find a short video showing water at its triple point freezing and boiling at the same time. Students will find it very interesting. Doing a keyword search on "triple point of water realization" will yield video of the process of preparing a triple point of water cell to be used for thermometer calibration.

 Formative Assessment:
Changes in Matter

1. Give the terms for phase changes from solid to liquid, liquid to gas, and solid to gas. *(melting, vaporization, and sublimation)*

2. Give the terms for phase changes when energy is removed from a substance, including the states involved. *(condensation [vapor to liquid], freezing [liquid to solid], and deposition [vapor to solid])*

3. Give examples of physical properties. *(color, texture, physical state, ductility, malleability, conductivity, luster)*

4. What type of substance might have the properties of ductility, malleability, conductivity, and luster? *(a metal)*

5. Give examples of chemical properties. *(reactivity, flammability)*

 Lab 2B: *Something Old, Something New*

Once students have an understanding of the material in Section 2D, they should be ready to work through Lab 2B.

Worldview Sleuthing Boxes

Throughout the Student Edition, select chapters will include worldview sleuthing boxes. These webquests put your students in a real-world role in which they have to research a topic related to the chapter. Students will present their research in a variety of formats.

Worldview Sleuthing: *Bulletproof!*

In this webquest, students will research ballistic materials that might be used to protect speed skaters from serious injury. Specify your expectations of them regarding any illustrations, the minimum number of sources, and any other details.

Webquest Rubric

Appendix G includes a reproducible rubric to assess this worldview sleuthing activity.

WORLDVIEW SLEUTHING: BULLETPROOF!

Bulletproof technology is an extension of the ancient art of armor making. Armor proved ineffective against modern weaponry, and metal suits are too heavy for the modern battlefield. Materials scientists started to investigate cloth materials. In the late 1800s, a doctor in Arizona discovered that silk could keep a bullet from penetrating the body, but silk was also extremely expensive. Over time scientists searched for other natural and synthetic (manmade) fibers with similar properties. Today scientists are looking at other applications for these materials.

TASK

You are working as a clothing designer for the nation's Olympic sports team. You have been assigned to research advances in ballistic materials to be used in clothing that will perform well and protect speed skaters from sharp skate blades during accidents. You are to write a one-page proposal for a material to be used for these uniforms.

PROCEDURE

1. Research the history and current information about bulletproof technology by doing keyword searches for "bulletproof," "ballistic materials," "Kevlar™," and "Twaron™."

2. Research other uses of similar materials by doing keyword searches for "aramid" and "para-aramid synthetic fiber."

3. Plan your proposal and collect any required photographs. Remember to cite your sources.

4. Write your proposal and show it to another person for feedback.

CONCLUSION

We are commanded in Scripture to serve others. We do this by meeting the needs of people around us. Developing materials such as these can protect the lives of athletes. Could they also help protect the lives of military and law enforcement personnel?

2D | REVIEW QUESTIONS

1. What is a physical property?

2. What is a chemical change?

3. Identify each of the following as a chemical or a physical change.

    a. Iron rusts and becomes iron (III) oxide.

    b. Sugar dissolves in iced tea.

    c. Paper burns.

4. Explain what is happening to the particles in a solid as you warm it to its melting point.

5. Define *boiling point*.

6. Compare evaporation and boiling.

7. State the law of conservation of matter.

8. List the changes of state that involve adding energy.

2D REVIEW ANSWERS

1. A physical property is a characteristic of a substance that can be observed without changing the chemical composition of the substance. *(p. 39)*

2. A chemical change is a change of one or more substances into new chemical substances. It is also known as a chemical reaction. *(p. 41)*

3. **a.** chemical

    b. physical

    c. chemical *(all pp. 40–41)*

4. The particles in a solid are vibrating while being held rigidly in place. As the substance is warmed, the particles vibrate faster and faster until they overcome the forces holding them rigidly in place. *(p. 42)*

5. The boiling point is the temperature at which a liquid starts to boil. *(p. 42)*

6. Evaporation and boiling are both forms of vaporization—the change of a liquid to a gas. Evaporation happens at any temperature between the melting and boiling points. It occurs when particles at the surface of the liquid gain enough energy from random collisions to escape the attractive forces of other particles. Boiling occurs at the boiling point anywhere in the liquid when the vapor pressure equals or exceeds the atmospheric pressure. Boiling generally happens faster than evaporation. *(p. 42)*

7. Matter can be neither created nor destroyed. *(p. 43)*

8. melting, vaporization (evaporating and boiling), and sublimation *(pp. 42–43)*

CHAPTER 2 REVIEW

UNDERSTANDING MATTER

- Matter is anything that has mass and takes up space. According to the particle model of matter, all matter is made of tiny particles (atoms and molecules) in constant random motion.

- The particle model of matter is very workable as it explains most of our observations about matter.

- Density is calculated by dividing the mass of an object by its volume.

- Mass is the amount of matter in an object, while weight is the force of gravity acting on that object.

2A Terms

matter	26
law of definite proportions	28
particle model of matter	28
atom	29
molecule	29
mass	29
volume	29
density	29
weight	30

2B

CLASSIFYING MATTER

- We classify matter by its physical and chemical properties.

- Matter is classified as either a pure substance or a mixture.

- Pure substances (elements and compounds) contain only one type of substance.

- Mixtures (heterogeneous or homogeneous) are physical combinations of two or more substances in changeable proportions.

2B Terms

pure substance	32
element	32
compound	33
mixture	34
heterogeneous mixture	34
homogeneous mixture	34

2C

STATES OF MATTER

- Particles in solids have low kinetic energy compared with the forces between particles. Solids have fixed shape and volume, high density, and low compressibility due to their close spacing and low energy.

- Liquid particles have enough kinetic energy to overcome some of the forces between particles. Liquids have fixed volume, high density, low compressibility due to their close spacing and low energy. Their ability to move makes them fluid and gives them a changeable shape.

- Gas particles have sufficiently high energy to overcome all the forces between them. The particles' wide spacing causes gases to have low densities and high compressibility. Their rapid motion allows them to fully occupy their container, no matter its shape or volume.

2C Terms

solid	35
liquid	35
gas	35
plasma	35

Recalling Facts

1. matter *(p. 26)*

2. Answers will vary. Students will probably list types of energy (e.g., light, sound, heat). *(p. 26)*

3. Students may list diffusion, the law of definite proportions, and Brownian motion. (Accept any two.) *(p. 28)*

4. The particle model of matter was accepted because it was more workable. It explained the observations of matter better than other models did. It also made more accurate predictions. *(pp. 28–29)*

5. Weight is the measure of Earth's gravity acting on an object. *(p. 30)*

6. A mixture is a combination of two or more substances physically combined in a changeable ratio. *(p. 34)*

7. homogeneous *(p. 34)*

8. Answers will vary. Examples: carbon (element), water (compound), trail mix (heterogeneous mixture), tea (homogeneous mixture) *(pp. 33–34)*

9. the particles' kinetic energy and the attraction between particles *(p. 36)*

10.

	Solid	Liquid	Gas
Volume	fixed	fixed	changeable
Shape	fixed	changeable	changeable
Compressibility	low	low	high
Particle Spacing	close	close	widely spaced
Density	high	between that of solid and gas	low
Particle Motion	vibrates in place	slides past each other	high speed
Fluid?	no	yes	yes

(p. 37)

11. gas (Accept plasma also.) *(p. 37)*

12. fluid *(p. 37)*

2D   CHANGES IN MATTER

- Physical properties are characteristics that can be observed without changing the chemical composition of a substance.
- Chemical properties of matter are characteristics that change the chemical composition of a substance when observed.
- The law of conservation of matter states that matter can't be created or destroyed but only changed in form.
- As a material warms, its particles move faster as it transitions from a solid to a liquid by melting. Liquids vaporize to gases as energy is added. Solids can change directly to gases by sublimation.
- Evaporation and boiling are different forms of vaporization that occur at different temperatures, different locations, and through different mechanisms.
- As a material cools, it transitions from a gas to a liquid by condensing. Liquids can change to a solid through freezing as energy is lost. Gases can change directly to a solid by deposition.
- Melting occurs at the melting point of a substance; boiling occurs at its boiling point.

2D Terms

physical property	39
physical change	41
chemical change	41
chemical property	41
melting	42
melting point	42
freezing	42
vaporization	42
evaporation	42
boiling	42
boiling point	42
law of conservation of matter	43
condensation	43
sublimation	43
deposition	43

13. A physical change is any change to a substance that does not alter its chemical composition. *(p. 41)*

14. A chemical property describes how a substance changes as it interacts with another substance. *(p. 41)*

15. **a.** melting

    b. vaporization

    c. sublimation *(all pp. 42–43)*

16. Answers will vary. (*Examples:* frost forming on a cold surface, soot coating a chimney) *(p. 43)*

Understanding Concepts

17. Aristotle's model says that matter is continuously divisible. The particle model says that matter consists of tiny indivisible particles. *(p. 27)*

18. Since the density doesn't change, the volume would also double. *(p. 29)*

19. Taking matter to space or another planet can alter the force of gravity, causing a change in weight. *(p. 30)*

REVIEW

CHAPTER REVIEW QUESTIONS

Recalling Facts

1. What has mass and takes up space?

2. Give three examples of non-matter.

3. Give two pieces of evidence that led to the acceptance of the particle model of matter.

4. Why was the particle model of matter accepted over competing models?

5. Define *weight*.

6. Define *mixture*.

7. What do we call a mixture that appears the same throughout?

8. Give an example of an element, a compound, a heterogeneous mixture, and a homogeneous mixture.

9. The state of matter of a substance depends on what two properties?

10. Copy the following table onto your paper and fill it in to compare the different states of matter. (See Appendix F for creating graphic organizers.)

	Solid	Liquid	Gas
Volume			
Shape			
Compressibility			
Particle Spacing			
Density			
Particle Motion			
Fluid?			

11. A substance with low density and changeable shape and volume that is highly compressible is in what state of matter?

12. What term is used to describe the ability of liquids and gases to flow?

13. What is a physical change?

14. Define *chemical property*.

15. Give the opposite change of state for the following processes.

   a. freezing

   b. condensation

   c. deposition

16. Give an example of deposition.

Understanding Concepts

17. How do Aristotle's model of matter and the particle model of matter represent matter?

18. Considering the formula for density, how will the volume of a piece of lead with a density of 11.34 g/mL change if you double the mass of the sample?

19. How can the weight of an object change even when its mass remains constant?

20. What mass of silver will occupy a volume of 87.75 mL? The density of silver is 10.49 g/mL.

21. Compare a pure substance with a mixture.

22. Draw a hierarchy chart using the terms *matter, pure substance, element, compound, mixture, heterogeneous mixture,* and *homogeneous mixture*. (See Appendix F.)

23. Describe a typical liquid.

24. Relate the spacing of particles in solids, liquids, and gases to their motion according to the particle model.

25. If particles in any substance are in constant motion, why can't solids flow?

26. What causes different liquids to vary in their viscosity?

27. Compare physical and chemical changes.

28. The mass of ash remaining after a log burns is much less than the original mass of the log. If the law of conservation of matter is true, how can you explain this?

29. Describe what happens to the particles in a material as it is warmed from a solid to a liquid and ultimately to a gas.

30. Explain how matter changes state as energy is removed.

Critical Thinking

31. A classmate tells you that air is not actually matter because you cannot see it. Do you agree? Explain.

32. Do we know for sure that the particle model is correct? Explain.

33. Water is unusual in that it is less dense in its solid form than in its liquid form. Hypothesize as to why this may be. Why might this be beneficial?

34. Given that solids have high density, liquids medium density, and gases low density, why do most types of wood float in water?

35. A gel is a combination of a solid network structure distributed throughout a liquid. It acts like a solid in most cases because the network structure keeps the liquid from flowing. An aerogel is a solid that is made by removing the liquid from a gel, leaving the solid network with air where the liquid had been. What properties would you expect an aerogel to have?

36. In the phase diagram on page 43, at the position marked by the x, the material is a gas. If you move vertically, increasing pressure at a constant temperature, the substance will change from a gas to a liquid. Why do you think this will happen?

---

20. What we know: $V = 87.75$ mL, $d = 10.49$ g/mL

Unknown: m

Write the formula and solve for the unknown:

$$d = \frac{m}{V}$$

$$dV = \frac{m}{\cancel{V}}\cancel{V}$$

$$m = dV$$

Evaluate:

$$m = \left(10.49\,\frac{g}{mL}\right)(87.75\,\cancel{mL})$$

$$= 920.5\text{ g } (p.\ 30)$$

21. A pure substance is made of only one type of material, whereas a mixture is two or more substances physically combined in a changeable ratio. *(pp. 33–34)*

22.

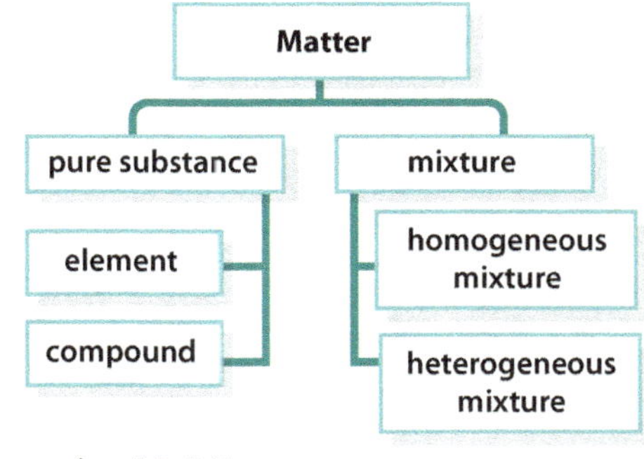

(pp. 33–34)

23. Liquid is the state of matter in which the particles are close together but have some freedom of movement. Liquids have medium to high densities and low compressibility. Liquids are fluids and

---

can change their shape, though their volume is fixed. *(p. 37)*

24. The particle model states that all matter is made of tiny particles in constant motion. Particles move faster as substances warm up, so they move slowest in a solid, faster in a liquid, and fastest in a gas. The faster the particles move, the more space between them, so solids are most closely spaced and gases are most widely spaced. *(p. 37)*

25. The particles in a solid move by vibrating but are held in place by the attractive forces between them. Thus a solid can't flow and has a fixed shape, even though the particles are moving. *(p. 37)*

26. the variation in attractive forces between particles *(p. 37)*

27. Both are changes to matter. Physical changes modify only the appearance of a substance. Chemical changes alter a substance's chemical makeup. *(p. 41)*

28. Conservation of matter states that matter cannot be created or destroyed. The mass decrease is due to some of the matter escaping in the smoke as gases and particulate matter. *(p. 43)*

29. As a solid is warmed, its particles vibrate faster until they have enough kinetic energy to overcome the forces holding them rigidly in place. The substance becomes a liquid. As the liquid is heated, the particles move faster until they have enough kinetic energy to overcome all the attractive forces between the particles and the substance becomes a gas. *(pp. 42–43)*

30. As energy flows out of a substance, the particles of the substance move more slowly. As the particles in a gas cool, they have less kinetic energy; when enough energy has been removed, the particles can't overcome the attractive forces and will stick together in a liquid. Condensation has occurred. As more energy is removed, the process continues until the particles get locked in place and the substance freezes to become a solid. *(pp. 42–43)*

Critical Thinking

31. Answers will vary. Matter is anything that takes up space and has mass. Air definitely takes up space. Air also has mass, as is evidenced by the presence of air pressure. *(p. 26)*

32. No. But we do know that the particle model is workable. It helps us understand what we observe in the world. It explains most of our observations and enables us to make useful predictions. *(pp. 28–29)*

33. Answers will vary. (Focus on students' reasoning more than on correctness.) The structure of the water molecule causes ice to arrange itself in a crystal structure that requires more space than liquid water needs. When water freezes, the particles move away from each other just a small amount, making the ice less dense. This is beneficial because lakes and streams freeze from the surface down. This prevents these bodies of water from completely freezing, which allows fish and other aquatic animals to survive winters in cold environments.

34. The comparison of densities is true for each substance. For example, solid metal is denser than the liquid form of that metal, which would be denser than the gas form of that metal. But not *all* solids are necessarily denser than *all* liquids. For example, most woods (solid) are actually less dense than water and therefore float.

35. Aerogels are solids, but they have very low densities because they take up space but include very little mass. They are rigid because of their solid network structure. They have a fixed shape and volume like other solids. They are poor conductors because they are mostly gas.

36. Increasing the pressure around the gas increases the boiling point of the substance. When the boiling point exceeds the actual temperature, the substance begins to condense and the material becomes a liquid. *(p. 43)*

REVIEW

CHAPTER REVIEW QUESTIONS

Recalling Facts

1. What has mass and takes up space?

2. Give three examples of non-matter.

3. Give two pieces of evidence that led to the acceptance of the particle model of matter.

4. Why was the particle model of matter accepted over competing models?

5. Define *weight*.

6. Define *mixture*.

7. What do we call a mixture that appears the same throughout?

8. Give an example of an element, a compound, a heterogeneous mixture, and a homogeneous mixture.

9. The state of matter of a substance depends on what two properties?

10. Copy the following table onto your paper and fill it in to compare the different states of matter. (See Appendix F for creating graphic organizers.)

	Solid	Liquid	Gas
Volume			
Shape			
Compressibility			
Particle Spacing			
Density			
Particle Motion			
Fluid?			

11. A substance with low density and changeable shape and volume that is highly compressible is in what state of matter?

12. What term is used to describe the ability of liquids and gases to flow?

13. What is a physical change?

14. Define *chemical property*.

15. Give the opposite change of state for the following processes.

  a. freezing

  b. condensation

  c. deposition

16. Give an example of deposition.

Understanding Concepts

17. How do Aristotle's model of matter and the particle model of matter represent matter?

18. Considering the formula for density, how will the volume of a piece of lead with a density of 11.34 g/mL change if you double the mass of the sample?

19. How can the weight of an object change even when its mass remains constant?

20. What mass of silver will occupy a volume of 87.75 mL? The density of silver is 10.49 g/mL.

21. Compare a pure substance with a mixture.

22. Draw a hierarchy chart using the terms *matter, pure substance, element, compound, mixture, heterogeneous mixture,* and *homogeneous mixture.* (See Appendix F.)

23. Describe a typical liquid.

24. Relate the spacing of particles in solids, liquids, and gases to their motion according to the particle model.

25. If particles in any substance are in constant motion, why can't solids flow?

26. What causes different liquids to vary in their viscosity?

27. Compare physical and chemical changes.

28. The mass of ash remaining after a log burns is much less than the original mass of the log. If the law of conservation of matter is true, how can you explain this?

29. Describe what happens to the particles in a material as it is warmed from a solid to a liquid and ultimately to a gas.

30. Explain how matter changes state as energy is removed.

Critical Thinking

31. A classmate tells you that air is not actually matter because you cannot see it. Do you agree? Explain.

32. Do we know for sure that the particle model is correct? Explain.

33. Water is unusual in that it is less dense in its solid form than in its liquid form. Hypothesize as to why this may be. Why might this be beneficial?

34. Given that solids have high density, liquids medium density, and gases low density, why do most types of wood float in water?

35. A gel is a combination of a solid network structure distributed throughout a liquid. It acts like a solid in most cases because the network structure keeps the liquid from flowing. An aerogel is a solid that is made by removing the liquid from a gel, leaving the solid network with air where the liquid had been. What properties would you expect an aerogel to have?

36. In the phase diagram on page 43, at the position marked by the x, the material is a gas. If you move vertically, increasing pressure at a constant temperature, the substance will change from a gas to a liquid. Why do you think this will happen?

The Chapter opener highlights a technique that scientists use to determine what atoms are made of. The colossal device pictured here is used to break atoms apart and detect the pieces that are emitted when particles collide.

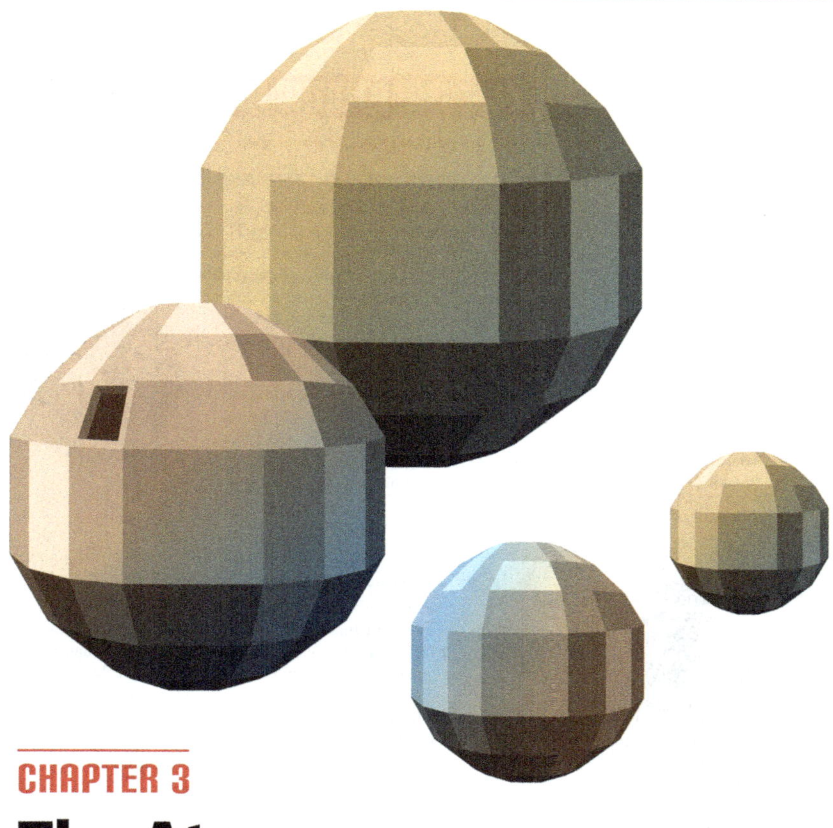

CHAPTER 3
The Atom

SIMPLY SMASHING!

What do you think is shown in the image on the left? It looks like something out of a sci-fi movie, right? Except the Large Hadron Collider isn't science fiction—it's science reality. The collider, built along the border between France and Switzerland, is located in a circular, underground tunnel whose circumference is a whopping *27 kilometers*! Incredibly, this world's-largest machine is used to explore the universe's tiniest particles, particles that are far smaller than atoms. By smashing these particles together at speeds near the speed of light, scientists hope to unlock the mysteries surrounding the basic properties of matter.

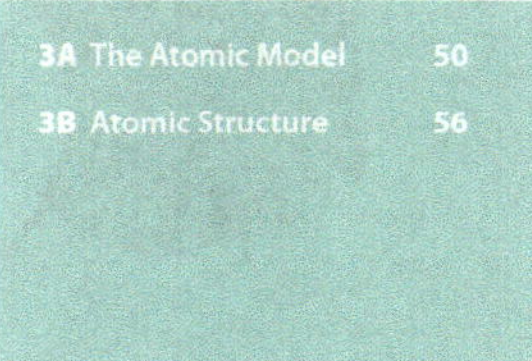

- Compare each of the atomic models.
- Evaluate the workability of each atomic model.
- Compare the properties of electrons, protons, and neutrons.
- Compare isotopes of the same element.

Overview

Chapter 3 is a foundational chapter that discusses the structure of atoms. Understanding this structure is key to understanding the chemical properties and the changes in matter that will be discussed in later chapters.

3A   **The Atomic Model**

3B   **Atomic Structure**

 ### Lab Activities

Lab 3A: *The Fiery Trial*—In this lab activity, students observe the characteristic colors that are emitted due to the arrangement of electrons in the metal of each salt.

Lab 3B: *Big Time*—In this lab activity, students create a model of an atom in order to make inferences regarding the relative sizes of different atoms and their properties.

Lab 3A: *The Fiery Trial*

Lab 3A is a good example of some of the evidence that led to the development of the Bohr model. It is a great activity to get students interested in the model. You should do the lab prior to beginning Section 3A.

Collaborative Learning

Preparation: Instead of teaching this section through lecture, consider dividing the class into learning groups and letting them rotate between five stations. Three of these stations should have a computer device for viewing video or simulations. Assign roles such as leader, recorder, timekeeper, and facilitator (to operate a device for video or simulations) and have them record the answers to the questions in a summary of what the group learned. If you want to make sure that students get all the details that you want them to have, you may prepare fill-in-the-blank notes for students to complete using the text, video, and simulations. The video and simulations referenced may be found through keyword searches. For a link to simulations, visit TeacherToolsOnline.com. *Under* Additional Resources, click on Chapter 3 Web Links. Then click on *PhET Simulations*. Locate the individual simulations using the website's search function. Other simulations and videos may be found on other sites by conducting an internet search using the subjects as keywords.

Performance:

Station 1—Study Aristotle, Democritus, and Dalton in the Student Edition (or other sources that you might provide). Answer the following: What is the law of definite proportions? What are the five points of Dalton's model? Are there any points that are contradicted by the modern model of the atom? How is Dalton's model different from Democritus's? How is it different from Aristotle's?

Station 2—View a short video regarding a Crookes tube that produces a cathode ray. Also view a Thomson experiment simulation. Answer the following: What is the law of electrostatic charges? How do electric and magnetic fields affect a cathode ray? How is this different from normal light? What is a cathode ray made of? What had Thomson discovered? Describe his model of the atom and how it was different from Dalton's model.

3A | THE ATOMIC MODEL

How have people thought about matter?

3.1 THE HISTORY OF THE ATOMIC MODEL

Many of the particles that are hurled together by the Large Hadron Collider are *subatomic particles*—particles that are smaller than atoms, which themselves were once thought to be the smallest possible bits of matter. But until very recently, people couldn't see atoms. Even now they are visible only when using the world's most powerful—and expensive—microscopes. And even with such advanced technology, atoms still appear only as blurry, indistinct spheres.

You can't see any of their internal structure. So how did we come to know so much about atoms? That story is a fascinating one. It shows both how scientific modeling works and how the work of scientists builds on the earlier work of others.

3A Questions

- Who were some of the key figures that contributed to our understanding of atoms?
- What were the significant discoveries that shaped atomic theory?
- If the Bohr model has been replaced, why do we still use it?
- How does the most current atomic model depict atoms?
- How does workability drive the development of models?

3A Terms

law of electrostatic charges, plum pudding model, nuclear model, quantum mechanics, Bohr model, energy level, quantum-mechanical model

The Ancient Greeks

Scientists had been thinking about atoms long before they were certain that atoms existed. You might be surprised to learn that ancient philosophers thought about the nature of matter. One of them, Democritus, a Greek, believed that matter existed in the form of very small particles, invisible to the human eye, which could not be further divided. He called these particles *atomos*, from a Greek word that means "indivisible." It's from that word that we get our modern English word *atom*.

It seems silly to us today, but many of the scholars of his time dismissed Democritus's notion of indivisible atoms. They believed instead that matter was continuous (i.e., it could be subdivided an unlimited number of times), an idea championed by Aristotle. No one bothered to test the two models—that way of thinking was still many centuries off.

C. 400 BC

Democritus suggested that atoms were tiny, indivisible particles. He also believed that their shapes and textures were related to the type of matter in which they were found: atoms in solids had hooks to hold them together, water atoms were slippery, and so on.

TEACHING THE MATERIAL

ESSENTIAL QUESTION

How have people thought about matter?

OBJECTIVES

- 3A1 Identify the key scientist(s) associated with each atomic model.
- 3A2 Summarize the key discoveries that shaped atomic theory.
- 3A3 Justify the continued use of the Bohr model of the atom. **BWS**
- 3A4 Summarize our current atomic model.
- 3A5 Explain how workability acted as the driving force in the development of atomic models.

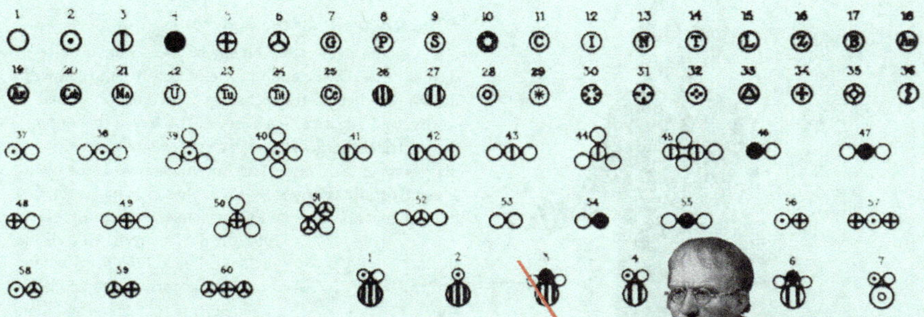

From Democritus to Dalton

By the turn of the nineteenth century, mounting evidence from the study of chemistry hinted that Democritus may have been right. The law of definite proportions, for instance, is what one would expect if matter were made of distinct particles, but not if it were continuous as Aristotle believed. It was time for some additions to the particle model of atoms.

An English schoolteacher, John Dalton, proposed such additions in 1803. Drawing in part on his own studies of the chemical reactions of gases, Dalton suggested several properties of atoms:

1. Elements are made of atoms.

2. Atoms are indivisible and cannot be destroyed.

3. The atoms of an element are all alike.

4. The atoms of one element are different from the atoms of all other elements, especially their masses.

5. Atoms combine chemically in small, whole number ratios.

Dalton's atomic model held up well for nearly a century. But scientists continued to delve into the mystery of atoms and developed better instruments for doing so. Along the way they discovered that some parts of Dalton's model were incorrect. Some changes needed to be made.

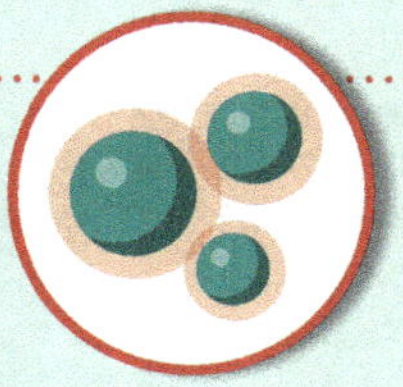

1803

Dalton imagined atoms as hard spheres of different sizes and weights, surrounded by heat envelopes.

Station 3—Study Nagaoka and Rutherford and their models and do a Rutherford gold foil experiment simulation. Answer the following questions: How are Nagaoka's and Rutherford's models different from Thomson's? How is Nagaoka's model described? What happened to the alpha particles as they passed through the gold foil? What kind of nucleus do these models have? How are electrons arranged?

Station 4—Do an emission spectra simulation and study the Bohr model. Answer the following questions: What was Bohr's explanation of spectral lines? How do the terms *photons* and *quanta* relate to atoms according to Bohr's model? How is Bohr's model different from Rutherford's and Nagaoka's?

Station 5—Study the quantum-mechanical model. Answer the following questions: How does this model differ from Bohr's model? What are orbitals? What do they look like? Is this the only model that we use now?

Seeing Atoms

The advanced microscopes mentioned in the Student Edition are Nion Hermes **S**canning **T**ransmission **E**lectron **M**icroscopes.

Odd Symbols

Dalton's element and compound symbols are shown in the background image.

More on John Dalton

Dalton's theories about atoms weren't published as a discrete body of work in a scientific journal or book as was usual for his time. Instead, his thoughts about atoms have been gleaned from his laboratory notebooks and other contemporary sources. Because of this, there is no standardized form of his atomic theory, and slight variations of it will exist from one source to another.

RESOURCES

Case Study: A World of Models

Demonstrating Electrostatic Charges, Demonstrating Cathode Rays; Demonstrating the Gold Foil Experiment

Lab 3A: *The Fiery Trial*—Using Flame Tests; Lab 3B: *Big Time*—Inquiring into Scale Models of Atoms

STRATEGIES

Class Opener: Ask students to write down a physical phenomenon that they know exists but have never seen. Have each student show it to a student nearby and discuss it. Then ask students to share their ideas with the class and explain how we know the phenomena exist. Students may mention, for example, radio waves, TV waves, cell signals, gravity, or oxygen. Use this discussion to introduce the model of the atom.

Active/Collaborative Learning: Use the Collaborative Learning teacher note on page 50 to help students develop an understanding of the evolution of the model of the atom.

(continued)

Demonstrating Electrostatic Charges

Preparation: There are an abundance of activities that can be used to demonstrate electrostatic charges using materials such as balloons, foam plates, rubber rods, plastic rods, PVC pipe, an electroscope, and more. Do a keyword search on "electrostatic charges demonstration" to find many to choose from.

Performance: First show that there are no attractions when the materials are electrically neutral. After rubbing materials together, try to show both attraction and repulsion.

Discussion:

1. Why is there no apparent attraction or repulsion initially? *(The materials have evenly distributed positive and negative particles; they are electrically neutral.)*

2. How do you think the rubbing causes attraction or repulsion? *(A buildup of electrons and therefore a buildup of charge are being produced on one of the materials. If this charge is the same as on another material, they repel each other. If the charges are opposite, they attract each other.)*

3. Why does the attraction or repulsion change if the charged material is touched? *(Electrons are redistributed.)*

4. What does this behavior tell us about electrons? *(They can be removed from atoms.)*

More on J. J. Thomson

Thomson was not the first to theorize that atoms were made of smaller fundamental particles, but earlier scientists had thought that these particles were about the same size as hydrogen atoms.

A plum pudding is a type of steamed dessert. In pre-Victorian England, *plum* was the usual word for "raisin."

Thomson was awarded the Nobel Prize in Physics in 1906.

The Discovery of the Electron

The existence of electrical charge has been known since ancient times. By the end of the eighteenth century it was also known that electrical charges existed in two forms, positive and negative. Scientists had observed that opposite electrical charges attract each other, while same charges repel each other. This became known as the **law of electrostatic charges**. By 1875, physicists had created a device that operated on this principle. Called a *Crookes tube*, the device produced a visible beam called a *cathode ray* when an electric current was applied to it. Scientists later learned that cathode rays could be deflected by both magnetic and electric fields.

On the basis of his research with Crookes tubes, university professor and physicist Joseph John Thomson theorized in 1897 that cathode rays were made of negatively charged particles. Using certain measurements, he was able to estimate the mass of these particles. This mass turned out to be extremely small—about 1/1800 of the mass of a hydrogen atom. Thomson called these tiny particles *corpuscles*, but scientists soon began calling them *electrons*. Thomson hypothesized that the electrons were coming from within the atom, which meant that the atom was *not* indivisible! Recognizing that the current atomic model wasn't workable, Thomson suggested a new model for atoms in 1904. Because atoms are electrically neutral (having neither a positive nor negative charge), Thomson proposed that the negatively charged electrons churned rapidly within a uniform positively charged matrix. This image of an atom reminded some scientists of a traditional English Christmas pudding, so Thomson's model became known as the **plum pudding model**.

Thomson's plum pudding model suggested negatively charged electrons embedded in a positive substance.

TEACHING THE MATERIAL

Demonstration: Use the Demonstrating Electrostatic Charges teacher note on this page to stimulate students' thinking about electrical charges.

Video: Do keyword searches on "Crookes tube," "Thomson experiment," "Rutherford's gold foil experiment," and "Bohr model emission spectra" for videos or simulations.

Demonstrations: Use the Demonstrating Cathode Rays and Demonstrating the Gold Foil Experiment teacher notes on page 53 to stimulate students' thinking regarding these experiments.

Formative Assessment: Use the Formative Assessment: The Bohr Model teacher note on page 54 to see how well students understand this model.

Ticket Out the Door: Write down a one-word goal of scientific modeling.

Alternate Ticket Out the Door: Imagine that you are sitting on the nucleus inside a neon atom. Describe what you are sitting on and what you see around you.

The Nuclear Model

But not everyone accepted Thomson's plum pudding model. A Japanese physicist, Hantaro Nagaoka, found fault with Thomson's notion that opposite electrical charges could mix in the manner that Thomson suggested. In that same year of 1904, Nagaoka proposed a new model. In his *Saturnian model*, the electrons orbited around a massive, positively charged center in a flat ring, much like the rings of Saturn orbit their planet.

New Zealander Ernest Rutherford had been one of Thomson's star pupils. By 1908 he had earned a Nobel Prize in Chemistry for his work on radioactive decay, but his greatest contribution to science was yet to come. Rutherford knew of the differences between the plum pudding and Saturnian models. Now a professor himself, he directed two of his students in a series of experiments to explore the nature of the inner atom. These experiments were performed between 1908 and 1913. In Thomson's model, an atom's electric charges were very spread out; Rutherford believed that a heavy and fast-moving type of particle, called an *alpha particle*, directed at an atom should be able to easily pass through it with little or no deflection.

To the surprise of all, a few of the alpha particles deflected away from the gold foil at very sharp angles. The Thomson model did not predict this. Such a result could be possible only if most of the atom's mass and positive charge were densely packed into a very small space. Rutherford called this dense, central portion of an atom the *nucleus*. Fascinatingly, Rutherford's discovery also showed that most of an atom consists of empty space. Again, the current model didn't explain the observations; it was not workable. These experiments later gave rise to a new model of the atom—the **nuclear model**—in which the atom was made up of a dense, positively charged central nucleus surrounded by negatively charged electrons.

1904

Nagaoka's atomic model had a massive and positively charged center surrounded by a flat ring of electrons.

1911

Rutherford's model of the atom had a tiny, but massive, positively charged nucleus surrounded by electrons.

Demonstrating Cathode Rays

Preparation: If you don't have a Crookes tube and appropriate power source, do a keyword search on "Crookes tube and magnet" to find a video or simulation of a Crookes tube in action. For a link to a simulation, visit TeacherToolsOnline.com. Under Additional Resources, click on Chapter 3 Web Links. Then click on *Crookes Tube*.

Performance: Show the effect of electric and magnetic fields on a cathode ray.

Discussion:

1. What do we call the beam going through the tube? *(a cathode ray)*

2. How is the cathode ray affected by the electric field? *(It is attracted to the positive side.)*

3. What does this indicate about the cathode ray? *(It is made of something that is negatively charged.)*

4. How is the cathode ray affected by the magnet? *(It is bent by it as well.)*

5. Would a beam of light be affected by a magnet in this way? *(No)*

6. What does this tell us about the cathode ray? *(It is not light; it is a type of matter composed of negatively charged particles.)*

Demonstrating the Gold Foil Experiment

Preparation: Do a keyword search on "Rutherford's gold foil experiment" for an animation, simulation, or video.

Performance: Show the simulation or video.

Discussion:

1. What was being passed through the gold foil? *(alpha particles)*

2. Why were so many of these particles passing through unaffected? *(The matter in the gold foil is mostly empty space.)*

3. Why were some of the particles bouncing off? *(Gold atoms have a very dense nucleus.)*

4. Why do you think gold foil was used for this research? (*Hint, if needed*: Gold has a property of matter that was discussed in the last chapter.) *(It is very malleable and can be hammered very thin. Thicker matter would have too many nuclei and would completely block the alpha particles.)*

Formative Assessment:
The Bohr Model

Use the light, helium, and neon spectra shown on this page to assess students' understanding of the Bohr model.

1. What wavelengths does the light spectrum show? *(all wavelengths of visible light)*

2. If each wavelength of light corresponds to a particular color and frequency of light, what determines the color, frequency, and wavelength? *(the amount of energy)*

3. Why do the helium and neon spectra have only lines of color? *(Each line represents an electron falling back to a lower energy level, and each line corresponds to a specific amount of energy.)*

4. Why is the helium spectrum different from the neon spectrum? *(Helium and neon have different arrangements of electrons.)*

5. Why do you think that neon has more spectral lines than helium? *(Neon has more electrons that move between various energy levels.)*

3.2 THE MODERN ATOMIC MODEL

As science began to unlock the secrets of the atom at the turn of the twentieth century, they saw things that classical physics—the work of Isaac Newton and others—could not explain. Classical physics was, and still is, fine for explaining the motion and energy of things like cars, rockets, and planets, but electrons were a whole new ball game. A new branch of physics was born, one that explored the behavior of matter and energy at the atomic and subatomic levels. The new branch became known as **quantum mechanics**.

The Bohr Model

Many of Rutherford's students became Nobel Prize winners in their own right. One of them, Niels Bohr from Denmark, turned his attention to learning more about how electrons move within atoms. In Rutherford's nuclear model, the electrons simply moved randomly about in the space surrounding the nucleus. If this were really happening, then atoms would constantly emit light and all matter would glow. So what was really happening? Bohr took a clue from the behavior of heated elements. When heated to high temperatures, each element produces an *emission spectrum* (p. spectra; see left) that shows only certain wavelengths of light and no others.

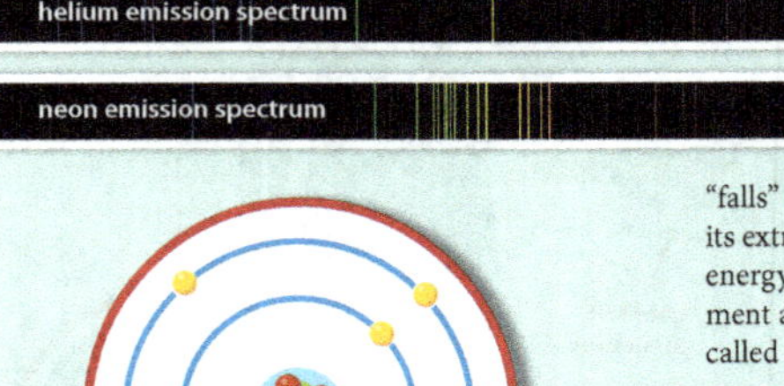

Bohr adjusted the nuclear model to explain this phenomenon. In the **Bohr model** of the atom, electrons do not travel randomly about the nucleus. Instead, electrons can move only in distinct **energy levels**, spherical regions located at fixed distances from the nucleus. If an electron absorbs energy, such as when being heated, it can "jump" to a higher energy level—one that is farther away from the nucleus. Almost instantly, though, it "falls" back to its original energy level. In the process, it loses its extra energy in the form of a *photon*, a packet of light energy that produces a single wavelength of color. An element absorbs and emits energy in very specific amounts only, called *quanta* (s. quantum).

Like the models before it, the Bohr model would soon need further refining. Even so, it still remains in widespread use today. Compared with later models, the Bohr model is fairly simple and easy to grasp. Though it is not completely correct in its depiction of atomic structure, it remains useful for showing how certain processes work, such as chemical bonding. It also represents energy levels in a way that is easier to visualize than the later improved model.

1913

Bohr's atomic model placed electrons in orbits at specific distances from the nucleus.

The Quantum-Mechanical Model

As their understanding of quantum mechanics grew, physicists soon realized that electrons didn't travel in neat, spherical orbits as predicted by the Bohr model. Nor could scientists even precisely pinpoint where an electron would be at any point in time. In the resulting **quantum-mechanical model** of the atom, the spherical orbits of the Bohr model were replaced by *orbitals*. Rather than being clearly traveled pathways, orbitals are indistinct and often oddly shaped regions where electrons are *likely* to be found. The quantum-mechanical model is still the primary model being used by physicists and chemists today.

The key factor that drove the development of new atomic models is *workability*. Could the model in use at a given time explain what scientists were seeing in their laboratories? As long as the answer was *yes*, the model remained useful. But if the answer ever became *no*, then the model would be revised or replaced in order to explain the new observations.

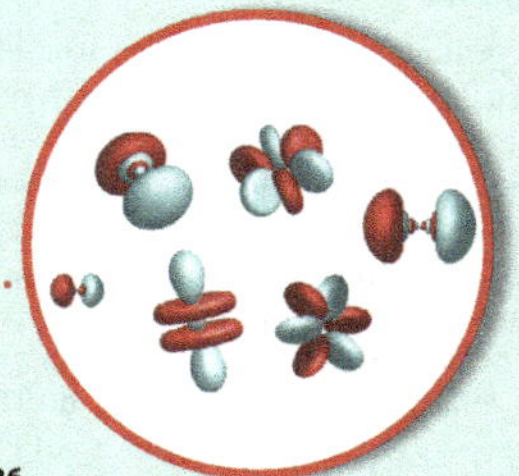

1926

The oddly shaped, overlapping orbitals of the quantum-mechanical model indicate areas where electrons in different energy levels are *likely* to be found.

CASE STUDY: A WORLD OF MODELS

What do you think of when you hear the word *model*? Perhaps you think of a plastic model car or airplane that is built from a kit. That's a kind of physical model, and scientists and engineers do make use of models like that. But there are lots of other kinds of models too. Mathematical formulas can model shapes or processes. A map models geographical features. Computer models analyze data and make predictions that are based on patterns discovered within the data.

What kind of model is being used in each of the following scenarios? Explain your reasoning for each choice.

1. Meteorologists predict the path of a developing hurricane.

2. Your class builds DNA molecules made of toothpicks and colored marshmallows.

3. A fish hatchery manager calculates how much food to feed to fish in a pond on the basis of the size of the pond and the amount of fish in it.

4. A student reads a book about mountain formation.

3A | REVIEW QUESTIONS

1. Name the scientist that is usually associated with each of the following:

   a. discovery of electrons

   b. electrons found in distinct energy levels

   c. first modern atomic model

   d. discovery of the nucleus

2. Why was Thomson's plum pudding model a significant departure from previous atomic models?

3. Why is the Bohr model still used, even though it is not completely accurate in its description of an atom's structure?

4. Following Ernest Rutherford's discovery of the atomic nucleus, what aspect of an atom's structure were scientists primarily trying to discern?

5. How is the quantum-mechanical model different from the Bohr model?

6. Why are newer models said to be more workable than older models?

Case Study: A World of Models

The questions in the case study can be discussed as a group. Use them to assess whether students have grasped what a scientific model is and is not.

Case Study Answers

1. Meteorologists use computer models that are based on current conditions and data collected during previous hurricanes.

2. These physical models are used to model a structure that is ordinarily too small to see.

3. The manager is using a mathematical formula. The size of the pond and amount of fish are variables in the formula.

4. The book is a type of descriptive model. It describes a process that happens too slowly to be seen.

Orbits Versus Orbitals

Students may be confused by the similarity of the terms *orbit* and *orbital*. Emphasize that an orbit is a fixed, distinct pathway for electrons in the Bohr model, whereas an orbital is a probable region for the location of electrons in the quantum-mechanical model.

Lab 3B: Big Time

Lab 3B should be done prior to introducing the material in Section 3B.

3A REVIEW ANSWERS

1. **a.** J. J. Thomson

   b. Niels Bohr

   c. John Dalton

   d. Ernest Rutherford *(all pp. 51–54)*

2. Prior to Thomson's discovery of the electron, atoms had been thought to be indivisible. *(pp. 50–52)*

3. The Bohr model is easy to grasp and is useful for explaining chemical processes and energy levels. *(p. 54)*

4. Scientists were trying to discern the location of electrons within the atom's structure. Some students may also reference the motion of electrons. *(pp. 54–55)*

5. The quantum-mechanical model includes orbitals that show where electrons are likely to be found, rather than showing electrons traveling in distinct paths or orbits. *(pp. 54–55)*

6. Models are said to be more workable if they better explain what is being observed and make better predictions. *(p. 55)*

 **Relative Masses and Sizes**

Hold up a paper clip and a 2 L bottle filled with water that has a mass of 1836 g and ask the following questions.

1. What subatomic particles could each item represent? *(paper clip—electron; 2 L bottle—neutron or proton)*

2. Where would the neutron or proton be located? *(in the nucleus)*

3. If the diameter of the bottle (about 11 cm) were its diameter in all directions, how far away would the electron be for a hydrogen atom? *(Pick out a place or two that all the students would be familiar with that is approximately 6 km away.)*

Scientific Notation

You may need to introduce or review scientific notation. The main takeaway here is that subatomic particles are very, *very* small!

Nucleons

Protons and neutrons are collectively known as *nucleons*. You may want to introduce this term.

- How are protons, neutrons, and electrons similar and different?
- What are some of the basic properties of atoms?
- How are the isotopes of an element similar and different?
- How can we identify specific isotopes?
- How do ions form?
- How can I tell which isotope of an element is most abundant?
- How can we protect people from radiation?

3B Terms

electron, proton, neutron, atomic number, isotope, mass number, ion, anion, cation, atomic mass

3B | ATOMIC STRUCTURE

What is it like inside an atom?

3.3 SUBATOMIC PARTICLES

The development of the modern atomic model revealed that atoms are made of three basic particles—protons, neutrons, and electrons. Each affects the structure and properties of the atom.

Electrons

Electrons (indicated by the symbol e⁻) are the smallest of the main subatomic particles, having a mass of 9.1094×10^{-31} kg. As you learned in the previous section, they were also the first subatomic particle to be discovered. They exist at a considerable distance outside the nucleus. Each electron carries a single negative electrical charge ($-1e$). Electrons are primarily responsible for all the chemical properties of atoms, including how bonds form between atoms. Many other properties of atoms are also determined by the number and arrangement of electrons around an atom's nucleus.

Protons

A **proton** (p⁺) is a subatomic particle located in the atom's nucleus. The discovery of the proton was not as straightforward as that of the electron. After Rutherford's famous experiments, scientists knew that the positive charge of an atom was concentrated in the nucleus. But the nature and identity of the particles that carried the positive charge revealed themselves only gradually. The name *proton* was first used in print in 1920. Protons have a mass of 1.6726×10^{-27} kg, which is about 1836 times the mass of an electron. They carry a single positive electrical charge ($+1e$).

If the nucleus of an atom were the size of a pea, the average atom would be the size of a football stadium!

TEACHING THE MATERIAL

ESSENTIAL QUESTION

What is it like inside an atom?

OBJECTIVES

- 3B1 Describe protons, neutrons, and electrons, including their masses, charge, and location.

- 3B2 Relate the concept of isotopes to variations within the nucleus.

- 3B3 Determine the number of protons, neutrons, and electrons in an atom using isotope notation.

- 3B4 Calculate atomic number, mass number, and charge given the number of protons, neutrons, and electrons in an atom.

- 3B5 Explain how an ion forms.

- 3B6 Predict the relative abundance of an isotope on the basis of its average atomic mass.

- 3B7 Explain how we can protect people from exposure to radiation. **BWS**

Neutrons

A second kind of subatomic particle found in the nucleus, the **neutron** (n), carries no electrical charge. It was therefore difficult to detect and was not discovered until 1932, though its existence had been predicted earlier. One of Ernest Rutherford's former students, James Chadwick, first isolated neutrons. A neutron has slightly more mass than a proton, about 1838 times that of an electron.

3.4 PROPERTIES OF ATOMS

Atomic Number

What makes silver unlike sulfur? Why is carbon so different from copper? The answer to these questions lies in the structure of the atoms that make up each element. Since each element is composed of its own kind of atom, something must make each kind of atom unique. That something is the distinctive number of protons in an atom's nucleus. For example, every gold atom has seventy-nine protons. If a proton could be removed from a gold atom, the remaining atom would no longer be a gold atom—it would be an atom of platinum because all atoms with seventy-eight protons are platinum atoms. As in gold and platinum, all atoms of a given element have the same number of protons. The unique number of protons in the atoms of each element is that element's **atomic number** and identifies that element. Atomic numbers are always whole numbers because atoms don't contain any partial protons.

The atomic number also tells us the number of electrons surrounding the nucleus in a *neutral atom*, an atom with balanced electric charges. Since protons carry a single positive charge and electrons each carry a single negative charge, an atom is neutral only if it has equal numbers of protons and electrons. For example, the atomic number of carbon is 6, so a neutral carbon atom has six protons and six electrons. The atomic numbers for all the elements are shown on a periodic table (pp. 80–81 and 570–71).

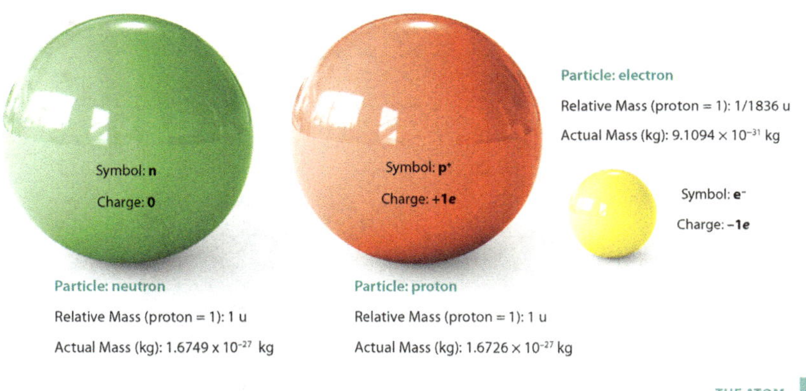

RESOURCES

Ethics: Strategies and Protecting People from Radiation (pp. 66–67)

Case Study: All Models Are Not Equal (p. 65)

Mini Lab: *Finding the Atomic Mass of Eggogen* (p. 63)

STRATEGIES

Class Opener: Use the Relative Masses and Sizes teacher note on page 56 to stimulate students' thinking regarding the relative masses of subatomic particles and the relative size of an atom.

Formative Assessment: Use the Formative Assessment: Subatomic Particles teacher note on page 58 to check students' understanding about these particles.

Biblical Worldview Shaping: Use the Sustainer of Matter teacher note on page 59 to remind students of God's power and might.

Formative Assessment: Use the Formative Assessment: Ions teacher note on page 60 to check whether students understand ions.

(continued)

Mass Number

On the basis of what you just read, what do you think makes every atom of carbon a carbon atom? If you said that it's the six protons in the nucleus of every atom of carbon, then you're correct! But there's more than just protons in the nuclei of carbon atoms—there are also neutrons. Does every carbon atom have the same number of neutrons? It turns out that the answer to that question is *no*. Most naturally occurring carbon atoms have six neutrons, but some carbon atoms have seven or even eight neutrons. These atoms of the same element with differing numbers of neutrons are called **isotopes**.

Each isotope of carbon is given a unique identifier consisting of its name followed by a number, called its *isotope name*. The isotope with six neutrons is called carbon-12. The number *12* is this isotope's **mass number**, the total number of particles in its nucleus. It indicates that there are twelve particles in each carbon-12 nucleus—six protons and six neutrons. The carbon isotope with seven neutrons is carbon-13, and the isotope with eight neutrons is carbon-14. Like atomic numbers, mass numbers are always whole numbers.

If you know the mass number for any isotope of an element, it is a simple matter to figure out how many neutrons that isotope has. To do so, one subtracts the element's atomic number from the isotope's mass number. The result is the number of neutrons present in the isotope. For example, how would one find the number of neutrons in each atom of sodium-23? By consulting a source such as a periodic table, we learn that sodium's atomic number is 11. Subtracting 11 from 23 shows us that each atom of sodium-23 has 12 neutrons.

TEACHING THE MATERIAL

Formative Assessment: Use the Formative Assessment: Properties of Atoms on page 61 to assess how well students understand atomic properties.

Discussion/Collaborative Review: Use the Collaborative Review teacher note on page 62 to encourage students to review atomic structure.

Ethics: Use the Ethics: Strategies and Protecting People from Radiation box on pages 66–67 to stimulate a discussion regarding how and why we should try to warn people about radon exposure.

Ticket Out the Door: What property of atoms determines the order of the periodic table?

Isotope Notation

An isotope can also be indicated by using *isotope notation*, which shows both the atomic and mass numbers of the element along with its chemical symbol. Starting with the element's symbol, which can be found in a periodic table (see 80–81 and pp. 570–71), the isotope's mass number is placed to the upper left and the element's atomic number is placed to the lower left. The general form is

$$^{A}_{Z}X,$$

where X is the chemical symbol of the element, A is the mass number, and Z is the atomic number. As a memory aid, always put the larger number (mass number) above the lower number (atomic number). Thus the isotope of sodium mentioned on the previous page can be written as

$$^{23}_{11}\text{Na}.$$

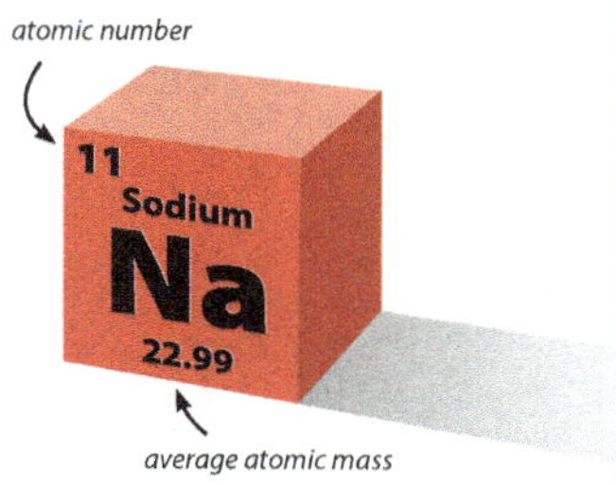

EXAMPLE 3-1: Isotope Notation

Chlorine-35 is the most common isotope of chlorine. In its pure form it is a toxic, greenish-yellow gas. Using the periodic table, write its isotope notation. How many protons, neutrons, and electrons are in one neutral atom of chlorine-35?

Answer: The symbol for chlorine is Cl and its atomic number is 17, making its isotope notation $^{35}_{17}\text{Cl}$. One neutral atom of chlorine-35 has seventeen protons, eighteen neutrons, and seventeen electrons.

Isotopes and Physical Properties

Isotopes of the same element can sometimes have quite different physical properties. Carbon-12, for example, is not only extremely common—it's also a prime building block of living things. As you might imagine, carbon-12 is a very stable sort of atom. Its cousin carbon-14 is not quite so stable. In fact, carbon-14 is radioactive (see Section 8A), a fact you might recall from studying life science. Scientists have used this property to try to determine the ages of some kinds of artifacts in a process known as *radiocarbon dating*.

Stable Isotopes

A general rule of thumb is that for atoms with low atomic numbers, stable isotopes have the same number of protons as neutrons. Generally, elements with fewer protons are more stable. As additional protons are added, neutrons act like a glue to hold the nucleus together in spite of the repulsion forces inherent in an increasingly concentrated positive charge.

Sustainer of Matter

Students may ask what holds subatomic particles together in an atom. Scientists describe forces such as the strong force, weak force, and the electromagnetic force. These are discussed in Chapter 8. But just as with the fourth fundamental force—gravity—scientists can only theorize regarding what causes these forces. Though this is not the ideal time to discuss these forces, mention them if students ask. Also remind them that Colossians 1:17 tells us that "by Him all things consist," or hold together. So whatever forces God uses to do it, He is the Ultimate Sustainer.

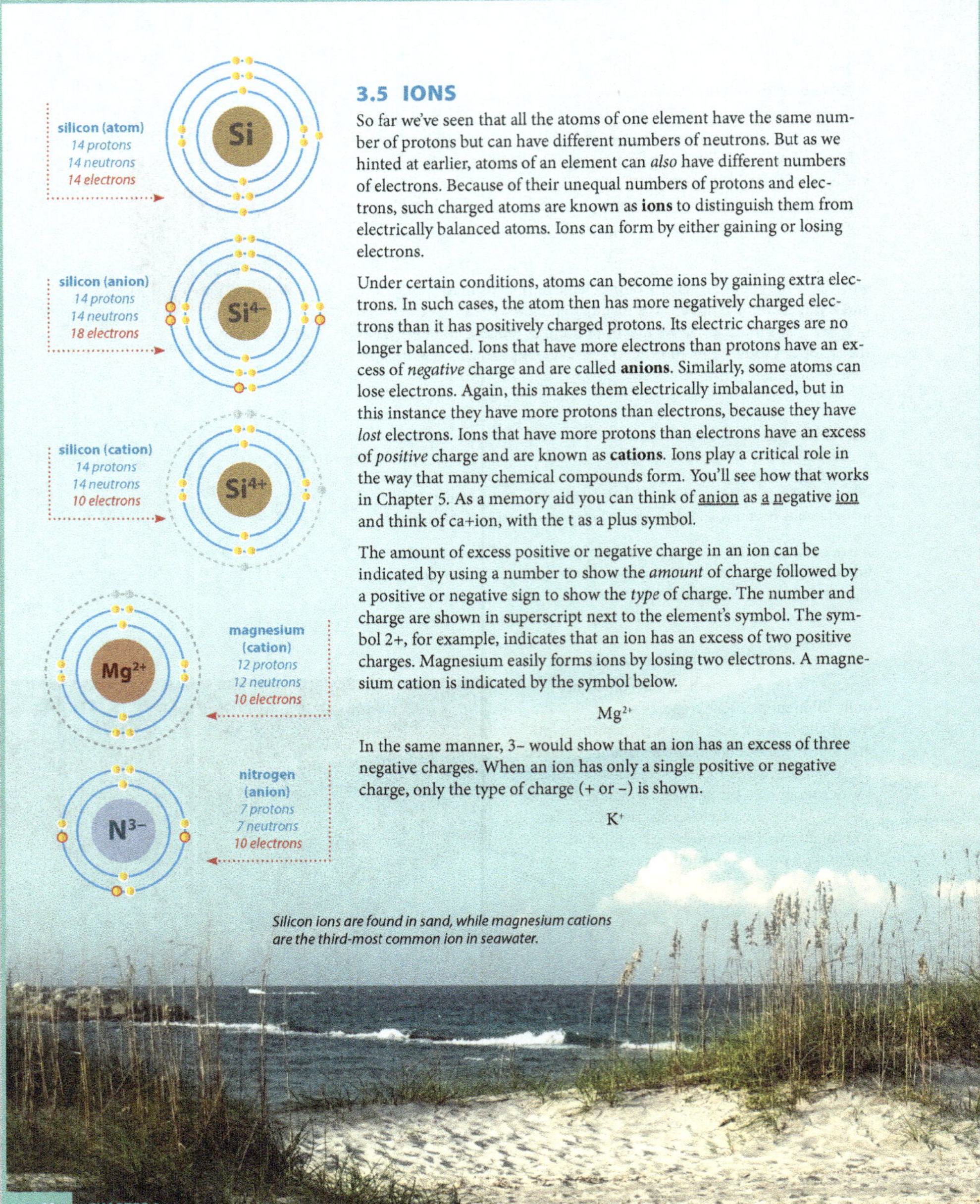

Silicon ions are found in sand, while magnesium cations are the third-most common ion in seawater.

3.5 IONS

So far we've seen that all the atoms of one element have the same number of protons but can have different numbers of neutrons. But as we hinted at earlier, atoms of an element can *also* have different numbers of electrons. Because of their unequal numbers of protons and electrons, such charged atoms are known as **ions** to distinguish them from electrically balanced atoms. Ions can form by either gaining or losing electrons.

Under certain conditions, atoms can become ions by gaining extra electrons. In such cases, the atom then has more negatively charged electrons than it has positively charged protons. Its electric charges are no longer balanced. Ions that have more electrons than protons have an excess of *negative* charge and are called **anions**. Similarly, some atoms can lose electrons. Again, this makes them electrically imbalanced, but in this instance they have more protons than electrons, because they have *lost* electrons. Ions that have more protons than electrons have an excess of *positive* charge and are known as **cations**. Ions play a critical role in the way that many chemical compounds form. You'll see how that works in Chapter 5. As a memory aid you can think of <u>anion</u> as <u>a</u> negative <u>ion</u> and think of ca+ion, with the t as a plus symbol.

The amount of excess positive or negative charge in an ion can be indicated by using a number to show the *amount* of charge followed by a positive or negative sign to show the *type* of charge. The number and charge are shown in superscript next to the element's symbol. The symbol 2+, for example, indicates that an ion has an excess of two positive charges. Magnesium easily forms ions by losing two electrons. A magnesium cation is indicated by the symbol below.

$$Mg^{2+}$$

In the same manner, 3– would show that an ion has an excess of three negative charges. When an ion has only a single positive or negative charge, only the type of charge (+ or –) is shown.

$$K^+$$

3.6 ATOMIC MASS

You've learned that both the atomic number and the mass number for any element are whole numbers. But you might notice something odd if you look at a periodic table. Take carbon, for example (see right). On a periodic table you'll see the letter C, the symbol for carbon, along with carbon's name and usually several numbers. There's a number 6, and we've already discussed that carbon has six protons, so that 6 in the box must be carbon's atomic number. But what about the number 12.01? That's not a whole number, so it can't be a mass number. So what's going on?

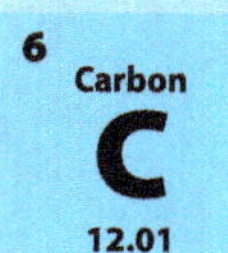

The decimal numbers that you see on a periodic table are the atomic masses of the elements. The **atomic mass** of an element is the weighted average of the masses of all the naturally occurring isotopes of that element. But before we consider how those weighted averages are determined, we first need to address the SI unit for atomic mass. Atomic masses are measured in *unified atomic mass units*, and the unit's symbol is the lowercase letter u. One u is not an exact quantity—it is defined as one-twelfth of the mass of a carbon-12 atom. Since a carbon-12 atom has six protons and six neutrons in its nucleus, 1 u approximates the mass of either one proton or neutron. As we saw earlier, protons and neutrons have slightly different masses, but the difference is very small. For most purposes, the difference in mass can be ignored. So 1 u is used to represent the mass of either particle.

But why is atomic mass called a "weighted average"? You already know how to calculate normal averages—you add up all the values in a set and then divide the total by the number of values. But how could you possibly add up *all the masses* of every carbon atom in existence? It's an absurdly impossible task! But what scientists *can* do is make measurements. They have made many measurements of the relative amounts of carbon's two stable isotopes, carbon-12 and carbon-13. These measurements have shown that about 98.93% of all the stable carbon on the earth is carbon-12. The remaining 1.07% is carbon-13. So most of the carbon on the earth is carbon-12, but not all of it.

Other Units

You will also see *Da* as the abbreviation for the unified atomic mass unit. *Da* is short for "Dalton." This is an acknowledgement of John Dalton's contribution to atomic theory.

Are Atomic Masses Constant?

The information in the third paragraph suggests that scientists can revise the atomic masses of elements if more, better, or more precise measurements of isotopic abundance are made, and that has indeed been the case. In addition, scientists have learned that not all the isotopes of an element are necessarily evenly distributed. One of bromine's two stable isotopes, for example, is more common in seawater than it is in organic substances. For this reason, IUPAC gives the atomic masses of bromine and some other elements as an interval rather than as a single value.

Formative Assessment:
Properties of Atoms

1. Which quantity best identifies the type of element? *(atomic number)*

2. From which quantity can the number of neutrons of an atom be determined? *(mass number)*

3. Which quantity best identifies an isotope? *(mass number)*

4. Which quantity is a weighted average? *(atomic mass)*

5. Which quantity tells the number of particles in the nucleus of an atom? *(mass number)*

Multiplying by Percents

Students may need a reminder that multiplying by a percentage requires shifting the decimal in the percentage two places to the left since they are actually multiplying by a fraction. Ninety-eight percent, for example, is 98/100, or 0.98.

Collaborative Review

Preparation: Consider dividing your class into as many small groups as you have devices. Groups of 2–3 students is suggested. The devices should have access to the simulations referenced in the procedure. For a link to the simulations, visit TeacherTools Online.com. Under Additional Resources, click on Chapter 3 Web Links. Then click on *PhET Simulations*. Locate the individual simulations using the website's search function.

Procedure:

1. Using the Build an Atom simulation, build neutral atoms of hydrogen-2, carbon-12, neon-22, a 2– ion of oxygen-16, and a 1+ ion of lithium-7.

2. Using the Isotopes version of the Isotopes and Atomic Mass simulation, determine the naturally occurring isotopes of hydrogen through neon, while noting which ones are not stable, and the relative amounts of each isotope in a normal sample.

3. Using the Mixtures version of the Isotopes and Atomic Mass simulation, determine which combination of isotopes comes closest to the atomic masses of lithium and chlorine.

Summary: Each group should discuss the answers to the following questions and put them on paper to be turned in.

1. Considering hydrogen through neon, which elements do you think have naturally occurring unstable isotopes, and in general, how abundant are they? *(H, Be, and C; Most samples contain only a trace.)*

2. Considering hydrogen through neon, which elements have three naturally occurring isotopes? *(H, C, O, and Ne)*

3. Considering hydrogen through neon, which isotope's atomic mass is expressed as a whole number? *(carbon-12)* Why is this true? *(Atomic mass units are based on carbon-12.)*

4. What were the ratios of lithium-6 to lithium-7 atoms and chlorine-35 to chlorine-37 atoms determined in Procedure Step 3 above? *(8 or 9 lithium-6 to 100 lithium-7; 75 chlorine-35 to 24 chlorine-37)*

If all Earth's carbon were carbon-12, then finding its average mass would be easy. Every carbon atom would have a mass of 12 u, so the average would also be 12 u. But since a small amount of carbon is carbon-13, the average mass should be more than 12 u—but how much more? That's where finding the weighted average comes in. To find the weighted average, we will multiply the mass of each isotope by its relative abundance and then add the resulting figures. The calculation is shown below.

$$12\,u(0.9893) + 13\,u(0.0107) = 12.01\,u$$

If you think about this process in reverse, you can see that the atomic masses listed on a periodic table give hints about which isotopes are most common. For example, lithium has two stable (nonradioactive) isotopes—lithium-6 and lithium-7—and its atomic mass is 6.94 u. Simply ask yourself, is the weighted average of 6.94 u closer to the mass of lithium-6 or to lithium-7? Since the value 6.94 is much closer to 7 than to 6, then most of the lithium on the earth must exist in the form of the lithium-7 isotope, and in fact it does. This determination gets a little trickier for elements that have more than two stable isotopes. Mercury, for instance, has seven stable isotopes and its average atomic mass value of 200.59 u is close to the atomic mass of mercury-201. But mercury-201 is *not* the most common isotope of mercury. Both mercury-202 (the most common isotope) and mercury-200 are more common than mercury-201.

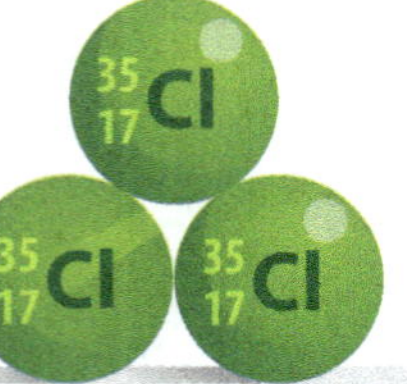

The weighted average of chlorine's two isotopes is 35.45 u.

3B | REVIEW QUESTIONS

1. Which type of subatomic particle is *not* found inside the nucleus?

2. Which two subatomic particles have approximately equal masses?

3. State the type of electric charge associated with each of the three main subatomic particles.

4. Lithium's atomic number is 3. How many electrons does a neutral lithium atom have? Explain.

5. Why does carbon-14 have a mass number of 14 if it has only six protons?

6. One isotope of tellurium, tellurium-123, has seventy-one neutrons. What must tellurium's atomic number be?

7. One isotope of sulfur has sixteen protons and eighteen neutrons. What is the name of this isotope?

8. What is the isotope notation for phosphorus-31?

9. How does an ion differ from an atom?

10. Fluorine's atomic number is 9. Indicate the type and amount of excess charge on a fluorine atom that has gained one electron.

11. The charge on a certain ion is indicated as 2+. Has the atom gained or lost electrons? Explain.

12. Why doesn't an atom's atomic mass necessarily have a whole-number value?

13. Bromine has two stable isotopes: bromine-79 and bromine-81. Bromine's atomic mass is 79.90 u. Which isotope of bromine is more common?

3B REVIEW ANSWERS

1. the electron *(p. 56)*

2. protons and neutrons *(pp. 56–57)*

3. electron: negative; proton: positive; neutron: no charge *(pp. 56–57)*

4. A neutral lithium atom has three electrons. To be neutral, a lithium atom must have the same number of protons as electrons. Since its atomic number indicates that it has three protons, a neutral atom must therefore have three electrons. *(p. 57)*

5. The other eight subatomic particles in a carbon-14 nucleus are neutrons. *(p. 58)*

6. 52 *(p. 58)*

7. sulfur-34 *(p. 58)*

8. $^{31}_{15}$P *(p. 59)*

9. An ion has gained or lost electrons compared with a neutral atom, resulting in either a positive or negative charge. *(pp. 57, 60)*

10. The fluorine atom has an excess of one negative charge. *(p. 60)*

11. It has lost electrons, resulting in an excess of positive charge. *(p. 60)*

12. The isotopes of an element have different atomic masses, so the atomic mass of the element is a weighted average of the masses of the element's stable isotopes. *(p. 61)*

13. bromine-79 *(p. 62)*

FINDING THE ATOMIC MASS OF EGGOGEN

Scientists have discovered a new element—eggogen! Your task in this lab activity is to help those scientists determine eggogen's atomic mass.

Procedure

A You and your team need to devise a plan for determining the atomic mass of eggogen. You will have only five minutes to come up with your plan. During those five minutes, you may examine the eggogen "atoms" in any manner you choose. But you may not open the atoms, and you may not make any mass measurements at this time.

B Your plan may include up to a *maximum of three* mass measurements.

1. How will your group determine the atomic mass of eggogen?

C Now carry out your plan. You will have five minutes to complete this task.

Conclusion

2. What is the atomic mass of eggogen?

3. How similar is your answer to that of other groups?

4. Describe at least one way in which you could improve the certainty of the value for eggogen's atomic mass.

5. You were not allowed to open the eggogen atoms. How is this similar to the limitations that scientists must work around when examining real atoms?

Going Further

6. Did your plan allow you to determine whether any "isotopes" of eggogen exist? If so, how many were there, and what is their ratio within your sample?

7. Suggest a plan for determining which eggogen isotope, if any, is most common and what the ratio of eggogen isotopes is in "nature" (i.e., all the samples in your class).

Essential Question:

How do scientists determine the atomic mass of an element?

Equipment
eggogen atoms (10)
laboratory balance

Finding the Atomic Mass of Eggogen

Preparing the Eggogen Atoms

Make eggogen atoms using plastic Easter eggs and marbles. Prepare enough atoms in advance to have ten atoms per group. Place one marble in 4/5 of the eggs and place two each in the remaining eggs. Although it would be beneficial to have all the eggs be the same color, if necessary just remind students that the color of the "atoms" is not important for the activity.

Make both triple-beam and electronic balances available. Also including balances with varying precision can add another element of uncertainty for students to think about.

It is suggested that you find the average mass of all of your eggogen atoms ahead of the activity so that you can have an accepted value for your students to compare with their calculated values.

Doing the Mini Lab

Steps 1 and 2 will need to be timed. Be prepared to tell students when to start and stop in both steps.

Adjust this lab activity to suit the abilities and interests of your students. You can vary the sample sizes or the number of allowed measurements or both. You can also throw in a third "isotope" by creating eggogen atoms with three marbles in them.

At your discretion, you may require students to record their measurements in a data table.

You can save class time by assigning the Going Further section as homework.

Group Discussion: Many Roads, One Destination

An excellent opportunity presents itself in this activity for you to discuss the variety of ways in which students could accomplish the task at hand. Include a discussion of which method obtained a result closest to the accepted value, along with suggestions for modifying students' methods in ways that might have improved their results.

Inquiry Labs

Students will find that this lab activity is not a traditional one in which they follow a prescribed procedure. Rather, it is an example of a guided inquiry activity in which the teacher initiates the activity with a question or problem, and students work together to plan and implement a procedure to answer the question or solve the problem.

Answers

1. Answers will vary.

2. Answers will vary.

3. Answers will vary.

4. Answers will vary. One way to improve the certainty would be to average the values across all groups.

5. Scientists cannot see into real atoms.

6. Answers will vary. There are two eggogen isotopes. They occur in a ratio of light to heavy of 4:1, but actual ratios within samples will vary.

7. Answers will vary.

Recalling Facts

1. b. *(p. 52)*

2. e. *(p. 54)*

3. a. *(p. 51)*

4. d. *(p. 53)*

5. c. *(p. 53)*

6. The formulation of the law of definite proportions, the law of electrostatic charges, and the discovery of protons, neutrons, and electrons all shaped the development of the modern atomic model. *(pp. 51–55)*

7. Protons and neutrons are found in the nucleus, while electrons are found outside the nucleus. *(pp. 56–57)*

8. Protons and neutrons are both about 1800 times more massive than an electron. *(pp. 56–57)*

9. An atom's identity is determined by its atomic number—the number of protons in its nucleus. *(p. 57)*

10. The number of electrons in a neutral atom is equal to the number of its protons. *(p. 57)*

Understanding Concepts

11. The plum pudding model explained cathode rays better than the indivisible atom of the Greek model or Dalton's model. The nuclear model provided a better explanation for the evidence from the gold foil experiments. The Bohr model explained the specific energy (colors of light) absorbed and emitted by atoms. The quantum-mechanical model explains the location of electrons better than previous models. In each case the model was more workable—it explained new observations better. *(pp. 51–55)*

12. Contrary to Dalton's theory, later discoveries showed that atoms were made of smaller parts, that is, not indivisible, and that atoms of the same element could be different from each other (e.g., isotopes and ions). *(pp. 51, 52, 58, 60)*

13. The Bohr model remains useful for showing how certain chemical processes work. It also provides a clear depiction of energy levels. *(p. 54)*

14. It should be modified to explain the new observation or be replaced. *(p. 55)*

15. According to the quantum-mechanical model of atoms, protons and neutrons are located within the nucleus of an atom, which in turn is surrounded by orbitals where electrons are likely to be found. *(pp. 56–57)*

16. Their lack of electric charge made neutrons difficult to detect. *(p. 57)*

17. 29 *(p. 57)*

18. 34 *(p. 58)*

19. 63 *(p. 58)*

20. 29 *(p. 57)*

21. protons (Some students may also correctly point out that they would have the same number of electrons.) *(p. 57)*

22. Atoms of the same element can have different numbers of neutrons, in which case they are isotopes, or they can gain or lose electrons, forming ions. *(pp. 58, 60)*

23. The term *mass number* describes the total number of protons and neutrons in an isotope and is a whole number. *Atomic mass* is the weighted average of the masses of all the naturally occuring isotopes of an element and typically is *not* a whole number. *(pp. 58, 61)*

24. a. $^{7}_{3}\text{Li}$

  b. $^{207}_{82}\text{Pb}$

  c. $^{56}_{26}\text{Fe}$

  d. $^{42}_{20}\text{Ca}$ *(all p. 59)*

CHAPTER 3

3A THE ATOMIC MODEL

- The Greek philosopher Democritus used the word *atomos* to describe indivisible particles of matter.

- John Dalton proposed a comprehensive theory of atoms in 1803.

- J. J. Thomson discovered electrons in 1897 and suggested a new model of the atom that came to be known as the plum pudding model.

- Ernest Rutherford and his students confirmed the existence of the nucleus, a small and dense central portion of an atom.

- Niels Bohr established that electrons move in distinct orbits with particular energy levels at fixed distances from the nucleus.

- Despite not being the most up-to-date model, the Bohr model remains useful for its ability to illustrate some chemical processes.

- The quantum-mechanical model of atoms replaced orbits with orbitals that showed where electrons were likely to be found.

- Workability was a key factor in the development of new atomic models.

3A Terms

law of electrostatic charges	52
plum pudding model	52
nuclear model	53
quantum mechanics	54
Bohr model	54
energy level	54
quantum-mechanical model	55

3B ATOMIC STRUCTURE

- Basic subatomic particles include positively charged protons and electrically neutral neutrons, both found in the nucleus, and negatively charged electrons found outside the nucleus.

- A neutron is slightly more massive than a proton, and both are roughly 1800 times more massive than an electron.

- Atoms of the same element always have the same number of protons, but the isotopes of an element have different numbers of neutrons.

- Isotope notation indicates the name of an element along with its atomic number and mass number.

- Ions exist as either positively charged cations that have lost electrons or as negatively charged anions that have gained electrons.

- The atomic mass of an element is the weighted average of the masses of the element's stable isotopes.

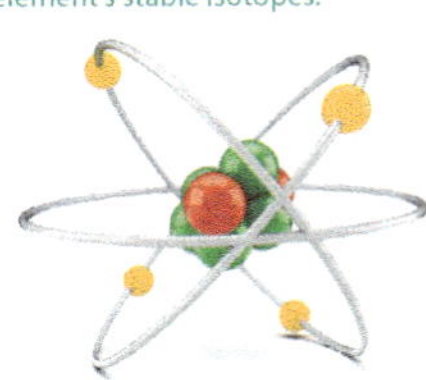

3B Terms

electron	56
proton	56
neutron	57
atomic number	57
isotope	58
mass number	58
ion	60
anion	60
cation	60
atomic mass	61

CHAPTER REVIEW QUESTIONS

Recalling Facts

For Questions 1–5, match each atomic model with the scientist most closely associated with it.

1. 

2.

3.

4.

5. 

a. John Dalton
b. J. J. Thomson
c. Hantaro Nagaoka
d. Ernest Rutherford
e. Niels Bohr

6. What key discoveries shaped the development of the atomic model?

7. Where are protons, neutrons, and electrons located in an atom?

8. How does the mass of a proton or neutron compare with that of an electron?

9. What property of an atom determines what element it is?

10. How is the number of electrons in a neutral atom related to the number of protons?

REVIEW

Understanding Concepts

11. Explain how each new atomic model was more workable than the last.

12. What parts of Dalton's atomic theory were disproven by later discoveries?

13. Why is the Bohr model still used today even though it is no longer believed to accurately represent the structure of an atom?

14. If an atomic model cannot explain a scientific observation, what should happen to it?

15. Summarize our current understanding of the locations of the basic subatomic particles within an atom.

16. Why were neutrons discovered later than either protons or electrons?

An atom of copper-63 contains 29 protons. Use this information to answer Questions 17–20.

17. What is copper's atomic number?

18. How many neutrons does copper-63 have?

19. What is copper-63's mass number?

20. How many electrons does a neutral copper-63 atom have?

21. Two of selenium's stable isotopes are selenium-78 and selenium-79. Which subatomic particle must each of these isotopes have the same number of?

22. Describe two ways that atoms of an element can be different from each other.

23. Explain the difference between *mass number* and *atomic mass*.

24. Write the isotope notation for the following isotopes.

    a. lithium-7   c. iron-56

    b. lead-207   d. calcium-42

A neutral atom of potassium (atomic number 19) loses one electron. Use this information to answer Questions 25–27.

25. Will the resulting ion have an excess positive or negative charge?

26. What kind of ion is formed by the loss of the electron?

27. Write the symbol for this ion using potassium's atomic symbol (K).

28. Hydrogen has three stable isotopes: hydrogen-1, hydrogen-2, and hydrogen-3. Hydrogen's atomic mass is 1.008. Which isotope do you think is most common? Explain.

Critical Thinking

29. Regarding his discovery of the nucleus, Ernest Rutherford said, "It was almost as incredible as if you had fired a 15-inch shell [an artillery shell] at a piece of tissue paper and it came back and hit you." Why was that so?

30. Why is it more proper to describe the quantum-mechanical model of atoms as *workable* rather than *correct*?

31. Could a magnesium cation, Mg^{2+}, form by the addition of two protons to the nucleus of a magnesium atom? Explain.

32. While helping a friend with his homework, you notice that he has written the isotope notation for bromine-81 as $^{35}_{81}Br$. What is wrong with your friend's notation, and why would the isotope that it suggests be physically impossible?

CASE STUDY:
ALL MODELS ARE NOT EQUAL

You may recall from a study of life science that scientists also model the beginnings of life on Earth. Until the mid-nineteenth century, most models of origins included God as the Creator of life, and some of these models also did not allow for change within created kinds of organisms. But around the turn of the nineteenth century, some scientists began to consider that living things might change, or *evolve*, over time. This idea became predominant after Charles Darwin introduced his theory of natural selection. Since then, most scientists have abandoned the Creation model despite the fact that it remains very workable.

Use the Case Study above to answer Questions 33–35.

33. How are the models of origins similar to the atomic models that you read about in this chapter?

34. One of the major differences between models of origins and models of atoms concerns the idea of testability. Explain how the two models are different in this regard.

35. Why do you think that most scientists have abandoned the Creation model of origins despite its workability?

ETHiCS — STRATEGIES AND PROTECTING PEOPLE FROM RADIATION

Do you remember the discussion of ethics in Chapter 1? If you are a little rusty, you should review pages 9–10 because we are going to study a strategy for evaluating the ethics of issues in physical science. Then we will apply this strategy to evaluate responses to radon gas exposure.

The Strategy

Strategies are helpful because they provide a way to deal with the complexities of ethics and help people organize their thoughts. This strategy is organized into six steps.

❶ What information can I get about this issue?

Often you may need to do some research on a topic before evaluating whether it is right or wrong. For this exercise, you can find the needed information in the example that follows.

❷ What does the Bible say about this issue?

Study the Bible to see what it says about this issue. Sometimes God's Word directly addresses an issue. Many times you have to study the Scriptures to learn how its principles apply to the issue. Consider the biblical principles of the image of God in man, the Creation Mandate, and God's whole truth.

❸ What are the acceptable and unacceptable options?

When trying to solve an ethical dilemma, you must look at all the options. Some of these potential solutions may not be obvious, while some may be talked about by almost anyone discussing the issue. Using the biblical teaching and principles that you considered in Step ❷, decide which of the options are acceptable from a biblical point of view. Discard those that are biblically unacceptable.

❹ What may be the consequences of the acceptable options?

Remember the biblical outcomes of human flourishing, a thriving creation, and God's glory? Some of the acceptable options may have consequences that fit with the biblical outcomes. Others may not. Reject the options that have consequences that are inconsistent with human flourishing, a thriving creation, and God's glory.

❺ What are the motivations of the acceptable options?

Remember the biblical motivations of faith in God, hope in God's promises, and love for God and others? Besides considering principles that inform our decisions and consequences biblically, we also need to be asking how this option will help me and others grow. We should also reject options that don't help us grow biblically in faith, hope, and love.

❻ What action should I take?

Suggest a course of action or form an opinion. Be prepared to justify the action that you suggest on the basis of the analysis in the previous steps.

The Issue: Radon Gas Exposure

❶ What information can I get about this issue?

Several times in this chapter, you've seen the word *stable* used to describe some isotopes. Unstable isotopes attempt to become stable by giving off excess particles or energy or both. These emissions are forms of *radiation* (see Chapter 8). Many people have some notion that radiation is a bad thing produced by nuclear power plants or atomic explosions, but did you know that there are naturally occurring radioactive isotopes? One source of these isotopes is the radioactive element uranium. Uranium is found throughout the earth's crust in small amounts. As uranium breaks down, it forms another radioactive element, radon, which is a gas. When radon gas seeps out of the earth's crust, it can be inhaled. This is not normally a problem because of the very small amounts produced. But radon gas can collect in places such as mines or in the basements of houses. In either of those places it has the potential to accumulate in amounts sufficient to cause health problems in humans.

❷ What does the Bible say about this issue?

Radioactive elements were unknown at the time the Bible was written, so exposure to radiation is not directly addressed in the Bible. But the Bible does teach us to have concern for our fellow image bearers, so dealing with radiation exposure is something we should certainly consider.

❸ What are the acceptable and unacceptable options?

An unacceptable option is to do nothing. Worse yet would be to allow people to be potentially harmed by radon gas exposure if you knew in advance that the danger existed. Exodus 22:5–7 contains some laws that show the importance that God places on the accidental or inadvertent damage of another person's property due to carelessness. How much more valuable is a human life compared with an item of property! Some acceptable options include mapping of areas at risk for radon exposure (see image on the facing page), requiring homeowners and businesses to disclose the presence of radon gas, and developing affordable ways to reduce radon gas exposure.

❹ What may be the consequences of the acceptable options?

One consequence would be the flourishing of human life. Knowing both where radon gas risks are highest and how best to reduce exposure to it when present would allow people to make informed decisions about where to live and how to build safe residences and businesses. Besides lowering the risks for exposure and illness to individuals, protection against radon gas exposure would lower the costs to society for treating and possibly compensating victims of exposure.

❺ What are the motivations of the acceptable options?

Alerting people to the possible hazards associated with exposure to radon gas in places where they live or work is an example of loving my neighbor as myself. Helping people take action against radon gas exposure can also demonstrate God's love and concern for His image bearers.

❻ What action should I take?

Now that I know about the risks of radon gas, I can make wise decisions for my future. For example, if I move to a different part of the country, I can consult resources that indicate where the risk of radon gas exposure is greatest and act accordingly. I can ask my employer whether my place of work has been tested for exposure. If I buy a house, I can request that the property be tested as a condition of the sale. This will help protect both me and my family.

Use the Ethics box above to answer Questions 36–40.

36. Which step(s) in this process explore(s) biblical principles?

37. Which step(s) in this process explore(s) biblical outcomes?

38. Which step(s) in this process explore(s) biblical motivation?

39. What might happen in Step ❷ if Step ❶ is not carried out thoroughly?

40. Why is Step ❻ necessary?

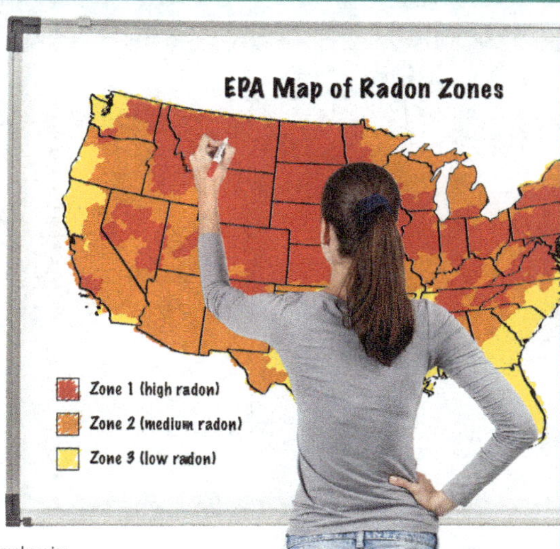

More About Radon

Radon gas is colorless and odorless. The most serious possible long-term effect is lung cancer. The EPA estimates that as many as 21,000 people die from radon-caused lung cancer in the United States each year. It most often enters a house through gaps in the walls or floors, especially over areas where the soil under a house is not covered by cement or some other material. It can even come through cracks in the concrete or around plumbing. Houses next to each other may have totally different levels of radon. Radon tests can be purchased from retail sources or may be available from state agencies. The sources of radon in a house can be dealt with to reduce the threat.

36. Step ❷

37. Step ❸

38. Steps ❺ and ❻

39. If you do not thoroughly understand an issue, you might incorrectly apply biblical principles to the issue.

40. It's not enough to decide the ethics of an issue. You must apply biblical ethics by turning them into action.

The photo of diamonds on this page highlights the amazing variety of forms of elements such as carbon in the earth. Thought to be made deep within the earth through intense heat and pressure, diamonds are now used to subject other materials to similar conditions. Using a diamond anvil cell (inset), scientists can study what may happen to substances deep underground .

DIAMOND ANVIL CELL
FORCE
particles
FORCE

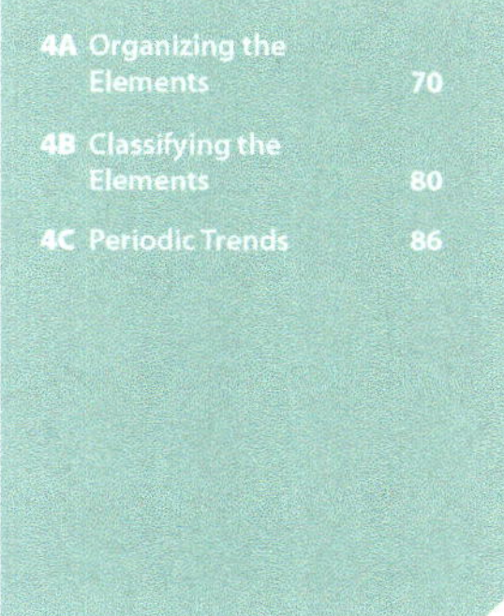

CHAPTER 4
The Periodic Table

DIAMONDS DELIVERING DATA

Diamonds are beautiful! Surprisingly, they are made of the same element as charcoal for your grill: carbon. We all know that diamonds are valued as gemstones, but they are valued for other reasons too. They are the hardest natural substance and have the highest thermal conductivity. These two characteristics make them great for industrial uses, such as cutting and grinding.

But did you know that diamonds are used for research also? A device called the *diamond anvil cell* enables scientists to create extreme pressures and temperatures. The design of the device (inset image left) is such that a sample is squeezed between the tips of two diamonds. Because of diamond's hardness and high thermal conductivity, scientists can apply immense pressures and high temperatures to a sample during testing. Diamonds are also transparent, which allows illumination of samples.

Diamond anvil cells are tiny laboratories in which scientists reproduce the most intense conditions in our universe—such as pressures and temperatures found in Earth's core. They "push" living cells to extremes to see whether they can survive. Diamonds may be a researcher's best friend.

💻 *Diamond Anvil Cells*

Some students may be interested in learning more about diamond anvil cells. Suggest that they do an internet search using the keywords "diamond anvil cell" or "hydrothermal diamond anvil cell."

- Summarize how the periodic table developed over time.
- Demonstrate how the periodic table predicted undiscovered elements.
- Relate the arrangement of the periodic table to atomic structure.
- Relate the arrangement of the periodic table to periodic trends.

Overview

Chapter 4 is foundational since a knowledge of the periodic table and an understanding of atomic structure are key to the chemistry chapters that follow.

🥽 *Lab Activities*

Lab 4A: *Bricks and Feathers*—In this lab activity, students measure the mass and volume of several different metal samples. They then calculate their densities and develop a mathematical model regarding how density changes across a row of the periodic table.

Lab 4B: *How Wide Is an Atom?*—In this lab activity, students investigate atomic radii to determine whether there is a pattern in the atomic radii of atoms in a period or group.

 ### Class Discussion:
Importance of Organizing

Begin class with a list of school events randomly ordered on the board. These could be events such as concerts, ball games, end of a quarter, and class trips. Ask students which events come first, which come last, which occur on a particular day of the week or occur on the same day, and other similar questions. Then ask them what would help in visualizing this schedule. After presenting a calendar showing a schedule of these events and their relationships to each other, show a list of randomly arranged names of elements. Use this to stimulate discussion regarding the best way to organize the elements.

Also emphasize that when we organize things in science, we do so on the basis of the order that God created.

✔ Key to Efficiency

Point out to students that since there are 118 known elements with such names as francium, scandium, technetium, protactinium, and tellurium, even scientists who work with these substances all the time need a means of keeping them straight. Organization is the key to using any large set of information effectively. For example, a calendar aids us in planning the next few months since it is organized into days, weeks, and months. In the same way, the periodic table gives us a visual means of analyzing the properties of the elements.

4A Questions

- Why do we need a periodic table?
- Who developed the periodic table?
- How has the periodic table changed over time?
- Can the periodic table predict new elements?
- Why is the periodic table arranged the way it is?

4A Terms

periodic law, periodic table, family, group, valence electron, period

Finding Elements on a Periodic Table

Throughout this chapter, the first reference to an element will include a number in parentheses. This number is the element's atomic number. Use atomic numbers to familiarize yourself with the location of elements on the table.

4A | ORGANIZING THE ELEMENTS

How does the periodic table relate to elements in the real world?

4.1 THE NEED FOR A TABLE

Chemical Discoveries

Today we know of 118 elements. Half of these were discovered only in the last 160 years, but some have been known since ancient times. The Bible refers to seven materials—gold (atomic number 79), silver (47), tin (50), copper (29), lead (82), iron (26), and brimstone, that is, sulfur (16)—that we know as elements today. While the Bible doesn't refer to these substances as elements, the concept that elements are the basis for matter is an ancient one.

Ancient Greeks thought that all matter was made of five elements, each with its own unique shape and characteristics. The original four elements were earth, air, fire, and water. Aristotle later added a fifth—aether. All other matter was thought to be made of combinations of these basic elements.

Interest in elements and other materials led to the rise of a group of scientists called *alchemists*. Much of their early work was aimed at turning low-value materials, such as lead, into high-value substances like gold. But some alchemists were interested in studying elements for purely scientific reasons.

TEACHING THE MATERIAL

ESSENTIAL QUESTION

How does the periodic table relate to elements in the real world?

OBJECTIVES

- 4A1 Identify the contributions of the key scientists associated with the development of the periodic table.

- 4A2 Explain how workability acted as the driving force in the development of the periodic table.

- 4A3 Evaluate the predictive power of the periodic table.

- 4A4 Identify periods, groups, and families on the periodic table.

- 4A5 Relate the arrangement of the periodic table to our understanding of atomic structure. **BWS**

Robert Boyle

In the seventeenth and eighteenth centuries, Robert Boyle and Antoine Lavoisier began a transition from alchemy to chemistry. Robert Boyle was an alchemist who believed that it was possible to change one element into another. His work solidified the position that matter consisted of indivisible particles of elements, rejecting the Greek concept of matter as mixtures of the five basic elements. In 1661, he published *The Skeptical Chemyst*, in which he called for experimentation to be the basis of science.

Antoine Lavoisier

Likewise, Antoine Lavoisier played a crucial role in our understanding of the elements. In 1779, while researching combustion reactions, he demonstrated that combustion required oxygen (8). Therefore fire formed as oxygen and other materials—the fuel—reacted. Fire couldn't be an element if it needed something else—oxygen—to form it. By this time, chemists had discovered many of the naturally occurring elements, and Lavoisier made one of the first lists of these elements. His list included the thirty-three elements known in his day.

Jacob Berzelius

As the list of known elements grew, scientists quickly recognized the need to organize all this data. Many scientists came up with their own symbols to represent each element. These different symbols made the sharing of information difficult. In 1814, Swedish chemist Jacob Berzelius came up with a system that would become the accepted standard. As Carolus Linnaeus did with animals, Berzelius gave each element a Latin name. He then represented each element with the capitalized first letter of its name. If needed, he used the first two letters of the name, with the second letter in lowercase. Berzelius represented oxygen with an O, and he used N to represent nitrogen (7).

Berzelius also combined these symbols to represent compounds. He would use the letters to represent the elements and *superscripts*, or other symbols, to represent how many atoms of each element were present in the compound. Berzelius came up with symbols for the forty-seven elements known to him.

✔️ Superscripts or Subscripts?

If students don't pick up on it themselves, point out the superscripts in the system that Jacob Berzelius used for chemical symbols. As you move on to writing chemical formulas, you can point out that subscripts act very similarly to mathematical exponents. For example, ammonium sulfate $(NH_4)_2SO_4$ contains two nitrogen atoms, eight hydrogen atoms, a sulfur atom, and four oxygen atoms. Note the number of hydrogen atoms. The subscript 2 on the (NH_4) signifies the same thing that the exponent 2 in $(xy^4)^2$ does. Just as $(xy^4)^2 = x^{1\times2}y^{4\times2} = x^2y^8$, this compound has four hydrogen atoms two times for a total of eight.

RESOURCES

🔵 Mini Lab: *Organizing Elements*; Lab 4A: *Bricks and Feathers*—Exploring Element Density in a Period

STRATEGIES

Class Opener: Use the Class Discussion: Importance of Organizing teacher note on page 70 to stimulate students' thinking regarding organizing information about elements.

Formative Assessment: Use the Formative Assessment: Early Chemists teacher note on page 72 to evaluate students' knowledge of chemistry history.

Resources: As you introduce the periodic table, conduct an internet search per the Periodic Table Resources teacher note on page 74.

Formative Assessment: Use the Formative Assessment: Developing Periodicity teacher note on page 75 to check students' knowledge of periodic table history.

Discussion: Use the Class Discussion: Atomic Mass or Atomic Number? teacher note on page 75 to help students recognize why Moseley reorganized the periodic table on the basis of atomic number instead of atomic mass.

(continued)

 Capitalization

Students must grasp the importance of writing elemental symbols correctly. As the Student Edition indicates, failing to use uppercase or lowercase letters changes the identity of the substances under discussion. Errors made in writing elemental symbols will be compounded when students are later required to write out chemical formulas and equations.

Three-Letter Symbols

While the periodic table included in the Student Edition has all the elements named, students are likely to see older periodic tables with unnamed elements. They will quickly notice that those elements have a three-letter symbol. Until an element is given its official name, its name is based on its atomic number. For instance, nihonium (113) was called ununtrium, meaning "one one three," and its symbol was Uut.

Formative Assessment:
Early Chemists

1. What aspect of the periodic table can we attribute to Berzelius? *(symbols of elements)*

2. Boyle's contribution to science was his insistence that science be done through what process? *(experimentation)*

3. Summarize Lavoisier's experiment that demonstrated that matter was not made of the five classical elements of the Greeks. *(Lavoisier was studying combustion reactions. He realized that oxygen was required for combustion to occur. Fire was a consequence of the combustion reaction and not an element involved in the process. Oxygen was an element involved in the process.)*

4. What did Berzelius add when representing compounds that we no longer use today? *(superscripts)* What is used instead? *(subscripts)*

5. Why do some of the symbols of elements, such as Na for sodium and Fe for iron, seem not to match their names? *(Some of the symbols are based on Latin names.)*

Modern Element Symbols

As the number of known elements has grown, the need for a standard way to represent them has become even greater than in the past. Our modern system is much like Berzelius's system, though it uses *subscripts* to indicate the number of atoms. The symbol for each element still consists of either one or two letters. In our modern system, twelve of the elements have a symbol that consists of a single capital letter, the first letter of the name. For instance, we use U to represent uranium (92). The rest of the elements use a capital letter followed by a lowercase letter. For example, Chlorine (17) has the symbol Cl, and Zn is the symbol for zinc (30). Remember that in two-letter symbols the first letter is always capitalized and the second letter is always lowercase. This helps us figure out whether we are dealing with an element or a compound. For example, are we using cobalt (27) with the symbol Co or carbon monoxide, a compound made of carbon (C) and oxygen (O) and whose *chemical formula* is CO?

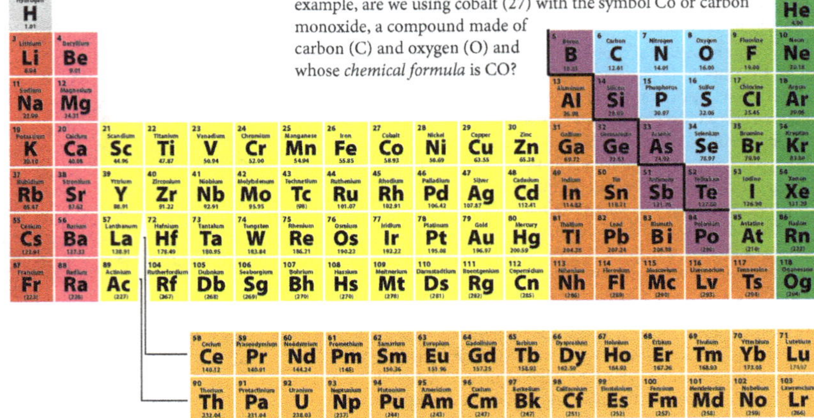

Tungsten (74) has the symbol W, which is the first letter in wolfram, tungsten's name in German.

At first glance, not all the symbols make sense, such as Na for sodium (11), Fe for iron, and Ag for silver. But remember that the elements were given names in Latin. Some of the symbols are based on the element's name in this original language. Na (sodium) came from *natrium*, Fe (iron) from *ferrum*, and Ag (silver) from *argentum*—all words from Latin.

4.2 DEVELOPMENT OF THE PERIODIC TABLE

Can you imagine trying to memorize the names, symbols, and characteristics of all known elements? In the seventeenth and eighteenth centuries, this wouldn't have been too difficult. But as the number of elements grew, many scientists looked for ways to arrange the information known about them.

TEACHING THE MATERIAL

Biblical Worldview Shaping: Use the "Why Is the Periodic Table So Orderly?" teacher note on page 76 to help students recognize that the order of the universe, including the periodic table, is a result of the orderliness of its Creator.

Differentiated Learning: Consider using the Differentiated Learning: Blocks teacher note on page 78 if you have advanced or curious students.

Formative Assessment: Use the Formative Assessment: Groups and Periods teacher note on page 78 to assess students' understanding of the structure of the periodic table.

Project-Based Learning: Use the Project-Based Learning: Periodic Table of . . . teacher note on page 78 to allow students the opportunity to creatively demonstrate their understanding of how the periodic table is organized.

Ticket Out the Door: Write a one-sentence explanation of why the periodic table is the best way to organize the elements.

ORGANIZING ELEMENTS

As scientists gathered more information, the need to organize their data increased. This became evident as chemists discovered more elements. How would you organize elements? In this activity, you will arrange fifteen "known" elements.

Procedure

A Study the data on each of the element cards.

   1. What data do you think will be most helpful for sorting the elements?

B Choose the property or properties by which you are going to sort.

C Arrange cards in the form of a table, starting with the first row. Start new rows when needed.

   2. Which property did you use for sorting the elements? Why did you make that choice? If you used more than one property for sorting, was one considered of first importance compared with the other?

   3. How did you decide to start a new row?

   4. Were any of the elements difficult to sort?

   5. Are there any gaps in your table of elements?

Conclusion

   6. What trends can you see in the properties?

Going Further

   7. What would a missing element imply?

   8. If you have an element missing, predict the characteristics of that element.

Equipment
Set of fifteen element cards

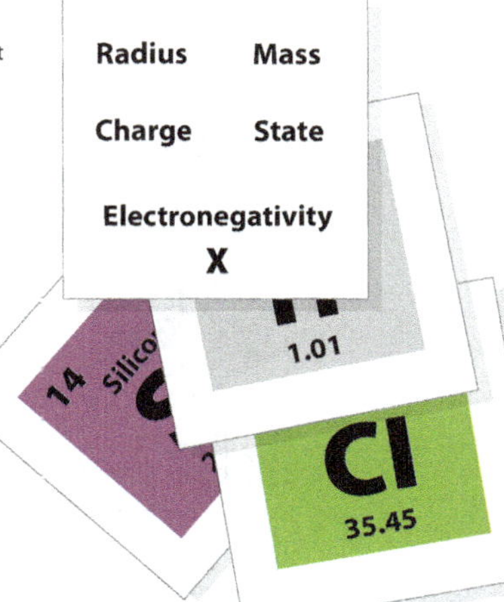

Periodicity

German chemist Johann Döbereiner made a key discovery while studying properties of the elements in the early 1800s. He noticed groups of elements, like chlorine, bromine (35), and iodine (53), with similar properties. Since these groups always seemed to have three elements, he called them *triads*. While most chemists didn't accept Döbereiner's work at the time, his efforts prompted a few to look for patterns in the properties of elements. These scientists started to notice repeating patterns when they arranged the elements in order by increasing atomic mass. This repeated pattern is called *periodicity*.

Preparation:

A reproducible copy of the element cards can be found in Appendix I. The cards are on a single page and will need to be cut out prior to beginning the activity. The expected organization of the elements is also provided in Appendix I. This solution assumes that students sort the elements by mass, with new rows according to radius. You will have to modify the answers if students take a different approach.

Answers

1. Answers will vary. The mass and radius could both be helpful. The state of matter may exhibit a pattern but is not likely to be very helpful. There are only four different charges, so they probably won't help. The electronegativity values seem to vary a lot. The chemical symbol won't be of any use.

2. Answers will vary. *Example*: I sorted the elements according to atomic mass. I started a new row when I noticed a sudden change in radius. I was unable to start new rows if I sorted by radius and used mass as secondary.

3. Answers will vary. *Example*: I noticed a sudden change in radius.

4. Answers will vary. *Example*: Lk and Dw were difficult to sort due to the closeness of their mass and radius values.

5. Answers will vary. *Example*: Yes. There seems to be a gap after Lt.

6. Answers will vary. *Examples*: The radii increase down the columns and decrease across the rows. The gases are all in the upper right corner.

7. an element that has yet to be discovered

8. The missing element should have a mass between 31.75 and 39.63 and a radius between 3.91 and 5.81. It should be a solid.

When discussing Newlands's law of octaves, you can use a piano keyboard or at least a photo of one to illustrate periodicity. Consider superimposing the symbols of elements in Newlands's table onto the keys in a photo of a piano keyboard.

Mistaken Correction

In 1869, Mendeleev published his first periodic table with elements in order of atomic mass in five vertical series. Some of the elements were lined up horizontally according to similar properties. He put iodine and tellurium in an order that was based on their properties but that was out of order according to their atomic masses. He put a question mark next to tellurium's atomic mass of 128, thinking that it might not be correct. When he published the table shown in the Student Edition in 1871, besides putting elements of similar properties in vertical groups, he "corrected" the atomic mass of tellurium to 125. Although he was mistaken in that, he was fairly accurate regarding his other corrections of properties of elements using their sequence in his table.

🖥 Periodic Table Resources

There are many internet resources related to the periodic table of the elements. These include interactive periodic tables, periodic table songs, and quizzes, flashcards, and tests. Do a keyword search for "interactive periodic table," "video periodic table," "element song," and "periodic table song."

In the 1860s, John Newlands expanded on the concept of periodicity. When he arranged the elements in order by atomic mass, he found pairs of elements with similar properties. These pairs of similar elements differed in their position by a multiple of eight. This led Newlands, in 1865, to develop his *law of octaves*, which states that when elements are arranged by atomic mass, the properties of every eighth element are similar. Newlands is now credited with the earliest form of the *periodic law*, which stated that the properties of the elements vary with their atomic masses in a regularly repeating way.

No.		No.		No.		No.		No.		No.		No.		No.	
H	1	F	8	Cl	15	Co & Ni	22	Br	29	Pd	36	I	42	Pt & Ir	50
Li	2	Na	9	K	16	Cu	23	Rb	30	Ag	37	Cs	44	Os	51
G	3	Mg	10	Ca	17	Zn	24	Sr	31	Cd	38	Ba & V	45	Hg	52
Bo	4	Al	11	Cr	18	Y	25	Ce & La	33	U	40	Ta	46	Tl	53
C	5	Si	12	Ti	19	In	26	Zr	32	Sn	39	W	47	Pb	54
N	6	P	13	Mn	20	As	27	Di & Mo	34	Sb	41	Nb	48	Bi	55
O	7	S	14	Fe	21	Se	28	Ro & Ru	35	Te	43	Au	49	Th	56

Mendeleev and Periodic Law

In 1869, Russian chemist Dmitri Mendeleev organized the elements in a table much like our modern periodic table. He arranged the elements in columns by increasing atomic masses. This arrangement displayed the periodic nature of the properties across the rows. Over time Mendeleev switched the rows and columns of the table. This change placed elements with similar characteristics in columns like our modern table.

Period	Group								
	I	II	III	IV	V	VI	VII	VIII	O
1	H 1								He 4
2	Li 7	Be 9.4	B 11	C 12	N 14	O 16	F 19		Ne 20
3	Na 23	Mg 24	Al 27.8	Si 28	P 31	S 32	Cl 35.5		Ar 39
4	K 39	Ca 40	— 44	Ti 48	V 51	Cr 52	Mn 55	Fe 56 Co 59 Ni 59 Cu 63	
5	Cu 63	Zn 65	— 68	— 72	As 75	Se 78	Br 80		Kr 83
6	Rb 86	Sr 87	Y 88	Zr 90	Nb 94	Mo 96	— 100	Ru 104 Rh 104 Pd 106 Ag 108	
7	Ag 108	Cd 112	In 113	Sn 118	Sb 122	Te 125	I 127		Xe 131
8	Cs 133	Ba 137	Di 138	Ce 140	—	—	—		
9	—	—	—	—	—	—	—		
10	—	—	Er 178	La 180	Ta 182	W 184	—	Os 195 Ir 197 Pt 198 Au 199	Rn (222)
11	Au 199	Hg 200	Tl 204	Pb 207	Bi 208	—	—		
12	—	—	—	Th 231	—	U 240	—		

The most striking features of Mendeleev's periodic table (bottom of previous page) were the blank spaces (unshaded cells) that he left between some elements. These spaces were for unknown elements that he predicted would be found in the future. He based his predictions on the periodic law. When Mendeleev ordered the elements on the basis of atomic mass, he recognized that in some cases there weren't elements with the expected properties. Therefore he concluded that there must be elements that would later fill those spots. At first, other scientists ignored these predictions, but the discoveries of gallium (31) and germanium (32) proved that Mendeleev was correct.

4.3 THE MODERN PERIODIC TABLE

Reordering the Table

English physicist Henry Moseley made an important discovery in 1914. He recognized that he was able to use a new technique called *X-ray spectroscopy* to determine the atomic number for each element. As a result, chemists rearranged the periodic table on the basis of increasing atomic number. This change corrected some misplaced elements from Mendeleev's table. Moseley's discovery also led to an update of the **periodic law**, which now states that the properties of the elements vary in a periodic way with their atomic numbers. This allowed scientists to produce the periodic table of the elements that we use today. The arrangement of the periodic table relates to the structure of the atoms themselves. Pages 80–81 show the modern **periodic table** of the elements—a table of the chemical elements arranged in a way that displays their periodic properties in relationship to their atomic numbers.

1	2	3	4	5	6	7	8	9	10	11	12	13	14	15	16	17	18
H																	He
Li	Be											B	C	N	O	F	Ne
Na	Mg											Al	Si	P	S	Cl	Ar
K	Ca	Sc	Ti	V	Cr	Mn	Fe	Co	Ni	Cu	Zn	Ga	Ge	As	Se	Br	Kr
Rb	Sr	Y	Zr	Nb	Mo	Tc	Ru	Rh	Pd	Ag	Cd	In	Sn	Sb	Te	I	Xe
Cs	Ba	La	Hf	Ta	W	Re	Os	Ir	Pt	Au	Hg	Tl	Pb	Bi	Po	At	Rn
Fr	Ra	Ac	Rf	Db	Sg	Bh	Hs	Mt	Ds	Rg	Cn	Nh	Fl	Mc	Lv	Ts	Og

Ce	Pr	Nd	Pm	Sm	Eu	Gd	Tb	Dy	Ho	Er	Tm	Yb	Lu
Th	Pa	U	Np	Pu	Am	Cm	Bk	Cf	Es	Fm	Md	No	Lr

The most updated versions of the periodic table include all the elements through atomic number 118. Four of these elements (113, 115, 117, and 118) were only recently added after a lengthy naming process by the International Union of Pure and Applied Chemistry (IUPAC), the group that is responsible for standardization in chemistry.

✔ Memorization

Teachers differ on the importance of memorizing the periodic table. Many consider the table a resource to be consulted when information is needed. Others believe that there are benefits to memorizing parts or even all of the table. It's your choice, but the general trend among teachers today is to have students memorize little to none of it. The reasoning is that in today's world we have this knowledge at our fingertips. Many would say that learning how to use and interpret the periodic table might be a more effective use of time than memorizing it.

Regardless of how much of the table your students memorize, encourage them to become familiar with the names, symbols, and location of at least the more commonly used elements. This will be a timesaver because they won't be hunting through the periodic table for each element of chemical formulas that they need to write. Many students will realize by the end of the course that they have indeed memorized many facts contained in the periodic table.

Formative Assessment: Developing Periodicity

1. How did Döbereiner contribute to the process of organizing the elements? *(Döbereiner was the first chemist to make note of the regularly repeating pattern in the properties of the elements.)*

2. What does periodicity mean? *(The properties of the elements repeat in a regular, or periodic, fashion when the elements are ordered by atomic mass or atomic number.)*

3. Why do you think Mendeleev gets the most credit for developing the periodic table despite other scientists developing similar arrangements around the same time? *(He left blanks that he predicted would be filled by yet-to-be-discovered elements, demonstrating the predictive power of his model.)*

4. Newlands is credited with the earliest form of what law? *(the periodic law)*

5. State the periodic law. *(The properties of the elements vary in a periodic way with their atomic numbers.)*

Class Discussion: Atomic Mass or Atomic Number?

Ask students to look at the periodic table on pages 80–81 and identify any pairs of elements that would be out of order if they were in order of increasing mass number. *(Ar and K, Co and Ni, Te and I)* Then ask students, "How did Moseley's discovery contribute to the development of the periodic table?" *(It helped chemists realize that the best order of elements to maintain periodicity is by atomic number.)*

Henry Moseley

Moseley was killed at the age of twenty-seven during the Gallipoli campaign of World War I. His death motivated the British government to enact policies that exempted from combat those scientists whose work could be considered vital for the nation's war effort.

Element Without a Family

Because it has only one electron in its only energy level, hydrogen is at the top of Group 1 despite being a gas and not a metal. It has only some properties similar to the alkali metals, while it has others that are very different—such as its ability to form diatomic molecules in gas form. It is sort of in a class by itself since its one electron half fills the first energy level. Since it does not truly belong to a particular family, it can be thought of as an element in a family of its own.

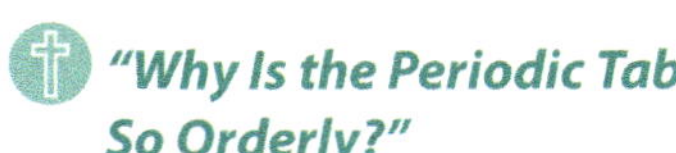 ### "Why Is the Periodic Table So Orderly?"

Discuss with students possible reasons why the periodic table is so orderly. The point to bring out is that orderliness is never a product of random processes. Rather the periodic table reflects a world with elements that, upon closer inspection, are very orderly because they were created that way.

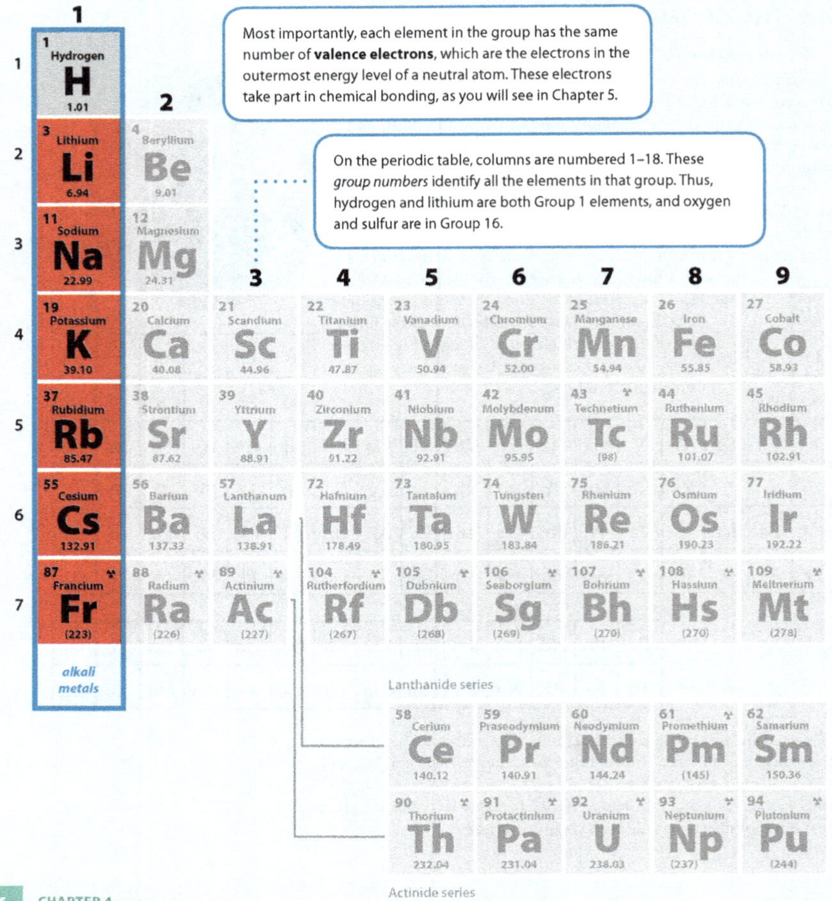

ORGANIZATION OF THE PERIODIC TABLE

FAMILY

You will notice that a periodic table is arranged in rows and columns. The first row starts with hydrogen (1), then helium (2), but then there is a new row. This is due to the periodic law. Lithium (3), needed to be placed below hydrogen because lithium and hydrogen have similar properties.

A **family**, or **group**, is a set of elements in the same column on the periodic table. Elements in the same group have similar structures, which result in their acting like each other. Just like members of your family have similarities, the elements within a group have similar physical and chemical properties.

Most importantly, each element in the group has the same number of **valence electrons**, which are the electrons in the outermost energy level of a neutral atom. These electrons take part in chemical bonding, as you will see in Chapter 5.

On the periodic table, columns are numbered 1–18. These *group numbers* identify all the elements in that group. Thus, hydrogen and lithium are both Group 1 elements, and oxygen and sulfur are in Group 16.

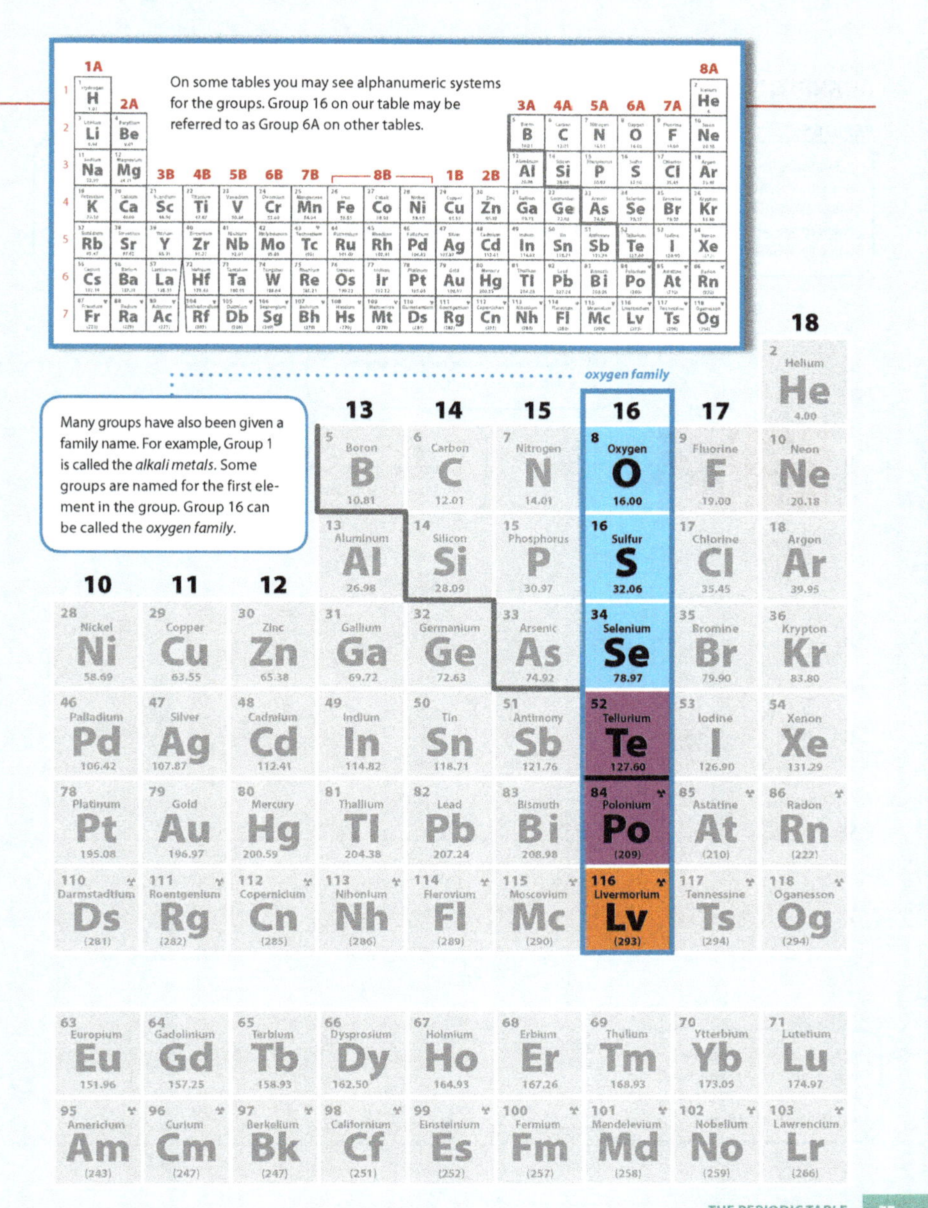

Group Numbers

Both IUPAC and the American Chemical Society (ACS) have adopted the Groups 1–18 naming convention. Some older tables will still use the alphanumerics, but since both the American and international standardization organizations have adopted the new numbers, we will use them also.

 **Differentiated Learning:** *Blocks*

Some students may be curious about the "stairstep" look of the periodic table, that is, why it appears that elements are missing from some rows. This particular aspect of the periodic table is related to the distribution of electrons within *energy sublevels*, a topic that is usually covered in a high-school chemistry class (e.g., see pages 88–97 in BJU Press CHEMISTRY 4th Edition). Within each energy level there are up to four sublevels, known as the *s*, *p*, *d*, and *f* sublevels. Elements in Groups 1 and 2 collectively form the *s* block. Similarly, Groups 3–12 constitute the *d* block, while Groups 13–18 form the *p* block. The lanthanide and actinide series are the *f* block.

 **Formative Assessment:**
Groups and Periods

1. Which periods contain the most elements? *(Periods 6 and 7)*

2. Which groups have the most elements? *(Groups 1 and 18)*

3. Which elements have properties that are similar to helium? Why would you think so? *(Neon, argon, krypton, xenon, radon, and oganesson are in the same family as helium.)*

4. What period is europium in? *(Period 6)*

5. What group is selenium in? *(Group 16)*

Project-Based Learning:
Periodic Table of …

To get students to really think about how the periodic table is organized, have them create their own periodic table. They could pick a topic (e.g., movies, cars, shoes) and create an organization scheme. Make sure that they explain the order that their objects are in, why they start a new row, and the common property for each column.

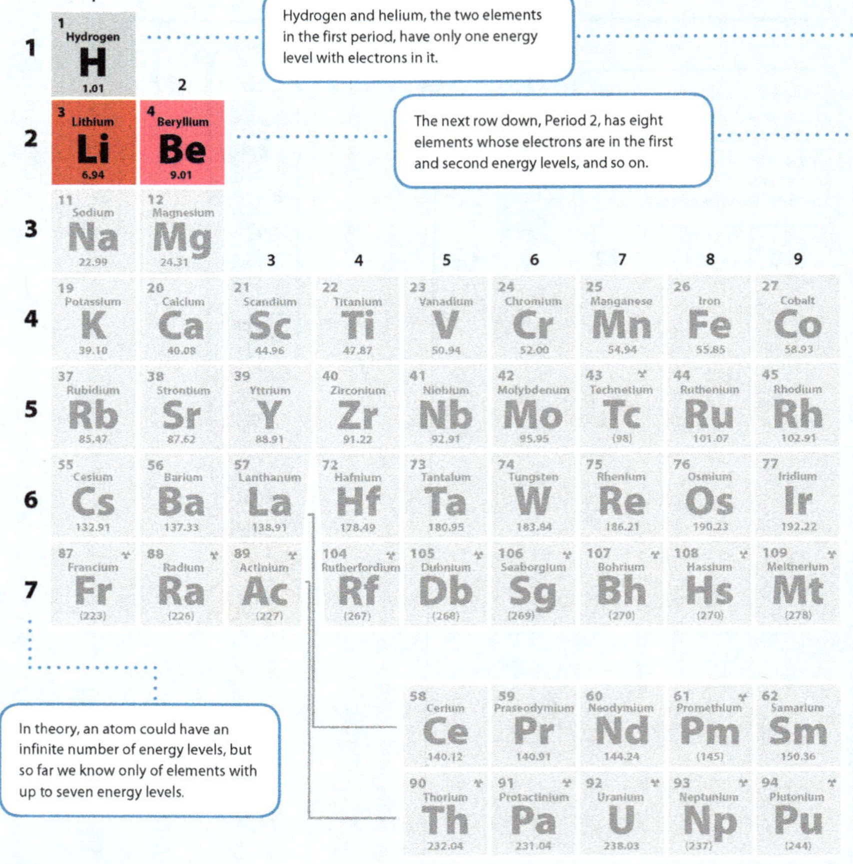

4A | REVIEW QUESTIONS

1. When did people first know about substances that we call elements?

2. Why were people driven to develop a way to organize the elements?

3. How did Lavoisier's study of combustion prove that fire was not an element?

4. Why do you think that potassium (19) has the symbol K?

5. What were Döbereiner's triads?

6. What was the most remarkable feature of Mendeleev's periodic table?

7. Give a definition for the periodic table.

8. What is a family or group of elements and how is it represented on the periodic table?

9. What is a period of the periodic table? What does it tell us about an atom?

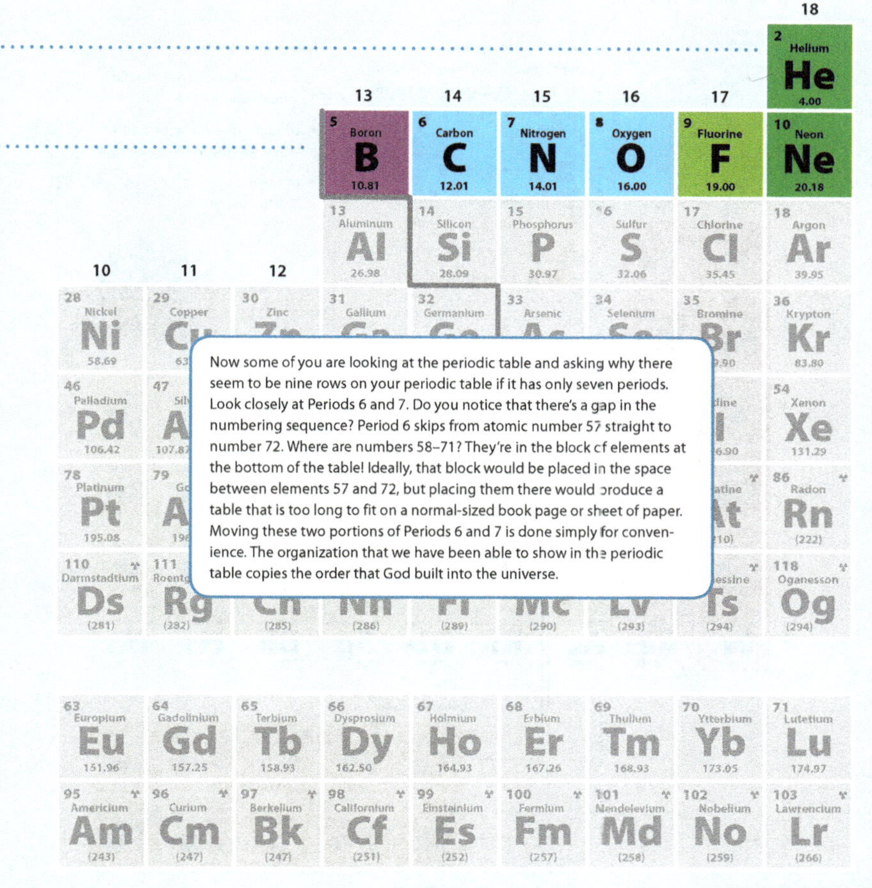

Lab 4A: Bricks and Feathers

After Section 4A has been completed, Lab 4A can be used to introduce trends across a period.

4A REVIEW ANSWERS

1. People have known about elements since Bible times. The Bible specifically mentions seven substances that we know are elements. *(p. 70)*

2. The amount of data was too much to memorize. The periodic table is a model that helps scientists explain their observations and make predictions about elements. *(p. 72)*

3. Lavoisier showed that since oxygen was required for combustion reactions, fire was not a separate element. *(p. 71)*

4. The symbol probably came from its name in a different language. Potassium was originally named *kalium*. *(p. 72)*

5. Döbereiner noticed that when he sorted elements by their masses, he found groups of three with similar properties. He called these groups of three *triads*. *(p. 73)*

6. When Mendeleev created his periodic table, he left open spaces that represented yet-undiscovered elements. Mendeleev predicted the discovery of these elements on the basis of atomic masses and periodic properties. These predictions proved accurate. *(pp. 74–75)*

7. The periodic table is a table of the chemical elements arranged on the basis of atomic structure, which displays their periodic properties in relationship to their atomic numbers. *(p. 75)*

8. A family or group of elements is a collection of elements with similar chemical and physical properties that are due to their similar structures. Families are in columns on the periodic table. *(p. 76)*

9. A period is a row of the periodic table. It indicates the energy level in which an atom's valence electrons are found. *(p. 78)*

 **"Looking Up" Elements**

If you have enough square ceiling tiles in your classroom, consider using them to display the symbol, atomic number, and atomic mass of each element as part of a periodic table.

Demonstrating Sodium's Properties

Students will love seeing you slice a piece of metal with a knife and then react it with water. Additionally, you could use a test tube and a candle to show that hydrogen gas is produced.

Preparation: You will need sodium metal, a knife, tongs, a medium-sized beaker about one-third full of water, and a laboratory safety shield. A small sliver of sodium is all that is needed to show its reactivity. Larger pieces reacting in water raises the likelihood of higher temperatures being produced and therefore more flames, sparks, or even explosions. As always, rehearse the demonstration before class without students present to ensure that you use an appropriately sized piece of sodium. If you are uncomfortable with this demonstration, do an internet search using the keywords "sodium in water demonstration" for videos showing this demonstration.

Performance/Discussion:
You will need the laboratory safety shield between the materials and your students, as the reaction can be mildly explosive.

Demonstrate sodium's softness by slicing a couple of pieces and then put the rest back in the container. Then, using tongs, drop a sliver in the beaker of water. The piece of sodium will move around on the surface of the water.

1. If the water is reacting with the sodium, what is being produced that burns so easily? *(hydrogen gas; Students may guess oxygen—a good guess but incorrect.)*

2. Why do you think that the metal is stored in mineral oil? *(Since sodium is so reactive, if exposed to the air it will easily react with water vapor in the air.)*

4B Questions

- Where are metals and nonmetals on the periodic table?
- How do metals, metalloids, and non-metals compare?
- How can we know the chemical family of an element?
- How many valence electrons does each element have?

4B Terms

metal, metalloid, nonmetal, alkali metal, alkaline-earth metal, transition metal, inner transition metal, mixed group, halogen, noble gas

4B | CLASSIFYING THE ELEMENTS

How is the periodic table useful?

4.4 TYPES OF ELEMENTS

Recall that the main point of the periodic table is to organize the elements to help us understand them better. Scientists arranged the table to group elements by their properties. One key characteristic of an element is how metallic it is. Elements range from highly metallic to nonmetallic as we move across the periodic table from left to right.

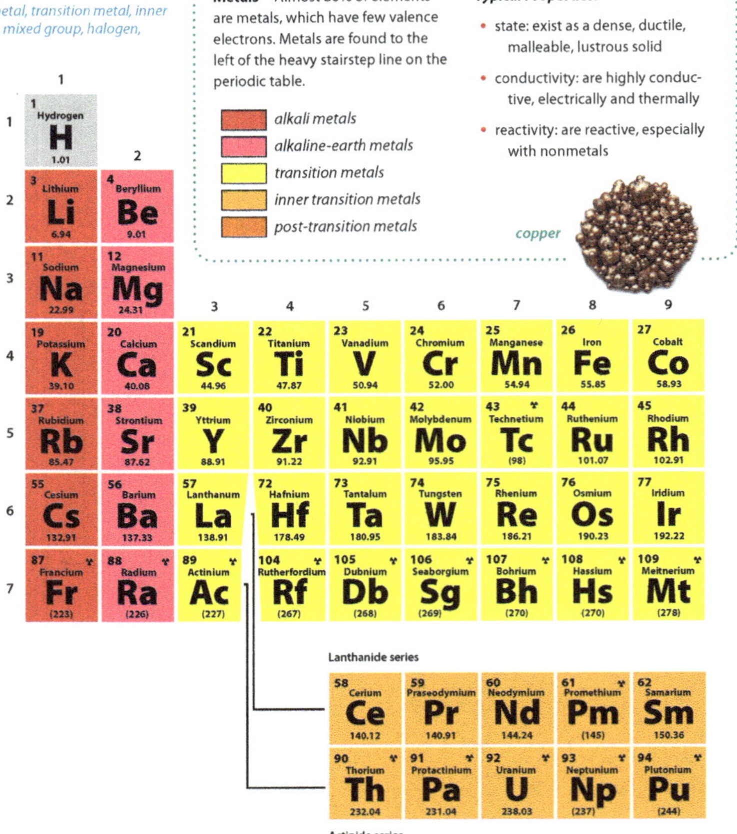

TEACHING THE MATERIAL

ESSENTIAL QUESTION

How is the periodic table useful?

OBJECTIVES

- 4B1 Classify elements as metals, nonmetals, and metalloids using a periodic table.
- 4B2 Compare the properties of metals, nonmetals, and metalloids.
- 4B3 Identify the families of elements in the periodic table.
- 4B4 Determine the number of valence electrons in an element according to its family.

RESOURCES

- Worldview Sleuthing: Giving Due Credit
- Demonstrating Sodium's Properties

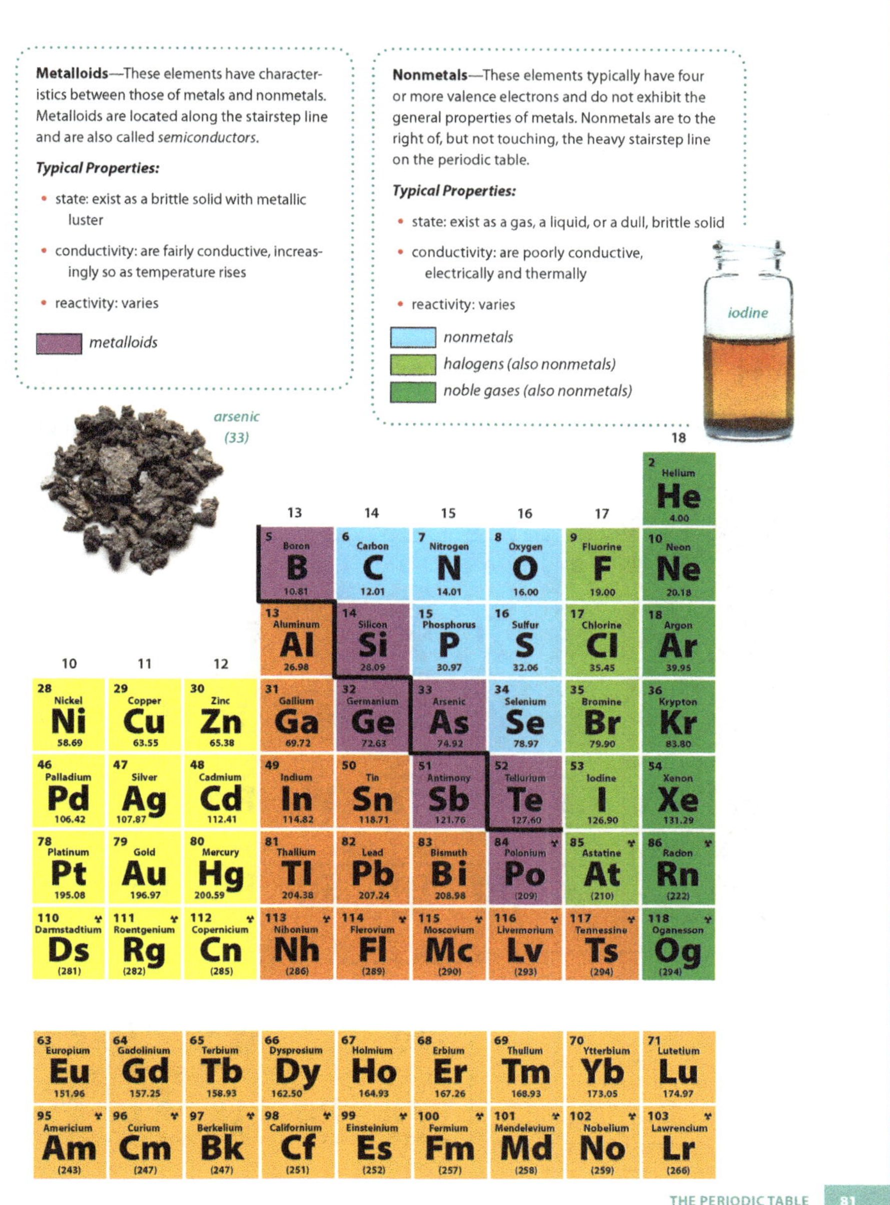

Metalloids—These elements have characteristics between those of metals and nonmetals. Metalloids are located along the stairstep line and are also called *semiconductors*.

Typical Properties:

- state: exist as a brittle solid with metallic luster
- conductivity: are fairly conductive, increasingly so as temperature rises
- reactivity: varies

metalloids

Nonmetals—These elements typically have four or more valence electrons and do not exhibit the general properties of metals. Nonmetals are to the right of, but not touching, the heavy stairstep line on the periodic table.

Typical Properties:

- state: exist as a gas, a liquid, or a dull, brittle solid
- conductivity: are poorly conductive, electrically and thermally
- reactivity: varies

nonmetals
halogens (also nonmetals)
noble gases (also nonmetals)

Post labels for the terms *metal*, *metalloid*, and *nonmetal* on the walls around the room. Quiz students by naming an element or by giving a chemical symbol or atomic number. Students will move to the area of the room with the appropriate label. Ask students to justify their selections and to give expected properties of the elements. You could have images ready to project for some of the elements.

As an alternative to the previous assessment method and if you don't have student response devices, put on the board the three element types labeled A, B, and C. Distribute a set of letter cards to each student. As you call out the name of an element, ask students to raise the appropriate card identifying its type.

Which Are Metalloids?

Many students will want a definitive list of which elements are metalloids. We have identified seven elements that are. Most scientists include six in the metalloids—boron, silicon, germanium, arsenic, antimony, and tellurium. Five others—carbon, aluminum, selenium, polonium, and astatine—are also included by some scientists.

STRATEGIES

Class Opener: Use the Demonstrating Sodium's Properties teacher note on page 80 to stimulate students' interest in the reactivity of alkali metals.

Formative Assessment: Use the Formative Assessment: Metal to Nonmetal teacher note on this page to assess students' understanding of the arrangement of types of elements in the periodic table.

Active Learning: Use the Active Learning: All in the Family teacher note on page 84 to help solidify students' grasp of families in the periodic table.

Worldview Sleuthing: Use the Giving Due Credit webquest activity on page 85 to get students thinking about properly recognizing the accomplishments of others.

Discussion: Use the Argumentation teacher note on page 85 to stimulate a student discussion regarding the worldview sleuthing box.

(continued)

The metal-to-nonmetal spectrum is a broad classification of elements, but remember that each column represents a family of elements. Just as in your family, there are distinctive traits in each of these families. Each family contains elements whose properties are similar. These properties arise because all the elements in the family have atomic structures, particularly the arrangements of their electrons, which are alike. Elements in each family have the same number of valence electrons. The descriptions of the families on this page and the next two pages specify how many valence electrons the elements have. Is there a pattern?

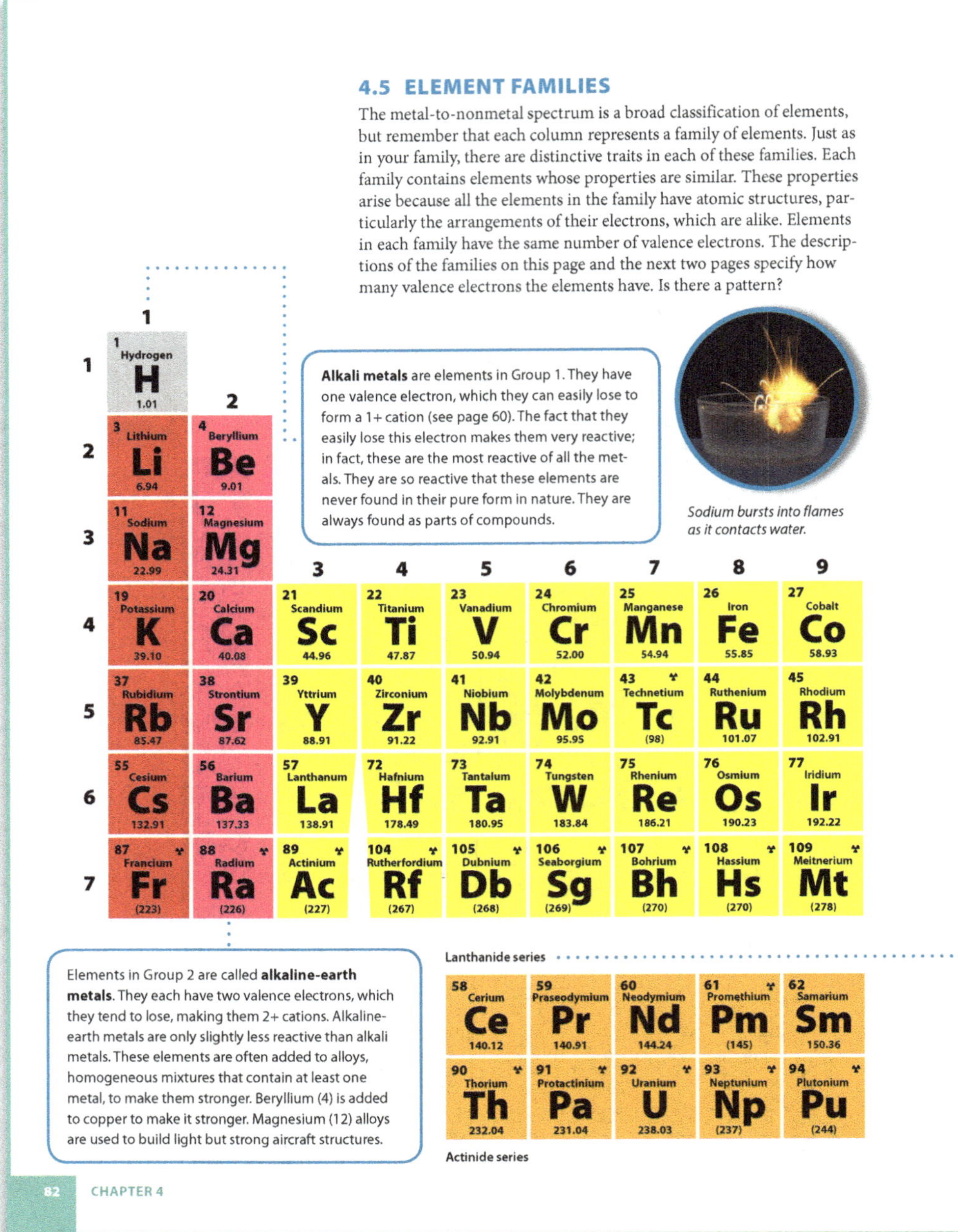

Alkali metals are elements in Group 1. They have one valence electron, which they can easily lose to form a 1+ cation (see page 60). The fact that they easily lose this electron makes them very reactive; in fact, these are the most reactive of all the metals. They are so reactive that these elements are never found in their pure form in nature. They are always found as parts of compounds.

Sodium bursts into flames as it contacts water.

Elements in Group 2 are called **alkaline-earth metals**. They each have two valence electrons, which they tend to lose, making them 2+ cations. Alkaline-earth metals are only slightly less reactive than alkali metals. These elements are often added to alloys, homogeneous mixtures that contain at least one metal, to make them stronger. Beryllium (4) is added to copper to make it stronger. Magnesium (12) alloys are used to build light but strong aircraft structures.

TEACHING THE MATERIAL

Ticket Out the Door: Using a periodic table that has only symbols, atomic numbers, and group numbers, write down three pairs of elements with the following characteristics:

1. a highly reactive metal and a highly reactive nonmetal

2. an element with two valence electrons and one with six

3. two noble gases

sterling silver

Groups 3–12 are called the **transition metals**. Most of these elements have one or two valence electrons. These electrons are easily lost, making cations with charges of 1+ or 2+. We often use these elements in alloys, such as sterling silver (a silver and copper alloy) or steel, an iron and carbon (6) alloy.

The two rows of elements shown below the main body on most periodic tables are the **inner transition metals**. They typically have two valence electrons, and so they form 2+ cations. The elements in the first row are rare and are used in lighting, lasers, superconductors, and magnets. The elements from the second row are all radioactive, and some are used as nuclear fuels.

This sample of holmium (67) shows the characteristic metal properties of inner transition metals.

Inner transition metals get their name from their proper placement—between the transition metals in the periodic table.

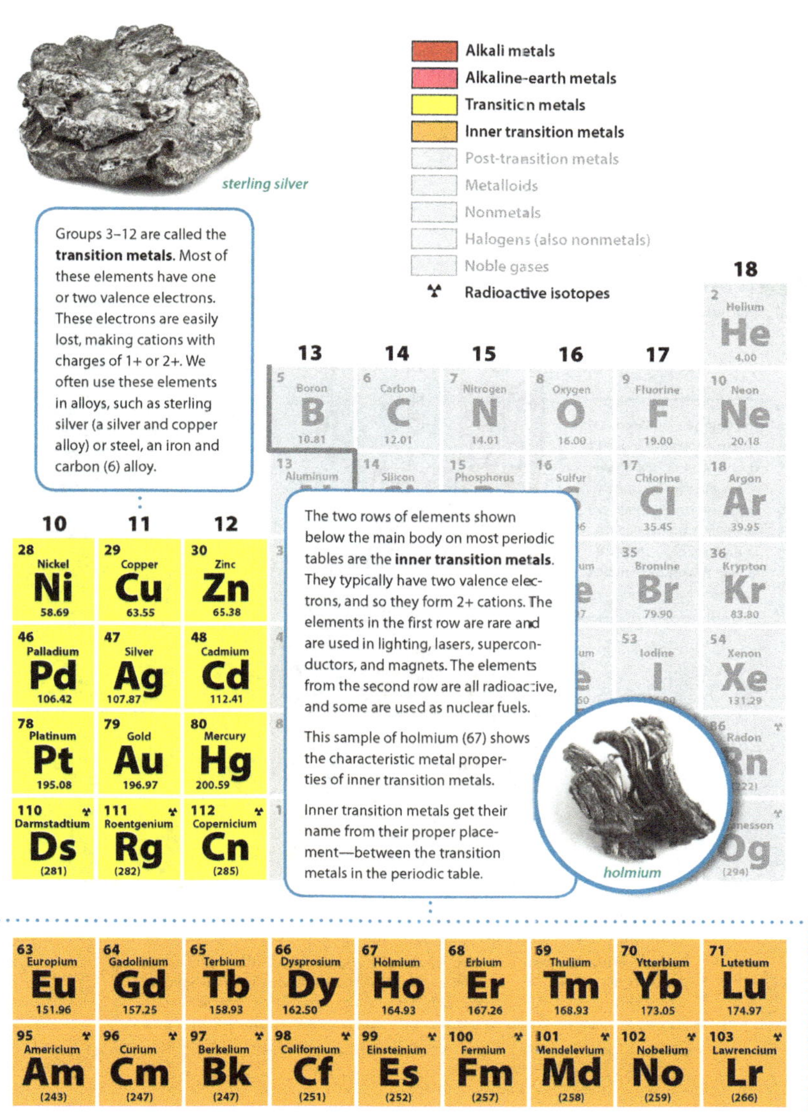

holmium

Valence Clue

Point out that with the exception of helium, the ones digit of Groups 1, 2, and 13–18 corresponds to the number of valence electrons in the elements of the group.

Halogens

The name *halogen* comes from the Greek words for "salt forming." In chemistry, a salt is a compound made from a metal and a nonmetal. Halogens readily combine with metals to make such salts, hence the name. Sodium chloride, or table salt, is just one example of a chemical salt containing a halogen.

Active Learning: *All in the Family*

Post labels of element family names on the walls around the room. Quiz students by naming an element, by giving a chemical symbol or atomic number, or by giving a description of one of the families. Students should move to the area of the room with the appropriate label. Ask them to justify their selection and to give expected properties of the elements to include the number of valence electrons. You could have images ready to project for some of the elements.

As an alternative to the previous assessment method and if you don't have student response devices, put on the board labels A–F for alkali metals, alkaline-earth metals, halogens, noble gases, transition metals, and inner transition metals. Distribute a set of letter cards to each student. As you call out the name of an element, ask students to raise the appropriate card identifying its type.

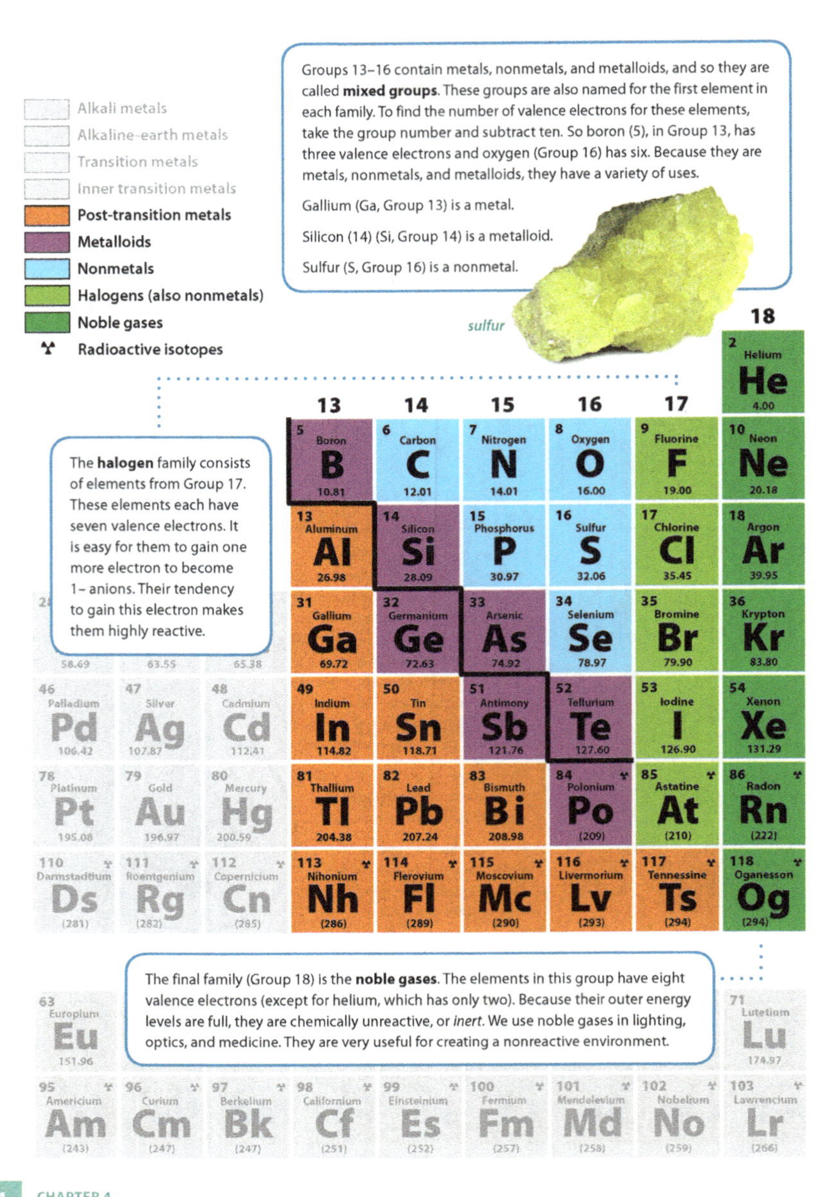

4B | REVIEW QUESTIONS

1. What is a metal?
2. Using the periodic table, would you expect oxygen or selenium (34) to act more like a nonmetal? Explain.
3. You have a sample of an element. At room temperature, it is a lightweight, brittle solid with a metallic luster. It has fair conductivity that increases with increased temperature. Do you think the material is a metal, nonmetal, or metalloid? Explain.
4. What do we call the family in Group 1?
5. Describe an alkaline-earth metal.
6. Why do most periodic tables show the inner transition metals below the main body of the table?
7. Describe a halogen.
8. How many valence electrons do the following elements have?
   a. magnesium
   b. antimony (51)
   c. chlorine

Worldview Sleuthing:
Giving Due Credit

In this webquest, students will research the discovery of element 113 as members of IUPAC. They will then take part in a debate regarding who should have the right to name this new element. You will need to assign each student the group that he is representing.

Worldview Sleuthing Rubric

A reproducible rubric to assess this worldview sleuthing activity can be found in Appendix G.

 Argumentation

This worldview sleuthing activity is good for giving students practice in formulating and supporting an argument. Expect students to support all claims with specific evidence on a daily basis. For example, if a student tells you that a carbon atom has six protons, expect him to tell you that he knows this because the atomic number of carbon is 6.

4B REVIEW ANSWERS

1. A metal is an element located to the left of the heavy stairstep line on the periodic table. Metals are typically solid, dense materials. Often they are ductile, malleable, conductive, reactive, and lustrous. They typically have one or two valence electrons. *(p. 80)*

2. Oxygen is farther from the stairstep, so it should have a stronger nonmetal character. *(p. 81)*

3. The material has properties of both metals (metallic luster, solid) and nonmetals (brittle solid, low conductivity). The material is probably a metalloid. *(p. 81)*

4. alkali metals *(p. 82)*

5. An alkaline-earth metal is a reactive metal with two valence electrons that are easily lost. *(p. 82)*

6. This placement is purely a space-saving decision. The table would be very long if the inner transition metals were properly placed. *(p. 83)*

7. Found in Group 17, a halogen is a nonmetal with seven valence electrons. Halogens easily gain one more electron, making them a 1– anion, and they are highly reactive. *(p. 84)*

8. a. 2
   b. 5
   c. 7 *(all pp. 82–84)*

Lab 4B: *How Wide Is an Atom?*

After completing Section 4B, students should work through Lab 4B as an introduction to Section 4C.

✔ Building Atoms

Make clear the connection between the atomic number and the arrangement of the periodic table. To draw a Bohr model of an atom, students should understand some facts. The number of protons and electrons comes from the atomic number, while the number of neutrons comes from the mass number minus the atomic number. The number of orbits, or energy levels, is the row number from the periodic table. Through element 20, the number of electrons in each energy level matches the number of elements in the corresponding row.

The Student Edition uses nitrogen as an example, but as an alternative, consider using Group 1 to explain the addition of energy levels with each period to accommodate more electrons. They all have one valence electron, so the increasing number of protons, and therefore electrons as well, requires the addition of another layer of electrons with each period. Use this concept to expand students' understanding of how an increase in the period number corresponds to increased sizes of atoms. This further corresponds to greater reactivity for metals: valence electrons are more loosely held the more distant they are from the nucleus. Remind students of the reactivity of sodium and ask them to suggest ramifications regarding properties of potassium and rubidium, for example.

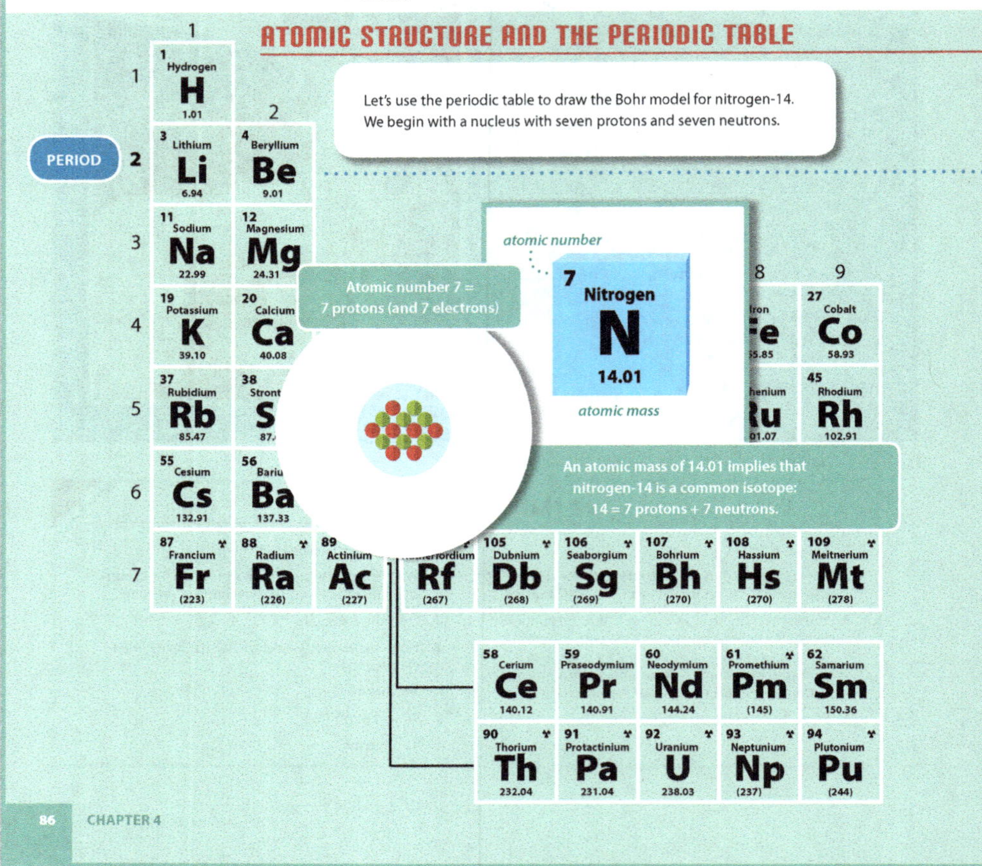

4C | PERIODIC TRENDS

What can an element's position on the periodic table tell us about the element?

4.6 ATOMIC STRUCTURE & THE PERIODIC TABLE

People often look at the periodic table and wonder why it is not a rectangle or some other regular shape. The shape of the periodic table is related to the structures of the atoms themselves. The periods, or rows, on the periodic table are related to the energy levels of the atoms. The groups are elements with similar electron arrangements, which is the reason for their similar properties. The arrangement of the periodic table tells us a lot about the composition and structure of each element's atoms.

TEACHING THE MATERIAL

ESSENTIAL QUESTION

What can an element's position on the periodic table tell us about the element?

OBJECTIVES

- 4C1 Write the electron dot notation for an element.
- 4C2 Explain why atomic radius changes as it does.
- 4C3 Arrange elements on the basis of atomic radius.
- 4C4 Explain the periodic trend in electronegativity.
- 4C5 Arrange elements on the basis of electronegativity.

RESOURCES

Case Study: Allotropes (p. 93)

Lab 4B: *How Wide Is an Atom*—Exploring Atomic Radii Trends

Let's look at oxygen on the periodic table. Oxygen is element number 8. It is to the right of the stairstep, has an average atomic mass of 16.00, is in Group 16, and is in the second period. What does this tell us about oxygen atoms?

Oxygen's being to the right of the stairstep tells us that it is a nonmetal. Its atomic number tells us that oxygen has eight protons, which means that as a neutral atom it must also have eight electrons. The average atomic mass implies that oxygen-16 is likely a common isotope with eight neutrons. The fact that oxygen is in Group 16 tells us that it has six valence electrons, which must be in the second energy level because oxygen is in the second period. And we can know all of this just by looking at where oxygen is located on the table! The periodic table is like a snapshot of the atomic structures of all the known elements.

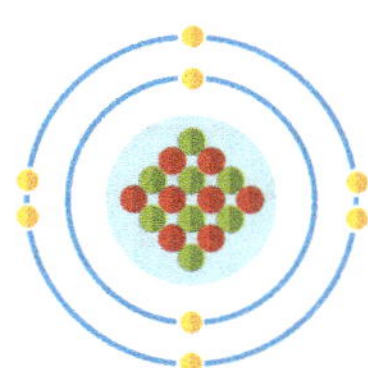

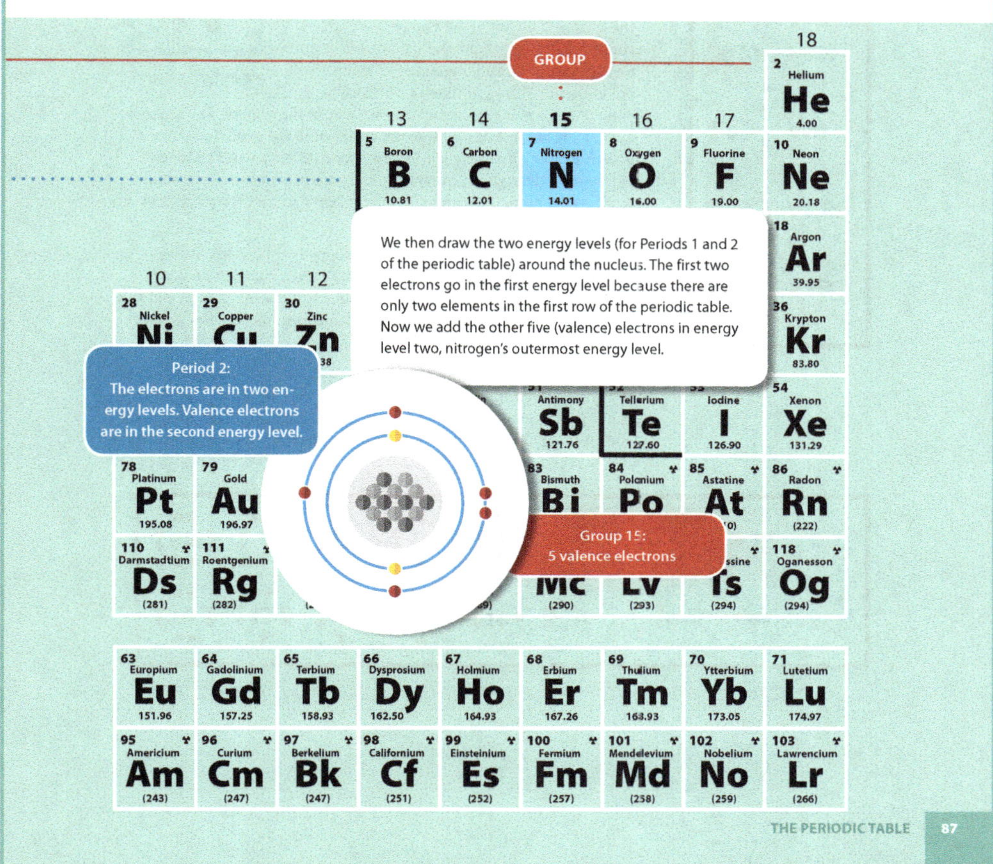

STRATEGIES

Lab: Use Lab 4B: *How Wide Is an Atom* as an introduction to section 4C.

Class Opener: Use the Building Atoms teacher note on page 86 to stimulate students' thinking regarding the trend of increasing numbers of energy levels as period number increases.

Formative Assessment: Use the Formative Assessment: Dot to Dot teacher note on page 88 to check students' understanding of electron dot notation.

Formative Assessment: Use the Formative Assessment: Less Than, Greater Than teacher note on page 89 to see whether students understand the trends in atomic radii.

Discussion: After students have been presented with the trends in the periodic table, have them propose the two elements that they think would most likely react together and defend their propositions on the basis of their locations in the periodic table.

Ticket Out the Door: Write down examples of elements that are stable and describe the electron arrangement that makes them so.

Formative Assessment: *Dot to Dot*

To assess students' understanding of electron dot notation, have them use only a periodic table to draw the electron dot notations for several elements. When they are finished, draw the correct representations on the board for students to check their answers.

So what happens to the atomic structure as we move from left to right across a period on the periodic table? Let's look at Period 2 on the table. As we move across the row, each element adds a proton, an electron, and typically some neutrons.

In many cases, the Bohr model shows more information than is needed. Therefore we can use a simpler drawing called an electron dot notation to represent atoms. **Electron dot notation** consists of the element's chemical symbol surrounded by the atom's valence electrons. The chemical symbol represents the nucleus and all of the *non-valence electrons*.

To write the electron dot notation, start with the chemical symbol. Then begin adding valence electrons. Add the first valence electron to the right side of the symbol. Add the second valence electron to form a pair on the right side of the symbol. The rest of the electrons are added counterclockwise around the symbol, placing one on each side before adding a second to any side. For elements with eight valence electrons, you will end up with a pair of electrons on each side of the symbol. The electrons are in pairs because of how they are arranged in the atom. In reality, it doesn't matter which side you start on; for consistency this textbook will start on the right side of the symbol and then add electrons counterclockwise around the atom. See the image above for the placement of valence electrons.

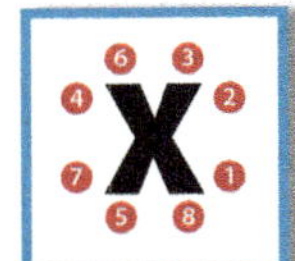

electron dot notation

Look at the electron dot notations for Groups 1 and 2 at left. As we move down each column, we add an energy level, but what do you notice about the number of valence electrons? It doesn't change, and this consistent structure is what causes the elements in the families to have similar properties.

Notice below that as we move across the rows (with the transition metals being the exceptions), we continue to add valence electrons. But also notice what happens as we move down a column (family). Even though we add an energy level, the number of valence electrons remains constant.

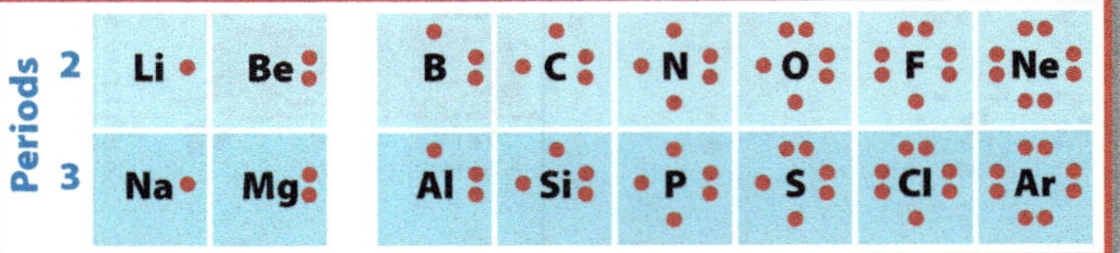

Atomic Radius and Electronegativity

Because of the changes in structure throughout the elements on the periodic table, there are trends in the properties of those elements as well. Two key trends are atomic radius and electronegativity.

TRENDS IN ATOMIC RADIUS

Atomic radius is simply the distance from the center of an atom's nucleus to the electrons in its outermost energy level. The size of an atom affects things such as how it will arrange itself with other atoms in molecules. It also affects how each atom will react and bond, since reactions and bonding depend on how close the valence electrons are to the nucleus. As we move down a column, the addition of energy levels, like layers on an onion, makes an atom bigger. As we move to the right across a period, adding protons and electrons causes the atoms to get smaller. This happens because opposite charges attract and more protons (positive) and electrons (negative) create a stronger force, pulling the electrons closer to the nucleus.

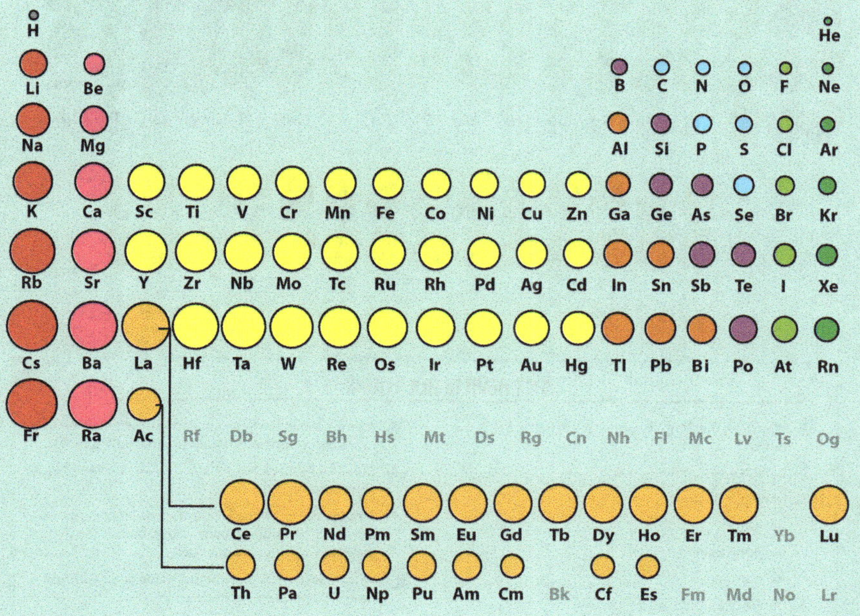

Effect of Adding Electrons

An attentive student may raise the concern that since electrons repel each other, adding electrons should make the atom larger. And he would be correct if the addition of electrons were all that happens. But as we move to the right across a period, we add both an electron *and* a proton. The addition of a proton increases the attractive force on all the electrons in the atom and therefore has a net effect of making the atom smaller.

Reasons for the Trends

Emphasize the importance of understanding the reasons for trends and not just memorizing them.

Getting a Handle on Atomic Radii

In order to engage students kinesthetically, have each student hold his hands apart from each other to indicate the atomic radii of elements in the table. As you move a pointer from *left to right*, have students move their hands closer together, and farther apart as you move the pointer from *right to left*. As you move a pointer *down* a group have them move their hands farther apart, and closer together as you move it *up* a group. Alternate moving vertically, horizontally, and finally diagonally.

Formative Assessment:
Less Than, Greater Than

To assess students' understanding of the trends of atomic radii and electronegativity, give students pairs of elements to write down indicating their relative atomic radius or electronegativity by putting a less than or greater than sign between them. After giving them six to eight element pairs, have them check their own responses.

As an alternative to this method and if you don't have student response devices, put on the board *A* for < and *B* for >. Distribute a set of letter cards to each student. As you call out pairs of elements regarding atomic radius or electronegativity, ask students to raise the appropriate cards.

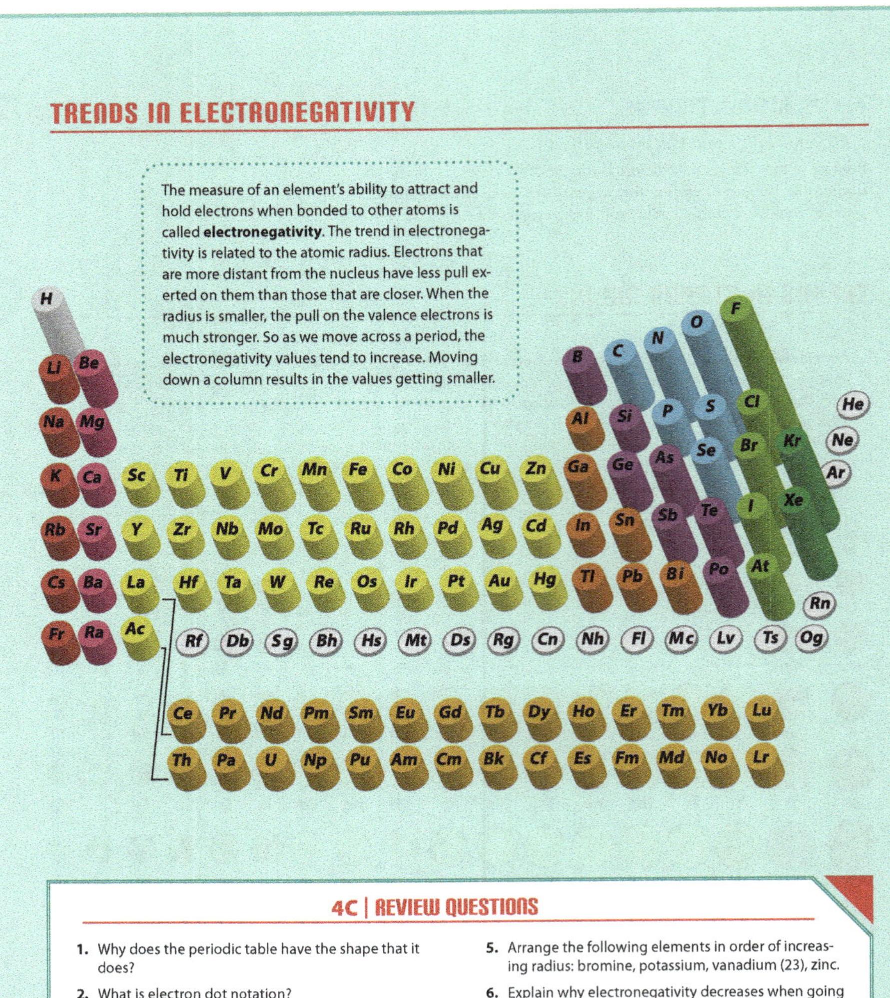

4C | REVIEW QUESTIONS

1. Why does the periodic table have the shape that it does?

2. What is electron dot notation?

3. Draw the electron dot notation for

   a. sulfur

   b. potassium

   c. nitrogen

4. Define *atomic radius*.

5. Arrange the following elements in order of increasing radius: bromine, potassium, vanadium (23), zinc.

6. Explain why electronegativity decreases when going down a column.

7. Arrange the following elements in order of expected increasing electronegativity values: barium (56), beryllium, calcium, magnesium.

8. Why do you suppose that most noble gases (Group 18) have little or no electronegativity?

4C REVIEW ANSWERS

1. The periodic table reflects the structure of atoms. The rows represent energy levels. So for example, the transition metals don't begin until the fourth row because transition metals all have four or more energy levels. *(p. 86)*

2. Electron dot notation is a way to depict atoms. It shows the chemical symbol with the valence electrons around the symbol. *(p. 88)*

3. a. ·$\overset{\cdot\cdot}{\underset{\cdot\cdot}{S}}$:   b. K·   c. ·$\overset{\cdot\cdot}{N}$: *(all p. 88)*

4. Atomic radius is the distance from the center of the nucleus to the electrons in the outermost energy level. *(p. 89)*

5. bromine, zinc, vanadium, potassium *(p. 89)*

6. As we move down a column, the valence electrons are farther from the nucleus. This means that there is less attraction on the valence electrons. The nucleus can't pull on bonded electrons as easily in a larger atom. *(p. 90)*

7. barium, calcium, magnesium, beryllium *(p. 90)*

8. The noble gases have completely filled outer energy levels.

CHAPTER 4 REVIEW . . . .

4A ORGANIZING THE ELEMENTS

- Scientists developed the periodic table to organize the data about the elements.

- Scientists noticed repeated patterns in the properties of elements, leading others to organize the periodic table.

- At first, chemists ordered the table on the basis of atomic mass, but this misplaced a few elements. Changing the table by using atomic numbers to order it made the model more workable.

- Dmitri Mendeleev predicted the discovery of elements that seemed to be missing from his periodic table.

- The periodic table's arrangement corresponds to atomic structure. The periods (rows) represent energy levels and the families (columns) represent similar electron arrangements.

4A Terms

periodic law	75
periodic table	75
family	76
group	76
valence electron	76
period	78

4B CLASSIFYING THE ELEMENTS

- The periodic table places metals on the left side and nonmetals on the right side, separated by a heavy stairstep line. Metalloids along the stairstep have both metallic and non-metallic properties.

- Metals tend to be lustrous, malleable, ductile, and reactive solids with high conductivity.

- Nonmetals can be solids, liquids, or gases, with varying reactivity and poor conductivity.

- The columns on the periodic table represent families of elements.

- Members of a family have the same number of valence electrons and properties.

4B Terms

metal	80
metalloid	81
nonmetal	81
alkali metal	82
alkaline-earth metal	82
transition metal	83
inner transition metal	83
mixed group	84
halogen	84
noble gas	84

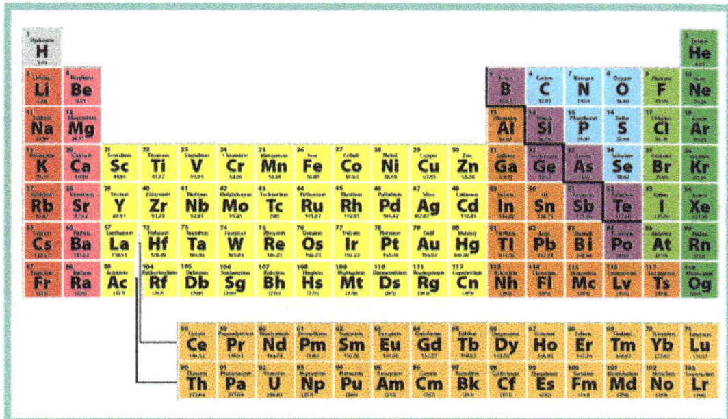

Recalling Facts

1. The ancient Greeks believed that all matter formed from variable mixtures of earth, water, fire, air (or wind), and aether. *(p. 70)*

2. Robert Boyle rejected the Greek concept of matter being mixtures of the five classical elements. He instead promoted the concept of matter consisting of indivisible particles of specific elements. He also promoted scientific inquiry and experimentation. Accept either. *(p. 71)*

3. Berzelius created a system of chemical symbols that were based on the name of the element and the number of atoms. This became the basis for our modern system. *(p. 71)*

4. Periodicity is the repeated pattern of properties exhibited by elements when put in order of atomic mass or atomic number. *(p. 73)*

5. Mendeleev originally put the elements in *columns* according to atomic mass. This resulted in elements with similar characteristics being in rows. Mendeleev changed his table to list the elements in *rows* according to their atomic masses. Later the table was ordered by atomic number instead of atomic mass. *(pp. 74–75)*

6. The periodic law states that the properties of the elements change in a repeating way with their atomic numbers. *(p. 75)*

7. A valence electron is an electron in the outermost energy level of an atom. *(p. 76)*

8. The pairs cobalt and nickel, argon and potassium, and tellurium and iodine would all have been out of order. *(pp. 76–77)*

9. A nonmetal is an element located to the right of and not touching the heavy stairstep line on the periodic table. It can be a gas, liquid, or solid. As a solid, it tends to be dull and brittle. It is a poor conductor. It tends to have four or more valence electrons. Its reactivity varies. *(p. 81)*

10. They are the most reactive metals. *(p. 82)*

11. one or two *(p. 83)*

12. The oxygen group has elements that are metals, nonmetals, and metalloids. *(p. 84)*

13. The noble gases have eight valence electrons and therefore tend to be non-reactive and inert. *(p. 84)*

14. **a.** mixed group (or carbon family)

    b. alkaline-earth metals

    c. inner transition metals *(all pp. 82–84)*

15. They are paired in the actual atom. *(p. 88)*

Understanding Concepts

16. Many chemists had created symbols to represent elements and compounds. Berzelius created his system of chemical symbols to standardize the symbols to make sharing information easier. *(p. 71)*

17. CN must represent a carbon-nitrogen compound. If this symbol were representing copernicium, then the second letter (n) would be lowercase. *(p. 72)*

18. While the specific form of a table may not be required, scientists recognized the need for a system to organize the information about chemical elements. Without some form of organization, the volume of information would be overwhelming. *(p. 72)*

4C PERIODIC TRENDS

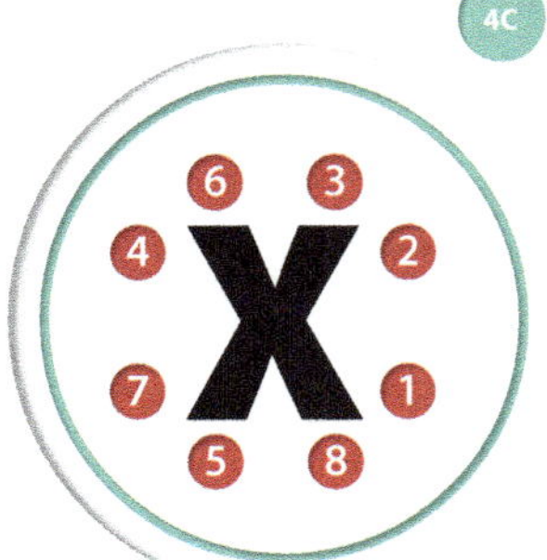

- Both the number of valence electrons that an element has and the energy level in which the valence electrons are found can be determined by the element's location on the periodic table.
- Electron dot notation is a simple way to represent atoms of elements and their valence electrons.
- The properties of elements change as we move across rows or down columns because of changes in atomic structure from one element to the next.

- Atomic radius *decreases across a row* as the attractive electric force between the nucleus and electrons increases. Atomic radius *increases down a column* due to the addition of energy levels.
- Electronegativity *increases across a row* as the valence electrons are nearer to the nucleus and *decreases down a column* as the valence electrons are farther from the nucleus.

4C Terms

electron dot notation	88
atomic radius	89
electronegativity	90

CHAPTER REVIEW QUESTIONS

Recalling Facts

1. Of what elements did the ancient Greeks believe all matter consisted?

2. Name a key contribution that Robert Boyle made to the study of chemistry.

3. What was Berzelius's contribution to the study of elements?

4. Define *periodicity*.

5. Name two changes made to Mendeleev's original periodic table to produce the one that we use today.

6. State the periodic law.

7. Define *valence electron*.

8. Using the periodic table, list two pairs of elements that would have been out of order on Mendeleev's periodic table (according to atomic mass) compared with our current table. Assume that scientists knew about all the elements up to and including lead.

9. Describe a nonmetal and list its properties.

10. What trait of alkali metals results from the fact that they easily lose their one valence electron?

11. How many valence electrons do the transition metals typically have?

12. Why is the oxygen group considered one of the mixed groups?

13. Which family's atoms generally have eight valence electrons? What property do these electrons produce?

14. Name the family for each of the following elements.

    a. lead

    b. calcium

    c. promethium (61)

15. Why are electrons paired in electron dot notation?

Understanding Concepts

16. Why did Berzelius establish his system of chemical symbols?

17. Does CN represent the element copernicium (112) or a compound of carbon and nitrogen? Explain.

18. Why do we need a periodic table?

19. Place in chronological order the following people who contributed to the development of the periodic table: Berzelius, Döbereiner, Mendeleev, Mosley, Newlands.

20. List the group number for each element.

    a. cadmium (48)

    b. cesium (55)

    c. carbon

21. How does the shape of the periodic table reflect the order that we see in atoms and in the universe?

22. Using the periodic table, would you expect copper or gallium to exhibit more metallic properties? Explain.

23. Using the periodic table, classify the following elements as metal, nonmetal, or metalloid.

    a. antimony

    b. strontium (38)

    c. bromine

24. What is similar about elements in a particular family?

25. What type of ion do the alkali metals form and why?

26. All of the transition metals have at least how many energy levels? Explain.

27. On the basis of its position on the periodic table alone, what do we know about sulfur?

28. What information about a neutral atom of silicon does the periodic table tell us?

29. Write the electron dot notation for the following elements.

    a. magnesium

    b. fluorine (9)

    c. indium (49)

30. Explain why atomic radius changes as we move to the right across a period.

31. Arrange the following elements in order of increasing radius: oxygen, polonium (84), selenium, sulfur.

32. Explain why electronegativity changes as it does across a period.

Critical Thinking

33. Why did people keep working on developing the periodic table? How are they continuing to work on the table now?

34. You want an element that is conductive, malleable, and ductile but not very reactive. Where on the periodic table would you look for your element? Explain.

35. What is wrong with the Bohr model of carbon shown below?

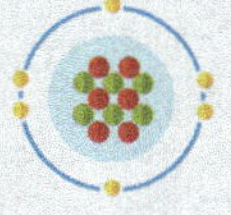

36. Ionization energy is the measure of how much work needs to be done to remove a valence electron from a neutral atom. Thinking about the trends in atomic radii and electronegativity, hypothesize as to how you think ionization energy would change across a row. Explain.

Use the Case Study at right to answer Questions 37–41.

37. What is an allotrope?

38. Of what element(s) are diamond and graphite made?

39. Considering your answer to Question 38, how can you explain the significant differences in properties between diamond and graphite?

40. Would a change from one allotrope to another be a chemical change or a physical change?

41. Why can we model the characteristics of elements in such an orderly way?

CASE STUDY: ALLOTROPES

As mentioned in the Chapter opener, diamonds are made of pure carbon, but so is the graphite in your pencil. Diamond and graphite are *allotropes* of carbon. Allotropes are different forms of the same element in the same state. Allotropes' internal structures differ, resulting in remarkably different characteristics.

Many elements exhibit allotropy. Metals such as polonium, iron, and cobalt each have allotropes. Some metalloids—antimony, arsenic, and silicon—exhibit allotropy. In addition to carbon, other nonmetals like oxygen (ozone and O_2) and phosphorus (15) can be found in different forms. Tin transforms from metallic tin into an allotrope that is a metalloid below 13.2 °C. Materials can change allotrope forms in response to the effects of temperature, light, and pressure.

While graphite and diamond are probably the two most well-known allotropes of carbon, there are a total of nine allotropes of carbon. Each allotrope has distinct properties, all stemming from varied internal structures. Graphite consists of thin sheets of carbon. The sheets easily separate, making graphite useful as pencil lead. Diamonds form from very strong crystal structures, giving them their extreme hardness. Carbon can even form exotic, soccer-ball shaped structures sometimes referred to as *buckyballs*.

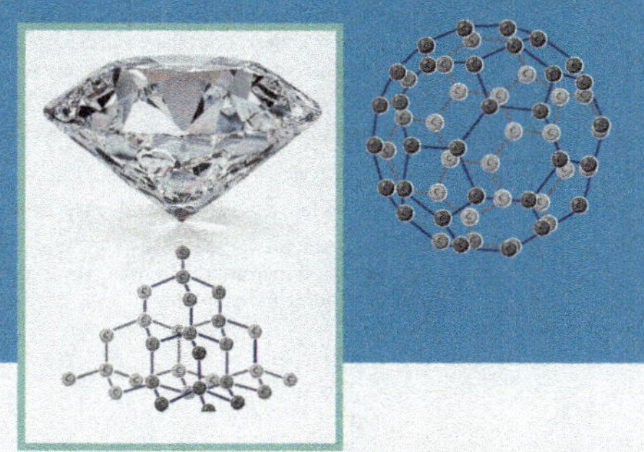

---

26. All transition metals have at least four energy levels. The transition metals are all in rows 4–7, and rows represent energy levels. *(pp. 82–83)*

27. Sulfur is a nonmetal in Period 3 and Group 16. It has six valence electrons, which are located in the third energy level. It is similar to oxygen, selenium, and other Group 16 elements. *(p. 84)*

28. Silicon is the fourteenth element, so it has fourteen protons and fourteen electrons. The average atomic mass implies that silicon-28 is a common isotope with fourteen neutrons. Since silicon is in Period 3, it has three energy levels. It's also in Group 14, so it has four valence electrons. *(pp. 86–87)*

29. a. **Mg:**

    b. **:F:**

    c. **In:** *(all p. 88)*

30. As we move to the right across a period, we add a proton and an electron to each additional element. This increases the attractive force on all the electrons, pulling them closer to the nucleus. This makes the radius smaller. *(p. 89)*

31. oxygen, sulfur, selenium, polonium *(p. 89)*

32. As we move to the right across a period, the atomic radii decrease. This means that the valence electrons are closer to the nucleus, and so there is a greater attraction on the valence electrons. The nucleus pulls more strongly on bonded electrons in this smaller atom. *(p. 90)*

Critical Thinking

33. People kept working on developing the periodic table to improve its ability to explain elements and make useful predictions about the elements. This process continues today as people discover new elements to add to the table with appropriate names. *(pp. 73–75)*

34. The properties that we are looking for are those of a metal. We need to look to the left of the stairstep. The metallic properties of metals decrease as we move to the right, so metals close to the stairstep would be best, such as aluminum, gallium, or tin. *(pp. 80–81)*

---

19. Berzelius, Döbereiner, Newlands, Mendeleev, Moseley *(pp. 71–75)*

20. a. 12

    b. 1

    c. 14 *(all pp. 76–77)*

21. The shape of the periodic table reflects our understanding of atomic structure. The rows represent energy levels in the atom. The groups represent similarities in properties caused by similar electron configurations. The order in the universe and in atomic structure clearly shows up in the organization of the periodic table. *(p. 80)*

22. Copper is farther to the left, so it should exhibit more metallic properties. *(pp. 80–81)*

23. a. metalloid

    b. metal

    c. nonmetal *(all pp. 80–81)*

24. Elements in a particular family have similar atomic structure, especially their arrangement of electrons. They also all have the same number of valence electrons. The similarities in atomic structure result in the elements having similar chemical and physical properties. *(p. 82)*

25. Alkali metals form a 1+ cation (positive ion) because they have one valence electron that they can easily lose. *(p. 82)*

REVIEW

24. What is similar about elements in a particular family?

25. What type of ion do the alkali metals form and why?

26. All of the transition metals have at least how many energy levels? Explain.

27. On the basis of its position on the periodic table alone, what do we know about sulfur?

28. What information about a neutral atom of silicon does the periodic table tell us?

29. Write the electron dot notation for the following elements.

  a. magnesium

  b. fluorine (9)

  c. indium (49)

30. Explain why atomic radius changes as we move to the right across a period.

31. Arrange the following elements in order of increasing radius: oxygen, polonium (84), selenium, sulfur.

32. Explain why electronegativity changes as it does across a period.

Critical Thinking

33. Why did people keep working on developing the periodic table? How are they continuing to work on the table now?

34. You want an element that is conductive, malleable, and ductile but not very reactive. Where on the periodic table would you look for your element? Explain.

35. What is wrong with the Bohr model of carbon shown below?

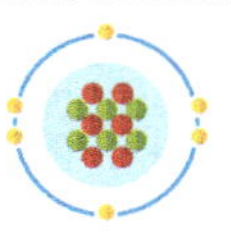

36. Ionization energy is the measure of how much work needs to be done to remove a valence electron from a neutral atom. Thinking about the trends in atomic radii and electronegativity, hypothesize as to how you think ionization energy would change across a row. Explain.

Use the Case Study at right to answer Questions 37–41.

37. What is an allotrope?

38. Of what element(s) are diamond and graphite made?

39. Considering your answer to Question 38, how can you explain the significant differences in properties between diamond and graphite?

40. Would a change from one allotrope to another be a chemical change or a physical change?

41. Why can we model the characteristics of elements in such an orderly way?

CASE STUDY: ALLOTROPES

As mentioned in the Chapter opener, diamonds are made of pure carbon, but so is the graphite in your pencil. Diamond and graphite are *allotropes* of carbon. Allotropes are different forms of the same element in the same state. Allotropes' internal structures differ, resulting in remarkably different characteristics.

Many elements exhibit allotropy. Metals such as polonium, iron, and cobalt each have allotropes. Some metalloids—antimony, arsenic, and silicon—exhibit allotropy. In addition to carbon, other nonmetals like oxygen (ozone and O_2) and phosphorus (15) can be found in different forms. Tin transforms from metallic tin into an allotrope that is a metalloid below 13.2 °C. Materials can change allotrope forms in response to the effects of temperature, light, and pressure.

While graphite and diamond are probably the two most well-known allotropes of carbon, there are a total of nine allotropes of carbon. Each allotrope has distinct properties, all stemming from varied internal structures. Graphite consists of thin sheets of carbon. The sheets easily separate, making graphite useful as pencil lead. Diamonds form from very strong crystal structures, giving them their extreme hardness. Carbon can even form exotic, soccer-ball shaped structures sometimes referred to as *buckyballs*.

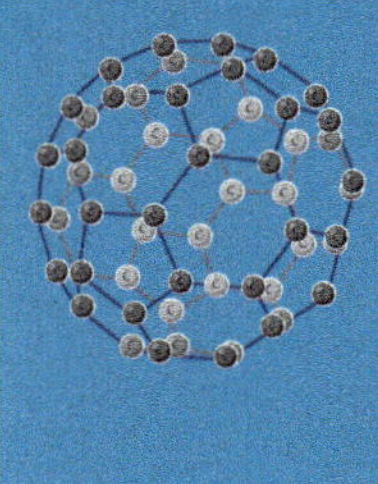

Latex from rubber trees is used to make rubber. These trees produce latex when they are six years old and continue to do so until they are almost thirty years old. Harvesting latex is an example of man wisely exercising dominion.

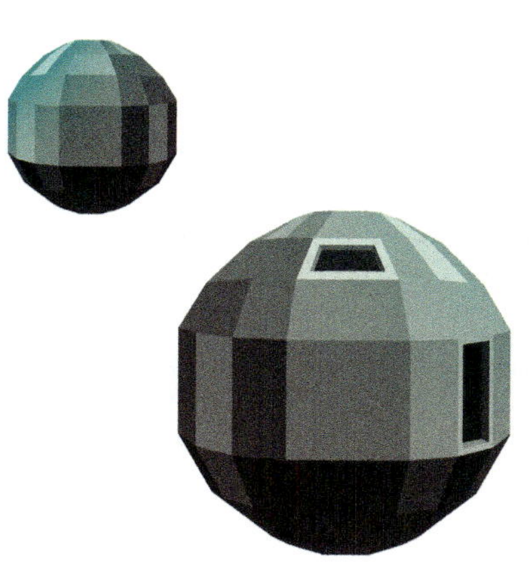

CHAPTER 5

Bonding and Compounds

TREE TRUNK TREASURE

What do you think is the link between these trees and the family car? Need a hint? The substance being harvested from these trees, once processed, can be used for a wide variety of products including garden hoses, waterproof rainwear, pencil erasers—and car tires. Have you guessed the connection yet? The trees in the picture are rubber trees, and the sticky, white latex from their bark is used to make natural rubber. Latex is just one of many naturally occurring chemical compounds that people depend on to meet everyday needs.

Why the Name "Rubber"?

The famous English chemist Joseph Priestly is credited with the discovery in 1770 that natural rubber was useful for rubbing away pencil marks on paper, hence the name *rubber*. In England, the term is still used to refer to a pencil eraser.

Overview

Since Chapter 5 involves bonding, naming compounds, and writing formulas, it is foundational to several chapters that follow. These formulas and names are necessary for understanding chemical reactions, solutions, and acid-base chemistry.

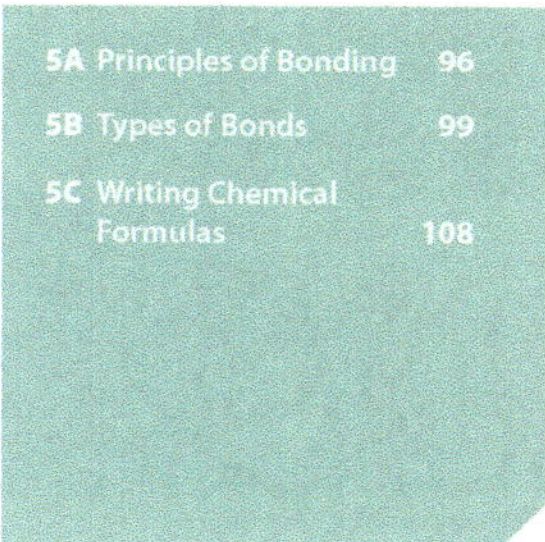 *Lab Activities*

Lab 5A: *The Solution to a Problem*—In this lab activity, students evaluate a substance's solubility on the basis of its bond type and predict a substance's bond type on the basis of its solubility.

Lab 5B: *Electric Lines*—In this lab activity, students evaluate a substance's conductivity on the basis of its bond type and predict a substance's bond type on the basis of its conductivity.

Chemical Bonds

This early in their budding chemistry careers, students will be tempted to think that breaking bonds releases energy and forming bonds absorbs it, which is not quite correct. Energy must actually be added to break bonds, and the amount of energy either released or absorbed during a chemical reaction is the net difference in energies between the reactants and products. This is covered in more detail in Chapter 7, but a more complete explanation is normally reserved for a regular chemistry course.

Demonstrating Hydrogen

Preparation: Do an internet search using the keywords "making hydrogen gas" for the demonstration method that suits your situation. You will find several to choose from.

Performance: Because several methods for making hydrogen gas involve a reaction of an acid or a strong base with a metal and produce heat, much caution should be used. Be sure to wear goggles and gloves and use glass containers instead of plastic. Collect the gas in a balloon on a bottle or in test tubes or gas-collecting bottles. Once the reaction has subsided and the gas has been collected, ignite small amounts of gas away from any dangerous chemicals and a sufficient distance from students.

Discussion:

1. Now that you've seen how the gas exploded, what kind do you think it is? *(hydrogen)*

2. Since water vapor contains hydrogen, would you expect it to react the same way? *(No)*

3. Since oxygen burns as well, why do you think oxygen and hydrogen burn separately but not when they are part of the same substance? *(Different substances have different properties. They are more stable together than apart.)*

4. What do you think is holding oxygen and hydrogen together? *(chemical bonds)*

5. So why do chemical bonds form? *(They make the elements involved more stable.)*

5A Questions

- What is a chemical bond?
- Why do chemical bonds form?
- Are molecules always made from more than one element?
- What determines how atoms bond?
- Are the properties of compounds the same as those of the elements of which they are made?

5A Terms

chemical bond, octet rule

5A | PRINCIPLES OF BONDING

How do compounds form?

5.1 CHEMICAL BONDS

Rubber, gasoline, starch, aspirin—these very different substances all have something in common. They are each made of compounds that are held together by chemical bonds. A **chemical bond** is an electrostatic attraction that forms between atoms when they share or transfer valence electrons. Chemical bonding is what makes it possible for the rather small number of naturally occurring elements to combine and form the great variety of compounds that exist in the world around us.

Chemical bonds store useful energy within molecules. The energy stored in the chemical bonds in gasoline powers your family's car. When chemical bonds in molecules are broken and new compounds are formed, some of that stored energy may become available to do work, like moving a car. Humans don't run on gasoline, but our food contains chemical bonds, just like gasoline. The energy stored in those bonds is what powers us too.

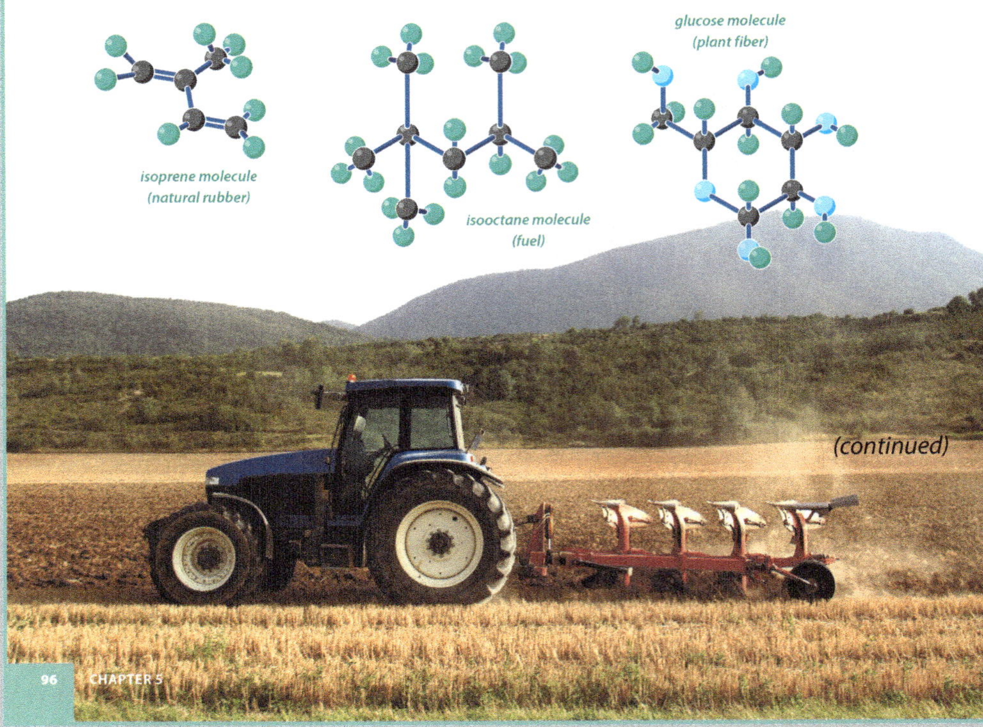

(continued)

TEACHING THE MATERIAL

ESSENTIAL QUESTION

How do compounds form?

OBJECTIVES

- 5A1 Define *chemical bond*.
- 5A2 Compare elements and compounds.
- 5A3 Explain how a molecule can be an element or a compound.
- 5A4 Explain how the octet rule guides chemical bonding.
- 5A5 Show how the properties of compounds can differ from the elements of which they are made.

RESOURCES

Demonstrating Hydrogen

Why Compounds?

Why do atoms form chemical bonds in the first place? It turns out that most atoms are chemically unstable when they are not bonded to other atoms. This is why few elements are found in their pure form in nature. Instead, they are usually found combined with other elements in the form of compounds. Groups of atoms bonded together as compounds are usually more stable than those individual atoms would be by themselves.

Element or Compound?

An atom of one kind of element doesn't always have to combine with an atom of a different element in order to gain stability. Some pure elements in nature occur as chemically bonded atoms of that element. Both oxygen and nitrogen in Earth's atmosphere exist mostly in the form of molecules that are made of two atoms of the same element bonded together. Sulfur is commonly found as ring-shaped molecules made of eight sulfur atoms.

5.2 THE OCTET RULE

The instability in atoms and their tendency to form chemical bonds is mainly due to incomplete valence energy levels (see Chapter 4). Atoms generally are most stable when they have a full eight electrons in their valence energy level. This principle is called the **octet rule** of bonding. There are a few exceptions to the octet rule. Hydrogen and helium, for example, have electrons only in their first energy level, which can have at most two valence electrons. The next three elements—lithium, beryllium, and boron—usually lose their valence electrons when bonding, again leaving electrons only in their first energy levels. You'll see later how the octet rule is applied in chemistry to determine how atoms will bond to other atoms. Extensive experimental evidence shows that an atom can fill an octet in its outer level by bonding with other atoms to form molecules and compounds.

Atoms achieve greater stability through bonding in one of two ways. First, atoms can share electrons. The shared electrons can fill the valence levels of all the atoms bonded together at the same time. Second, atoms can gain or lose electrons through a process of electron transfer to gain a full octet. Atoms with few valence electrons, which are loosely held, can easily lose them, exposing the full octet of the next lower energy level. On the other hand, if an atom needs a few electrons to complete an octet, it can acquire them from other atoms or its surroundings. Ions formed by losing or gaining electrons are pulled together by their opposite charges. No matter the type of bonding, the end result is the same—atoms are more stable with completely filled outer energy levels.

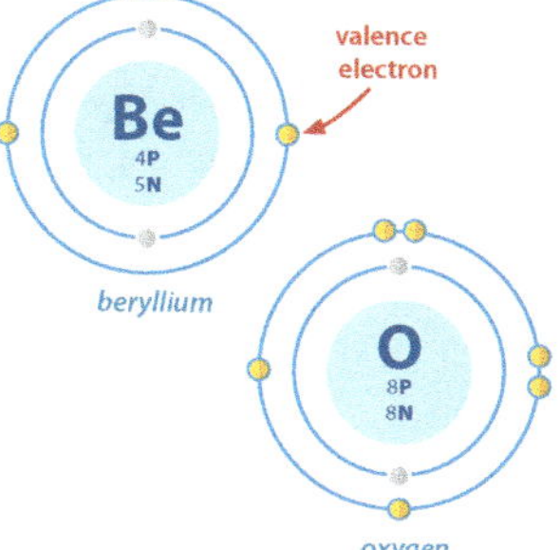

STRATEGIES

Class Opener: Use the Demonstrating Hydrogen teacher note on page 96 to stimulate students' thinking about the reasons why chemical bonds form.

Formative Assessment: Use the Formative Assessment: Losers, Gainers, or Sharers teacher note on this page to help students understand how elements react in order to achieve a full outer shell or stable octet.

Ticket Out the Door: Why are so few elements found by themselves rather than combined with other elements in compounds?

 ### Sulfur Dioxide

The Lewis structure for sulfur dioxide shows a single and a double bond. You could also draw the molecule with the single and double bonds switched. This is known as *resonance*. In reality, sulfur dioxide molecules have two identical bonds that are between a single and a double bond. This concept is beyond the scope of the Student Edition, but we point it out because if you do an internet search for sulfur dioxide, you may see either Lewis structure mentioned above or even a structure that appears to show two double bonds. In the latter case, the two double bonds actually represent the two 1.5 bonds of the resonance structure.

The Octet Rule and Periodic Trends

Connect the tendency of atoms to gain or lose electrons to the periodic trends learned in Chapter 4. As the radius gets smaller across a row, the more tightly the electrons are held. Thus, elements to the right of the table will tend to gain electrons because it is hard to remove their valence electrons. Elements to the bottom and left of the periodic table easily lose electrons because these atoms are relatively large, so their valence electrons are loosely held.

It should be obvious to students that the elements in Group 18 all have full valence levels and therefore do not need to form bonds, but they might need reminding of this.

Formative Assessment:
Losers, Gainers, or Sharers

To help students see the importance of the octet rule in bonding, consider projecting or drawing on the board Bohr models of several elements. Following are suggested elements and questions to ask regarding them.

For noble gases such as neon and argon:

1. What do these have in common? *(They are chemically stable and have eight valence electrons, that is, full outer shells.)*

(continued)

For metals such as lithium, sodium, or potassium:

2. What do these have in common? *(They are very reactive metals, have low electro-negativities, and have only one valence electron.)*

3. What are two ways that those metals could become like noble gases? *(They could lose one electron or they could gain seven electrons.)*

4. Which option do you think would be easier for them to do, considering that it involves energy for each electron? *(lose one electron)*

For nonmetals such as chlorine, bromine, and fluorine:

5. What do they have in common? *(They are reactive nonmetals, have high electronegativities, and have seven valence electrons.)*

6. What are two ways that they could become like noble gases? *(They could lose seven electrons, or they could gain one electron.)*

7. Considering that adding or taking away electrons involves energy for each electron, which do you think is easier? *(adding one electron)*

For sodium and fluorine:

8. If these two were together, what might each do to become more stable? *(Sodium could lose an electron and fluorine could gain an electron.)*

9. What type of bond would this be, and why is there attraction between the atoms? *(ionic; Because the atoms form ions that are oppositely charged, the sodium would be positive and the fluoride would be negative.)*

Hydrogen and Oxygen Gas

Both hydrogen and oxygen occur in their pure forms as diatomic molecules (H_2 and O_2). These diatomic molecules are more stable than individual hydrogen or oxygen atoms but less stable in turn than a water molecule.

Have you ever tried to set water on fire? Silly question, right? Water, as you probably know, is a compound made of the elements hydrogen and oxygen. At room temperature, water is a liquid that is useful for drinking and bathing. It's essential for life as we know it. It's also not flammable, so it's terrible for starting fires but great for putting them out. Pure hydrogen and oxygen are another story. Both are colorless, odorless, and extremely reactive gases that easily, and sometimes explosively, combine with other elements. Hydrogen and oxygen are chemically reactive precisely because their atoms are unstable. When they combine with each other, the resulting compound—water—is much more stable than its parent elements. Chemical bonding changes the physical and chemical properties of all the substances involved.

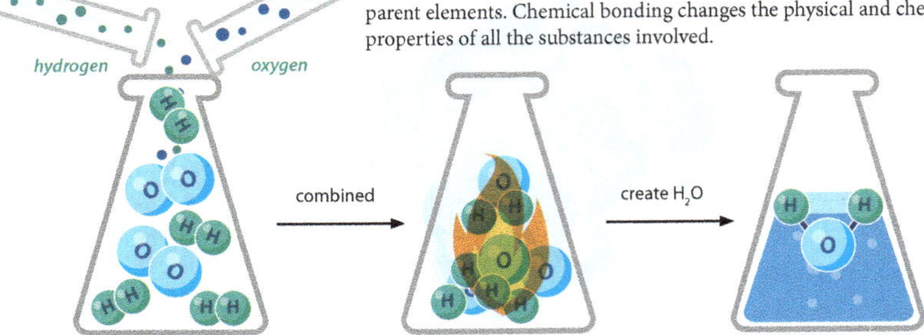

What's true for water is usually true for other compounds as well. They tend to be quite different from the elements of which they are made. Think about pencil lead for a moment. We call it "lead," but it's actually a mixture of clay and *graphite*, a form of pure carbon (see page 93). Would you consider stirring some powdered graphite into your coffee or sprinkling it on a bowl of oatmeal? Of course not! But if those same carbon atoms found in graphite are combined and bonded in a certain way with hydrogen and oxygen atoms, a surprising change takes place. They produce a white, crystalline, and sweet-tasting solid—table sugar. Sugar tastes great, but it's useless for writing. Its physical and chemical properties are just far too different from those of its parent element, carbon.

5A | REVIEW QUESTIONS

1. A chemical bond forms when atoms _____ electrons.
   a. share
   b. transfer
   c. either share or transfer
   d. both share and transfer

2. Why do atoms form chemical bonds?

3. Are molecules always made of atoms of different elements? Explain.

4. What is the octet rule?

5. What are two ways that an atom can meet the octet rule requirement?

6. Describe two properties of water that are different from the properties of the elements from which it is formed.

5A REVIEW ANSWERS

1. c. *(p. 96)*

2. The atoms in compounds produced by bonding are more stable than the individual atoms that form them. *(p. 97)*

3. No. Some elements exist naturally as molecules. *(p. 97)*

4. The octet rule is a principle stating that most atoms are stable when they have a full valence energy level, typically with eight valence electrons. *(p. 97)*

5. by sharing electrons or by transferring (gaining or losing) electrons *(p. 97)*

6. Answers will vary. Unlike both hydrogen and oxygen, water is relatively unreactive at room temperature. Water is a liquid at room temperature, while hydrogen and oxygen are both gases. *(p. 98)*

Have students turn back to the Trends in Electronegativity chart on page 90. Ask them to consider the following combinations of elements and speculate what might happen to valence electrons.

1. a fluorine atom and a francium atom *(The fluorine atom would pull electrons away from the francium atom.)*

2. two chlorine atoms *(The atoms would pull on the electrons equally.)*

3. several iron atoms *(The atoms would pull on the electrons equally.)*

4. a carbon and an oxygen atom *(The oxygen would pull the electrons from the carbon but not completely away.)*

Formative Assessment: *Covalent Bonds*

1. What arrangement of electrons makes most atoms more stable? *(having eight valence electrons, that is, a full outer shell)*

2. In general, how can an atom achieve that arrangement of electrons? *(by transferring [gaining or losing] electrons or by sharing them with another atom, thereby forming a bond)*

3. What property of two atoms determines the type of bond that forms between them? *(electronegativity)*

4. If both atoms have high electronegativities, what type of bond forms between them? *(covalent)*

5. What do we call a unit of a compound that forms through covalent bonding? *(a molecule)*

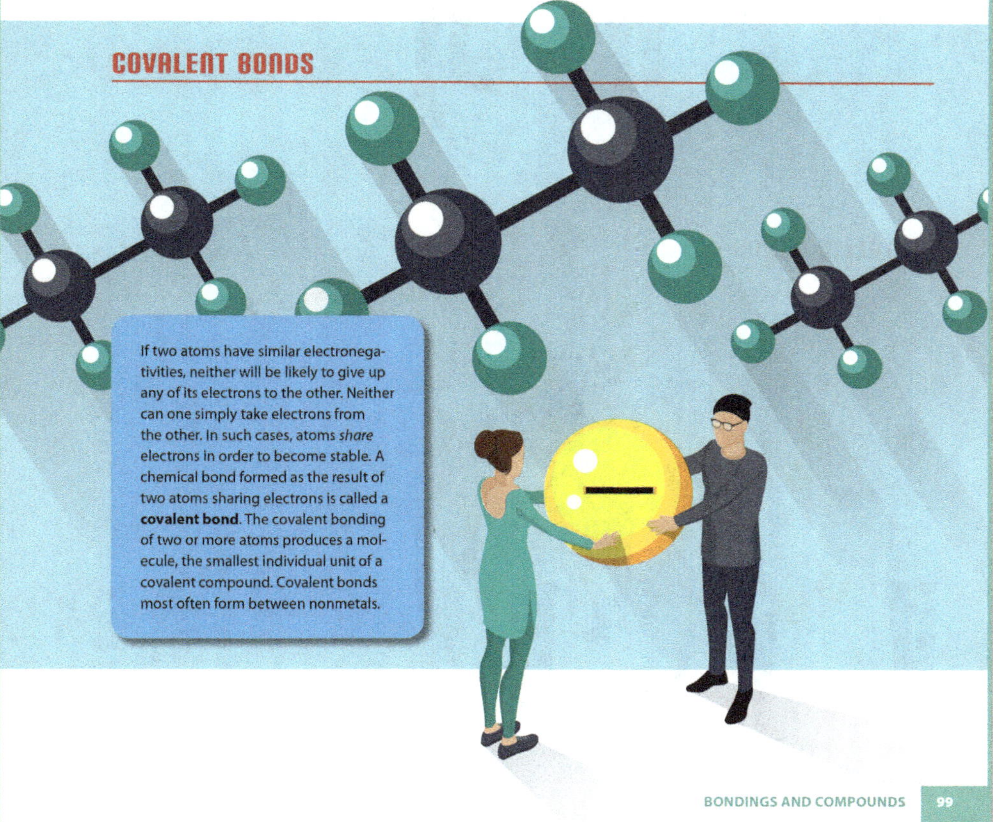

5B | TYPES OF BONDS

Why do atoms bond in different ways?

5.4 TYPES OF BONDS

Remember, most atoms are seeking to have eight valence electrons, and they can accomplish this by gaining, losing, or sharing electrons. As you saw in Chapter 4, not all atoms exert the same amount of tug on their electrons. Atoms with high electronegativities hold on to their valence electrons tightly and are eager to acquire more. Those with low electronegativities, on the other hand, pull weakly on their valence electrons and easily give them up. The difference (or lack thereof) between the electronegativities of two atoms governs what sort of bond they will make.

5B Questions

- How does the role of electrons differ in covalent, ionic, and metallic bonding?
- What's an easy way to illustrate simple compounds?
- Can two atoms form more than one bond between them?
- How does the polarity of a bond compare with that of a molecule?

5B Terms

covalent bond, ionic bond, formula unit, metallic bond, diatomic molecule, Lewis structure, polarity

COVALENT BONDS

If two atoms have similar electronegativities, neither will be likely to give up any of its electrons to the other. Neither can one simply take electrons from the other. In such cases, atoms *share* electrons in order to become stable. A chemical bond formed as the result of two atoms sharing electrons is called a **covalent bond**. The covalent bonding of two or more atoms produces a molecule, the smallest individual unit of a covalent compound. Covalent bonds most often form between nonmetals.

TEACHING THE MATERIAL

ESSENTIAL QUESTION

Why do atoms bond in different ways?

OBJECTIVES

- 5B1 Compare the role of electrons in ionic, covalent, and metallic bonding.
- 5B2 Interpret the Lewis structures for simple compounds.
- 5B3 Explain why double and triple bonds form using the octet rule.
- 5B4 Relate polarity to bonds and molecules.

RESOURCES

- Demonstrating Properties of Ionic and Metallic Substances
- Mini-Lab: *Modeling Bonds in Three Dimensions* (p. 106); Lab 5A: *The Solution to a Problem—Solubility and Chemical Bonds*; Lab 5B: *Electric Lines—*Conductivity and Chemical Bonds

(continued)

Preparation:

You will need a hammer, a laboratory safety shield, rock salt, copper, iron, or aluminum metal, and something solid such as a brick or cinder block.

Performance:

Use the hammer to smash a piece of rock salt on the solid surface behind the laboratory safety shield. You could even have a student volunteer don goggles and do it. Then use the hammer to beat on the metal.

Discussion:

1. What difference do you observe? *(The ionic substance breaks into pieces. The metal may change shape but remains intact.)*

2. What accounts for this difference? *(The ionic substance is held together by ionic bonds, which hold the ions rigidly in place until the force violently breaks them apart. But the atoms of metals are held by metallic bonds, which allow the metal atoms to rearrange when struck by the hammer.)*

🔬 *Formative Assessment:*
Ionic and Metallic Bonds

1. What properties of two elements would result in ionic bonds forming between them? *(They would have very different electronegativity values, and typically one would be a metal and the other a nonmetal.)*

2. What do metals usually do with their valence electrons? *(They give them up.)*

3. If a nonmetal atom needed two valence electrons and a metal atom had only one to donate, where would the nonmetal get another electron? *(from another metal atom)*

4. What are some of the physical properties of a metal that result from its metallic bonds? *(ductility, malleability, luster, conductivity)*

5. What is being shared between atoms involved in metallic bonds? *(a "sea" of electrons)*

IONIC BONDS

When atoms with very different electronegativities bond, one typically loses one or more valence electrons and the other gains the lost electrons. The atom with low electronegativity, which loses electrons, becomes a cation. The atom with high electronegativity, which gains electrons, becomes an anion. The opposite electrical charges on these ions attract each other to form an **ionic bond**. Ionic bonds normally form between a metal and a nonmetal.

Ionic substances form crystals rather than molecules (see Chapter 2). Each ion is equally attracted to a number of oppositely charged ions within a lattice. Crystals of a particular ionic solid always form with the same ratio of cations to anions. Thus, the chemical formula for an ionic substance represents a **formula unit**, the smallest ratio of the ions within the compound.

METALLIC BONDS

Many metals, especially transition metals, exhibit a third type of chemical bond. These elements have similar low electronegativities. This causes atoms of those metals to hold their valence electrons very loosely. As a result, the electrons are able to move about easily from one atom to another. They are not shared or exchanged between any two atoms in particular in the manner of either a covalent bond or an ionic bond. The result of this metallic bonding is a "sea" of freely moving electrons within the lattice structure of the metal. A **metallic bond** is the attraction between metal atoms and their sea of shared electrons. Most of the typical properties of metals are due to these metallic bonds.

TEACHING THE MATERIAL

STRATEGIES

Class Opener: Use the Discussion: Electron Tug of War teacher note on page 99 to stimulate students' thinking regarding what happens to electrons between atoms.

Formative Assessment: Use the Formative Assessment: Covalent Bonds teacher note on page 99 to assess students' understanding of covalent bonds.

Formative Assessment: Use the Formative Assessment: Ionic and Metallic Bonds teacher note on this page to assess students' understanding of these types of bonds.

Ticket Out the Door: Write the names of a covalent compound and an ionic compound and draw the Lewis structure for each.

5.5 HOW COVALENT BONDING WORKS

Diatomic Molecules

The simplest example of two atoms with similar electronegativities sharing electrons in order to satisfy the octet rule occurs when the atoms are of the same element. Molecules made of two atoms, whether of the same element or not, are called **diatomic molecules**. Let's look at how a molecule of chlorine (Cl_2) forms. Both chlorine atoms have seven valence electrons and need one more to complete an octet (see right, top). The electronegativity of chlorine is high, so it is not willing to give up an electron. And since each atom has the same electronegativity, neither can remove an electron from the other. But if each chlorine atom shares one of its electrons with the other, both atoms can have an octet. The shared electrons count toward meeting the octet rule requirement for both atoms. The two shared electrons, called a bonding pair, form a *single covalent bond* (right, bottom) between the two chlorine atoms.

The example just given uses Bohr models to show how electrons are shared. But there's a simpler way to model a covalent bond using **Lewis structures**. Chemist Gilbert Lewis introduced this system in 1916. By rotating the unpaired dots in the electron dot notation for two chlorine atoms, the single dots can be shown forming a bonding pair. The pair of dots between the symbols identifies a covalent bond. The Lewis structure for the bonding of chlorine is shown below.

Covalent bonds may be represented by a pair of dots, but more often a single dash between the bonded atoms is used to distinguish between the electrons involved in bonding and those that are not. When dashes are used in Lewis structures, the nonbonding electrons may or may not be shown.

Seven elements are found in nature as diatomic molecules. You'll need to know these for Chapter 7. The seven diatomic elements are hydrogen, nitrogen, oxygen, fluorine, chlorine, bromine, and iodine. A memory aid for these is the *Hydrogen Seven*. Notice on the periodic table below that the diatomic elements have been shaded. Hydrogen is in the upper left corner, but the other six form the numeral seven starting at nitrogen, element number 7.

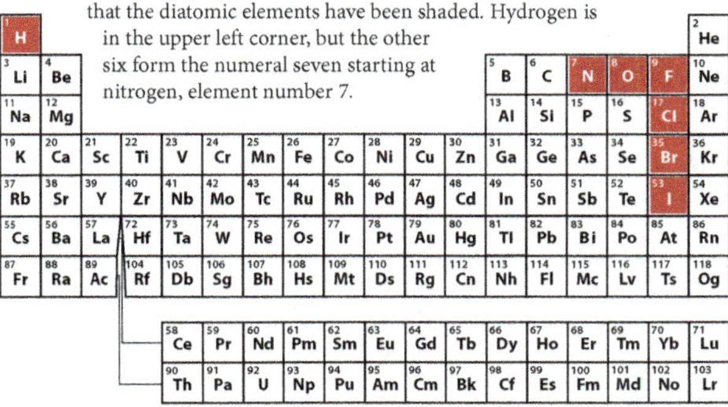

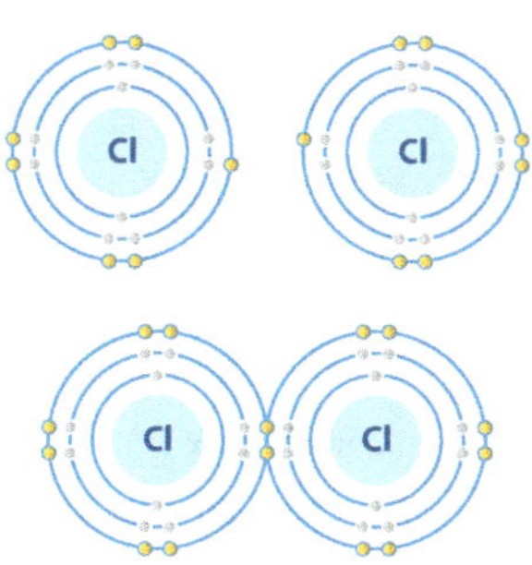

Chlorine forms a covalent bond by sharing a single electron with another chlorine atom. When counting valence electrons, the shared electrons count for both atoms.

Diatomic Mnemonics

Use the Hydrogen Seven memory aid to help students remember the seven diatomic elements. *Hydrogen* reminds us that hydrogen is one of them, and *Seven* tells us to go to element number 7 (nitrogen) and imagine a large number seven on the periodic table—covering nitrogen, oxygen, fluorine, chlorine, bromine, and iodine.

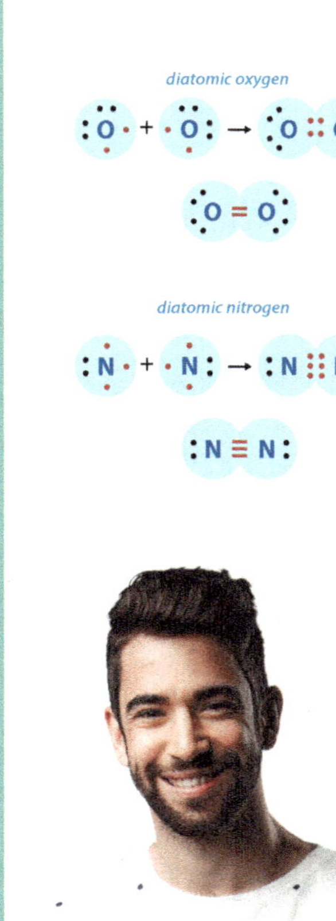

Shifting Electrons?

Some students may ask why the unbonded electrons around the oxygen atoms shifted their position. While this gets into VSEPR theory, for now ask them about the charge on electrons and what we know about objects with like charge. These electrons move because they are repelling each other.

Quadruple Bonds?

Students may wonder whether more than three bonds can be made between atoms. Quadruple bonds can occur between atoms of certain transition metals, but none are known to occur among main block elements (Groups 1, 2, and 13–18).

Multiple Bonds

Like chlorine, pure oxygen occurs in the form of diatomic molecules. But there's a difference. An oxygen atom has six valence electrons, so it lacks two electrons to make a complete valence octet. To address the shortage, an oxygen atom needs to share not just one, but two, of its electrons with another oxygen atom. Sharing two pairs of electrons forms a *double covalent bond*. Double bonds are stronger than single bonds. In Lewis structures they are represented by two dashes instead of two pairs of dots. Even stronger *triple covalent bonds* form when atoms must share three electrons to complete their octets. Diatomic nitrogen is one example.

Polyatomic Molecules

Of course, matter in our world consists of much more than just diatomic elements. Most substances are made of polyatomic molecules (Gk. *poly*, meaning "many") that contain different kinds of atoms. Such molecules are also held together by covalent bonds. Let's look at water as one example. As you probably already know, water consists of one part oxygen for every two parts hydrogen. We've just seen that each oxygen atom needs two valence electrons to complete an octet. That gives us a hint as to why a molecule of water contains two hydrogen atoms.

If you locate hydrogen on the periodic table, you'll see that its atomic number is 1 and that it is in Group 1 and in the first period. That means that a neutral hydrogen atom has one valence electron (Group 1) and only one energy level (Period 1). You'll recall from Chapter 4 that the first energy level of every atom can hold only two electrons. This means that there are a few elements, including hydrogen, that are exceptions to the octet rule. They can be stable if their first energy level is at full capacity—just two electrons.

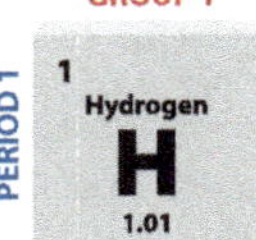

Now we can see how forming a molecule of water satisfies the electron needs of each atom in the molecule. Each hydrogen atom shares its one electron with the oxygen atom, and the oxygen atom shares two electrons, one with each hydrogen atom. This forms two single covalent bonds—one between each hydrogen atom and the oxygen atom.

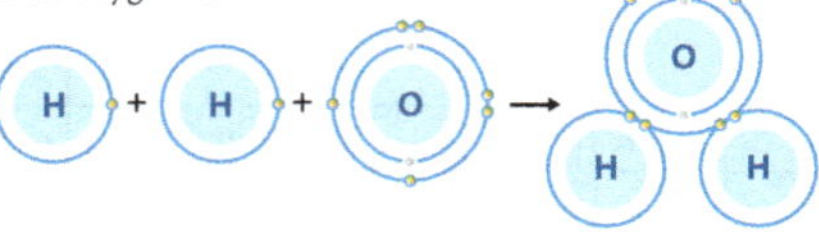

Covalent bonds within other polyatomic molecules are formed in similar fashion. Take a look at the Lewis structures for two common substances and see whether you can identify the kinds of atoms and bonds found in them.

Polarity

Do atoms share their electrons "nicely" with each other? It turns out they don't! It's more like a tug of war. Remember that elements vary in their electronegativity—their pull on electrons. Unequal pulling results in unequal sharing, which results in an unequal distribution of electric charge. **Polarity** is the name we use for this unequal distribution of electric charge. Remember the quantum mechanical model of atoms? Recall that electrons inhabit a region that looks like a cloud. The cloud surrounding a bonded pair of atoms is shifted toward the more electronegative atom of the pair.

Let's see how this works in a molecule of water. Oxygen is the second-most electronegative element in existence. Hydrogen's electronegativity is much lower. As a result, the oxygen atom in a water molecule (top right) pulls on its shared electrons much more than the hydrogen atoms do. This means that the shared electrons in a water molecule spend more of their time near the oxygen atom than they do near the hydrogen atoms. We call this unequal sharing of electrons a *polar covalent bond*. Not all covalent bonds are polar. If the bond exists between two atoms of the same element, such as in oxygen or nitrogen gas, then each atom exerts an equal pull on the shared electrons. Equal sharing like this produces a *nonpolar bond* (middle right).

The unequal sharing of electrons can produce more than just polar bonds. It can produce *polar molecules* as well. This results when the unequal sharing of electrons causes some portion of a molecule to have a negative charge while another portion is positive. Again, let's see how this happens in a water molecule (bottom right). When the shared electrons in a water molecule are near the oxygen atom, they create a temporary region of excess negative electrical charge around the oxygen atom. At the same time, their absence from the hydrogen atoms creates a region of temporary excess positive charge around those atoms. Overall, this causes the oxygen side of the molecule to have a slight negative charge and the hydrogen side to have a slight positive charge. The polarity of water molecules causes them to be slightly "sticky." The positive side of one water molecule is attracted to the negative sides of other water molecules.

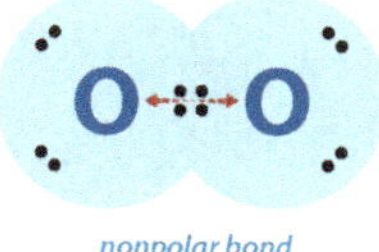

If these two dogs and their tug toy represent a covalent bond between oxygen and hydrogen, which dog is oxygen?

VSEPR

Many students are probably aware that a water molecule has a bent shape. Of course, there's a reason why water is shaped that way. The geometry of various molecules is explained by **v**alence **s**hell **e**lectron **p**air **r**epulsion theory, or VSEPR (pronounced "vesper"). Molecular geometry is examined in the mini lab activity for this chapter, but a full treatment of the subject is usually reserved for a course in chemistry. It is sufficient at this grade level that students recognize that a bent configuration such as that found in water produces a polar molecule.

Answer to Question in Caption

The caption asks students to consider which of the two dogs represents oxygen. The larger dog is able to pull the tug toy more, so it represents oxygen with its higher electronegativity value.

Properties of Water

The statement "Life as we know it could not exist without water" involves more than water being a liquid at room temperature. The polar nature of water molecules explains why water is such a good solvent. Since this will be discussed further in Chapter 9, it is important that students understand the polarity concept. In addition, the hydrogen bonds that are involved with polarity are responsible for other properties of water, including its high specific heat, its surface tension, and its ability to freeze from the top down.

Ionic or Covalent?

Students may wonder whether there's a cutoff line somewhere along the periodic table where elements transition from making ionic bonds to making covalent bonds. In reality, most chemical bonds are neither completely one nor the other. As the difference in the electronegativities between two bonded elements increases, so does the tendency of their bonds to be ionic in nature. Conversely, as the difference decreases, the bonds tends to be more covalent.

The cohesion and surface tension that allow spilled water to bead on a tomato are both caused by the polarity of water molecules.

The polarity of water accounts for many of its physical and chemical properties, including its being a liquid at room temperature. If water were not polar, life as we know it could not exist!

5.6 HOW IONIC BONDING WORKS

In the last section, we saw how atoms with similar electronegativities can share electrons to fill their valence energy levels. But if two atoms have very different electronegativities, they will form a different kind of bond. In ionic bonding, atoms don't share valence electrons. Instead, the atoms transfer electrons. One of them, the donor, will give away one or more of its electrons. The other, the receptor, will accept the electrons. Another way to think about this is as very unequal sharing—so much so that the nonmetal takes the metal's valence electrons. We'll see how this works by looking closely at another very common substance—table salt.

Table salt, or sodium chloride, is an ionic compound made of equal parts, or a one-to-one ratio, of sodium and chlorine. Chlorine's electronegativity is very high, the third highest on the periodic table in fact. Chlorine holds on to its valence electrons very tightly. Sodium, on the

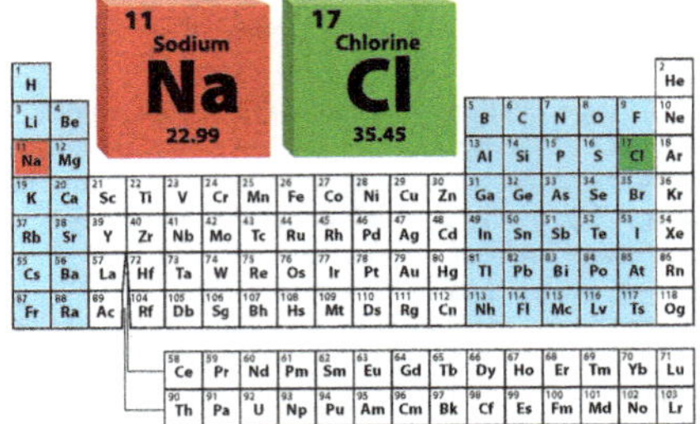

The weakly electronegative metals in Groups 1 and 2 easily form salts with the strongly electronegative nonmetals in Groups 13–17.

other hand, has only one valence electron. It's in the third energy level, so it's rather far from the nucleus. Therefore, the positive charge in sodium's nucleus attracts its one valence electron only weakly. Chlorine's strong pull on electrons, coupled with sodium's weak hold on its valence electron, allows an atom of chlorine to remove the valence electron from an atom of sodium. In so doing, chlorine completes an octet of valence electrons. When sodium loses its one valence electron, its third energy level is emptied. Since its second energy level already has an octet of electrons, the exchange of electrons fulfills the octet rule for both atoms.

Transferring sodium's one valence electron to chlorine exposes the eight electrons in sodium's already-full second energy level.

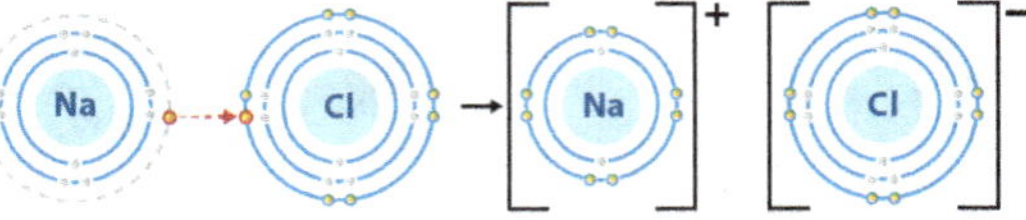

But more than just a simple exchange of an electron has taken place in this scenario. By losing an electron, the sodium atom has become a sodium cation. The chlorine atom, by accepting the electron, has become a chloride anion (the switch to the *-ide* suffix is a naming convention—an agreed upon way to name substances). The attraction between oppositely charged particles is what holds the pair together.

As we did with covalent bonds, we can use Lewis structures to show ionic bonds. There are a couple of differences though. The Lewis structures for ionic compounds do not show all of the valence electrons that you might expect. Take a look at the Lewis structure for sodium chloride.

$$\left[\text{Na} \right]^{+} \left[:\overset{\cdot\cdot}{\underset{\cdot\cdot}{Cl}}: \right]^{-}$$

Can you spot the differences? Recall from Chapter 4 that electron dot notation, and thus Lewis structures, show only valence electrons. Where is sodium's one valence electron? It has been transferred over to chlorine, which is now shown with eight electrons instead of its original seven. Also note that the resulting electric charge is shown with each ion produced by the electron exchange.

Ionic bonding commonly occurs between highly electronegative nonmetals on the right-hand side of the periodic table and the weakly electronegative metals in Groups 1 and 2. The compounds formed in this manner are referred to as salts, and table salt is just one example. Other salts that you may be familiar with include Epsom salt (magnesium sulfate), found in bath salts, and baking soda (sodium bicarbonate). As you see, chemical bonds are very important for giving us some of the materials and energy that we need to live.

5B | REVIEW QUESTIONS

1. Describe the role of electrons in covalent, ionic, and metallic bonds.

2. What do the red electrons below represent?

3. Under what circumstances will a triple bond form?

Use the Lewis structure for chloromethane and the table of electronegativites below to answer Questions 4–5.

Element	Electronegativity
Carbon	2.6
Chlorine	3.2
Hydrogen	2.2

4. How many of the bonds shown in the Lewis structure for chloromethane are polar? Explain.

5. Is chloromethane a polar molecule? Defend your answer.

6. Do ionic bonds tend to form between atoms with similar or dissimilar electronegativities?

7. If the atoms in an ionic bond are not sharing electrons, what keeps the atoms together?

Once students have an understanding of the material in Section 5B, they should be ready to work through Labs 5A and 5B.

5B REVIEW ANSWERS

1. In covalent bonding, electrons are shared between atoms. In ionic bonding, one atom loses electrons while the other gains them. In metallic bonding, an electron "sea" of freely moving electrons is shared by all the metal atoms in a lattice. *(pp. 99–100)*

2. the four shared electrons in a double bond between two oxygen atoms *(p. 102)*

3. Triple bonds form when atoms must share three pairs of electrons to satisfy the octet rule. *(p. 102)*

4. All the bonds are polar because each bond is between two atoms with different electronegativities. *(p. 103)*

5. Yes. The polarity of the C–Cl bond is different from that of the C–H bonds, causing an unequal distribution of charge on the molecule. *(p. 103)*

6. dissimilar *(p. 104)*

7. They are held together by the opposite electric charges on the ions that are formed. *(p. 105)*

MODELING BONDS IN THREE DIMENSIONS

Essential Question:

What does a methane molecule actually look like?

Equipment

foam ball, 5 cm.

foam balls, 2.5 cm. (4)

toothpicks (4)

protractor

methane

Methane is a colorless, odorless gas at normal temperatures. One source of methane is the decomposition of dead plant matter in wetlands; there it is an important component of *swamp gas*. Methane is also considered a greenhouse gas.

The Lewis structure for methane (below left) is, of course, a model. Carbon and hydrogen don't look like the letters *C* and *H* in real life. Real bonds don't look like straight lines either. But there's another big difference between this model and real-life methane. The Lewis diagram is a flat, two-dimensional model, whereas methane is three-dimensional. In this lab activity, you'll try to figure out the shape of a methane molecule.

1. Describe what you think the three-dimensional shape of methane might be.

PROCEDURE

Ⓐ Take a moment to think about the Lewis structure for methane. Each line that connects two atoms represents a shared electron pair. Electrons all have the same negative charge, and negative charges repel each other. The slight positive charge on each hydrogen atom, a result of its polar bond with carbon, also repels its hydrogen neighbors. The result of all this is that each bonding pair in a methane molecule tries to get as far away as possible from every other bonding pair.

Ⓑ With the 5 cm. foam ball as the central carbon atom, use toothpicks to connect the four 2.5 cm. balls (representing hydrogen) to the carbon atom.

👓 Modeling Bonds in Three Dimensions

Answers

1. Answers will vary. Methane's actual molecular geometry is tetrahedral. The four hydrogen atoms are at the four corners of a tetrahedron, a four-sided polyhedron.

2. bonding pairs of electrons or chemical bonds

3. They should all be the same.

4. 109°; Accept some variation to allow for the difficulty of measuring the model.

5. Answers will vary. Make sure that students' answers are consistent with their responses elsewhere.

6. The negative charge has no effect. The fluorine atoms in tetrafluoromethane repel each other in the same manner as the four hydrogens in methane do. This produces the same tetrahedral shape.

2. What do the toothpicks represent in this model?

C Move the hydrogen atoms around until you think you have them as far apart as possible.

ANALYSIS

D Use the protractor to measure the *bond angle* between two adjacent hydrogen atoms. The bond angle is the angle formed by two bonds with the central carbon atom at the vertex of the angle.

3. If the hydrogen atoms are as far apart as possible, what should be true about all the bond angles within the molecule?

E Using your answer in Question 3, make adjustments to your model, if needed.

CONCLUSION

4. What bond angle measurement produces the desired spacing of the hydrogen atoms?

5. How does your final version of methane compare to your prediction in Question 1?

GOING FURTHER

6. Replacing the four hydrogen atoms in methane with fluorine atoms creates a molecule of tetrafluoromethane. Unlike hydrogen, fluorine is more electronegative than carbon, so a fluorine atom bonded with carbon has a negative charge rather than a positive one. How do you think this negative charge affects the three-dimensional shape of tetrafluoromethane compared with that of methane?

Common Chemical Formulas

Ask students to write two chemical formulas for common chemical compounds. Have them share with another student and interpret the meaning of each other's formulas. Ask a few students to share their examples with the class. Most will probably come up with the formulas for water, carbon dioxide, and maybe sodium chloride. Use one of their examples to remind them of the meaning of the symbols and subscripts.

5C | WRITING CHEMICAL FORMULAS

How do you know how many atoms are in a chemical formula?

5.7 CHEMICAL FORMULAS FOR BINARY IONIC COMPOUNDS

Chemical Formulas: A Quick Review

You're probably already somewhat acquainted with chemical formulas. A **chemical formula** is a shorthand way of identifying a chemical compound. Chemical formulas also give us information about the composition of the compounds they identify. The chemical formula for water, H_2O, is familiar to most people. H is the symbol for hydrogen and O is the symbol for oxygen. The subscript 2 after the H tells us that each water molecule has two hydrogen atoms. There is no subscript after the O, so it's understood that a water molecule has only a single oxygen atom. Formulas for other compounds are written in a similar way. Each formula includes the element symbols that identify the kinds of atoms in the compound, along with subscripts to indicate the number of atoms of each element.

TEACHING THE MATERIAL

ESSENTIAL QUESTION

How do you know how many atoms are in a chemical formula?

OBJECTIVES

• 5C1 Predict the ratios of ions or atoms in ionic and covalent compounds.

• 5C2 Write the chemical formulas for ionic and covalent compounds.

• 5C3 Name ionic and covalent compounds.

• 5C4 Justify the use of medications. **BWS**

RESOURCES

Ethics: Pseudoephedrine (p. 118)

Case Study: Sticky Situation (p. 121)

Predicting Formulas for Binary Ionic Compounds

It's one thing to already know the formula for a chemical compound, but can we also predict how different atoms will bond together to form them? How do we know, for instance, that sodium will bond with fluorine in a one-to-one ratio to make sodium fluoride, the stuff in your toothpaste? For many simple **binary compounds**—those made from only two elements—formulas can be predicted on the basis of where those elements are found on the periodic table. This is especially true for ionic compounds—those made from oppositely charged ions. The number of electrons that each atom can donate or accept determines the ratio of elements in the compound. We'll soon see how this works, but it helps to first understand how ionic compounds are named.

Naming Ionic Compounds

How would you name an ionic compound, such as that formed by bromine and potassium? A hint is found in some of the names that you've already seen in this chapter, such as sodium chloride. Sodium, the cation, is named first, followed by the name of the anion, chlorine. Other ionic compounds are named in like manner. But notice that there is a slight twist. The names of the anions below are modified to include the suffix *–ide*. Thus, chlor*ine* is changed to chlor*ide*. Similarly, when bromine forms an anion, it becomes brom*ide*. So by giving the name of the metal cation first followed by the nonmetal anion with its modified ending, we can see that the compound of bromine and potassium is called potassium bromide.

COMMON ANIONS

Name	Symbol	Name	Symbol
fluoride	F^-	oxide	O^{2-}
chloride	Cl^-	sulfide	S^{2-}
bromide	Br^-	nitride	N^{3-}
iodide	I^-	phosphide	P^{3-}

Got that? Okay, then let's get back to predicting formulas.

 From Names to Formulas

In Section 5C we demonstrate that the names of compounds are determined by their chemical formulas. You may also desire to show that the reverse is true—that chemical formulas can be determined from the names of compounds. This is especially true for covalent compounds. If you choose to do so, it would be good to have students practice by giving them names of compounds and asking them to determine the compounds' formulas.

STRATEGIES

Class Opener: Use the Common Chemical Formulas teacher note on page 108 to review the meaning of chemical formulas.

Formative Assessment: Use the two formative assessments on page 116 to check students' understanding of writing chemical formulas.

Ticket Out the Door: Use a periodic table and your textbook to write formulas for aluminum sulfide, barium phosphate, and dinitrogen pentoxide.

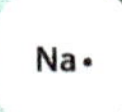

First, locate the cation (the metal) on the periodic table. Sodium is a Group 1 alkali metal with one valence electron. It needs to lose the electron to fulfill the octet rule.

Next, find the anion (the nonmetal). Fluorine is a Group 17 halogen, so we know that it is one valence electron short of fulfilling the octet rule.

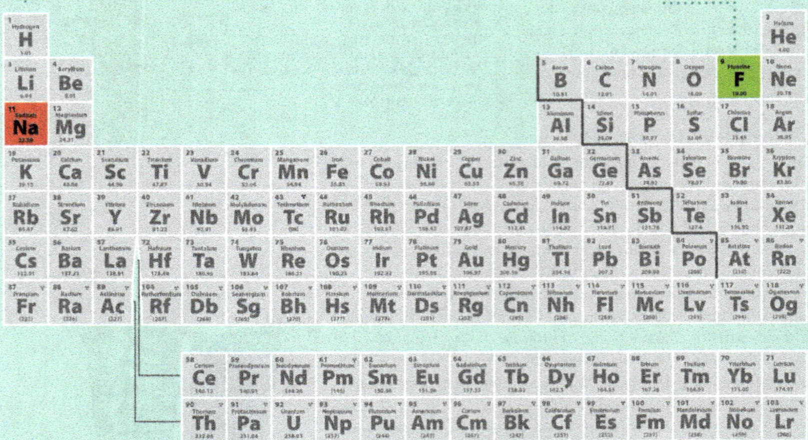

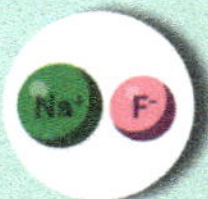

Because sodium is a metal and fluorine is a nonmetal, we know that the two will form an ionic bond. And since each needs to lose or gain only one electron to satisfy the octet rule, we know that sodium and fluorine will bond in a one-to-one ratio.

In the formula for sodium fluoride, the symbol for the cation is written first, followed by the symbol for the anion. There is only one of each ion, so no subscripts are needed. The formula is therefore NaF.

Sodium can form many kinds of chemical salts, and not all of them are made in a one-to-one ratio with a nonmetal. For example, sodium can also combine with oxygen to form sodium oxide. In what ratio do you think sodium and oxygen will combine? Use the technique described above to find out.

In which group is sodium? How many electrons can it donate?

Sodium is in Group 1. It has one electron to donate.

In which group is oxygen? How many electrons can it accept?

Oxygen is in Group 16. It can accept two electrons.

How many sodium atoms are needed to fulfill the electron requirements for oxygen?

Since each sodium atom has only one electron to donate, two are necessary to fulfill oxygen's need for two electrons.

What is the ratio of sodium to oxygen in sodium oxide?

The ratio of sodium to oxygen in sodium oxide must be two atoms of sodium for every atom of oxygen, or 2:1.

What is the formula for this compound?

Since there are two sodium cations for every oxide anion, the formula is Na_2O.

Predicting chemical formulas isn't always as straightforward as this example suggests. For instance, magnesium can combine with nitrogen to form magnesium nitride. We can use the same procedure demonstrated in Example 5-1, but we'll need to add an extra step.

EXAMPLE 5-2: Predicting and Writing the Formula for Magnesium Nitride

In which group is magnesium? How many electrons can it donate?

Magnesium is in Group 2. It has two electrons to donate.

In which group is nitrogen? How many electrons can it accept?

Nitrogen is in Group 15. It can accept three electrons.

How many magnesium atoms are needed to fulfill the electron requirements for nitrogen?

Each nitrogen atom needs three electrons, but each magnesium atom has only two to donate. It will take at least two magnesium atoms, with four total electrons to donate, to meet nitrogen's need for three electrons. But that will leave one electron left over, requiring at least one more nitrogen atom to accept it. Atoms can bond only in whole number ratios, so we need to find the smallest number of each atom to combine so that no electrons are left over. To do this, we'll borrow a skill from mathematics—finding a *least common multiple*.

What is the least common multiple of two and three (the numbers of electrons donated and received by magnesium and nitrogen, respectively)?

The least common multiple of two and three is six. Two will divide into six three times, and three will divide into six two times. Therefore, every three magnesium atoms (with two electrons to donate each) can fulfill the electron needs of two nitrogen atoms (three electrons to receive each).

What is the ratio of magnesium to nitrogen in magnesium nitride?

Since three magnesium atoms are required to fulfill the electron needs of two nitrogen atoms, the ratio of magnesium to nitrogen in magnesium nitride must be 3:2.

What is the formula for this compound?

Since there are three magnesium cations for every two nitride anions, the formula is Mg_3N_2.

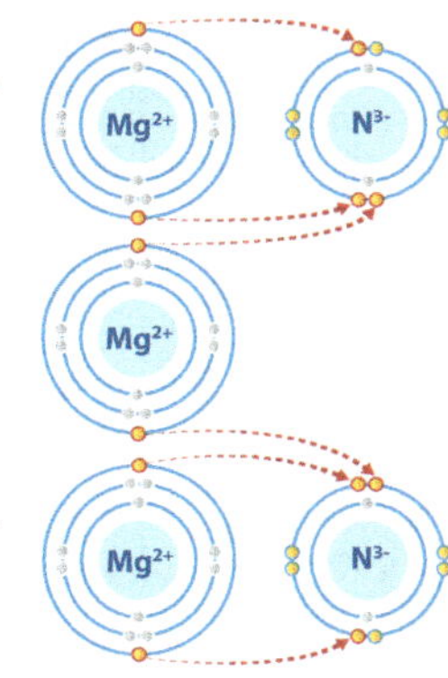

Trisodium phosphate is the name that students may see in lists of ingredients for cleaners and many processed foods. The actual IUPAC name is simply *sodium phosphate*. It would be good to stress to students that in industry the normal system of naming substances is not always followed. Another substance that they have probably heard of before is monosodium glutamate, or MSG. Despite the use of a prefix in trisodium phosphate, students should be encouraged to add prefixes only to covalent binary compounds.

Some teachers may want to have students memorize the commonly used polyatomic ions listed on this page in the Student Edition.

Subscripts

Connect the subscripts on polyatomic ions in ionic compounds to the power of powers property in mathematics. For example, $Mg_3(PO_4)_2$ contains three atoms of magnesium and two phosphate ions, each with a phosphorus and four oxygens. This means that there are two phosphorus atoms and eight oxygen atoms altogether. This is similar to $z^3(xy^4)^2$, which has three of the z variable multiplied together, two of the x, and eight of the y variable multiplied as well.

Predicting and Writing Formulas for Ionic Compounds Containing Polyatomic Ions

So far, all of the ionic compounds that we have looked at have been made of monatomic ions, such as Mg^{2+} or Cl^-. But some ions consist of more than one atom. A **polyatomic ion** is a group of covalently bonded atoms that *together* have gained or lost electrons. The bonded atoms act as a single ionized particle. One such polyatomic ion is known as phosphate. It consists of a single atom of phosphorus covalently bonded to four atoms of oxygen. Together they act as a single ion that has accepted three electrons, giving it a charge of 3−. One phosphate ion can bond with three sodium atoms, which have each lost one electron and have a charge of 1+. The compound formed is called *trisodium phosphate*, a common cleaning agent. A table showing the names and chemical formulas of common polyatomic ions is shown on the left. Note that you'll see both subscript *and* superscript numbers in their symbols. Remember, subscripts indicate the number of atoms and superscripts indicate charge. The amount of charge is equal to the number of electrons that each polyatomic ion has donated or accepted. A positive charge indicates that the ion has lost electrons, while a negative charge shows that electrons have been gained.

When predicting the formulas for compounds containing polyatomic ions, the polyatomic ion is treated as a single particle. The same goes for writing formulas for compounds with polyatomic ions. The only difference occurs when more than one polyatomic ion is in the formula. In such instances, the polyatomic ion is shown in parentheses with a subscript number outside of them.

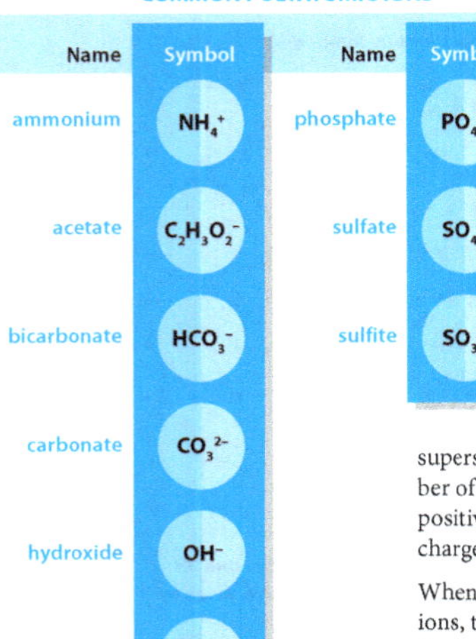

EXAMPLE 5-3:
Predicting and Writing the Formula for Magnesium Hydroxide

In which group is magnesium? What is its charge when it forms ions?

Magnesium is in Group 2. It has two electrons to donate; therefore ionized magnesium will have a charge of 2+.

What is the charge on a hydroxide ion?

1−

How many of each ion are needed to balance their positive and negative charges?

Since magnesium's charge is 2+ and hydroxide's is 1−, each magnesium cation can balance the electrical charge of two hydroxide anions.

What is the ratio of magnesium to hydroxide in magnesium hydroxide?

1:2

What is the formula for this compound?

Since there are two hydroxide anions for every magnesium cation, the formula is $Mg(OH)_2$.

Predicting Formulas Using Oxidation States

The method that we just looked at works well for predicting the formulas of some ionic binary compounds but not for all of them. For instance, copper and oxygen can form ionic bonds. Copper has one valence electron to donate. Oxygen needs two. It seems easy enough to predict that two copper atoms will donate their electrons to one oxygen atom, forming Cu_2O. That combination meets the electron needs of all three atoms. And that is in fact one way that the two elements can bond—but it's not the only way. Copper can also bond with oxygen in a one-to-one ratio, forming CuO. This formula isn't suggested by a straightforward analysis of the location of each element on the periodic table.

To reflect these varied ways that two elements can bond with each other, chemists have developed another way to predict chemical formulas. On the basis of many years of observing the ways that elements bond with each other, each element has been assigned one or more oxidation states. An element's **oxidation state** shows the electric charge gained or lost by that element when it forms a compound. This definition treats all chemical bonds as though they are ionic in nature, even though they may be covalent in reality. Oxidation states are whole numbers whose values can be positive, negative, or zero. Usually one or two oxidation states for an element, called primary oxida-

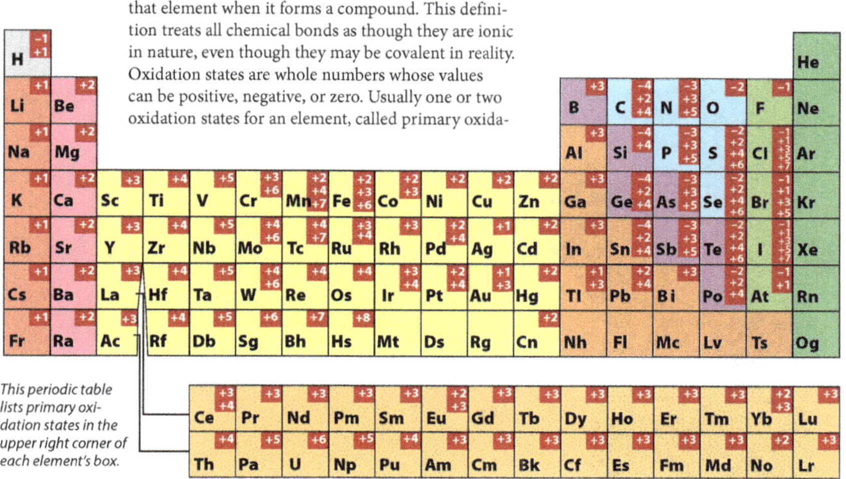

This periodic table lists primary oxidation states in the upper right corner of each element's box.

tion states, are more common than the others. For main block elements (those in Groups 1–2 and 13–18), one of the common oxidation states is usually the same as the number of electrons predicted to be gained or lost on the basis of an element's location on the periodic table. If shown on the table, common oxidation states are normally printed in bold type. When predicting a chemical formula using oxidation states, the oxidation states for all the elements involved must add up to zero. Let's look at some examples.

Why Oxidation State?

The concept of oxidation originally described chemical reactions involving oxygen. Much later, chemists realized that substances that bonded with oxygen lost electrons. Subsequently, the term *oxidation* was used to describe *any* reaction in which a substance lost electrons, not just those reactions that involved oxygen.

Curious students may wonder where the second copper electron in copper(II) oxide comes from in the example given in this section, given that copper has only one valence electron. The first bonding electron is found in copper's 4s sublevel; the second electron available for bonding comes from the copper atom's 3d sublevel.

The use of roman numerals to indicate the oxidation state of a metal is called the *Stock system*.

It's not necessary to have students memorize all the oxidation states. It may be best to have them just use the oxidation states that are easily deduced from the elements' locations in the periodic table (Groups 1, 2, and 13–17 or as indicated by a roman numeral in the name of the compound). You may choose to add those of carbon and nitrogen.

The Stock System

Since the primary reason for the Stock system is to distinguish between compounds made of the same metal but with differing oxidation states, it is often not used for metals such as silver and zinc, which are rarely found in more than one particular oxidation state.

The familiar green patina on copper statues is copper(II) carbonate.

EXAMPLE 5-4:
Predicting the Ratio of Cesium to Oxygen in Cesium Oxide Using Oxidation States

What is cesium's most common oxidation state?

+1

What is oxygen's most common oxidation state?

−2

On the basis of its most common oxidation state, how many cesium atoms will bond with one oxygen atom?

Since the oxidation states of all the bonded atoms must add up to zero, two cesium atoms at +1 each are required to balance the −2 oxidation state of one oxygen atom.

What ratio of cesium to oxygen do these oxidation states indicate in cesium oxide?

On the basis of these oxidation states, the ratio of cesium to oxygen in cesium oxide will be two atoms of cesium for every atom of oxygen, or 2:1.

Notice that in Example 5-4, the predicted ratio of atoms is the same as what we would predict according to the location of both elements on the periodic table. The compound produced in this case is ionic and is named accordingly. The cation, *cesium*, is written first, followed by the anion, *oxide*. But the copper compounds we mentioned earlier require an extra step in the naming process. The names of compounds containing transition metals, like copper, add a roman numeral in parentheses between the names of the ions. The roman numeral indicates the oxidation state of the metal cation. Thus, Cu_2O is copper(I) oxide because copper's oxidation state in this compound is +1. CuO is copper(II) oxide.

Let's look at another example of using oxidation states to predict the formula for a compound that includes a transition metal.

EXAMPLE 5-5: Predicting the Formula and Name for a Compound of Iron and Oxygen If Iron's Oxidation State is +3

Which oxidation state for oxygen should I use?

Oxygen doesn't have a −3 oxidation state to balance iron's +3; its most common oxidation state is −2.

On the basis of oxygen's most common oxidation state, what is the most likely ratio of iron atoms to oxygen atoms?

Since 3 is not a whole number multiple of 2, we'll need to use least common multiples again. The least common multiple of 3 and 2 is 6; therefore the ratio of iron atoms (+3 oxidation state) to oxygen atoms (−2 oxidation state) is 2:3.

What is the formula for this compound?

Again, start with the cation, iron, followed by the anion, oxide. This yields the formula Fe_2O_3.

What is the name of this combination of iron and oxygen?

Using the naming process for ionic compounds, the name of the metal cation is given first, followed by the name of the anion. Since iron is a transition metal, its oxidation state must be included in parentheses. The full name of the compound is thus iron(III) oxide.

5.8 CHEMICAL FORMULAS FOR BINARY COVALENT COMPOUNDS

To this point we have considered only ionic compounds. Now let's look at some examples that show how oxidation states can be used to predict the formulas for covalent compounds. We'll look at a group of compounds called *carboxides*—compounds made of carbon and oxygen.

carbon monoxide

carbon dioxide

EXAMPLE 5-6:
Predicting the Ratio of Carbon to Oxygen in Carboxides

What are the most common oxidation states for carbon? What is the most common oxidation state for oxygen?

Carbon's most common oxidation states are −4 and +4. We've seen that oxygen's most common oxidation state is −2. Since the oxidation states in a compound must add up to zero, we'll need to use the +4 oxidation state for carbon.

What is the ratio of carbon to oxygen according to these oxidation states?

To balance the +4 oxidation state of one carbon atom, two oxygen atoms are needed. The ratio of carbon to oxygen is therefore 1:2. The formula for this compound is CO_2.

Does carbon have any other oxidation states that are positive multiples of 2?

Yes! Carbon also has a +2 oxidation state.

What would be the ratio of carbon to oxygen using a +2 oxidation state for carbon?

Since the +2 oxidation state of a carbon atom can be balanced by a single −2 oxygen atom, the ratio of carbon to oxygen for this compound would be 1:1. Its formula is simply CO.

Naming Covalent Compounds

As you just saw, carbon can covalently bond with oxygen in multiple ways. Neither element is a metal, so how do chemists know which element to list first in the names of these compounds? How can chemists distinguish between the 1:2 carbon-to-oxygen compound and the 1:1 version? It turns out that chemists have a system for doing this. In some ways it is similar to the system for naming binary ionic compounds, but there are a few key differences.

Carbon monoxide (a key component of smog) or carbon dioxide? The difference is kind of important!

When to Use Prefixes

The names of ionic compounds do not include prefixes because the numbers for each ion in a formula unit are easily deduced from their electric charges. Nonmetals generally have more oxidation states than most metals, so prefixes are necessary to determine formulas of covalent compounds.

1. What type of compound forms between lead(IV) and oxygen? *(ionic)*

2. What are the oxidation states of lead and oxygen? *(+4, −2)*

3. What is its formula? *(PbO_2)*

Practice! Practice!

Once you have presented naming and writing formulas for ionic and covalent compounds, it is important that you give students ample opportunity to practice. This can be done with a variety of forms of worksheets, flashcards, or collaborative review. It would be good to start off with similar sets, such as naming binary ionic compounds, then naming compounds that contain polyatomic ions, then naming covalent compounds, then writing formulas for binary ionic compounds, and so on. Finally, give students a mixture of these to make sure that they can switch from one to the other.

Practice with a Neighbor

One way for students to practice is to have each student write two names of substances (one ionic and one covalent) and write two formulas of substances (one ionic and one covalent). Then have each give them to another student to write formulas for the names and names for the formulas.

Formative Assessment: *Writing Formulas from Names of Compounds*

1. When determining the name of a compound from a chemical formula, what must you decide first? *(whether it is ionic or covalent)*

2. How can you determine from a formula whether a compound is ionic or covalent? *(A compound is ionic if the formula is binary and contains a metal and a nonmetal or if it contains a polyatomic ion. If it contains two nonmetals or it contains prefixes, it is usually covalent.)*

3. What type of compound is nitrogen tribromide? *(covalent)*

4. What is its formula? *(NBr_3)*

5. What type of compound is iron(III) chloride? *(ionic)*

6. What are the oxidation states of iron and chloride? *(+3, −1)*

7. What is its formula? *($FeCl_3$)*

The names of binary covalent compounds indicate both the elements involved and the number of atoms of each element. To name a binary covalent compound, follow the few simple rules below.

❶ The element with the *lower group number* is written first.

❷ If both elements are in the same group, then the element with the *higher period number* is written first.

❸ The second element to be named is modified using the *-ide* suffix, just as you saw when naming ionic compounds (e.g., ox*ide*, sulf*ide*, nitr*ide*).

❹ A system of *Greek prefixes* (see below) is used to indicate the number of atoms of each element in the compound. The exception to this rule is when there is only one atom of the first element; in such cases, no prefix is used.

Let's apply these rules to carboxides. For both of them, carbon has the lower group number (14 compared with 16 for oxygen), so carbon will be named first ❶. In each case, there is only one carbon atom, so carbon, being named first, requires no prefix ❹. Since oxygen will be named second, it requires the *-ide* suffix and will be written as *oxide* ❸. The first example had two oxygen atoms, which calls for the prefix *di-* ❹. This compound is thus *carbon dioxide*, the stuff in your carbonated drinks. The second example had only one oxygen, which will be indicated with the prefix *mon-*. This compound is *carbon monoxide*. Recall that carbon monoxide is a toxic gas that is a component of smog (page 115).

As you can see from the examples just shown, the names of chemical compounds are not random. A lot of thought went into giving them names that not only serve as unique identifiers, but also provide information about the number and kinds of atoms in each compound as well.

BINARY COVALENT COMPOUND PREFIXES

Prefix	Number of atoms	Example
mon–, mono–	●	nitrogen monoxide (NO)
di–	● ●	silicon dioxide (SiO_2)
tri–	● ● ●	boron trifluoride (BF_3)
tetr–, tetra–	● ● ● ●	carbon tetrachloride (CCl_4)
pent–, penta–	● ● ● ● ●	diphosphorus pentoxide (P_2O_5)
hex–, hexa–	● ● ● ● ● ●	sulfur hexafluoride (SF_6)

Millions of chemical compounds can be made from the less than 120 elements found on the periodic table. The many different ways that atoms can bond make this great variety possible. Part of exercising wise dominion over God's earth is seeking to learn as much as we can about how the world works, including its compounds. The knowledge that we gain from studying compounds and bonding can change lives for the better. From finding new materials to discovering new medicines, our efforts to understand compounds have helped God's image bearers to thrive.

Of course, the study of chemical compounds has had a dark side too. Chemists have produced many less noble compounds—for example, addictive drugs, weapons of war, and toxins that pollute our world. These chemicals are often the very same ones that can be put to more beneficial uses. Godly wisdom and compassion for others are needed to decide how best to use the compounds that chemistry makes possible.

5C | REVIEW QUESTIONS

1. Using tellurium's location on the periodic table, predict one of its common oxidation states.

2. Your friend writes down the formula for a compound of oxygen and lithium as OLi_2. Is this correct? Explain.

3. What is the name of the compound in Question 2?

4. Predict the formula and give the name for a compound of calcium and fluorine.

5. When naming binary compounds, how is determining which element is named first in an ionic compound different than doing so for a covalent compound?

6. What is a polyatomic ion?

7. Predict the formula and give the name for a compound of magnesium and phosphate.

8. What is the difference between manganese(II) oxide and manganese(III) oxide? How will this difference be reflected in the chemical formulas of the two compounds?

9. How is the naming of the second element in a binary ionic compound similar to that for a covalent compound? How is it different?

10. For the compound whose formula is NCl_3, is the correct name *nitrogen trichloride*, *nitrogen chloride*, or *trichlorine mononitride*? Explain.

11. (See the Ethics box on the next page.)

117

✅ Don't Miss Question 11

Be sure that students know that Question 11 refers to the ethics box Question 11 on page 118.

5C REVIEW ANSWERS

1. −2 *(p. 110)*

2. Lithium forms a cation in a compound with oxygen and should be shown first in the formula, not second. *(p. 110)*

3. lithium oxide *(p. 109)*

4. CaF_2, calcium fluoride *(pp. 109–10)*

5. In ionic compounds, the cation is named first. In covalent compounds, the element in the lower-numbered group (or higher-numbered period if both are in the same group) is named first. *(pp. 109, 116)*

6. a group of covalently bonded atoms that acts as a single ionized particle *(p. 112)*

7. $Mg_3(PO_4)_2$, magnesium phosphate *(p. 112)*

8. The oxidation state of manganese in the first is +2, while that of the second is +3. The ratio of elements in the two compounds will thus be different, resulting in two different chemical formulas, MnO and Mn_2O_3. *(p. 114)*

9. In both cases, the second element is modified by adding the –ide suffix. In covalent compounds, Greek numeric prefixes are used. The prefixes are not used in naming ionic compounds. *(pp. 109, 116)*

10. The correct name is nitrogen trichloride. Since nitrogen and chlorine are both nonmetals, the compound is covalently bonded. Therefore, the Greek prefixes are used. Since nitrogen is in a lower-numbered group, it is named first. *(p. 116)*

11. See page 118.

Pseudoephedrine

Pseudoephedrine (PSE) was originally intended to be used as a decongestant, and it is very effective in that capacity. But it is also a stimulant and has been used as a performance enhancing drug in the world of sports. Moreover, PSE is a precursor molecule used in the production of illegal methamphetamine. These latter two uses have led federal and state governments to restrict sales of the drug. A few states have gone as far as requiring a prescription for it.

Potential substitutes for PSE, some of which are commonly available, have not proven to be as effective as PSE for alleviating sinus congestion. The ethical issue, then, is how to allow the continued use of the drug by those who genuinely need it, while at the same time preventing its illicit uses. Government monitoring of PSE sales brings up the related issue of privacy since the government is not, in effect, tracking actual criminals but rather persons who have only the *potential* for committing a crime. This has led to the unintended consequence of some people being prosecuted simply because they purchased more than the legally allowable amount of decongestant.

11. Answers will vary. Check to see that the student has (1) identified the principles of Scripture related to the sale and use of pseudoephedrine; (2) identified possible outcomes and motivations of heeding or ignoring these principles; and (3) proposed one or more actions to take in response to the issue. Some suggested ideas for the three paragraphs follow.

Paragraph 1: First Corinthians 6:12–20 teaches an important principle that applies to drugs. The things that we can partake of that are not proscribed in Scripture are lawful as long as they do not dominate us. Our bodies belong to God, and we may not use them as we want. Since Paul told Timothy in 1 Timothy 5:23 to use "a little wine for thy stomach's sake," implying a medicinal function, a biblically acceptable option includes using PSE to mitigate symptoms of congestion. It certainly is not acceptable to use the drug for illicit purposes that often lead to addiction.

Paragraph 2: Heeding Scriptural principles would result in a person avoiding uses of PSE that might lead to addiction. This would allow that person to be controlled by the Holy Spirit and God's Word rather than by the fleshly desires that result from substance abuse. This should come from our love for God and faith in His promises to lead us and direct us through His Word.

Also, a Christian's love for others would cause him to be mindful of his testimony and how his misuse of PSE might cause another brother to stumble into addiction. Using PSE properly would help a person to lessen the symptoms of congestion in order to sleep better and allow his body to fight off the illness that may be hindering his productiveness.

Paragraph 3: A student should use PSE according to its instructions only. He could advocate for policing the misuse of PSE and even find an organization committed to this cause to promote or donate to.

 **Ethics Rubric**

A rubric to grade this essay is available in Appendix H.

CHAPTER 5 REVIEW

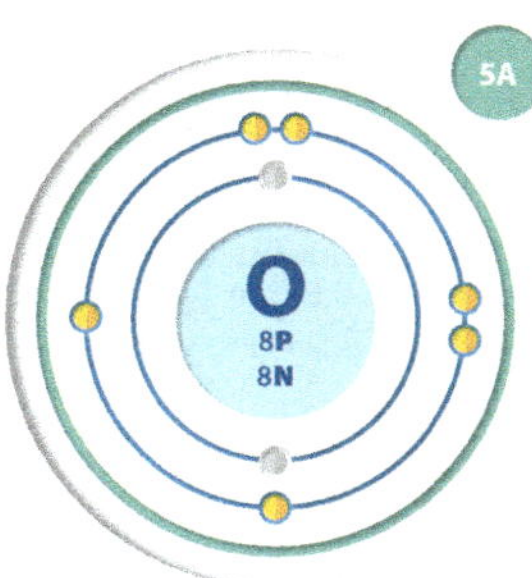

5A PRINCIPLES OF BONDING

- A chemical bond is an electrostatic attraction that exists between atoms due to the sharing or exchanging of valence electrons.

- Molecules form when atoms bond with other atoms in an attempt to gain stability.

- With few exceptions, atoms are most stable when they have eight valence electrons; this is known as the octet rule.

- The properties of compounds are usually very different from the properties of the elements from which they are formed.

- Atoms of some elements will bond with each other to form molecules, such as the diatomic elements.

5A Terms

chemical bond	96
octet rule	97

5B TYPES OF BONDS

- Covalent bonds are produced when atoms must share electrons due to having similar electronegativities. A group of two or more covalently bonded atoms is a molecule.

- When atoms have very different electronegativities, the more weakly electronegative atom donates electrons to the stronger. The resulting ions are held together in a crystal structure in a ratio represented as a formula unit.

- The atoms of metals bond by mutually sharing all of their valence electrons, forming an electron sea.

- A covalent bond consists of a bonding pair of electrons. Atoms can share up to three pairs of electrons, forming either single, double, or triple bonds.

- Atoms having different electronegativities share electrons unequally, resulting in polar bonds.

- An uneven distribution of electric charge on a molecule produces a polar molecule.

5B Terms

covalent bond	99
ionic bond	100
formula unit	100
metallic bond	100
diatomic molecule	101
Lewis structure	101
polarity	103

5C WRITING CHEMICAL FORMULAS

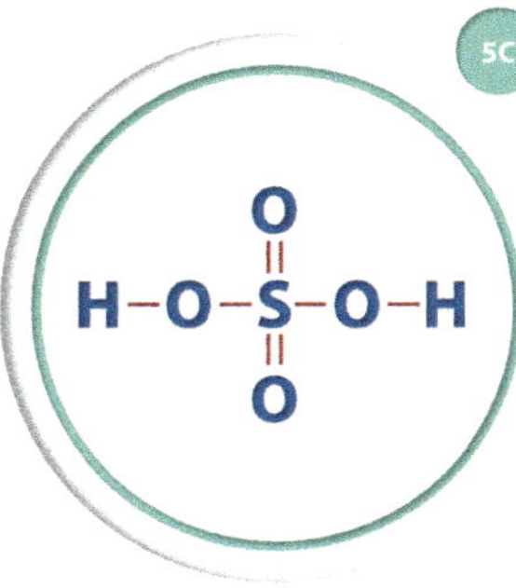

- The chemical formulas for some ionic compounds can be predicted by the location of the component elements on the periodic table.

- Oxidation states are used to predict chemical formulas for both ionic and covalent compounds.

- In naming and writing binary ionic compounds, the cation is identified first, followed by the anion.

- Ionic compounds may contain polyatomic ions, which are groups of covalently bonded atoms that act as single ionized particles.

- The naming and writing of binary covalent compounds take into account the locations of the elements on the periodic table as well as the number of atoms of each element.

5C Terms

chemical formula	108
binary compound	109
polyatomic ion	112
oxidation state	113

CHAPTER REVIEW ANSWERS

Recalling Facts

1. A chemical bond is an electrostatic attraction that exists between atoms resulting from the sharing or transferring of valence electrons. *(p. 96)*

2. False. *(p. 98)* Typically, compounds have significantly different properties than the elements that they contain.

3. A covalent bond is the attraction between two atoms due to the sharing of valence electrons. *(p. 99)*

4. a formula unit *(p. 100)*

5. A metallic bond occurs between metals. Because metals have low electronegativities, they easily release their valence electrons. The metal atoms form a lattice structure and share all their valence electrons. *(p. 100)*

6. Multiple (double or triple) bonds exist when a pair of atoms shares more than one pair of electrons. *(p. 102)*

7. False. *(p. 102)* A triple bond is the sharing of three pairs of electrons by two atoms.

8. A polar covalent bond forms between two atoms with moderately differing electronegativities. The valence electrons between the atoms are shared but the sharing is unequal. The electrons are held closer to the atom with the higher electronegativity. *(p. 103)*

9. ionic *(p. 104)*

10. A chemical formula tells the identity and number of atoms of each element in the compound. *(p. 108)*

11. The oxidation state indicates the electric charge gained or lost by that element when it forms a compound. *(p. 113)*

Understanding Concepts

12. No. Atoms of the same element can sometimes become stable by bonding with each other, such as in diatomic molecules of oxygen or nitrogen. *(p. 97)*

13. Answers will vary. Hydrogen can become stable by either gaining one electron or losing its one valence electron. Helium already has a full first energy level. Lithium, beryllium, and boron can all become stable by losing the electrons in their second (valence) energy level. *(p. 97)*

14. **a.** gain

 b. lose

 c. lose *(all pp. 97, 100)*

15. The electronegativity of two iodine atoms is the same, therefore the two atoms must share electrons in a covalent bond. *(pp. 99, 101)*

16. In students' diagrams, the single and double bonds between sulfur and both oxygen atoms should be circled. *(p. 102)*

17. Oxygen is more electronegative than sulfur and will exert a greater pull on the shared electrons, so all the bonds shown will be polar. *(p. 103)*

18. Sulfur dioxide is polar since the oxygen atoms will pull shared electrons toward their side of the molecule, resulting in that side having a negative charge. The sulfur side of the molecule will be positive. *(p. 103)*

19. Each sulfur atom needs two additional valence electrons to satisfy the octet rule, so they will share two pairs of electrons and form a double bond. *(p. 102)*

20. Since the electronegativity of the two atoms is the same, they will share electrons equally, producing a nonpolar molecule. *(p. 103)*

21. fluorine; strontium *(p. 104)*

CHAPTER REVIEW QUESTIONS

Recalling Facts

1. What is a chemical bond?

2. (True or False) Compounds often have physical and chemical properties that are similar to those of the elements of which they are made.

3. What is a covalent bond?

4. What is used to represent the ratio of ions in a group of ionically bonded atoms?

5. Describe a metallic bond.

6. Define the term *multiple bond*.

7. (True or False) A triple bond holds three atoms together.

8. What is a polar covalent bond?

9. If two atoms have very different electronegativities, will they form an ionic bond or a covalent bond?

10. What does a chemical formula tell about a compound?

11. What does an oxidation state indicate about an element?

Understanding Concepts

12. Must atoms of one element always bond with atoms of a different element in order to become stable? Explain.

13. Describe an example of an element that does not need to meet the octet rule requirement in order to become stable.

14. Identify whether each of the following elements would be more likely to gain or lose electrons in order to form an ionic bond.

 a. bromine

 b. rubidium

 c. strontium

15. Will two iodine atoms form an ionic or covalent bond? Explain.

Refer to the Lewis structure for sulfur dioxide below to answer Questions 16–18.

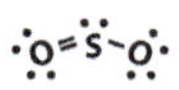

$$:\!\ddot{O}\!=\!\ddot{S}\!-\!\ddot{O}\!:$$

16. Copy the Lewis structure for sulfur dioxide onto your paper, then circle all of the shared pairs of electrons.

17. Knowing that the electronegativity of oxygen and sulfur are 3.4 and 2.6 respectively, do you think that sulfur dioxide contains polar or nonpolar bonds? Explain.

18. Actual sulfur dioxide molecules are bent, just as the Lewis structure shows. Are molecules of sulfur dioxide polar or nonpolar? Explain.

19. If two sulfur atoms bond with each other, will they form a single bond or a double bond? Explain.

20. Will the sulfur molecule in Question 19 be polar or nonpolar? Explain.

21. In a compound of strontium and fluorine, which element forms an anion? Which forms a cation?

22. Which of the following compounds are salts? Explain.

 a. selenium dioxide

 b. barium iodide

 c. sulfur dibromide

 d. triphosphorus pentanitride

23. Name the elements and the number of atoms of each element in the chemical formula $C_5H_{11}NO_3$.

24. Which of the following Lewis structures is the correct one for beryllium chloride? Explain your choice.

 a. 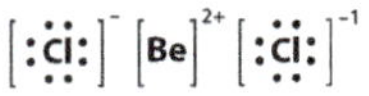 $\left[:\!\ddot{Cl}\!:\right]^{-}\ \left[Be\right]^{2+}\ \left[:\!\ddot{Cl}\!:\right]^{-1}$

 b. 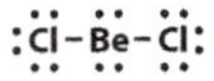 $:\!\ddot{Cl}\!-\!\ddot{Be}\!-\!\ddot{Cl}\!:$

25. Predict the names and formulas for compounds containing the following elements or polyatomic ions.

 a. sodium and bromine

 b. calcium and chlorine

 c. barium and nitrate

 d. nickel and fluorine

26. What is (are) the primary oxidation state(s) for each of the following elements?

 a. scandium

 b. sulfur

 c. zirconium

 d. iodine

27. Using primary oxidation states (see table on page 113), predict the names and formulas for two different compounds composed of nitrogen and oxygen.

28. Are the compounds you predicted in Question 27 the only possible compounds of nitrogen and oxygen? Explain.

REVIEW

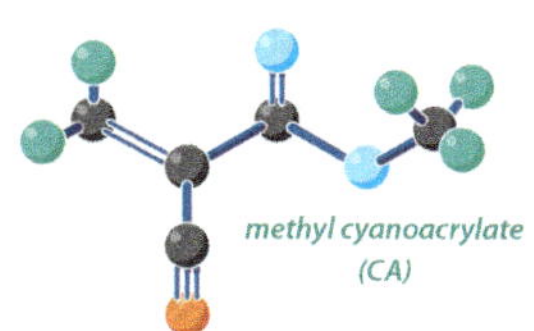

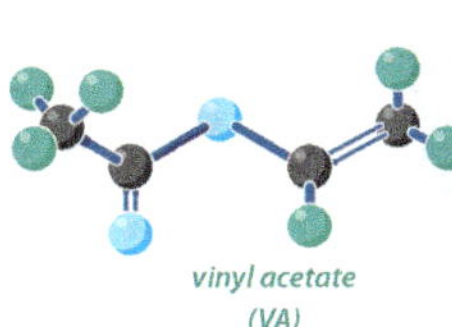

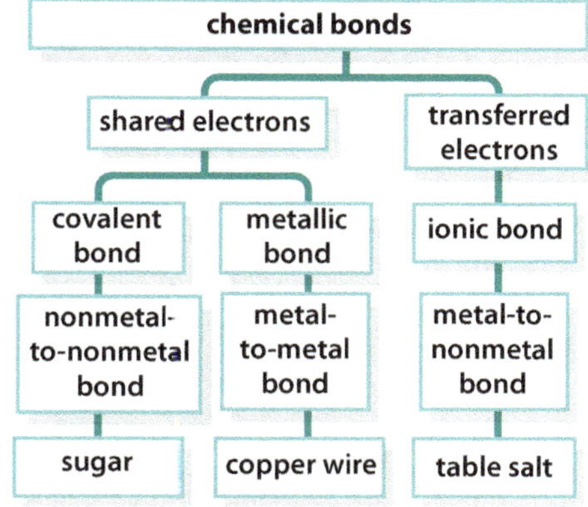

---

22. b.; Salts are formed when a metal and nonmetal form an ionic compound. Only barium iodide is an ionic compound. *(p. 105)*

23. five atoms of carbon, eleven atoms of hydrogen, one atom of nitrogen, and three atoms of oxygen *(p. 108)*

24. a.; The compound is ionic and the beryllium cation should show no valence electrons. *(pp. 105, 111)*

25. **a.** sodium bromide, NaBr

   b. calcium chloride, $CaCl_2$

   c. barium nitrate, $Ba(NO_3)_2$

   d. nickel(II) fluoride, NiF_2
   (all pp. 109–12)

26. **a.** +3

   b. +6, +4, +2, −2

   c. +4

   d. +7, −5, +3, +1, −1 *(all p. 113)*

27. dinitrogen trioxide (N_2O_3) and dinitrogen pentoxide (N_2O_5) *(pp. 113, 115)*

28. No. Since these compounds were predicted from only the primary oxidation states, there could be other compounds that are based on the other oxidation states. *(pp. 113, 115)*

Critical Thinking

29. Magnesium and chlorine are highly reactive, but together they form magnesium chloride, a very stable salt. Magnesium is a solid gray metal and chlorine is a poisonous yellow-green gas. When combined, they become a hard, white crystalline solid.

30. See hierarchy chart below. Other arrangements are possible. Check to make sure that students' charts correctly link terms in a logical manner. *(pp. 96–100)*

31. Because the molecules in cooking sprays are nonpolar, they are not attracted to the surfaces of cookware. *(p. 103)*

32. two atoms of aluminum, three atoms of sulfur, and twelve atoms of oxygen (*Note:* Students may answer only one sulfur atom and four oxygen atoms, but there are three sulfate ions.) *(pp. 108, 112)*

33. If the electronegativities of all the elements were the same, then any atom could bond at random with any other atom. The formation of compounds would be random and unpredictable.

34. The differences in properties are due to the different arrangements of the atoms and bonds within the three molecules.

35. Answers will vary. Students may note, for example, that the molecules are roughly the same size, have approximately the same number of atoms, and are made of similar elements. They may notice that each molecule contains a chain of five atoms, one of which is oxygen.

36. Answers will vary. Students may note that the arrangement of the atoms in each molecule is different and that methyl cyanoacrylate contains a triple-bonded nitrogen atom that vinyl acetate does not. They may notice that the oxygen in CA is near an end carbon, whereas in PVA it is in the middle of the chain.

The striking black-and-white pattern of a skunk's fur serves as a warning of its second level of defense: its extremely malodorous spray. Warning coloration such as the skunk's is called aposematism.

One of the most distinctive smells is that of a skunk. A collection of organic chemicals produces its pungent odor.

CHAPTER 6

The Chemistry of Life

SKUNK-SAVING SCENT

The phrase "lifesaving chemical" may bring many things to mind. Some of you may know people who have undergone chemotherapy. Most of you have taken medicine when you were sick, and everyone needs the compounds in foods to stay healthy. Some people need to carry lifesaving medicines, such as insulin (for diabetes) or epinephrine (for allergic reactions), with them at all times.

Skunks carry a supply of lifesaving chemicals with them too. God designed skunks with a highly effective defense system—skunk scent. A skunk warns a threat by hissing, stomping its feet, and doing a little dance. If the warning is not heeded, a skunk will spray from two glands at the base of its tail, aiming at the face of an assailant. In addition to its telltale odor, skunk scent is also an eye irritant that can temporarily blind an attacker. The mixture is so effective that most predators avoid skunks at all costs. The great horned owl is the only predator that routinely preys on skunks. While almost everyone seeks to sidestep its stench, skunks certainly savor the sweet scent of this lifesaving chemical.

 **Lifesaving Molecules**

Ask students whether they know anyone who takes insulin or who carries an epinephrine delivery system. Ask how these medications are related to a skunk's smell. *(They are all organic molecules.)*

- Relate molecular structure to physical and chemical properties.
- Use models to understand the structures of organic substances.
- Compare classes of organic compounds.

Overview

Chapter 6 is an enrichment chapter that covers organic compounds. This subject will be studied again in chemistry. Consider covering this chapter if you have sufficient time and have students who are interested in a career in chemistry or medicine.

Lab Activities

Lab 6A: *Sticky Business*—In this inquiry lab activity, students work to improve a glue recipe.

Lab 6B: *Milking Chemistry*—In this lab activity, students test food for the presence of proteins.

6A | ORGANIC COMPOUNDS

Why is carbon so important?

6.1 DEVELOPMENT OF ORGANIC CHEMISTRY

Most of us have seen the term *organic* in the grocery store. Up and down the aisles we are surrounded by signs for organic carrots, lettuce, tomatoes, and much more. In this context, the term is meant to make us think that these are healthier options. This may be true because often these products are grown through organic farming methods, such as using only pesticides and fertilizers that come from natural, living sources—other plants and animals.

In chemistry, the term *organic* began with a similar meaning because all the then-known organic compounds came from living things. Over time we have changed our definition, but we are getting ahead of ourselves. Today organic chemistry touches every aspect of our lives. All around your house you can find many products that are the work of organic chemists. Plastics, fuels, dyes, drugs, and many other goods are made using modern organic chemistry. But this is not a new field of science.

Since ancient times, people have used many compounds from plants and animals. They used these as medicines, poisons, and dyes. Because

TEACHING THE MATERIAL

ESSENTIAL QUESTION

Why is carbon so important?

OBJECTIVES

- 6A1 Define *organic compound* and *hydrocarbon*.
- 6A2 Explain how isomers are formed.
- 6A3 Compare saturated and unsaturated hydrocarbons.

RESOURCES

Demonstrating Straight Chains

Mini Lab: *Modeling Hexane Isomers* (p. 133); Lab 6A: *Sticky Business*—Inquiring into Glues

they believed that these products came from organisms, scientists called them *organic compounds*. The study of these compounds became known as *organic chemistry*, a term coined by Jacob Berzelius in the early 1800s.

Scientists thought that these compounds contained the "vital force," and that only living organisms could make them. While many scientists studied these compounds, no one had been able to produce them. In 1828, German chemist Friedrich Wöhler inadvertently made urea, an organic compound, in the laboratory. Through this accidental discovery, Wöhler refuted the vital force theory and started a period of synthesizing many organic compounds.

Today we still divide chemistry into organic and inorganic chemistry, but our definitions have changed. Chemists originally defined an organic compound as any compound produced by a living organism that therefore contained the vital force. Today, scientists define an **organic compound** as one that contains carbon. While some compounds (e.g., carbon dioxide and carbonates) have carbon in them but are not considered organic, all organic compounds contain carbon. So today, organic chemistry is the study of carbon-containing compounds.

 ### Models

Some people see the history of science as throwing off old superstitions such as the vital force theory and embracing the truth. But it's important to remember that a lot of good work was done by chemists who believed in the vital force theory. It's also important to remember that current models are not so much true as they are workable. They may very well be replaced one day by models that explain observations better and have greater predictive power.

⁉️ Why the Exceptions?

Some students might wonder why carbon dioxide and carbonates are not considered organic compounds. While the reason may be partly based on historical convention, it also reflects the fact that nearly all organic compounds contain carbon and hydrogen. Since these compounds do not contain hydrogen, they, along with cyanide compounds, are generally considered inorganic compounds.

STRATEGIES

Class Opener: Use the Lifesaving Molecules teacher note on page 123 to begin a discussion of organic chemistry.

Biblical Worldview Shaping: Use the Models teacher note on this page to lead students to understand that new scientific models are created to be more workable and are not to be seen as a progression toward greater truth.

Discussion: Use the Active Learning: Carbon's Four Bonds teacher note on page 126 to introduce a discussion on how carbon's number of bonding sites makes it ideal for forming many different molecules.

Demonstration: Use the Demonstrating Straight Chains teacher note on page 126 to give a good visual of a straight chain hydrocarbon.

Visual: Use the Visualizing Organic Compounds teacher note on page 127 to help students get an understanding of the structure of organic compounds.

(continued)

As a way to show the effect of carbon's four bonds on molecular structure, select three volunteers to represent carbon and three to represent oxygen. Give each "carbon" a length of ribbon approximately 2 m long to tie loosely around his waist so that the ribbon forms two tails at opposite sides of his body. Inform the other students that they represent hydrogen atoms.

Have one carbon and one oxygen stand in the center of the room. Instruct the hydrogens to bond where possible. Carbon bonding sites are represented by the hands and the two ribbon tails of each carbon. Count with students the number of "atoms" in each "molecule."

Repeat the process with two carbons bonded and two oxygens bonded. Then compare three-carbon molecules with three-oxygen molecules.

After the students return to their seats, discuss how the number of bonding sites of the primary molecules (carbon and oxygen) affected the complexity of the molecule. Students should note that while a three-carbon chain has twice as many open bonding sites as a single carbon atom (eight versus four), a three-oxygen chain has the same number of bonding sites as a single oxygen atom (two).

Straight but Not Straight

Many students may struggle at first with the fact that a straight chain can have turns in it and doesn't necessarily look straight. Over the course of the chapter, you may need to emphasize several times the description of a straight chain as having a single continuous chain of carbon atoms.

Demonstrating Straight Chains

Preparation: Prepare a straight chain of carbon atoms from a molecular modeling kit. (You can substitute a string of beads.) Prepare another chain that has at least one branch.

Performance: Show students the twisted chain and then straighten it out to show them that the chain was straight the entire time. Show them the branched chain and point out that it cannot be pulled into a straight chain.

(continued)

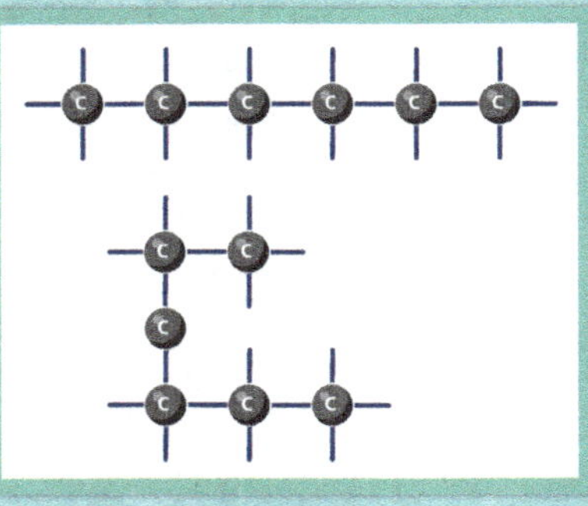

6.2 VERSATILE CARBON

Bonding

There are millions of organic compounds. Just as a small number of types of Lego™ blocks is the basis for many structures, so the atoms of one element—carbon—provide the framework for all organic matter. This is because the structure of carbon atoms makes them extremely versatile, allowing for great variety in the compounds that are possible.

Carbon is a Group 14 element, which means that it has four valence electrons. These electrons allow carbon atoms to bond with up to four other atoms. Most organic compounds form on a framework of carbon atoms bonded to each other, like a carbon backbone, with bonds that are quite strong. The structure of carbon, with its four valence electrons, also allows it to form single, double, and triple bonds. These three properties of carbon atoms—four valence electrons that can bond with

MOLECULAR ARRANGEMENT

STRAIGHT CHAINS

The most basic form of an organic molecule is a *straight chain*.

Straight chains consist of a single continuous series of any number of carbon atoms bonded to each other. The term "straight" doesn't imply a straight line, just a single path of carbon atoms from one end of the molecule to the other. Straight chains will often have angles and bends in them.

While the chain at top left is clearly a straight chain, the one at bottom left is also a straight chain, even though it has two turns in it. In both cases there is a single continuous path of six carbon atoms.

BRANCHED CHAINS

Even if the types and number of atoms remain the same, changing how the atoms are arranged will result in molecules with different properties.

Organic molecules can be made more complex by including branches. A *branched chain* has carbon atoms that connect to other carbon atoms that are not on the ends of a straight chain. So there is more than one path that you could follow when counting the carbon atoms.

Like the straight chains, these branched chains (right) still contain six carbon atoms. But one of the carbon atoms has been removed from the continuous chain and now branches off from the main chain. Because the branched arrangement is different from a straight chain, it behaves differently than the straight chain from which it derives.

TEACHING THE MATERIAL

Formative Assessment: Use the Formative Assessment: Numerical Prefixes teacher note on page 128 to check students' knowledge and understanding of this basic feature of organic compound names.

Discussion: Use the Active Learning: Modeling Hydrocarbons teacher note on page 130 to introduce a discussion on how the ratio of carbon atoms and hydrogen atoms differs among the three types of hydrocarbons.

Formative Assessment: Use the Formative Assessment: Alkenes teacher note on page 130 to check students' knowledge and understanding of alkenes.

Formative Assessment: Use the Formative Assessment: Alkynes teacher note on page 131 to see how well students understand alkynes.

Ticket Out the Door: Instruct students to draw a three-carbon alkane and correctly write its name.

four other atoms, strong carbon–carbon bonds, and the potential for multiple bonds—enable it to combine in many ways with other atoms to form many substances. While other members of Group 14 have similar characteristics, carbon's smaller size allows it to form stronger C–C bonds compared with Si–Si bonds, for example. But silicon's closeness to carbon makes it a great candidate for building blocks of life in science fiction stories!

Arrangement

Most organic compounds have a structure that is based on multiple carbon atoms. In Chapter 4 we learned that the structure of an atom determines the properties of the elements. Similarly, the atoms that make up an organic molecule and the arrangement of those atoms dictate the properties of the molecule. The framework of carbon atoms can take one of three main forms—straight chains, branched chains, or rings.

RINGS

The third main form for organic molecules is a *ring*, which is made by connecting the two ends of a continuous chain.

Notice at right that a straight chain of six carbon atoms has been turned into a ring of six carbon atoms by connecting the two carbon atoms at the ends of the chain. Rings can also have branches (bottom left), or they can be a branch attached to a chain (bottom right).

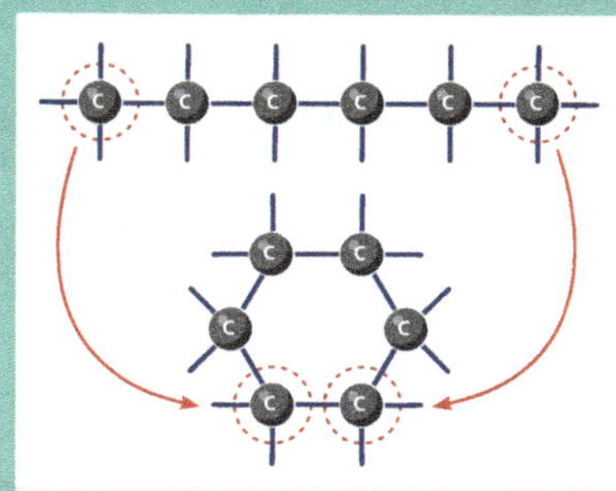

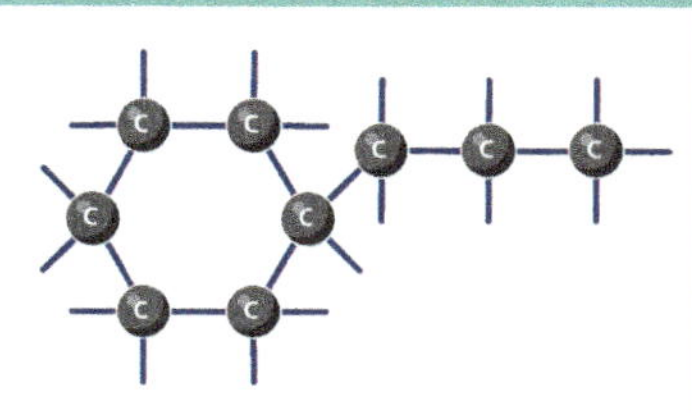

Above is a three-carbon branch attached to one of the carbon atoms in a six-carbon ring.

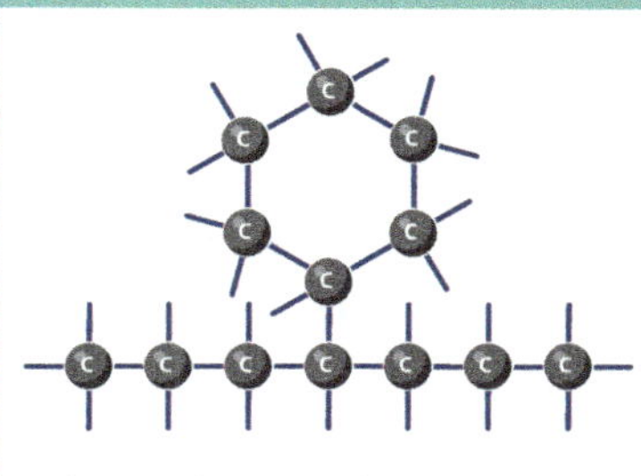

In the image above, a six-carbon ring is a branch off the fourth carbon atom in a seven-carbon chain.

1. Did either chain look straight at the beginning? *(No)*

2. Could either chain be pulled into a straight chain? *(Yes, one but not the other)*

3. Give the organic chemistry definition of a straight chain. *(A straight chain is a continuous series of carbon atoms.)*

Differentiated Instruction

Many students may benefit from multiple ways of discussing and visualizing the different types of hydrocarbons and substituted hydrocarbons. Having students write down basic formulas (e.g., a two-carbon example from each class) will help many students remember them better.

Visualizing Organic Compounds

Many students will be helped by seeing 3D models of different hydrocarbon molecules. A molecular modeling kit is invaluable for this. If not available, polystyrene balls and toothpicks or toy construction kits can be substituted.

Write the following chemical names on the board (except for Questions 4 and 5) as you ask the questions.

1. State the number of carbon atoms in heptane. *(seven)*

2. State the number of carbon atoms in nonene. *(nine)*

3. State the number of carbon atoms in butyne. *(four)*

4. What prefix indicates that a hydrocarbon molecule has two carbon atoms? *(eth-)*

5. What prefix indicates that a hydrocarbon molecule has six carbon atoms? *(hex-)*

6. Other than the prefixes, what do you notice about the names *heptane*, *nonene*, and *butyne*? *(They also have different endings.)*

7. What do you think the endings on those three names indicate? *(Answers will vary. Students should hypothesize that the endings indicate something about the structure of the molecules.)*

Formulas

To help students distinguish between the different ways to refer to hydrocarbons, show them the name, molecular formula, and structural formula of some of the simpler hydrocarbons.

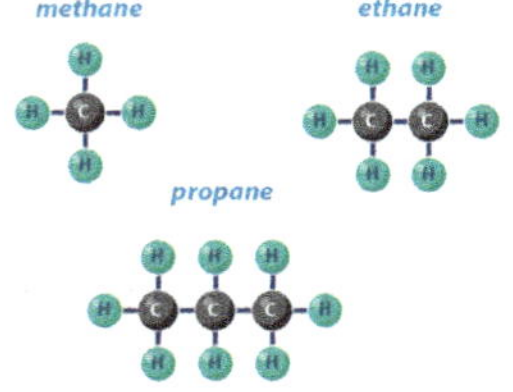

6.3 HYDROCARBONS

Organic compounds always contain carbon. Notice that carbon atoms, with their C–C bonds, are the framework for organic molecules. In addition to carbon, hydrogen is the most common element in organic compounds. Compounds made of only carbon and hydrogen atoms are called **hydrocarbons**.

We can refer to hydrocarbons by their names, molecular formulas, or structural formulas. A hydrocarbon's name starts with a prefix for the number of carbon atoms in the longest chain or ring. Some of these prefixes will be familiar to you; in the table at left, you likely know *pent-*, *hex-*, *oct-*, and *dec-*. Do you recognize *eth-*, *prop-*, or *but-* (BYOOT)? The ending of a hydrocarbon's name indicates the types of bonds between the carbon atoms; we'll discuss that a little later.

Molecular formulas are the same chemical formulas that you learned in Chapter 5. A *structural formula* is a drawing that shows not only the atoms but also the bonds and their arrangement in the molecule. Though more complex, the structural formulas are similar to the Lewis structures that you learned about in Chapter 5. The examples shown on pages 126–27 for straight chains, branched chains, and rings were just the carbon-atom frames for structural formulas; to emphasize the carbon backbones and basic shapes, they didn't include any other atoms bonded to the carbon atoms.

CLASSIFYING HYDROCARBONS

One way to classify hydrocarbons is by the types of bonds between the carbon atoms.

SATURATED HYDROCARBONS (ALKANES)

If only single bonds exist between the carbon atoms in the molecule, we call it a **saturated hydrocarbon**. Because the carbon atoms have only single bonds, these hydrocarbons have the greatest number of hydrogen atoms possible. So a saturated hydrocarbon is saturated with hydrogen.

Alkanes are saturated hydrocarbons. The names of alkanes end with *-ane*. The simplest alkane, methane, consists of one carbon atom bonded to four hydrogen atoms. Its name comes from *meth-* (one carbon) and *-ane* (only single bonds).

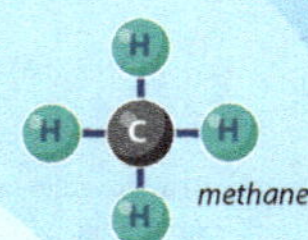

Alkanes have low reactivity. Low-number alkanes are natural gases and liquid fuels. We use high-number alkanes to produce plastics, waxes, and asphalt.

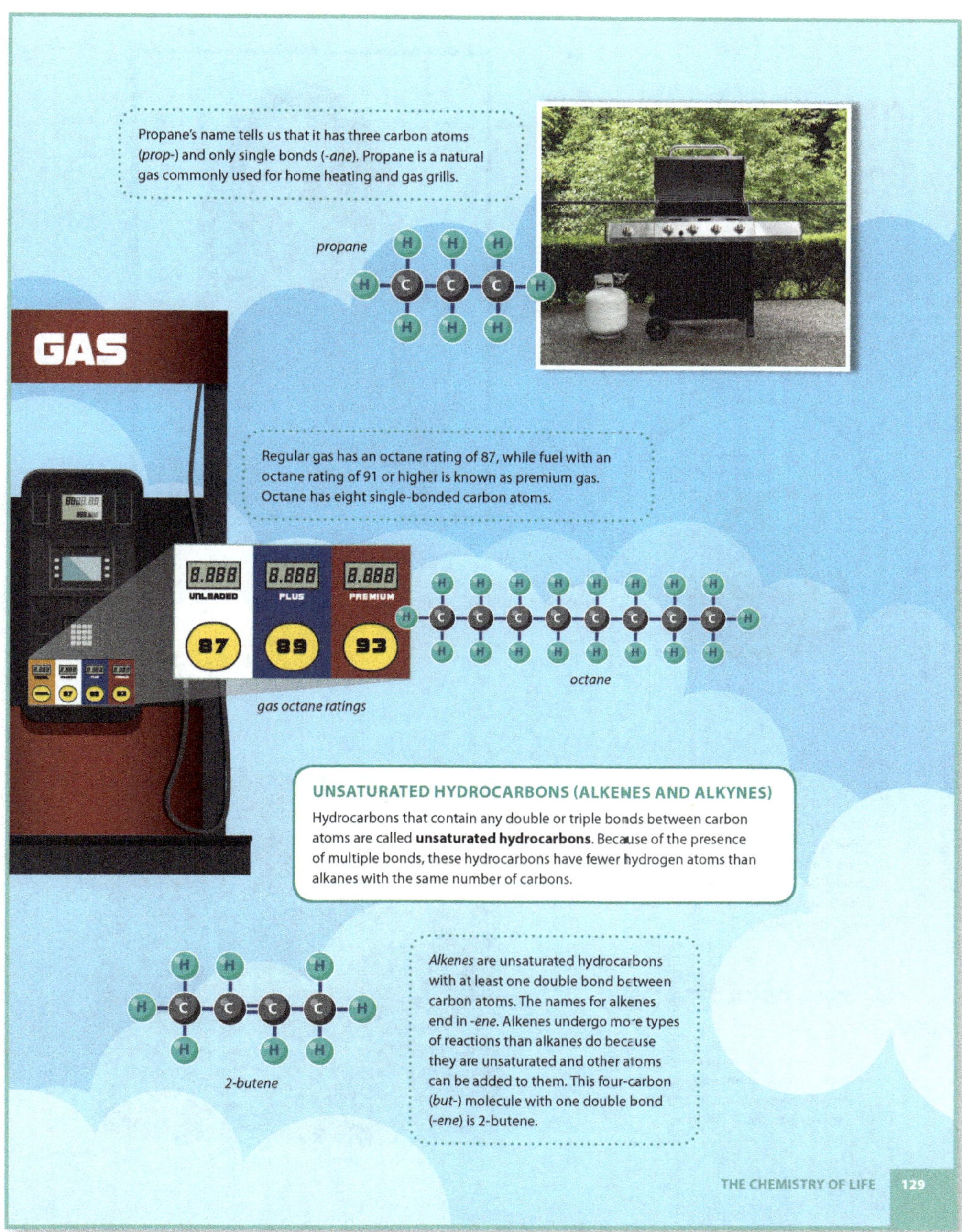

UNSATURATED HYDROCARBONS (ALKENES AND ALKYNES)

Hydrocarbons that contain any double or triple bonds between carbon atoms are called **unsaturated hydrocarbons**. Because of the presence of multiple bonds, these hydrocarbons have fewer hydrogen atoms than alkanes with the same number of carbons.

Alkenes are unsaturated hydrocarbons with at least one double bond between carbon atoms. The names for alkenes end in *-ene*. Alkenes undergo more types of reactions than alkanes do because they are unsaturated and other atoms can be added to them. This four-carbon (*but-*) molecule with one double bond (*-ene*) is 2-butene.

Active Learning: Modeling Hydrocarbons

Use the same setup as described for the molecules containing carbon in the Active Learning: Carbon's Four Bonds note on page 126. Have students form a model of a butane molecule, then one of a butene molecule. (The position of the double bond is not important.) Finally, have them form a model of a butyne molecule.

After the students return to their seats, lead a discussion about the ratio of hydrogen atoms to carbon atoms in each type of molecule. Lead them to recognize that each double bond decreases the number of hydrogens in that molecule by two and that each triple bond decreases the number of hydrogens in that molecule by four.

Formative Assessment: *Alkenes*

1. How many carbon atoms are in pentene? *(five)*

2. How many hydrogen atoms are in pentene? *(ten)*

3. How many single carbon–carbon bonds are in pentene? *(three)*

4. How many double carbon–carbon bonds are in pentene? *(one)*

5. How many triple carbon–carbon bonds are in pentene? *(none)*

6. Describe a heptene molecule. *(A heptene molecule has seven carbon atoms with one double bond. Heptene molecules have fourteen hydrogen atoms.)*

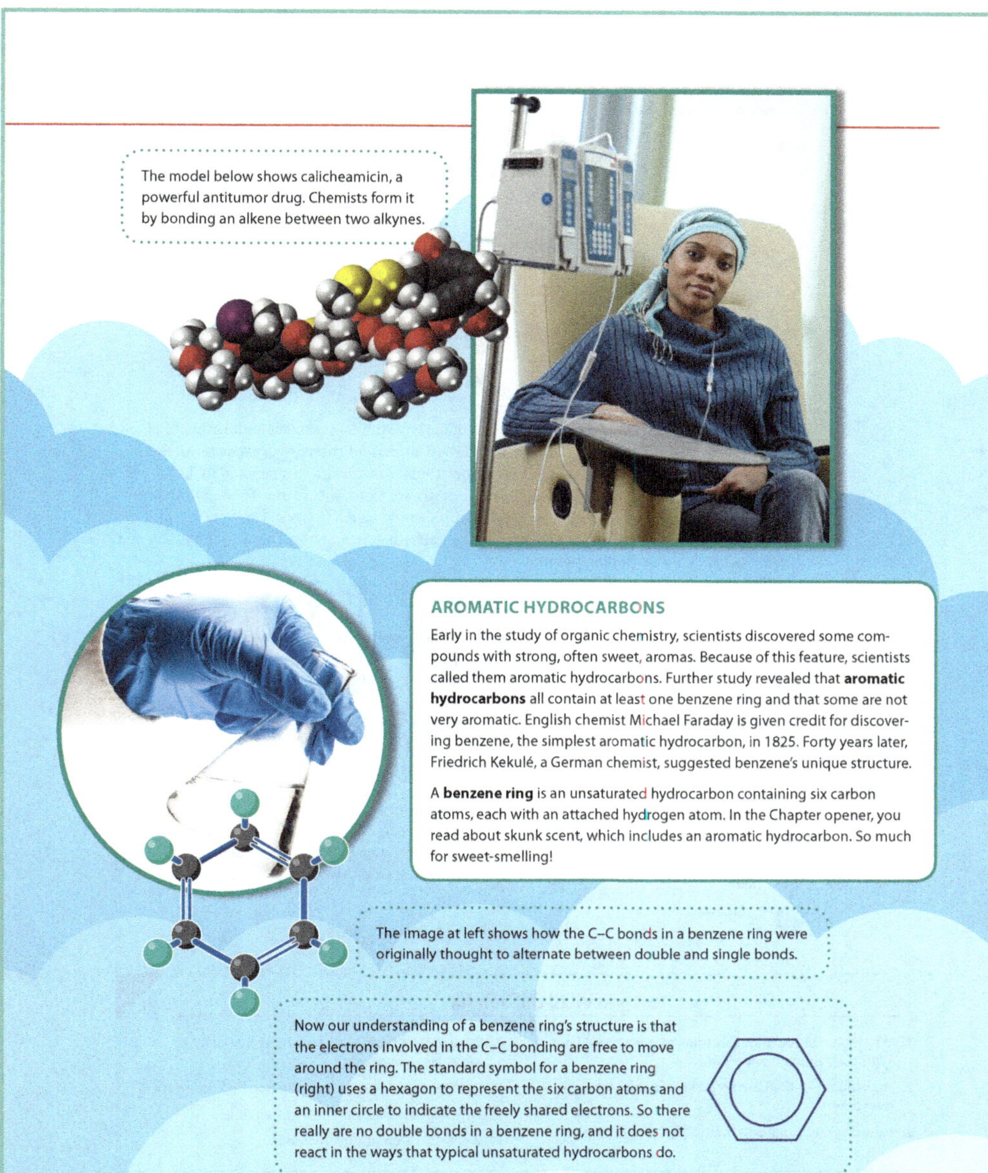

The model below shows calicheamicin, a powerful antitumor drug. Chemists form it by bonding an alkene between two alkynes.

AROMATIC HYDROCARBONS

Early in the study of organic chemistry, scientists discovered some compounds with strong, often sweet, aromas. Because of this feature, scientists called them aromatic hydrocarbons. Further study revealed that **aromatic hydrocarbons** all contain at least one benzene ring and that some are not very aromatic. English chemist Michael Faraday is given credit for discovering benzene, the simplest aromatic hydrocarbon, in 1825. Forty years later, Friedrich Kekulé, a German chemist, suggested benzene's unique structure.

A **benzene ring** is an unsaturated hydrocarbon containing six carbon atoms, each with an attached hydrogen atom. In the Chapter opener, you read about skunk scent, which includes an aromatic hydrocarbon. So much for sweet-smelling!

The image at left shows how the C–C bonds in a benzene ring were originally thought to alternate between double and single bonds.

Now our understanding of a benzene ring's structure is that the electrons involved in the C–C bonding are free to move around the ring. The standard symbol for a benzene ring (right) uses a hexagon to represent the six carbon atoms and an inner circle to indicate the freely shared electrons. So there really are no double bonds in a benzene ring, and it does not react in the ways that typical unsaturated hydrocarbons do.

 ### Formative Assessment: *Alkynes*

1. How many carbon atoms are in propyne? *(three)*

2. How many hydrogen atoms are in propyne? *(four)*

3. How many single carbon–carbon bonds are in propyne? *(one)*

4. How many double carbon–carbon bonds are in propyne? *(none)*

5. How many triple carbon–carbon bonds are in propyne? *(one)*

6. Describe an octyne molecule. *(An octyne molecule has eight carbon atoms with one triple bond. Octyne molecules have fourteen hydrogen atoms.)*

 ### Benzene

Although scientists no longer think that benzene has alternating double and single bonds, the ball-and-stick model showing the older concept is still sometimes used to represent benzene rings. While the circle representing the delocalized electrons inside the carbon hexagon ring represents the current model, both images are commonly used in scientific and educational literature.

Do you recall what determines the properties of molecules? If you thought of the atoms that make up the molecule and the arrangement of those atoms, then you're correct. If we rearrange the atoms in a hydrocarbon, thus changing its form, we change the way it behaves. Two molecules with the same molecular formula but whose structures differ are called **isomers**.

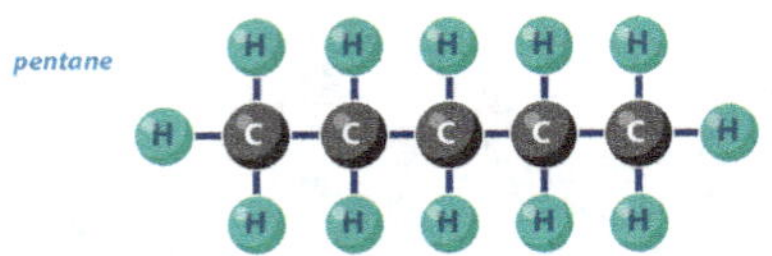

Both pentane and isopentane have the same molecular formula, C_5H_{12}—they each contain five carbon atoms and twelve hydrogen atoms. By moving one of the carbon atoms from an end of a pentane molecule and connecting it to form a branch, we form isopentane. This small change in structure produces significant changes in physical and chemical properties. Isopentane is more stable than pentane and has lower melting and boiling points.

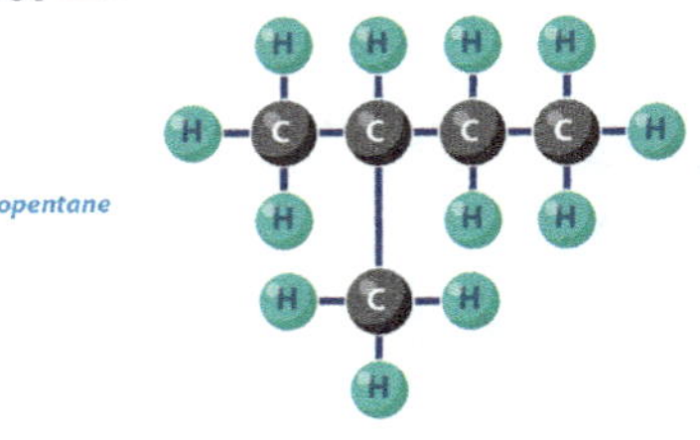

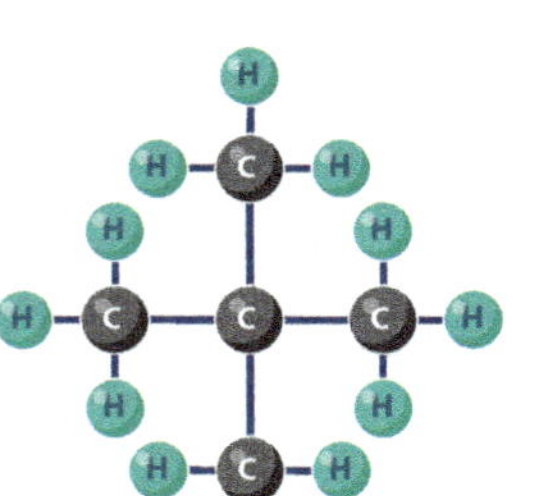

We can move another carbon atom from the end of isopentane and attach it as a second branch to form neopentane. This makes an isomer that is even more stable than isopentane. The varied properties of isomers allow scientists to select a particular isomer for a particular function.

6A | REVIEW QUESTIONS

1. Why did scientists originally think that man could not synthesize organic compounds?

2. How did the definition of *organic compound* change over time?

3. How do you recognize a branched chain molecule?

4. What are the three basic structures that organic molecules can have?

5. Compare saturated and unsaturated hydrocarbons.

6. Why does *benzene* have the *-ene* ending?

7. What is an aromatic hydrocarbon?

Lab 6A: *Sticky Business*

Once students have an understanding of the material in Section 6A, they should be ready to work through Lab 6A.

6A REVIEW ANSWERS

1. Scientists believed that the vital force, an aspect of nature, was needed. *(p. 125)*

2. Originally scientists defined organic compounds as compounds formed by living organisms. Today we define them as compounds containing carbon. *(p. 125)*

3. If you start at one end of a branched chain and count carbon atoms, you will come to a point at which there is more than one path to follow. *(p. 126)*

4. straight chain, branched chain, and ring *(pp. 126–27)*

5. Saturated hydrocarbons have only single bonds between carbon atoms, while unsaturated hydrocarbons have at least one double or triple bond between carbon atoms. *(pp. 128–29)*

6. The suffix *-ene* indicates that scientists originally believed that benzene contained double bonds. *(pp. 130–31)*

7. An aromatic hydrocarbon is a hydrocarbon with at least one benzene ring in it. *(p. 131)*

MODELING HEXANE ISOMERS

Many of us have used rubbing alcohol (isopropyl alcohol, bottom right) as a disinfectant or hand sanitizer. It also has a distinct smell. Isopropyl alcohol is an isomer of propanol. Isomers have the same molecular formula, but their structures differ. In this lab activity, you will build models of the isomers of hexane.

1. What does the name hexane tell us?

Procedure

(A) Draw the structural formula for hexane.

2. What is the molecular formula for hexane?

(B) Using the large foam balls to represent carbon atoms, the small foam balls to represent hydrogen atoms, and the toothpicks as bonds, create a model of hexane. Remember that each carbon atom can have only four bonds and each hydrogen atom can have only one.

(C) Rearrange your atoms to try to form an isomer of hexane.

(D) Draw the structural formula for the isomer.

3. How many carbon atoms and hydrogen atoms does this isomer have?

(E) Rearrange your atoms again to try to form as many isomers of hexane as you can.

Conclusion

4. How many isomers did you form?

Going Further

(F) Rearrange your atoms again to form a hexane ring. Remember that each carbon atom can have only four bonds.

5. Is this an isomer of hexane? Explain.

6. Why are isomers important?

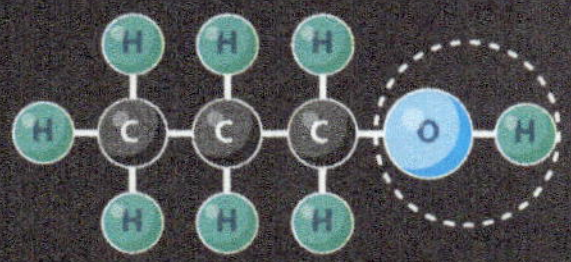

Essential Question:

How can we rearrange the atoms in a molecule?

Equipment

foam balls, 5 cm (6)

foam balls, 2.5 cm (14)

toothpicks (19)

The OH at the end of this propane molecule makes this propanol.

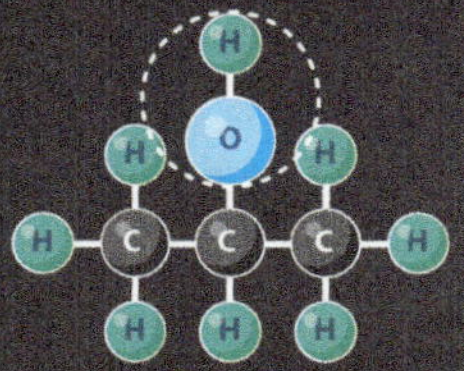

By moving the hydroxyl group (OH) to the middle of the molecule, we turn propanol into isopropyl alcohol.

Modeling Hexane Isomers

This lab activity introduces students to the basics of modeling hydrocarbons. The foam balls and toothpicks work well for this quick activity. Molecular modeling kits can also be used to good effect.

Answers

1. The name tells us that hexane is made of six carbon atoms with only single bonds between them. It also implies that there are fourteen hydrogen atoms.

2. C_6H_{14}

3. If done correctly, it will still have six carbon and fourteen hydrogen atoms.

4. Answers will vary. (There are five isomers of hexane.)

5. No. When the ring was formed, two of the hydrogen atoms had to be removed. Therefore, the molecular formula is C_6H_{12}.

6. Since isomers have different structures, they have different properties. This allows scientists more flexibility in choosing needed compounds on the basis of their properties.

E85

Ask students how many of them have seen E85 at the gas station. Then ask whether they know what it means. *(It is a mixture of 85% ethanol and 15% gasoline.)*

1. What about the name ethanol is familiar and what does that tell us? *(The eth- prefix tells us that there are two carbon atoms in ethanol. Some students may recognize the -an- root indicating an alkane (all single bonded carbon atoms).)*

2. What is new in the name ethanol? *(The -ol suffix is new).*

The *-ol* suffix on ethanol indicates that this is an alcohol, a substituted hydrocarbon. Use this as an introduction to substituted hydrocarbons in general.

Other Functional Groups

There are a number of functional groups in addition to those listed in the table, which shows three of the more common. Students will learn about some of the others in chemistry.

Effects of Functional Groups

Discuss some of the differences between propane and propanol. The most obvious difference is that propane is a gas and propanol is a liquid. Many other different characteristics can be found online. Search using the keywords "propane versus propanol."

6B Questions

- Can organic compounds contain other elements?
- What makes various substituted hydrocarbons different from each other?
- What are substituted hydrocarbons used for?

6B Terms

substituted hydrocarbon, functional group, alcohol, aldehyde, ketone

6B | SUBSTITUTED HYDROCARBONS

What other elements can be found in organic compounds?

6.5 SUBSTITUTION

Modern food processors are similar to blenders, but they can also do the work of five or six other machines. How can they do this? These machines have attachments that allow them to do different tasks. They have blades for slicing, shredding, chopping, grating, and mixing. One can attach other tools for juicing or even kneading bread dough. Just substitute one blade for another and you have a machine with vastly different properties.

Chemists can do the same thing with hydrocarbons. By replacing, or *substituting*, one of the hydrogen atoms with a different atom or group of atoms, they can create a **substituted hydrocarbon**. This new substance has properties that differ from those of the original hydrocarbon. The atom or group of atoms that scientists insert to form the substituted hydrocarbon is called a **functional group**, or a *substituent*. The same compounds that chemists make may also be found occurring naturally, however, and are not always man-made. Notice the common functional groups in the table on the left.

COMMON FUNCTIONAL GROUPS		
Name	hydroxyl	carbonyl	halogen
Structure	O – H	C = O	Cl , F , I , Br
Hydrocarbon	alcohol	ketone or aldehyde	alkyl halide
Example	propan-2-ol	methanal	chloroethane

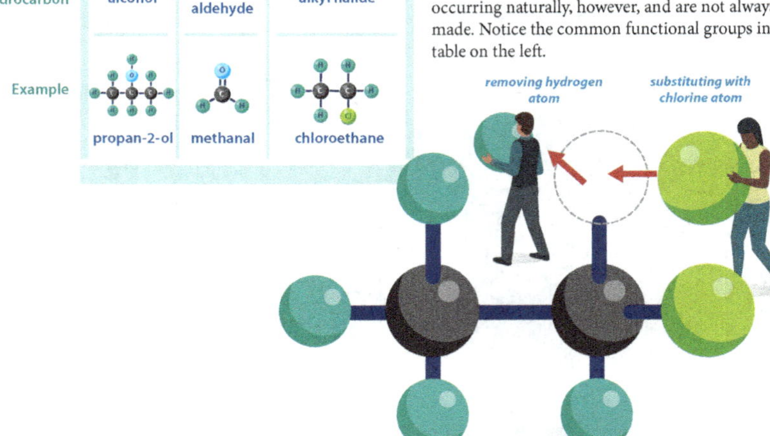

6.6 COMMON SUBSTITUTED HYDROCARBONS

Substituted hydrocarbons form when a hydrogen atom in a hydrocarbon is replaced with a functional group. While many functional groups contain oxygen, others include nitrogen, sulfur, or one of the halogens.

TEACHING THE MATERIAL

ESSENTIAL QUESTION

What other elements can be found in organic compounds?

OBJECTIVES

- 6B1 Classify substituted hydrocarbons on the basis of their functional group.
- 6B2 Give examples of how common substituted hydrocarbons are used.

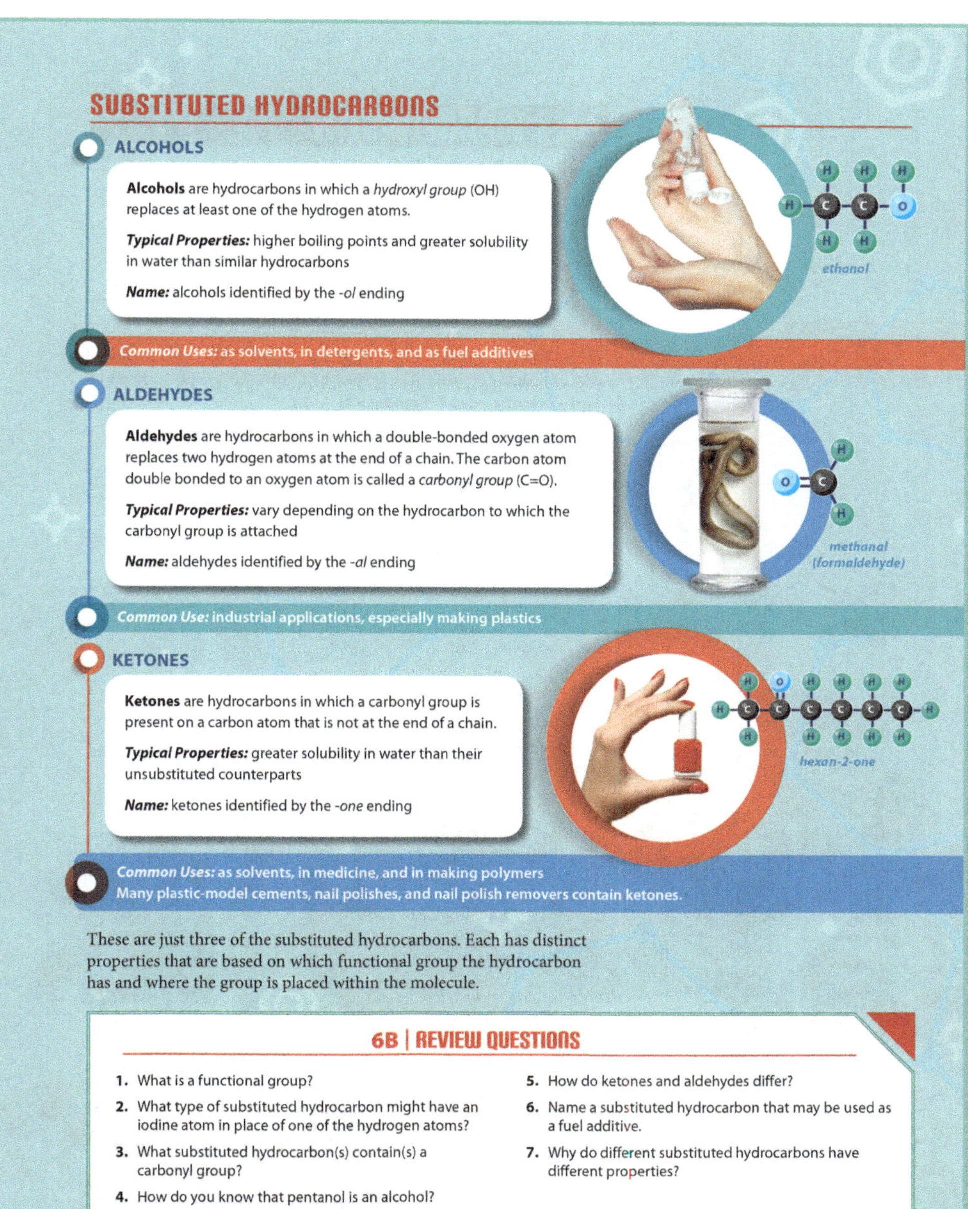

These are just three of the substituted hydrocarbons. Each has distinct properties that are based on which functional group the hydrocarbon has and where the group is placed within the molecule.

6B | REVIEW QUESTIONS

1. What is a functional group?
2. What type of substituted hydrocarbon might have an iodine atom in place of one of the hydrogen atoms?
3. What substituted hydrocarbon(s) contain(s) a carbonyl group?
4. How do you know that pentanol is an alcohol?
5. How do ketones and aldehydes differ?
6. Name a substituted hydrocarbon that may be used as a fuel additive.
7. Why do different substituted hydrocarbons have different properties?

THE CHEMISTRY OF LIFE   135

Discussion:
Naming Substituted Hydrocarbons

Use the hexan-2-one from the infographic (see Ketones box on the student page) to discuss the naming of substituted hydrocarbons. The 2 in the name indicates where the carbonyl group (second carbon) is attached. This naming convention also applies to the position of the double or triple bonds in alkenes and alkynes. For example, oct-2-ene is an eight carbon alkene with the double bond between the second and third carbon.

Formative Assessment:
Functional Groups

1. What functional group forms an alkyl halide? (a halogen atom)

2. How does butanol compare with butane? (They both have the same number of carbon and hydrogen atoms, but butanol has an oxygen atom bonded to one carbon and one hydrogen atom.)

3. Why is it not possible to determine the difference between an aldehyde and a ketone by the ratio of carbon, hydrogen, and oxygen atoms? (Both types of molecules can have the same ratio of atoms. The difference lies in the location of the carbonyl group.)

6B REVIEW ANSWERS

1. A functional group is an atom or a group of atoms that replaces a hydrogen atom to form a substituted hydrocarbon. (p. 134)

2. an alkyl halide (p. 134)

3. ketones and aldehydes (p. 134)

4. The -ol ending indicates that the compound is an alcohol. (p. 135)

5. Aldehydes have the carbonyl group at one end of the hydrocarbon, while the ketone has the carbonyl group attached to an interior carbon atom. The names of ketones have an -one ending, while aldehydes end with -al. (p. 135)

6. alcohol (Some students may specifically mention ethanol.) (p. 135)

7. The properties of substituted hydrocarbons are determined by the functional group(s) present and their location within the molecule. (p. 135)

STRATEGIES

Class Opener: Use the E85 teacher note on page 134 to introduce the topic of substituted hydrocarbons.

Discussion: Use the Effects of Functional Groups teacher note on page 134 to compare propane and propanol. As you do so, help students recognize that adding a functional group to a hydrocarbon can greatly alter its properties.

Discussion: Use the Discussion: Naming Substituted Hydrocarbons teacher note on this page to help students better understand the IUPAC naming conventions.

Formative Assessment: Use the Formative Assessment: Functional Groups teacher note on this page to assess students' understanding of substituted hydrocarbons and their functional groups.

Ticket Out the Door: What is the difference between an aldehyde and a ketone?

Common Polymers

Hold up a piece of nylon fabric and a potato. Ask students what they have in common. The answer of course is that nylon is a polymer and potatoes are rich in the polymer starch. Use this to introduce the subject of polymers.

6C | BIOCHEMISTRY

What molecules are needed for life?

6C Questions

- What are polymers?
- How do our bodies use polymers?
- Can food be fast *and* healthy?

6C Terms

polymer, monomer, carbohydrate, protein, amino acid, lipid, nucleic acid, nucleotide, DNA

6.7 POLYMERS

Have you ever watched as a long train passes a railroad crossing? Car after car rumbles by. Most of the world's products move around on freight trains. The conductor at the stockyard created the train by connecting smaller units, or boxcars.

Chemistry has its own version of these trains—gigantic molecules formed by linking many smaller molecules together. These *macromolecules* are called **polymers**. The smaller molecules that are put together to form a polymer are called **monomers**.

Polymers are made either by linking identical monomers or by connecting different ones. The properties of polymers are determined by the monomers that make them and how those smaller units are arranged. While many polymers form in nature, others are synthetic.

Glutamine (left) is one of the amino acids that form proteins, which are critical macronutrients for our bodies. Many foods, especially from animals, are good sources of proteins.

136

TEACHING THE MATERIAL

ESSENTIAL QUESTION

What molecules are needed for life?

OBJECTIVES

- 6C1 Define *polymer*.
- 6C2 Compare carbohydrates, lipids, and proteins.
- 6C3 Explain the role of nucleic acids in cell reproduction.
- 6C4 Explain the importance of nutrition in a fast-food society. **BWS**

RESOURCES

- Ethics: Can Fast Food Be Nutritious? (p. 143)
- Demonstrating Superabsorbent Polymers
- Serving as a Food Chemist
- Lab 6B: *Milking Chemistry*—Proteins in Food

6.8 SYNTHETIC POLYMERS

As scientists studied hydrocarbons, they realized that they could link many small hydrocarbons to form polymers. Many of the products that we use daily are made from these manmade polymers. While working with a methane derivative, scientists accidentally made a long chain of ethane molecules. Today many food containers and plastic bottles are made from this material, called polyethylene (see page 130). We use polyethylene because of its flexibility and high-impact strength.

Other scientists worked to create manmade fibers to replace natural fibers that are costly or difficult to obtain. The desire to replace silk and cotton led to the discovery of nylon. By linking a couple of different substituted hydrocarbons, scientists made the first nylon. Today there are many forms of nylon, each with distinct properties. Scientists also created other fibers such as Nomex® for flame-resistant clothing, bulletproof Twaron LFT® for ballistic vests, and neoprene for water-resistant cloth.

We don't always think about the amazing properties of synthetic polymers such as nylon. Do you think the skydiver above is thinking about the nylon that made his parachute?

In the image above you can see nylon forming at the boundary between two liquids.

Demonstrating Superabsorbent Polymers

Preparation: Prepare a cup with about 1 mL of sodium polyacrylate (available from science supply companies but also found in the linings of baby diapers). Measure 100 mL of warm water into another cup.

Performance: Pour the water into the cup *containing* the sodium polyacrylate. If necessary, stir it until the polymer stops expanding. Then tip the cup to show that the water has been absorbed.

Discussion:

1. What happened to the water? *(It was absorbed by the polymer.)*

2. What are some common products that contain this polymer? *(Examples: baby diapers, bandages, pet pads)*

3. How could you find out whether it is possible to remove the water and reuse the polymer? *(Setting it in the sun to dry would drive the water off and leave the dry powder.)*

Note: You can find many video clips of this demonstration. Search using the keywords "sodium polyacrylate and water demonstration."

STRATEGIES

Class Opener: Use the Common Polymers teacher note on page 136 to introduce the subject of polymer chemistry.

Demonstration: Use the Demonstrating Superabsorbent Polymers teacher note on this page to show a common but little-known polymer. As an alternative to the demonstration, do a keyword search for a video on "sodium polyacrylate and water demonstration."

Formative Assessment: Use the Formative Assessment: Biomolecules teacher note on page 139 to check students' knowledge and understanding of these necessary molecules.

Biblical Worldview Shaping: Use the Can Fast Food be Nutritious? ethics box on page 143 to stimulate a discussion on why and how we should make choices about what we eat.

Ticket Out the Door: What monomers make up the polymer starch?

Organic chemistry was originally thought to be the chemistry of living things. Today this branch of chemistry is defined as the chemistry of carbon-containing compounds. But the chemistry of living things is still a major field of study and one that we call *biochemistry*.

MACRONUTRIENTS

CARBOHYDRATES

Carbohydrates are compounds made of carbon, hydrogen, and oxygen atoms, with the hydrogen and oxygen usually in a 2:1 ratio. Carbohydrates are often polymers of sugars in our food, and the most common sources are fruits, vegetables, and grains.

Starches are more complex carbohydrates and are formed by linking glucose molecules to form large polymers, some of which contain branches of smaller glucose chains off the main chain. Plants make starch for energy storage. Starches are a key source of energy for humans.

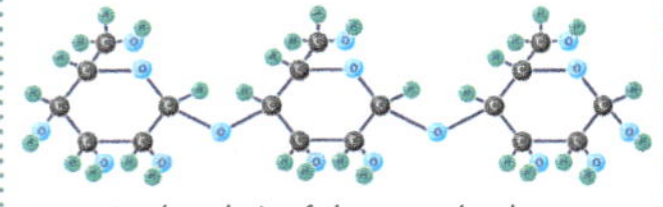

Sugars are simple carbohydrates that provide quick energy to people and animals. Common sugars include glucose and fructose. Plants make glucose during photosynthesis—to produce energy during cellular respiration, to convert to starches for storage, or to change into cellulose for plant structures. Glucose circulates in the blood of animals and humans and is called *blood sugar*. It is the primary form of energy for humans and animals. Animals and people convert excess glucose into a polymer called *glycogen* for short-term storage. The sugar we are probably most familiar with is sucrose, or table sugar, which consists of a glucose and a fructose molecule bonded together.

⁉️ *Relatively Simple*

Some students might wonder why the term *simple* is applied to sugars, since most of them are composed of approximately twenty-four atoms. While these are very complex molecules, they are still relatively simple carbohydrates when compared with extremely complex polysaccharides like starch and cellulose.

The Most Abundant Carbohydrate

The most abundant carbohydrate in the world is cellulose. Plants create it from glucose and use it to make their rigid cell walls. Many vertebrates (including humans) cannot digest cellulose in their food because they lack the enzyme necessary to break apart the bonds between the glucose monomers in cellulose polymers. But ruminants such as cows have microorganisms in their stomachs that are capable of digesting cellulose, allowing them to gain nutrition from the resulting glucose monomers. Though humans cannot digest cellulose, it forms insoluble fiber, or roughage, which is important to the health of the intestines.

➗ *Addition and Subtraction*

An attentive student might notice that the equation represented in the image showing the formation of sucrose is not balanced. For simplicity, we did not show the water molecule released in this process. When two sugar monomers bond, they release a water molecule. So while glucose and fructose both have the chemical formula $C_6H_{12}O_6$, sucrose's chemical formula is $C_{12}H_{22}O_{11}$.

PROTEINS

Another major group on a nutrition label is proteins—the building blocks of muscles, blood, skin, tendons, and hair. **Proteins** are polymers formed by linking amino acids. **Amino acids** are organic molecules with both an amine (NH_2) and a carboxyl group (C=OOH). Proteins contain hundreds of amino acids, formed by the amine group of one monomer bonded to the carboxyl group of the next. About twenty amino acids are needed to form all of the proteins that are essential for our bodies. Our bodies can produce some of these amino acids, but others must come directly from our food. Even in cases where we can produce the amino acids, it is more efficient to get them from food sources. Common sources of proteins are meats, dairy products, eggs, beans, and nuts.

LIPIDS

The third important group of compounds in biochemistry is lipids. **Lipids** are organic compounds that include fats, oils, waxes, and cholesterol. Fats get a lot of negative attention, but we need fats to maintain good health. They help with many processes within our cells as well as with brain function. Like hydrocarbons, fats can be saturated or unsaturated.

Saturated fats have only single bonds between carbon atoms within a portion of the molecule called a *fatty acid chain*. They are typically solid at room temperature, come from animal sources, and are simply called *fats*.

Unsaturated fats contain one or more double bonds in their fatty acids. They typically come from plants, remain liquid at room temperature, and are called *oils*. To increase the shelf life of foods, manufacturers will add hydrogen atoms to unsaturated fats—breaking the double bonds and changing them into saturated fats. There is concern about the long-term effects of these modified fats.

1. What type of molecule is glucose? *(carbohydrate)*

2. What are the building blocks of proteins? *(amino acids)*

3. Name two types of lipids. *(Accept any two of the following: fats, oils, waxes, cholesterol.)*

4. What are the three parts of a nucleotide? *(a sugar, a phosphate group, and a nitrogen-containing base)*

5. An unknown biomolecule has nitrogen in it. Which type of biomolecule is it likely to be? *(Accept either a protein or a nucleic acid.)*

Trans Fats

The modified fats mentioned in the last box on this page are often referred to as *trans fats*, which were originally created because they were thought to be healthier than saturated fats. But concerns were raised that they were in fact even worse. As a result of increased public awareness and new FDA guidelines, trans fats have been largely replaced in processed foods, often by saturated fats.

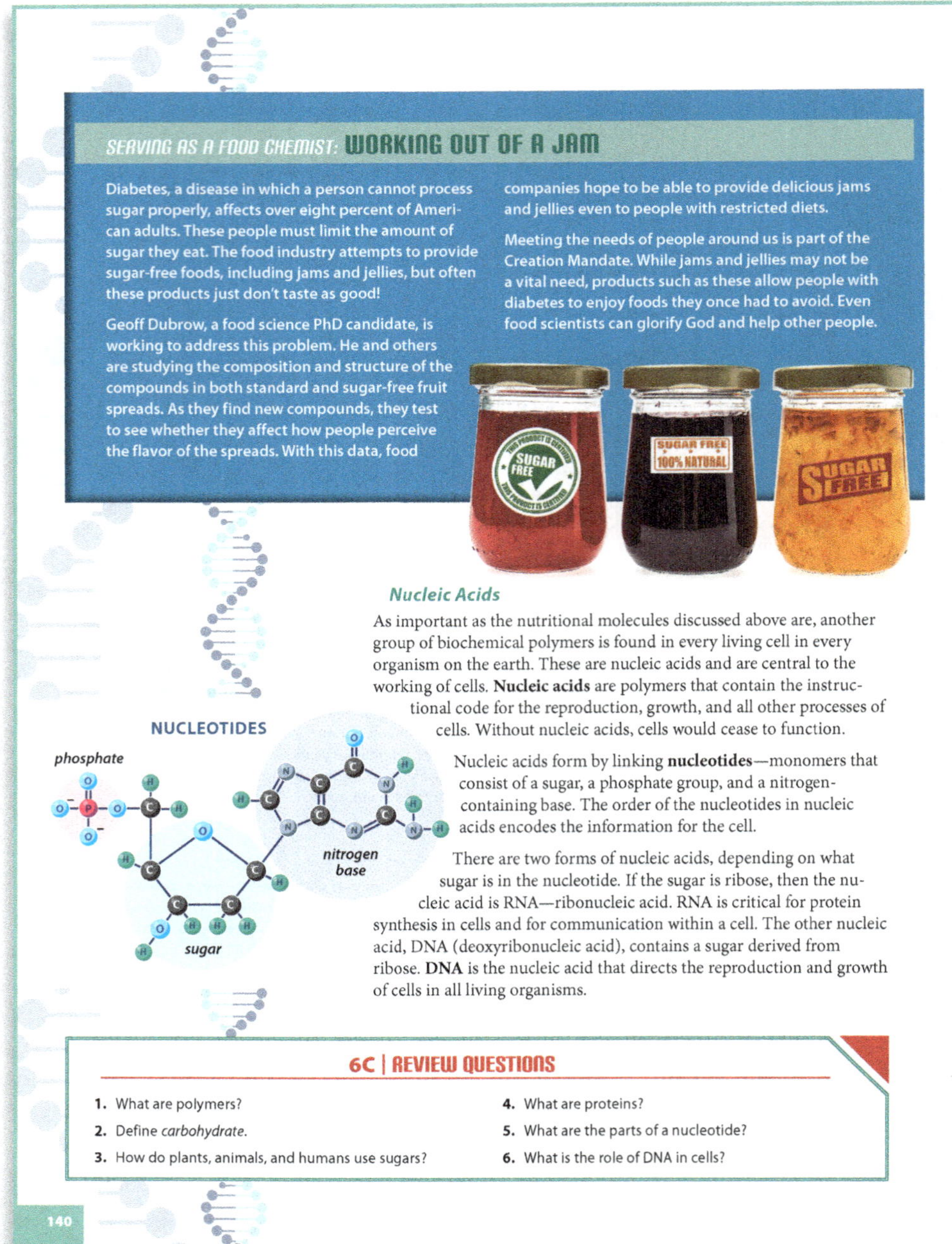

Nucleic Acids

As important as the nutritional molecules discussed above are, another group of biochemical polymers is found in every living cell in every organism on the earth. These are nucleic acids and are central to the working of cells. **Nucleic acids** are polymers that contain the instructional code for the reproduction, growth, and all other processes of cells. Without nucleic acids, cells would cease to function.

Nucleic acids form by linking **nucleotides**—monomers that consist of a sugar, a phosphate group, and a nitrogen-containing base. The order of the nucleotides in nucleic acids encodes the information for the cell.

There are two forms of nucleic acids, depending on what sugar is in the nucleotide. If the sugar is ribose, then the nucleic acid is RNA—ribonucleic acid. RNA is critical for protein synthesis in cells and for communication within a cell. The other nucleic acid, DNA (deoxyribonucleic acid), contains a sugar derived from ribose. **DNA** is the nucleic acid that directs the reproduction and growth of cells in all living organisms.

6C | REVIEW QUESTIONS

1. What are polymers?
2. Define *carbohydrate*.
3. How do plants, animals, and humans use sugars?
4. What are proteins?
5. What are the parts of a nucleotide?
6. What is the role of DNA in cells?

Lab 6B: *Milking Chemistry*

Once students have an understanding of the material in Section 6C, they should be ready to work through Lab 6B.

6C REVIEW ANSWERS

1. Polymers are massive molecules formed by linking many monomers. *(p. 136)*

2. Carbohydrates are molecules made of carbon, hydrogen, and oxygen atoms. *(p. 138)*

3. Plants produce glucose during photosynthesis. Plants can use the glucose to generate energy in cellular respiration, convert it to starches, oils, and fats for storage, or convert it to cellulose for plant structures. Animals and humans use glucose as one of their primary energy sources. They can convert unused glucose to glycogen for storage. *(p. 138)*

4. Proteins are the building blocks of muscles, blood, skin, tendons, and hair. They are polymers composed of amino acids. *(p. 139)*

5. a sugar, a phosphate group, and a nitrogen-containing base *(p. 140)*

6. DNA directs the reproduction and growth of cells. *(p. 140)*

CHAPTER 6 REVIEW . . . .

6A ORGANIC COMPOUNDS

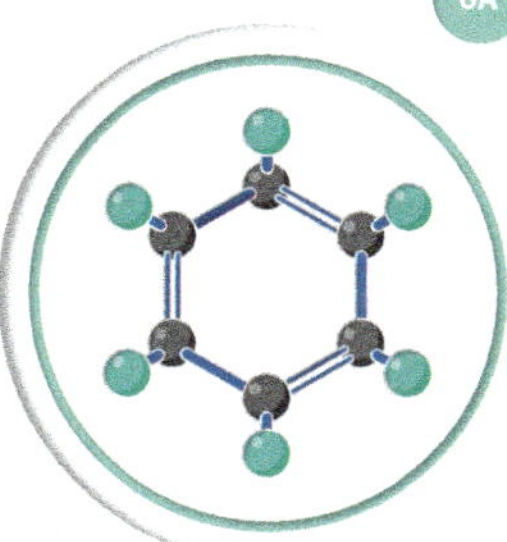

- Organic chemistry is the study of the composition, structure, and properties of carbon-containing compounds.

- Organic compounds form as straight chains, branched chains, and rings.

- Hydrocarbons are organic compounds consisting of only carbon and hydrogen atoms.

- Saturated hydrocarbons contain only single bonds between carbon atoms; unsaturated hydrocarbons have double or triple bonds.

- Aromatic hydrocarbons contain six-carbon benzene rings.

- Molecules with the same chemical formula but whose atoms are arranged differently are called isomers.

- Differences in the structures of molecules result in differences in the properties of compounds.

6A Terms

organic compound	125
hydrocarbon	128
saturated hydrocarbon	128
unsaturated hydrocarbon	129
aromatic hydrocarbon	131
benzene ring	131
isomer	132

6B SUBSTITUTED HYDROCARBONS

- Substituted hydrocarbons are organic compounds where one or more hydrogen atoms have been replaced with other atoms or groups of atoms called functional groups.

- Functional groups substituted into hydrocarbons change the composition and therefore the properties of the original molecule.

6B Terms

substituted hydrocarbon	134
functional group	134
alcohol	135
aldehyde	135
ketone	135

6C BIOCHEMISTRY

- Polymers are huge molecules formed by linking monomers together.

- Many manmade products are synthetic polymers, such as plastics, nylon, and synthetic rubber.

- Carbohydrates, proteins, lipids, and nucleic acids are organic compounds involved in biochemical processes.

- Carbohydrates, sugars, and starches provide most energy for living organisms. They are also building blocks for other structures.

- Examples of lipids are the fats, oils, and waxes in our foods.

- Proteins, which are made from amino acids, are the building blocks for many tissues in our bodies.

- Nucleic acids encode and store information for cells. They direct functions such as reproduction, growth, and development in the cell.

6C Terms

polymer	136
monomer	136
carbohydrate	138
protein	139
amino acid	139
lipid	139
nucleic acid	140
nucleotide	140
DNA	140

CHAPTER REVIEW ANSWERS

Recalling Facts

1. Friedrich Wöhler accidentally made the organic compound urea, which meant that man could produce organic compounds. *(p. 125)*

2. Since carbon has four valence electrons, it can bond with up to four different atoms. When carbon atoms bond with each other, they form very strong, durable bonds. Carbon's four valence electrons also enable them to form single, double, and triple bonds. *(pp. 126–27)*

3. Hydrocarbons are molecules that contain only carbon and hydrogen atoms. *(p. 128)*

4. Isomers are molecules with the same molecular formula but with a different arrangement of atoms. *(p. 132)*

5. Substituted hydrocarbons are formed by replacing a hydrogen atom with an atom of another element or with a group of atoms. *(p. 134)*

6. A hydroxyl group is a hydrogen atom bonded to an oxygen atom. It is the functional group of alcohols. *(p. 134)*

7. A carbonyl group is a carbon atom and an oxygen atom bonded with a double bond. It is the functional group in ketones and aldehydes. *(p. 134)*

8. Alcohols are commonly used as solvents, in detergents, and as fuel additives. (Accept any two.) *(p. 135)*

9. Answers will vary. Examples include polyethylene for water bottles and food containers, nylon as a silk and cotton replacement, Nomex® in flame-resistant clothing, and Twaron LFT® in ballistic-resistant clothing. (Accept any two.) *(p. 137)*

10. carbohydrates, proteins, and lipids *(pp. 138–39)*

11. amino acids *(p. 139)*

12. Answers will vary. Examples include meats, eggs, beans, peanuts, tree nuts, and legumes. *(p. 139)*

13. Lipids are polymers made from carbon and hydrogen. We know them as fats, oils, waxes, and cholesterol. *(p. 139)*

14. Nucleic acids are polymers made from nucleotides. They encode, store, and provide instructions for processes in all living cells. *(p. 140)*

CHAPTER REVIEW QUESTIONS

Recalling Facts

1. What event changed scientists' thinking about organic compounds and chemistry?

2. Name three characteristics of carbon atoms that allow them to form so many different organic compounds.

3. What are hydrocarbons?

4. What are isomers?

5. How do substituted hydrocarbons differ from hydrocarbons?

6. What is a hydroxyl group, and what is its relationship to alcohols?

7. What is a carbonyl group? In which group(s) of substituted hydrocarbons would you find one?

8. Name two uses of alcohols.

9. Give two examples of synthetic polymers and how we use them.

10. What biochemical macromolecules do our bodies use?

11. What monomers link to form proteins?

12. What are good food sources of proteins?

13. What are lipids?

14. What are nucleic acids?

Understanding Concepts

15. Why is the term *straight chain* somewhat misleading?

16. Compare alkanes, alkenes, and alkynes.

17. Identify each of the following as an alkane, alkene, alcohol, aldehyde, or ketone.

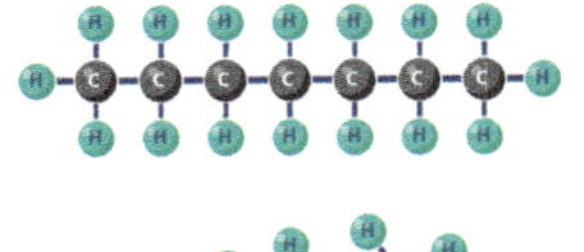

18. Create a concept map using the following terms: *hydrocarbon, saturated hydrocarbon, unsaturated hydrocarbon, single bond, double bond, triple bond, alkanes, alkenes,* and *alkynes.*

19. Why is butyne much more reactive than butane?

20. What changes the properties of hydrocarbons?

21. Explain how monomers and polymers are related.

22. Explain how starches and sugars are related.

23. How do plants, animals, and humans use starches?

24. Evaluate the statement, "For a healthy diet, we must avoid eating fats."

25. Create a concept map using the following terms: *molecule, polymer, monomer, carbohydrate, starch, sugar, lipid, protein, energy, cellulose,* and *glycogen.*

26. Compare the structure and function of RNA and DNA.

Critical Thinking

27. What is the mathematical relationship between the number of hydrogen atoms and the number of carbon atoms in alkanes?

28. Why does methene not exist?

29. Are the two molecules shown below isomers of each other? Explain.

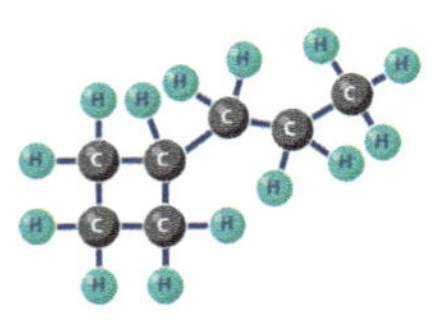

30. Change the propane molecule shown below into

a. a ketone.

b. an alcohol.

c. an aldehyde.

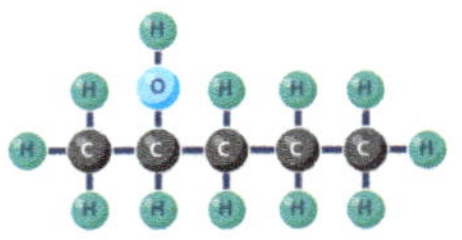

31. The IUPAC name for the isomer shown below is pentan-2-ol. Why do you think it is so named?

32. Compare carbohydrates, proteins, and lipids by copying the following table on your paper and filling in the cells.

	Carbo-hydrates	Lipids	Proteins
Made of chains of	sugars	carbon and hydrogen atoms	amino acids
Uses in the body	energy	energy	building body structure
Examples in food	fruit, grains	oils, lard	meats, nuts

Understanding Concepts

15. The term suggests that the chain always forms a straight line, but straight chains can have bends and turns in them. They are straight in the sense that there is one path of carbon atoms from end to end. *(p. 126)*

16. All three are hydrocarbons. Alkanes are saturated, while alkenes and alkynes are unsaturated. Alkanes have only single bonds between carbon atoms, alkenes have at least one double bond, and alkynes have at least one triple bond. *(pp. 128–30)*

17. **a.** aldehyde *(pp. 134–35)*

b. alcohol *(pp. 134–35)*

c. alkene *(pp. 129–30)*

18. See the concept map below. *(pp. 128–30)*

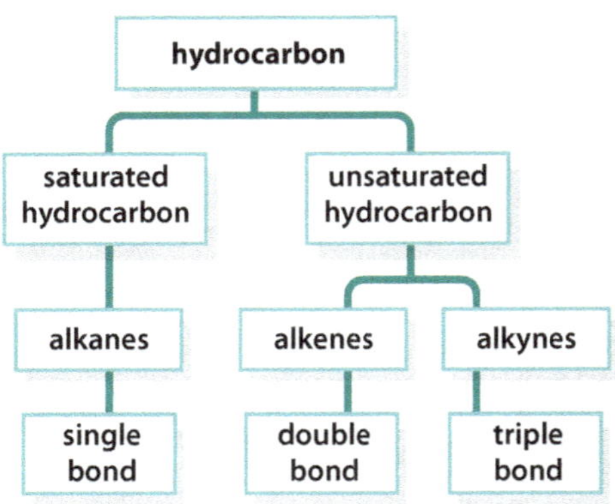

19. Butyne has a triple bond. *(p. 130)*

26. RNA and DNA are both nucleic acids. They differ in the composition of their nucleotides. RNA has ribose as the sugar in its nucleotide, while DNA has a ribose-derived sugar—deoxyribose—as its sugar. RNA functions in protein synthesis, while DNA contains the genetic information that governs the growth, function, and reproduction of cells. *(p. 140,*

Critical Thinking

27. Alkanes are named for the number of carbons in them: each carbon has two hydrogens attached to it with one more hydrogen attached on each end of the chain. Therefore the pattern follows the formula C_nH_{2n+2}. *(p. 128)*

28. An *-ene* ending indicates that a substance is an alkene with at least one double bond between carbon atoms. But the *meth-* prefix indicates the presence of only one carbon atom, thus no carbon–carbon bonds are possible. *(pp. 128–29)*

29. These are not isomers of each other because they have different numbers of hydrogen atoms. *(p. 132)*

30. See examples below. Other answers are possible.

a.

$$H-\underset{\underset{H}{|}}{\overset{\overset{H}{|}}{C}}-\overset{\overset{O}{||}}{C}-\underset{\underset{H}{|}}{\overset{\overset{H}{|}}{C}}-H$$

b.

$$H-\underset{\underset{H}{|}}{\overset{\overset{H}{|}}{C}}-\underset{\underset{H}{|}}{\overset{\overset{H}{|}}{C}}-\underset{\underset{H}{|}}{\overset{\overset{H}{|}}{C}}-O$$

c.

$$H-\underset{\underset{H}{|}}{\overset{\overset{H}{|}}{C}}-\underset{\underset{H}{|}}{\overset{\overset{H}{|}}{C}}-\overset{\overset{H}{|}}{C}=O$$

(all pp. 134–35)

31. The hydroxyl group is attached to the second carbon atom in the chain. *(p. 135)*

32. See reduced student page on facing page. *(all pp. 138–39)*

---

20. The properties of a hydrocarbon are changed by altering its composition, internal structure, or both. *(p. 132)*

21. Polymers are massive molecules formed from linking smaller molecules called monomers. *(p. 136)*

22. Starches and sugars are both carbohydrates. Starches are polymers made from the linking of sugar molecules. *(p. 138)*

23. Plants produce starches from glucose and use it to store energy. Animals and humans use starch as a source of energy. *(p. 138)*

24. This statement is false. Fats help many cellular functions as well as brain function. *(p. 139)*

25. See the concept map below. *(pp. 136–39)*

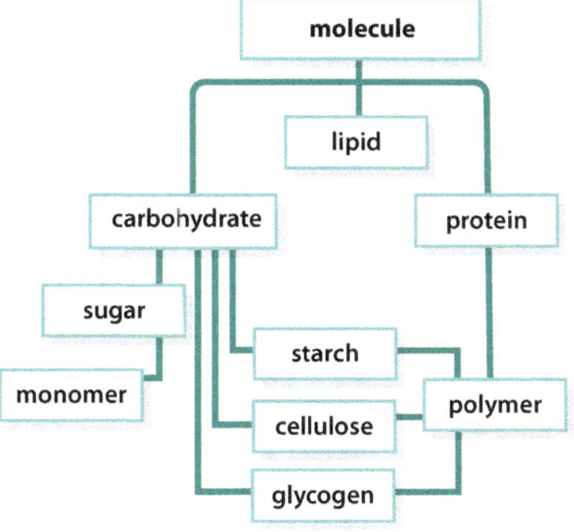

The reduced student page:

REVIEW

ETHiCS — CAN FAST FOOD BE NUTRITIOUS?

We live in a fast-paced world. People run from event to event without time for a breath, much less time to think about what they are eating. Twenty-five percent of Americans eat at least one fast-food meal each week, and Americans eat twenty percent of their meals in their cars. Drive through many commercial districts and you may see five, ten, even fifteen fast-food restaurants. Watch an hour of television, and you will see at least one fast-food commercial. The food looks delicious, and it will be ready . . . well, fast, so that you can keep moving. And it is food, right?

But is all food the same when we consider its macronutrients—proteins, lipids, and carbohydrates? What about when we start to look at its micronutrients, such as vitamins and minerals?

THE ISSUE: EATING WELL

33. What information can you get about this issue? Do a keyword search for "fast food nutrition" and "fast food's impact on health." Write a summary of what you have learned.

34. What does the Bible say about this issue? Does the Bible specifically address what people eat? Can you find general principles in Scripture about what we eat?

35. What are the acceptable and unacceptable options of fast food? Should we never eat any fast food? Should you eat and drink whatever you feel like? And there are many options between those two. What might some of those options be?

36. What might be the consequences of adopting the acceptable or unacceptable options? What might be the result of never eating any fast food? What could result from eating and drinking whatever you feel like?

37. What are the motivations for the acceptable options? How do the acceptable options that you've identified demonstrate things such as respect for God's laws and growing in Christlike character?

38. What action should you take? Using all the information that you have found and your answers to the previous questions, think about what sorts of actions you should take with regard to fast food.

	Task completed	Task partially completed	Task not completed
Additional Information			
Biblical Principles			
Acceptable and Unacceptable Options			
Consequences			
Biblical Outcomes			
Biblical Motivations			
Opinion Statement			

THE CHEMISTRY OF LIFE 143

33. Answers will vary. Fast food generally is high in calories, fat, and sodium. The balance of macronutrients typically doesn't match the recommendation of nutrition experts and many of the foods lack needed micronutrients. While most fast food menus include healthier options, the general nutritional value of the food has not changed over time.

34. Answers will vary. Scripture does not specifically address fast food but does provide general principles. In 1 Corinthians 6:19–20 we read that the body is the temple of the Holy Spirit. Our bodies are not our own to do with whatever we want; we are to glorify God with our bodies. In 1 Corinthians 10:31 we are told to do all, even eating and drinking, to the glory of God.

35. Answers will vary. *Unacceptable*: We can eat whatever we want and ignore the consequences. *Acceptable*: We should always eat wisely. We should seek to strengthen our bodies, minds, and souls to best serve God and others. Eating fast food occasionally is acceptable, but we may want to seek out the healthier options on the menu.

36. Answers will vary. By adopting an acceptable option we could maintain our health, enabling us to better serve God and others. Adopting the attitude that we can eat and drink whatever we want contradicts what Scripture teaches about the ownership of our bodies. To decide never to eat fast food could place us in a position with no options. The occasional fast food meal won't have a lasting, unhealthy effect.

37. Answers will vary. The best motivation for adopting the acceptable option is that we recognize that everything we do can glorify God. We can glorify God by taking care of ourselves physically and spiritually.

38. Answers will vary. God created food for both our enjoyment and our nourishment. I am commanded to do everything to the glory of God. I therefore should seek to glorify God when I make food choices. While the occasional fast food meal won't have a lasting, negative effect, I should limit how often and how much I consume. (*Note:* Be sure that students do not just write, "I will eat less fast food." While that may be a good choice, a student's answer should include his motivation for that choice.)

REVIEW

ETHiCS   CAN FAST FOOD BE NUTRITIOUS?

We live in a fast-paced world. People run from event to event without time for a breath, much less time to think about what they are eating. Twenty-five percent of Americans eat at least one fast-food meal each week, and Americans eat twenty percent of their meals in their cars. Drive through many commercial districts and you may see five, ten, even fifteen fast-food restaurants. Watch an hour of television, and you will see at least one fast-food commercial. The food looks delicious, and it will be ready . . . well, fast, so that you can keep moving. And it is food, right?

But is all food the same when we consider its macronutrients—proteins, lipids, and carbohydrates? What about when we start to look at its micronutrients, such as vitamins and minerals?

THE ISSUE: EATING WELL

33. What information can you get about this issue? Do a keyword search for "fast food nutrition" and "fast food's impact on health." Write a summary of what you have learned.

34. What does the Bible say about this issue? Does the Bible specifically address what people eat? Can you find general principles in Scripture about what we eat?

35. What are the acceptable and unacceptable options of fast food? Should we never eat any fast food? Should you eat and drink whatever you feel like? And there are many options between those two. What might some of those options be?

36. What might be the consequences of adopting the acceptable or unacceptable options? What might be the result of never eating any fast food? What could result from eating and drinking whatever you feel like?

37. What are the motivations for the acceptable options? How do the acceptable options that you've identified demonstrate things such as respect for God's laws and growing in Christlike character?

38. What action should you take? Using all the information that you have found and your answers to the previous questions, think about what sorts of actions you should take with regard to fast food.

	Task completed	Task partially completed	Task not completed
Additional Information			
Biblical Principles			
Acceptable and Unacceptable Options			
Consequences			
Biblical Outcomes			
Biblical Motivations			
Opinion Statement			

The Unit opener quote emphasizes the importance of pure science. Discoveries are made
because we strive to learn more about the
world around us. Often we later realize that
these discoveries are also useful.

J. Robert Oppenheimer was the director of the
Los Alamos Laboratory during the Manhattan
Project. He is sometimes called the Father of
the Atomic Bomb.

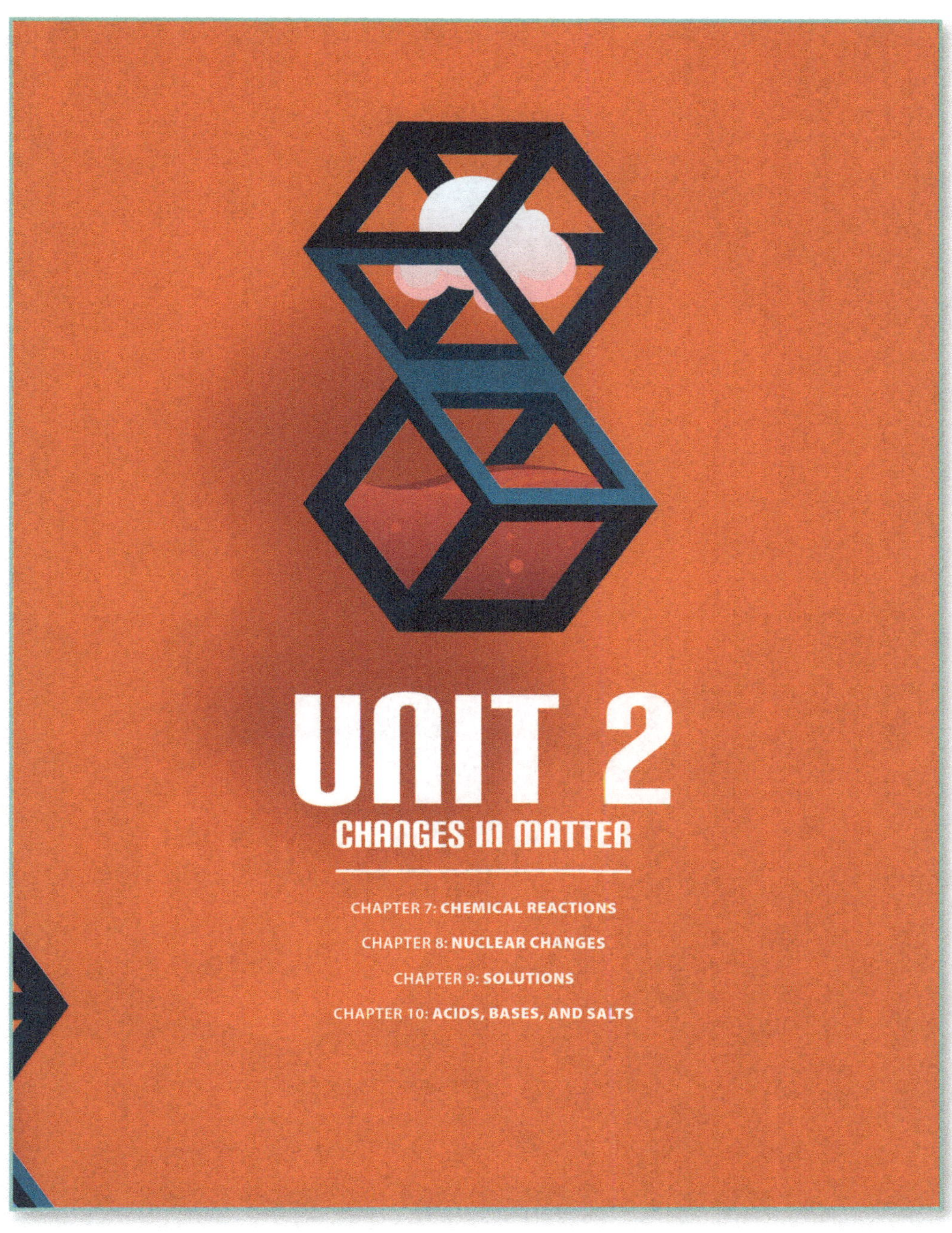

UNIT 2
CHANGES IN MATTER

Main photo: *Explosions are rapid combustion reactions and can be devastating when uncontrolled.*

Inset: *Residents comb through the wreckage after the* Mont Blanc *explosion destroyed most of Halifax.*

CHAPTER 7

Chemical Reactions

LITTLE CHANGES, BIG EFFECTS

In Halifax, Nova Scotia, the tranquility of Sunday morning, December 6, 1917, was suddenly shattered by an enormous blast. A French cargo ship, the SS *Mont Blanc*, fully laden with explosives for the Allied war effort in Europe, collided with another ship, caught fire, and exploded. The energy released was the equivalent of 2900 metric tons of TNT being detonated—the largest manmade explosion in history before the advent of atomic weapons. About 2000 people lost their lives.

What makes the Halifax tragedy even more astonishing is that all that energy was released when unimaginably tiny atoms within the explosive compounds swiftly rearranged themselves. Happily for us, not all chemical reactions are as powerful or as deadly as this example. God designed chemical reactions to sustain life on Earth.

Halifax Explosion

Do an internet search for a video about the Halifax explosion. This can be used as an opener to stimulate students' thinking and interest in chemical processes.

Overview

Because so many living and nonliving phenomena involve chemical processes, this chapter is foundational to an understanding of many of the changes we see in the physical world around us.

Lab Activities

Lab 7A: *Science Fair Revisited*—In this lab activity, students evaluate the evidence of chemical reactions, classify the types of reactions on the basis of their equations, and distinguish between reactants and products.

Lab 7B: *It's in the Bag*—In this lab activity, students observe a chemical reaction to determine which reactants have the greatest effect on the reaction.

This demonstration can be done as an alternative to doing Lab 7B. If you are planning on doing Lab 7B, using this demonstration will reduce the effectiveness of that lab activity.

Preparation:

1. Prepare aqueous solutions of silver nitrate and potassium chromate by stirring in approximately 2 g of silver nitrate in one beaker and 1 g of potassium chromate in another, each containing approximately 50 mL of water.

2. In a test tube or small flask, put about 15 g of baking soda, and in another put about 100 mL of vinegar.

   Note: You could substitute a mixture of about 15 mL of dry yeast, a pinch of sugar, and 45 mL of warm water in a small beaker. Then in another large beaker pour 150 mL of hydrogen peroxide. Using hydrogen peroxide solution that is more concentrated than the household variety (3%) will produce a more vigorous reaction. Protective clothing, gloves, and goggles should be worn. Mixing some liquid dishwashing soap with the concentrated hydrogen peroxide beforehand will produce foam as an added effect.

3. Obtain an unused glow stick.

4. Measure out about 10 g of potassium permanganate in a metal or ceramic container behind a laboratory safety shield. Measure out about 5 mL of glycerin.

Performance:

1. Pour some of the potassium chromate solution into the silver nitrate solution.

2. Mix the baking soda with the vinegar or add the yeast mixture to the hydrogen peroxide. If using the former, a balloon can be placed over the mouth of the test tube or flask. If using the latter, a thermometer can be placed in the flask; have a student read the temperature before and after the reaction.

3. Turn off the lights and snap the glow stick.

(continued)

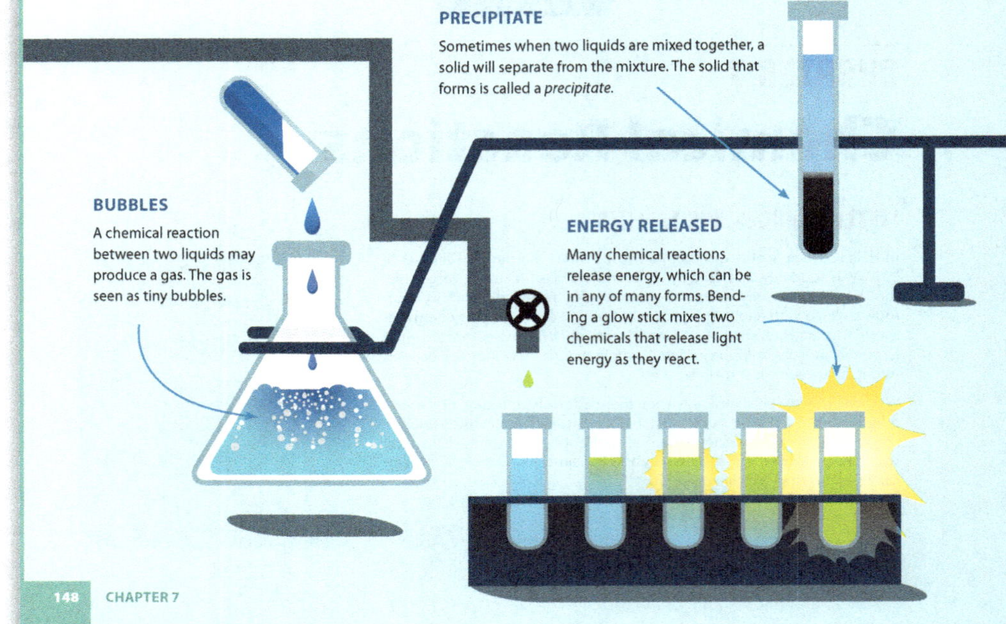

TEACHING THE MATERIAL

ESSENTIAL QUESTION

How can I tell whether a chemical change has taken place?

OBJECTIVES

• 7A1 List evidences for a chemical change.

• 7A2 Explain the use of a chemical equation as a model of a chemical reaction.

• 7A3 Write chemical equations from word equations.

• 7A4 Balance chemical equations by using the law of conservation of matter.

• 7A5 Solve simple mole-mass conversion problems.

RESOURCES

Demonstrating Chemical Reaction Indicators

Serving as a Toxicologist

Mini Lab: *Balanced Diet* (p. 155)

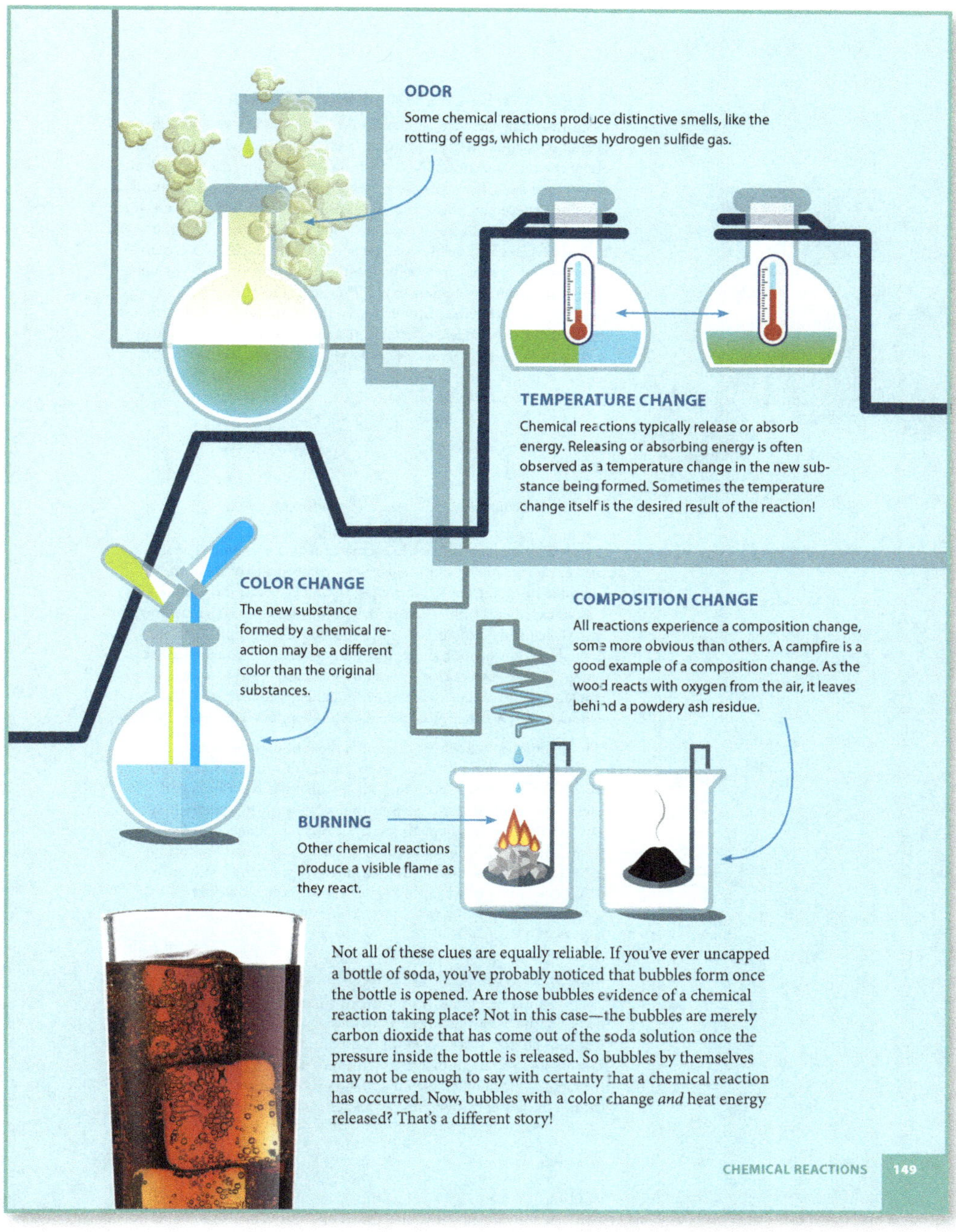

Not all of these clues are equally reliable. If you've ever uncapped a bottle of soda, you've probably noticed that bubbles form once the bottle is opened. Are those bubbles evidence of a chemical reaction taking place? Not in this case—the bubbles are merely carbon dioxide that has come out of the soda solution once the pressure inside the bottle is released. So bubbles by themselves may not be enough to say with certainty that a chemical reaction has occurred. Now, bubbles with a color change *and* heat energy released? That's a different story!

4. Positioning yourself behind the laboratory safety shield and wearing gloves and goggles, add the glycerin to the container of potassium permanganate on a fire-safe surface and step back to observe. Within thirty seconds to a minute, there should be a vigorous reaction-producing fire.

Discussion:

1. In the first reaction, what were two indicators that a chemical reaction had occurred? *(the formation of a precipitate and a color change)*

2. In the second reaction, what indicator of a reaction was observed? *(Bubbles were formed and—if using the yeast and hydrogen peroxide—there was a temperature change.)*

3. In the third reaction, what indicator of a reaction was observed? *(Energy in the form of light was given off.)*

4. In the fourth reaction, what indicator of a reaction was observed? *(There was a temperature change and fire was produced.)*

5. What happened in each of these instances that caused the particular indicator? *(New substances with different properties were being produced.)*

STRATEGIES

Class Opener: Use the Halifax Explosion teacher note on page 147 to stimulate students' thinking and interest. Do an internet search for a video on the Halifax Explosion.

Demonstration: Use the Demonstrating Chemical Reaction Indicators teacher note on page 148 to show students examples of chemical reaction indicators.

Think, Pair, Share: Use the Think, Pair, Share: Balancing Practice teacher note on page 152 to reinforce students' abilities to balance equations.

Think, Pair, Share: Use the Think, Pair, Share: Mole Conversion Practice teacher note on page 153 to reinforce the mole concept.

Formative Assessment: Use the Formative Assessment: Interpreting Coefficients teacher note on page 154 to check students' understanding of coefficients used in chemical equations.

Ticket Out the Door: To check students' understanding regarding the coefficients in an equation, use the Formative Assessment: Interpreting Coefficients teacher note on page 154.

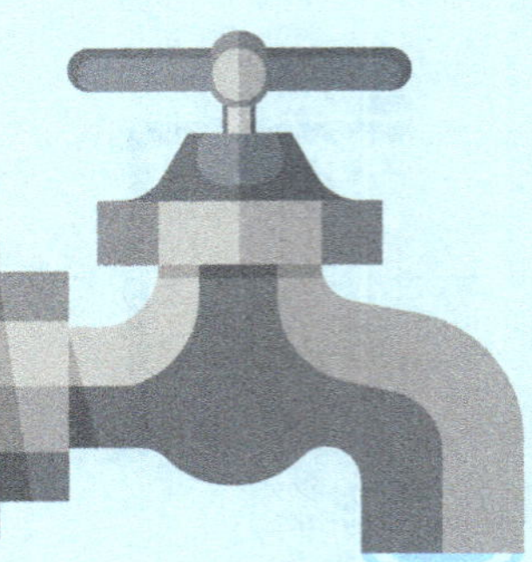

7.2 WRITING CHEMICAL EQUATIONS

Chemists—and chemistry students—do more than just look for signs of chemical reactions. They also study, create, describe, and communicate about chemical reactions. But describing a chemical reaction with words alone is not a wholly satisfactory method. Take for instance a chemical reaction between hydrogen and oxygen to form water. We could create a word equation such as, "Two units of hydrogen gas react with one unit of oxygen gas to yield two units of water." That's a lot of words to describe a fairly simple reaction—most reactions are far more complex. Chemists need a simpler way to model chemical reactions. For this purpose they use a **chemical equation**, a combination of chemical formulas and symbols that models a chemical reaction. Let's look at what the hydrogen and oxygen reaction looks like when it's written as an equation.

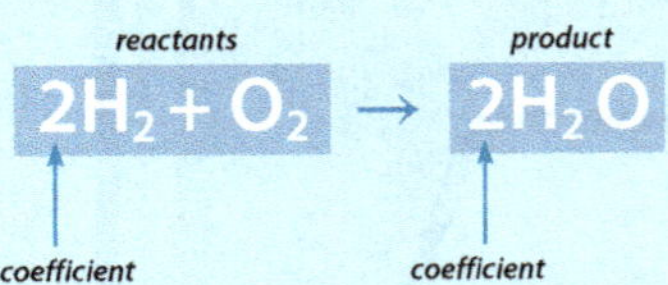

A chemical equation is much like a mathematical equation. The substances on the left side of the equation, hydrogen and oxygen, are indicated by their chemical formulas, H_2 and O_2 (don't forget your diatomic elements from Chapter 5). The plus sign shows that hydrogen and oxygen will be combined in this reaction. The formula for water on the right of the equation shows that water is what is formed by the reaction. The substances that enter into a chemical reaction are called the **reactants**. The substances that they combine to produce are called the **products**. The arrow symbol, like the equal sign in math, means "yields."

What about the number "2" in front of the formulas for hydrogen and water? Go back to the original description of this reaction. Notice the words "two units" and "one unit." The number placed in front of a chemical formula within a chemical equation is called a **coefficient**, similar to coefficients in math class. It shows how many units (e.g., atoms, molecules, or formula units) of each reactant and product are in the chemical equation. Hydrogen and oxygen always combine in a ratio of two units of hydrogen to one unit of oxygen to form two units of water. A coefficient is not shown if only one unit of a substance is indicated in an equation.

7.3 BALANCING CHEMICAL EQUATIONS

There's another reason why coefficients are needed for chemical equations. On the basis of many observations, chemists know that chemical reactions do not create any new matter, nor do they destroy any matter. The number and kinds of atoms in the products of a reaction are always the same as the number and kind found in the reactants. This relates back to the law of conservation of matter concept that you learned in

Chapter 2. To see how this works, let's look at our equation again, but this time without the coefficients.

$$H_2 + O_2 \rightarrow H_2O$$

This new equation isn't completely wrong. After all, it does show that hydrogen combines with oxygen to yield water, and that is true. But let's count up the number of each kind of atom shown in both the reactants and the product. On the reactants side, we can see two atoms of hydrogen and two atoms of oxygen. But on the product side there are two hydrogen atoms and only one oxygen atom. We call this an *unbalanced* chemical equation. We know from the law of conservation of matter that the missing oxygen atom couldn't have been destroyed during the reaction. We need to account for that second oxygen atom, and to do so we need to balance the equation. Let's see how that's done.

EXAMPLE 7-1: Writing Balanced Chemical Equations

Step 1: Write the word equation for the reaction.

Hydrogen plus oxygen yields water.

Step 2: Write the chemical equation for the reaction.

$$H_2 + O_2 \rightarrow H_2O$$

Step 3: Balance the chemical equation with coefficients, if necessary. Here we see right away that the number of oxygen atoms on both sides of the equation is not balanced. We need to double the number of oxygen atoms on the right side (product side) of the equation. We'll do that by placing a 2 in front of the formula for water. That coefficient doubles the number of units of water, which are molecules in this case.

$$H_2 + O_2 \rightarrow 2H_2O$$

Notice that we did not place a subscript 2 after the symbol for oxygen in H_2O in order to double the number of oxygen atoms! The reason for this is simple and important: adding subscripts changes the identity of a chemical formula. H_2O is water; H_2O_2 is not!

Step 4: Check to see whether the addition of the coefficient balances the equation. In this case, adding the coefficient for water corrected the imbalance for oxygen but unbalanced the number of hydrogen atoms on each side.

Step 5: Repeat Steps 3 and 4, if necessary. Adding the coefficient 2 for water doubled the number of hydrogen atoms on the product side (note that it is 2 units of water × 2 hydrogen atoms in each unit = 4 total hydrogen atoms). We need to double the number of hydrogen atoms on the reactant side to balance the new number on the product side.

$$2H_2 + O_2 \rightarrow 2H_2O$$

A quick check of this equation shows that it is now balanced. It has the same number of hydrogen and oxygen atoms on each side of the equation.

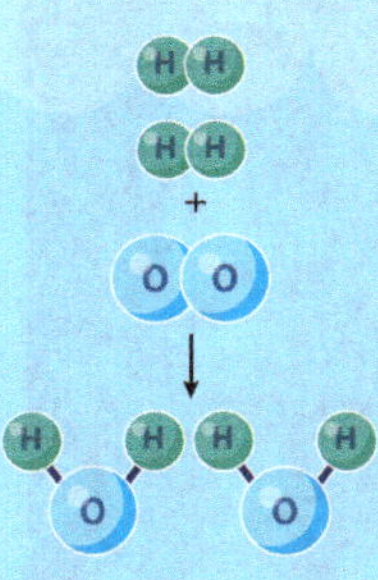

Diatomic Elements

Some students may wonder why the hydrogen and oxygen in Step 2 each have a subscript of 2. This is a great time to remind them about the diatomic elements. These elements—the Hydrogen Seven (hydrogen, nitrogen, oxygen, fluorine, chlorine, bromine, and iodine)—exist as two-atom molecules when they are by themselves.

 Naming Conventions

If necessary, refer back to page 114 to review naming conventions for compounds of transition metals.

 Tips for Balancing Equations

Instruct students to start by trying to balance elements that are in the most complex formulas on both sides of the arrow. Save elements without subscripts for last.

Some students find balancing equations easier if they have a way to track the atoms on each side of the equation. A table is helpful for this purpose. In the first column, enter the elements involved in the reaction. The next two columns are for the reactants and products. As coefficients are added, update the numbers until the equation is balanced.

	Reactants	Products
Fe	1 4	2 4
O	2 6	3 6

 Think, Pair, Share:
Balancing Practice

Some students will need practice to balance equations consistently. Give them an equation and time to balance it on their own. Then have them share with another student and discuss between themselves whether each has the equation balanced correctly. After a couple minutes, ask for a student to give the correct answer and verify. Some sample equations with their solutions are given below.

1. $Fe_2O_3 + C \rightarrow Fe + CO_2$

   $2Fe_2O_3 + 3C \rightarrow 4Fe + 3CO_2$

2. $Na + H_2O \rightarrow NaOH + H_2$

   $2Na + 2H_2O \rightarrow 2NaOH + H_2$

3. $FeCl_3 + NaOH \rightarrow Fe(OH)_3 + NaCl$

   $FeCl_3 + 3NaOH \rightarrow Fe(OH)_3 + 3NaCl$

4. $C_3H_8 + O_2 \rightarrow CO_2 + H_2O$

   $C_3H_8 + 5O_2 \rightarrow 3CO_2 + 4H_2O$

5. $Pb(OH)_2 + HCl \rightarrow H_2O + PbCl_2$

   $Pb(OH)_2 + 2HCl \rightarrow 2H_2O + PbCl_2$

Rust is a combination of iron and oxygen that we first saw back in Chapter 5 as iron (III) oxide. Let's see how the chemical equation for this reaction is balanced.

Step 1: *Write the word equation for the reaction.*

Iron plus oxygen yields iron (III) oxide (rust).

Step 2: *Write the chemical equation for the reaction.*

$$Fe + O_2 \rightarrow Fe_2O_3$$

Step 3: *Balance the chemical equation with coefficients, if necessary.*

The presence of two oxygen atoms on the left and three on the right shows that least common multiples will be necessary. The least common multiple of 2 and 3 is 6, so a coefficient of 3 will need to be added to the reactant oxygen, and a 2 will be necessary for the product.

$$Fe + 3O_2 \rightarrow 2Fe_2O_3$$

This balances the number of oxygen atoms on each side but leaves the iron atoms unbalanced. Since there are four iron atoms in the product, the addition of a 4 as the coefficient for the iron reactant easily fixes the problem.

$$4Fe + 3O_2 \rightarrow 2Fe_2O_3$$

A quick check shows that the equation is now balanced. The balanced equation reveals that four atoms of iron react with three molecules of oxygen to produce two formula units of iron (III) oxide.

7.4 WORKING WITH CHEMICAL EQUATIONS

Why do we bother with balancing chemical equations? First of all, we balance equations so that they will conform to the law of conservation of matter. But we also need to think about how chemists work. Do chemists work with individual atoms and molecules? Of course not! Even very small masses of atoms contain enormous numbers of particles. One nanogram (a billionth of a gram) of hydrogen gas contains over 597 *trillion* hydrogen atoms! To make working with these gigantic numbers easier, scientists use a quantity called a *mole*. Just as we can talk about a *pair* (2) of shoes, a *dozen* (12) doughnuts, or a *case* of copier paper (5000 sheets), a mole is a defined quantity of items. One **mole** (mol) is 6.02×10^{23} particles, either atoms, ions, or molecules.

Balanced chemical equations don't just tell us how many individual atoms, ions, or molecules take part in a reaction. They also tell us how many moles of those particles will react together. We've seen that carbon and oxygen can combine in a one-to-one ratio to form carbon monoxide. So if a chemist wanted to combine one mole of carbon atoms

---

with oxygen to form carbon monoxide, how many oxygen atoms would be needed? The answer is one mole. Similarly, the chemical equation we saw in the previous section, $4Fe + 3O_2 \longrightarrow 2Fe_2O_3$, can be read as, "Four moles of iron atoms react with three moles of oxygen molecules to produce two moles of iron (III) oxide formula units."

Mole-Mass Conversions

Still, chemists don't have time to count out moles of atoms. No one does! But there is a way to *measure* out moles of atoms! You saw in Chapter 5 that the atomic mass of an element as shown on the periodic table represents the average mass of an atom of that element. For example, an average carbon atom's mass is 12.01u. But 12.01 also happens to be the mass in grams of one mole of carbon atoms. So 12.01g of carbon contain 6.02×10^{23} atoms of carbon. This relationship is true for every other element on the periodic table as well. The mass of one mole of a substance is its **molar mass**. The molar mass of oxygen atoms is 16.00 g, of calcium atoms is 40.08 g, and of tin atoms is 118.71g.

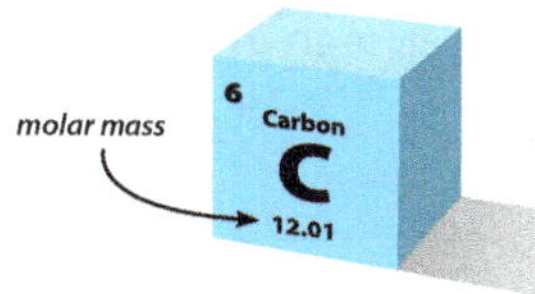

Suppose a chemist wished to carry out the iron (III) oxide reaction described in the last section. To produce 2 mol of iron (III) oxide, he needs 4 mol of iron and 3 mol of oxygen. With the help of conversion factors, he can easily calculate how many grams of each substance he needs. The relationship between moles of iron and grams of iron is shown in the conversion factors below.

$$\frac{1 \text{ mol Fe}}{55.85 \text{ g Fe}} \text{ and } \frac{55.85 \text{ g Fe}}{1 \text{ mol Fe}}$$

Likewise, the conversion factors for moles and grams of oxygen are as follows.

$$\frac{1 \text{ mol O}_2}{32.00 \text{ g O}_2} \text{ and } \frac{32.00 \text{ g O}_2}{1 \text{ mol O}_2}$$

Note that these conversion factors show oxygen as O_2 because pure oxygen exists as diatomic molecules. Therefore 1 mol of oxygen molecules consists of 2 mol of oxygen atoms. Now let's see how these conversion factors are used to solve the chemist's problem.

The wildlife personnel shown below are not just splashing and having fun. They're applying a chemical called *rotenone* to a pond. The rotenone will kill all the fish in the pond. That sounds like a terrible thing—why would they do it? It's usually done to eliminate invasive species. Rotenone degrades rapidly once applied, so native fish can be stocked back into the pond after several days. Rotenone is derived from plants. Indigenous peoples discovered that crushing these plants and adding them into a body of water would cause fish to come to the surface for air, where they could easily be caught.

The science of identifying and studying toxic substances, such as rotenone, is called *toxicology*. Toxicologists decode the chemical structure of toxins and figure out how they work. Many of these poisonous substances are found in nature. Like rotenone, some of them can be put to beneficial uses, such as helping to manage the environment or treat disease. There are still lots of toxins out there waiting to be discovered—maybe you'll be the first to identify one of them.

More on the Mole

Like some other fundamental concepts in chemistry, the understanding of the mole developed gradually over a fairly long period of time. Currently, the definition of a mole is a count that contains exactly $6.022\,140\,76 \times 10^{23}$ particles.

 Think, Pair, Share:
Mole Conversion Practice

Some students will need some practice to solve mole-mass conversions consistently. Give them a problem to solve and time to solve it on their own. Then have them share with another student and discuss between themselves whether each has solved it correctly. After a couple minutes, ask for a student to give the correct answer and verify. Some sample problems with their solutions are given below.

1. Give the molar mass of NaOH. *(40.00 g)*

2. Give the molar mass of NH_3. *(17.04 g)*

3. Give the molar mass of $Mg(OH)_2$. *(58.33 g)*

 Toxicology: *Organic Chemistry Connection*

Make a connection for your students between the name of the chemical being applied (rotenone) and your study of organic chemistry. See whether anyone can identify the type of organic compound and the functional group from the name *rotenone*. *(a ketone with a carbonyl group; p. 135)*

Checking the Calculation

To show students that the calculation at the end of the section is in fact correct, show them from Example 7-3 that making 2 mol of Fe_2O_3 requires 223.4 g of iron plus 96.00 g of oxygen, for a total of 319.4 g. Half that value, or 1 mol, is 159.7 g.

Formative Assessment: *Interpreting Coefficients*

Display the following equation:

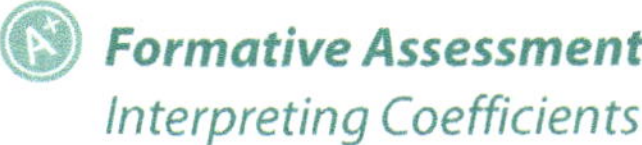

$$4FeS + 7O_2 \rightarrow 2Fe_2O_3 + 4SO_2$$

Ask, "If 6 mol of FeS reacted with an unlimited supply of oxygen, how many moles of Fe_2O_3 would be produced? *(3)*

EXAMPLE 7-3

How many grams of iron and oxygen are needed to make 2 mol of iron (III) oxide?

Use conversion factors to convert 4 mol of iron and 3 mol of oxygen to grams.

$$4\ \cancel{\text{mol Fe}} \times \frac{55.85\ \text{g Fe}}{1\ \cancel{\text{mol Fe}}} = 223.4\ \text{g Fe}$$

$$3\ \cancel{\text{mol O}_2} \times \frac{32.00\ \text{g O}_2}{1\ \cancel{\text{mol O}_2}} = 96.00\ \text{g O}_2$$

To make 2 mol of iron (III) oxide, the chemist needs 223.4 g of iron and 96.00 g of oxygen.

In the same manner as you saw for oxygen, the molar masses of compounds can be found by adding together the molar masses of the compound's elements. Remember, the formula Fe_2O_3 tells us that each formula unit of iron (III) oxide contains two atoms of iron and three atoms of oxygen. So 1 mol of Fe_2O_3 will contain 2 mol of iron and 3 mol of oxygen. So the calculation for the molar mass of iron (III) oxide will be as shown below.

$$\left[\left(\frac{2\ \cancel{\text{mol Fe}}}{1\ \text{mol Fe}_2\text{O}_3}\right)\left(\frac{55.85\ \text{g Fe}}{1\ \cancel{\text{mol Fe}}}\right)\right] + \left[\left(\frac{3\ \cancel{\text{mol O}}}{1\ \text{mol Fe}_2\text{O}_3}\right)\left(\frac{16.00\ \text{g O}}{1\ \cancel{\text{mol O}}}\right)\right] = \frac{159.7\ \text{g Fe}_2\text{O}_3}{1\ \text{mol Fe}_2\text{O}_3}$$

7A | REVIEW QUESTIONS

1. Why is melting ice not an example of a chemical change?
2. List five evidences that suggest a chemical change may have occurred.
3. How does a chemical equation model a chemical reaction?
4. Write a chemical equation that is based on the following word equation: "Carbon combines with oxygen to form carbon dioxide." You do not need to balance the equation.
5. Why do chemical equations need to be balanced?
6. Determine whether each of the following equations is balanced. Write "balanced" if the equation is balanced. If it is not, then add the coefficients needed to balance it.
   a. $N_2 + 3H_2 \rightarrow 2NH_3$
   b. $C + S_8 \rightarrow CS_2$
   c. $K + O_2 \rightarrow K_2O$
7. Write the balanced chemical equation for the decomposition of hydrogen peroxide (H_2O_2) into water and oxygen.

Use the equation below to answer Questions 8–9.

$$CH_4 + 4Cl_2 \rightarrow CCl_4 + 4HCl$$

8. A chemist wants to react 1 mol of methane (CH_4) with chlorine gas (Cl_2) to produce carbon tetrachloride (CCl_4). How many moles of chlorine gas will he need for this reaction?
9. How many grams of chlorine gas will the chemist need to react with 1 mol of methane?

7A REVIEW ANSWERS

1. Chemical changes change the identity of substances. Melting does not change the identity of a substance. *(p. 148)*
2. Accept any five of the following: the formation of a precipitate, the presence of bubbles, the release of energy, a temperature change, a color change, odor, burning, a change in composition. *(pp. 148–49)*
3. A chemical equation shows the kinds and amounts of substances that enter into a chemical reaction as well as the kinds and amounts of products that are formed. *(p. 150)*
4. $C + O_2 \rightarrow CO_2$ *(p. 150)*
5. Because atoms are neither created nor destroyed during a reaction, equations must show that the types and number of atoms in the reactants are equal to the types and number of atoms in the products. *(p. 150)*
6. a. balanced
   b. $4C + S_8 \rightarrow 4CS_2$
   c. $4K + O_2 \rightarrow 2K_2O$ *(all pp. 150–52)*
7. $2H_2O_2 \rightarrow 2H_2O + O_2$ *(pp. 151–52)*
8. 4 mol *(pp. 152–53)*
9. 283.60 g *(p. 154)*

BALANCED DIET

Want to make balancing chemical equations more appetizing? Then give this visual method a try!

Procedure

A Choose one color of candy to represent atoms of carbon, another for hydrogen, a third for oxygen, and a fourth to represent any other element.

B Set the ruler on the desk in front of you, pointing away. Set two cups to the left of the ruler and one to the right.

C Then have a look at the following chemical equation.

$$C_2H_2 + H_2 \rightarrow C_2H_6$$

D Use your colored candies to fill the cups, representing the reactants and product of the reaction. It will help if you place your "models" before you in the same order as they occur in the reaction.

1. What do the cups represent?

2. What does the ruler represent?

E Carefully count up the number and kind of each atom.

3. Does your model show the same number and kind of each atom on both the reactant and product sides of the equation?

F If necessary, add more cups of candies representing additional molecules of one or more of the substances in the equation until you have the same number and kind of atoms on each side of the equation. Remember, you can't add only part of a molecule! Molecules enter into reactions only as whole units.

4. What did you have to add in order to balance the equation?

5. What is the balanced form of the equation?

Going Further

G Use this edible model technique to balance the following equations. After finishing each model, write the balanced equation.

6. $NH_3 + O_2 \rightarrow N_2O + H_2O$

7. $C_4H_8O_4 + O_2 \rightarrow CO_2 + H_2O$

Essential Question:

How can I tell whether a chemical equation is balanced?

Equipment

colored candies, such as gumdrops
small cups
ruler

Edible Options

If you're trying to keep your students away from sugary treats, try substituting any kind of small, colored, but durable food items. You can of course use nonfood items as well—but they're not as much fun to eat afterward!

For each group, regardless of what you use, you will need at least four different colors and at least twenty-four of each color.

Balanced Diet

Answers

1. molecules

2. the yield arrow

3. No

4. It was necessary to add one more molecule of hydrogen (H_2).

5. $C_2H_2 + 2H_2 \rightarrow C_2H_6$

6. $2NH_3 + 2O_2 \rightarrow N_2O + 3H_2O$

7. $C_4H_8O_4 + 4O_2 \rightarrow 4CO_2 + 4H_2O$

 Think, Pair, Share: *Why Classify?*

Ask students to write down reasons why scientists classify chemical reactions. Have each of them share with another student and discuss. Then ask students to volunteer reasons that they agreed on. (*Possible reasons: It helps us understand better what is happening in a reaction and allows us to predict what substances might be produced.*)

Chemical Equations

Chemical equations often include extra symbols and abbreviations that indicate the physical states of the reactants and products, the presence of solutions, whether energy is needed, and other bits of information. For this introductory-level course, we have chosen to omit these extra details, though you may introduce them to students at your discretion. They will learn about these in a standard chemistry course.

Cracking

The refining of crude oil is actually a two-part process. The first part, *distillation*, sorts the molecules found in crude oil according to their boiling points, with longer hydrocarbon chains boiling at higher temperatures. This process yields a high proportion of longer, less valuable hydrocarbon chains. The second step, which breaks these long chains into smaller ones, is known as *cracking*.

Demonstrating Types of Chemical Reactions

Each of the following example reactions can be demonstrated to some extent.

Preparation:

1. *Synthesis*: Obtain two containers, a bottle of carbonated water, and two pieces of pH or litmus paper.

2. *Decomposition*: If an electrolysis apparatus is not available, do a keyword search for "electrolysis apparatus" or "electrolysis of water." You can find several possible do-it-yourself versions of this apparatus. Most involve two test tubes, a 9 V battery, and wires to act as leads. Some type of substance will need to be added to the water to carry the current. Baking soda is probably one of the safest.

3. *Single Replacement*: Dissolve about 5 g of silver nitrate in 50 mL of water. Obtain a 4–6 in. length of copper wire.

(continued)

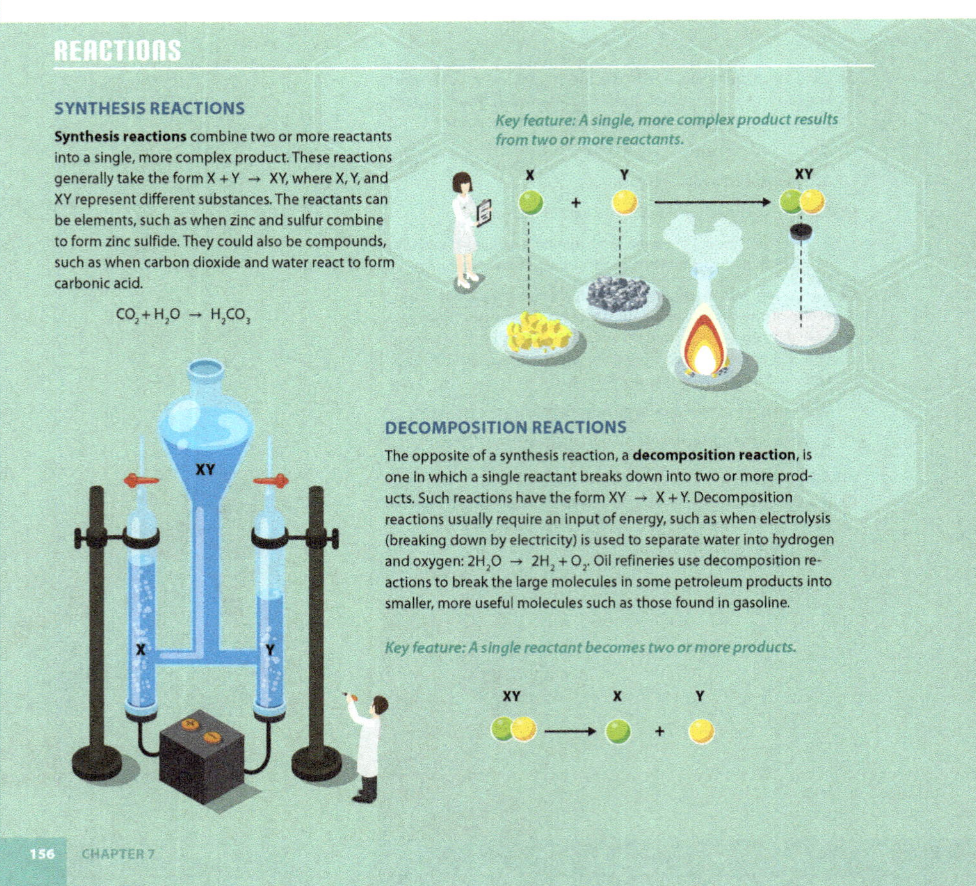

TEACHING THE MATERIAL

ESSENTIAL QUESTION

Are there different kinds of chemical reactions?

OBJECTIVES

- 7B1 Classify chemical reactions.
- 7B2 Compare oxidation and reduction.

RESOURCES

 Demonstrating Types of Chemical Reactions

Lab 7A: *Science Fair Revisited*—Types of Chemical Reactions

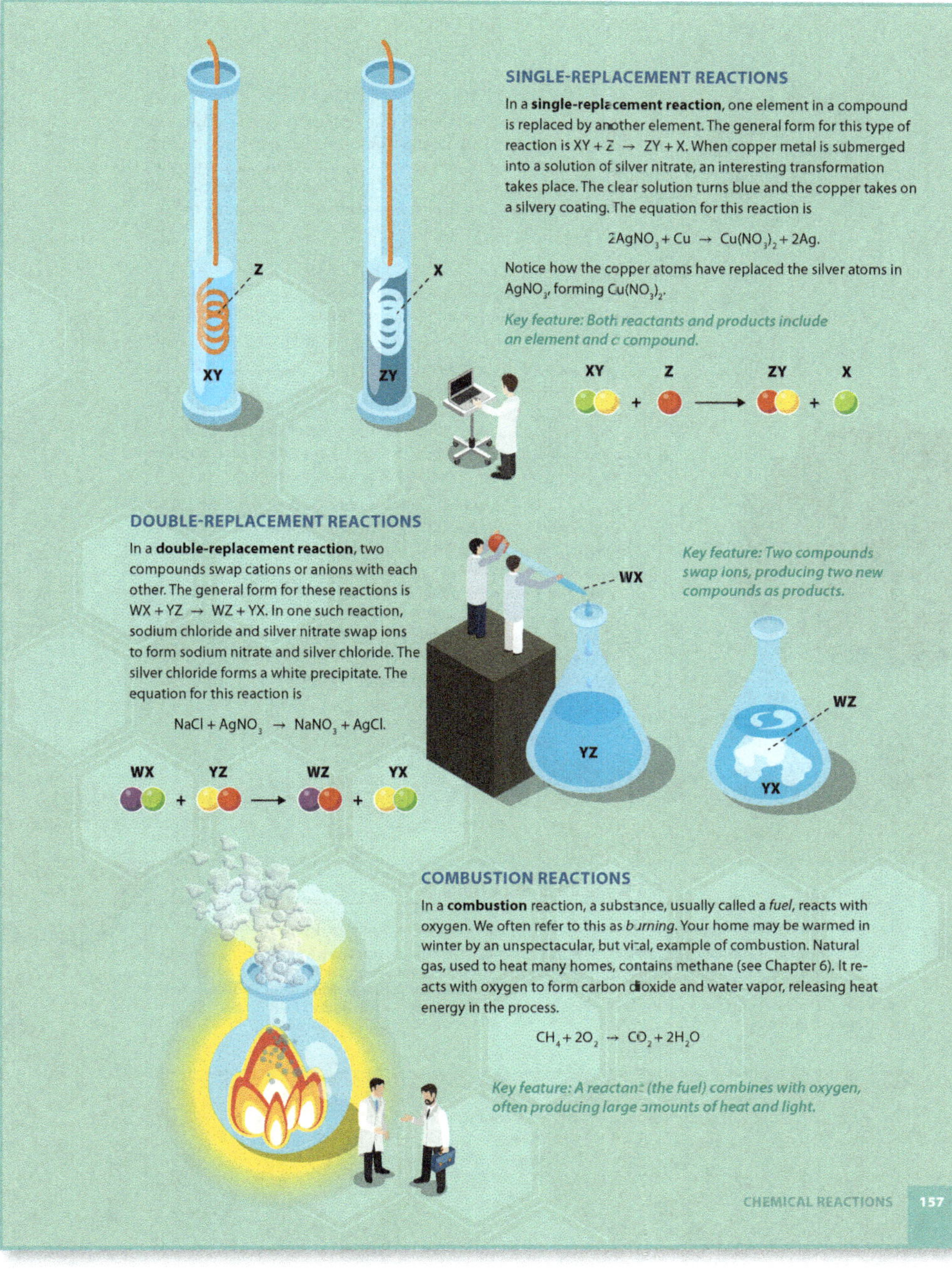

4. *Double Replacement*: Prepare aqueous solutions of silver nitrate and sodium chloride by stirring in approximately 2 g of silver nitrate and 1 g of salt each in separate beakers containing approximately 50 mL of water.

5. *Combustion*: If your laboratory does not have natural gas, you can substitute a propane burner.

Performance/Discussion:

Synthesis: Pour some carbonated water into a beaker. Pour the same amount of distilled water into a beaker. Dip a piece of blue litmus paper, pH paper, or the probe of a pH meter into the pure water and one into the carbonated water.

1. Why do you think that the litmus paper turned pink in the carbonated water and not in the pure water? (Modify the question for pH paper or a pH meter.) *(because there is an additional substance in the water—an acid)* Some students may not know that pink litmus means that the solution is acidic since acids have not yet been studied.

2. Where did the acid come from? *(It was formed when the water and carbon dioxide combined.)*

Decomposition: If you have an electrolysis apparatus, you can show the decomposition of water. Put a large pinch of baking soda in water in a container in which you can insert the leads for the electrolysis apparatus. Connect the leads to the battery and insert them into the water underneath the tubes in the beaker.

3. How can you tell which tube the hydrogen is collecting in? *(There are twice as many hydrogen atoms in each water molecule as there are oxygen. So there should be twice as many hydrogen molecules and twice as much gas produced compared with oxygen.)*

Single Replacement: Partially fill a test tube with the silver nitrate solution and insert the copper wire and let it sit while going on to the next reaction. This may take some time, and if you are short on time in class, you can have two test tubes with silver nitrate solution. Put a piece of copper wire in one test tube the night before and put a piece of copper wire in the other one during class.

4. If copper is replacing the silver, what new compound is being formed? *[copper(II) nitrate]*

(continued)

Class Opener: Use the Think, Pair, Share: Why Classify? teacher note on page 156 to stimulate students' thinking regarding types of chemical reactions.

Ticket Out the Door: Write a balanced equation for a redox reaction involving calcium. (2Ca + O₂ → 2CaO or Ca + S → CaS are possibilities.)

Double Replacement: Pour some of the salt water into the silver nitrate solution.

5. If the resulting precipitate is silver chloride, what is the other compound produced? *(sodium nitrate)*

Combustion: Light a burner to show combustion of either mostly methane (natural gas) or propane.

6. If combustion forms water, why don't we see the water being produced? *(Because of the amount of heat involved, the water is in vapor form and quickly mixes with the air.)*

Alternative Term
Replacement reactions are sometimes called *displacement reactions*.

⁉️ More on Combustion
There is some disagreement on what exactly is meant by the term *combustion*. We usually think of heat and flames when we think of combustion, but the term can also be used to describe smoldering that occurs slowly, at low temperatures, and without flames. Students may also note that combustion reactions are a subset of redox reactions, which are treated in the next subsection.

⁉️ What Is Reduced?
It is important to help students realize that reduction is referring to the lowering of the oxidation number of the element. Adding negatively charged electrons reduces the oxidation number or state. Oxidation, on the other hand, refers to the increase of an oxidation state. Although it most commonly occurs through the action of oxygen, it can also happen through the action of elements such as sulfur, fluorine, bromine, or chlorine.

🥽 Lab 7A: *Science Fair Revisited*
Once students have an understanding of the material in Section 7B, they should be ready to work through Lab 7A.

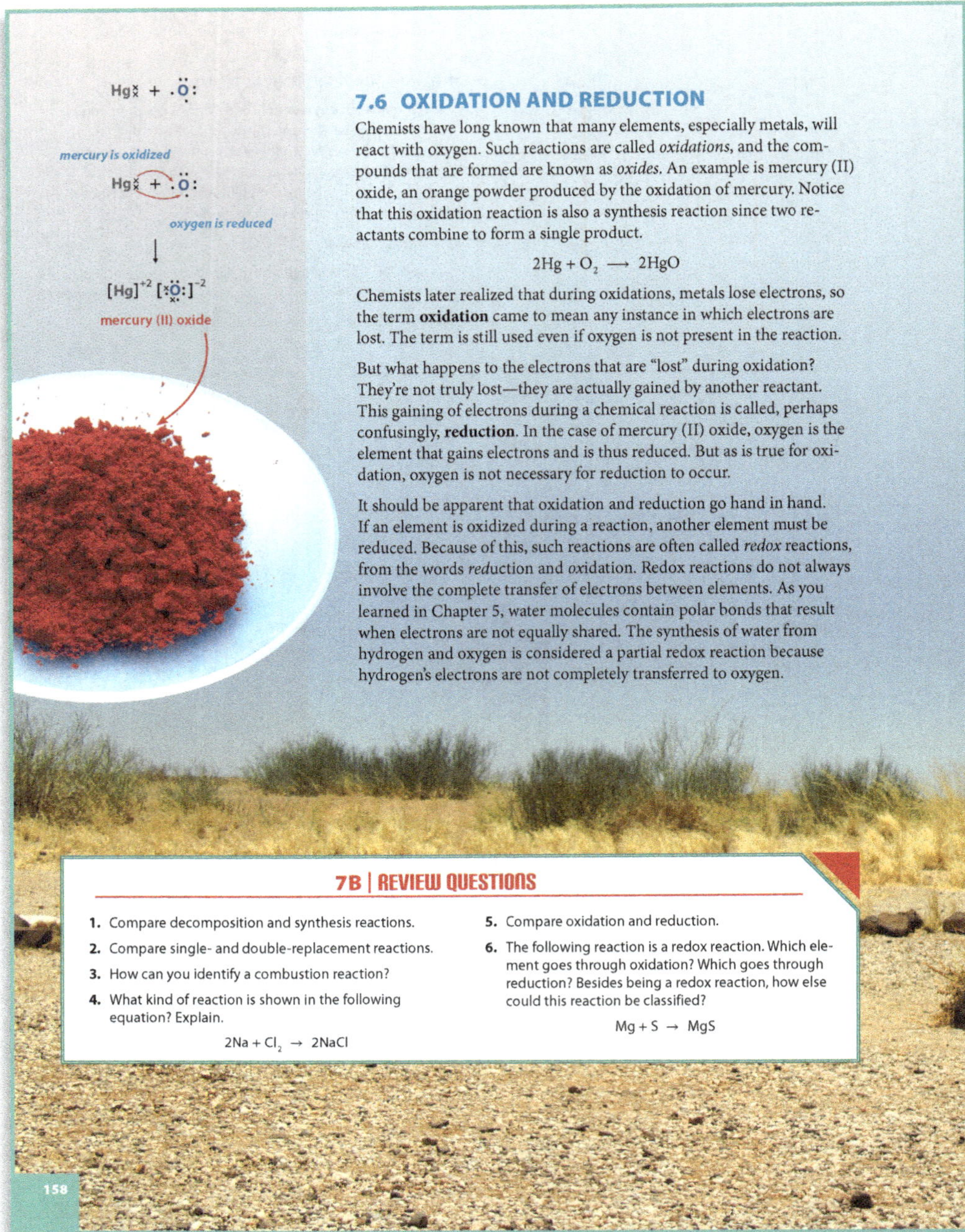

7.6 OXIDATION AND REDUCTION
Chemists have long known that many elements, especially metals, will react with oxygen. Such reactions are called *oxidations*, and the compounds that are formed are known as *oxides*. An example is mercury (II) oxide, an orange powder produced by the oxidation of mercury. Notice that this oxidation reaction is also a synthesis reaction since two reactants combine to form a single product.

$$2Hg + O_2 \longrightarrow 2HgO$$

Chemists later realized that during oxidations, metals lose electrons, so the term **oxidation** came to mean any instance in which electrons are lost. The term is still used even if oxygen is not present in the reaction.

But what happens to the electrons that are "lost" during oxidation? They're not truly lost—they are actually gained by another reactant. This gaining of electrons during a chemical reaction is called, perhaps confusingly, **reduction**. In the case of mercury (II) oxide, oxygen is the element that gains electrons and is thus reduced. But as is true for oxidation, oxygen is not necessary for reduction to occur.

It should be apparent that oxidation and reduction go hand in hand. If an element is oxidized during a reaction, another element must be reduced. Because of this, such reactions are often called *redox* reactions, from the words *reduction* and *oxidation*. Redox reactions do not always involve the complete transfer of electrons between elements. As you learned in Chapter 5, water molecules contain polar bonds that result when electrons are not equally shared. The synthesis of water from hydrogen and oxygen is considered a partial redox reaction because hydrogen's electrons are not completely transferred to oxygen.

7B | REVIEW QUESTIONS

1. Compare decomposition and synthesis reactions.

2. Compare single- and double-replacement reactions.

3. How can you identify a combustion reaction?

4. What kind of reaction is shown in the following equation? Explain.

$$2Na + Cl_2 \rightarrow 2NaCl$$

5. Compare oxidation and reduction.

6. The following reaction is a redox reaction. Which element goes through oxidation? Which goes through reduction? Besides being a redox reaction, how else could this reaction be classified?

$$Mg + S \rightarrow MgS$$

7B REVIEW ANSWERS

1. Decomposition reactions break a single reactant down into two or more products. Synthesis reactions combine two or more reactants into a single product. *(p. 156)*

2. In a single-replacement reaction, one element in a compound is replaced by a lone element. The reaction has the form XY + Z → ZY + X. In a double-replacement reaction, two compounds swap cations or anions with each other. The general form for these reactions is WX + YZ → WZ + YX. *(p. 157)*

3. Combustion reactions include oxygen as a reactant and often release large amounts of energy in the form of heat and light. *(p. 157)*

4. This is a synthesis reaction because two reactants combine to form a single product. Some students may also recognize it as a redox reaction. *(pp. 156, 158)*

5. In oxidation, an element loses electrons. In reduction, an element gains electrons. During a redox reaction, one reactant is oxidized while another is reduced. *(p. 158)*

6. Magnesium is oxidized and sulfur is reduced. This is also a synthesis reaction. *(pp. 156, 158)*

1. While holding up a match before the students, ask, "What would it take to get this match burning?" *(Strike the match.)*

2. Ask, "Could it burn without striking it?" *(Yes)* "How?" *(if it got hot enough)*

3. "In either case, how would that start the match burning?" *(by providing energy to break chemical bonds)*

4. Ask, "What is happening energy-wise when the match is burning?" *(Energy is being released.)*

5. "Overall, is more energy given off or put into the reaction?" *(given off)*

6. "What is the primary use for matches?" *(to start fires or get something burning, that is, to start other combustion reactions)*

7. "Besides fire or heat, what are some other sources of energy for chemical reactions? Give an example." *(light for photosynthesis, glucose [or ATP] for cellular respiration, a spark from an engine spark plug, or ATP for other cellular processes)*

7C | ENERGY IN CHEMICAL REACTIONS

Do chemical reactions always give off energy?

7.7 UNDERSTANDING ENERGY EXCHANGES

Sure, something like a rocket launch is an exciting chemical reaction to see, and it obviously releases lots of energy too. Smoke, noise, heat, flames—there's plenty of evidence that a dramatic reaction is happening. But what about this old, rusting truck? It's experiencing a chemical reaction, too, only it takes years to happen instead of a few minutes. Does rusting release energy? Yes, it does, but it sure doesn't seem like it!

As you learned in Chapter 5, energy is stored in the chemical bonds that form between atoms. To start a chemical reaction, additional energy is required to break the chemical bonds within reactants, but when bonds are reformed afterward in products, energy is released. This relationship between absorbing and releasing energy determines whether the overall reaction itself will give off or absorb energy.

7C Questions

- How does energy change during a chemical reaction?
- Can energy be created or destroyed?

7C Terms

exothermic reaction, endothermic reaction, activation energy, law of conservation of energy

TEACHING THE MATERIAL

ESSENTIAL QUESTION

Do chemical reactions always give off energy?

OBJECTIVES

- 7C1 Explain the flow of energy in exothermic and endothermic reactions.
- 7C2 Relate exothermic and endothermic reactions to their energy graphs.
- 7C3 Define *activation energy*.

RESOURCES

Demonstrating Endothermic and Exothermic Reactions

(continued)

Exo or Endo?

Students should recall the prefixes *exo-* and *endo-* from their study of life science. *Exo-* means "out," as seen in the word *exoskeleton*, which is on the outside of an organism. *Endo-* means "in." *Endoskeletons* are on the inside of an organism.

⁉️ Light or Heat?

Observant students may notice that photosynthesis is described here as endothermic, even though the energy being absorbed is light, not heat. Strictly speaking, such reactions are described as being endergonic (the opposite is exergonic), but it is not necessary to make this fine a distinction at this grade level.

Demonstrating Endothermic and Exothermic Reactions

This demonstration can be done if you are not planning to do Lab 7B, which also involves endothermic and exothermic reactions. Otherwise, you may choose to skip this demonstration. Note that this demonstration provides a visual, quantitative illustration, whereas Lab 7B allows students to feel the temperature change.

Preparation:

Set up a small beaker with about 10 mL of baking soda solution in it and a beaker containing about 50 mL of water. Put a thermometer or temperature probe in each. Some type of digital display would be ideal. If you have only laboratory thermometers, have a student read aloud the temperatures of the mixtures. Measure out about 2 g of calcium chloride and about 30 grams of ammonium nitrate.

Performance:

Mix the 2 g of calcium chloride with the baking soda solution and allow students to see the rise in temperature.

Mix the ammonium nitrate with the water and allow students to see the decrease in temperature.

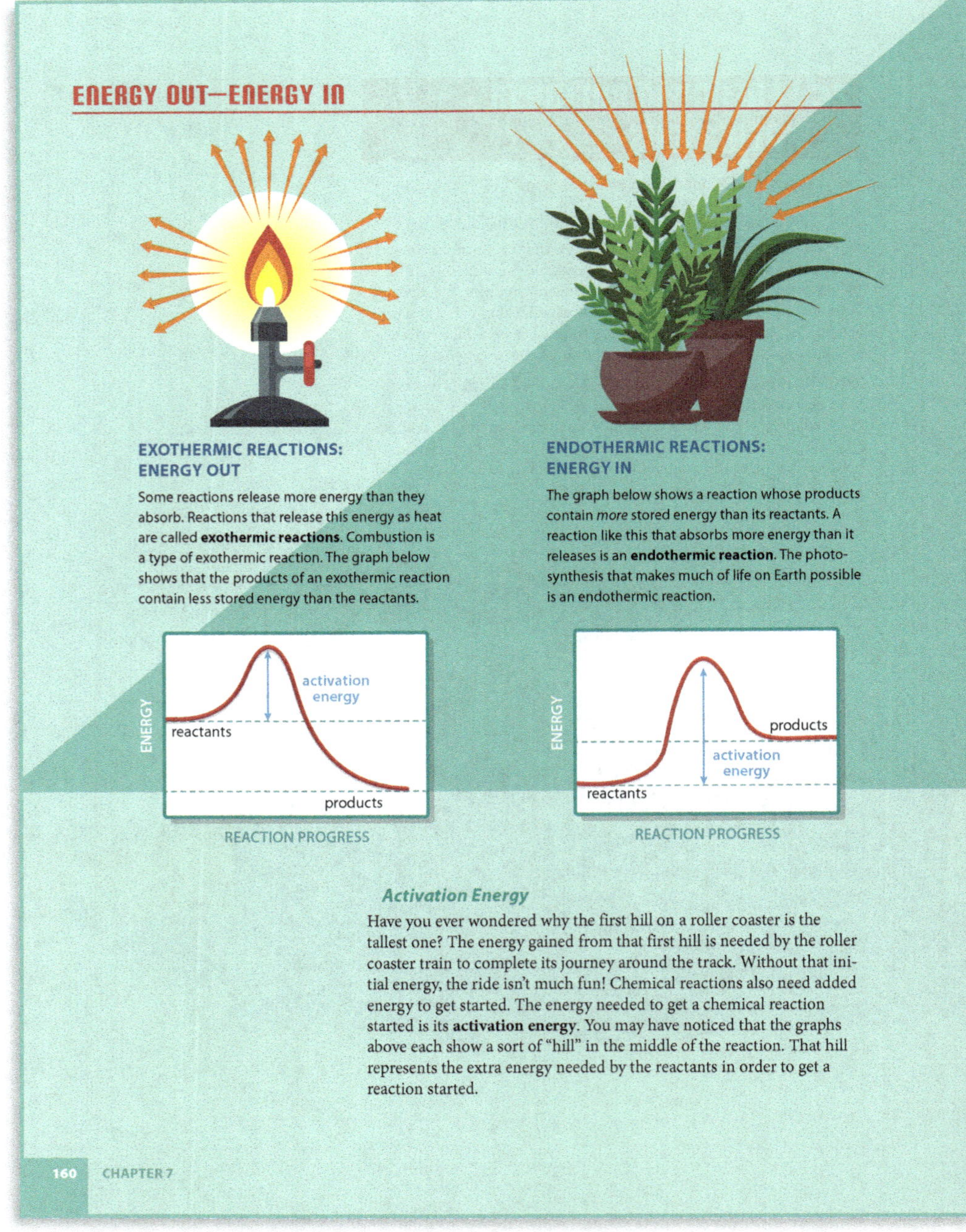

TEACHING THE MATERIAL

STRATEGIES

Class Opener: Use the Discussion: Striking Up a Reaction teacher note on page 159 to stimulate students' thinking regarding energy involved in chemical reactions.

Demonstration: Use the Demonstrating Endothermic and Exothermic Reactions teacher note on this page to differentiate between endothermic and exothermic reactions.

Ticket Out the Door: With your book closed, draw on a piece of paper a basic graph for an exothermic reaction and one for an endothermic reaction. Include labels.

7.8 CONSERVATION OF ENERGY

A campfire almost completely consumes the wood that is burned, while photosynthesis creates sugars in plants. It's tempting to think that exothermic reactions destroy energy while endothermic reactions create it. Such thinking would be a serious error, though, for it is contrary to one of the fundamental laws of science. The **law of conservation of energy** states that during a chemical reaction no energy is created or destroyed, but only changed from one form to another. The diagrams at right show how this works.

As the stored chemical energy in a piece of firewood is changed into thermal energy, where does it go? It is radiated out into and absorbed by the environment (including you, if you are standing near enough). This energy that seems to be "lost" is actually shown on the graph (above, right) as the difference in the *y*-axis values for the energy found in the reactants and products. Where does the stored chemical energy in a plant come from? It started out as light energy produced by the sun. Similar to the exothermic example, the difference in energy between reactants and products can also be seen in the graph (far right) of an endothermic reaction. In this case the difference between the *y*-axis values represents the energy that is gained during the reaction. So even though some reactions may *look* like violations of the law of conservation of energy, they clearly are not.

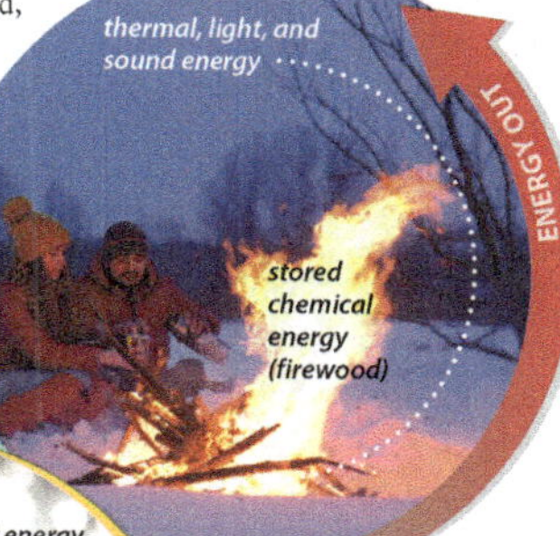

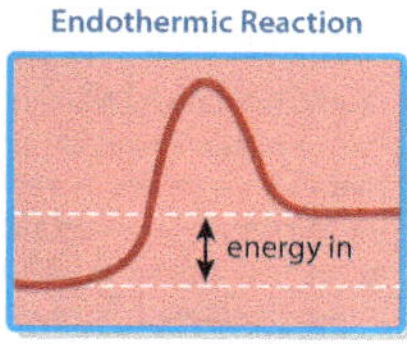

7C | REVIEW QUESTIONS

1. A reaction whose products have more energy than its reactants is a(n) ______ reaction.

2. Is combustion a type of exothermic or endothermic reaction?

3. What is activation energy?

Use the information in the graph at right to answer Questions 4–5.

4. Label parts A, B, and C of the graph.

5. Does the graph represent an exothermic or endothermic reaction?

6. What must be true about the energy given off by an exothermic reaction and the energy remaining in the products according to the law of conservation of energy?

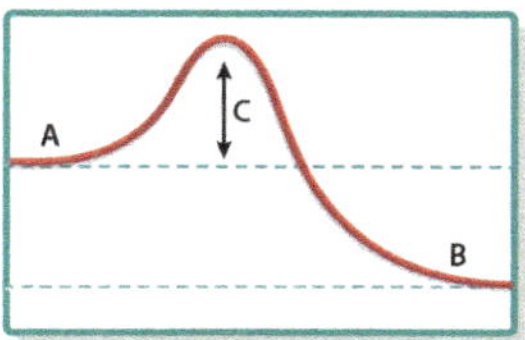

7C REVIEW ANSWERS

1. endothermic *(p. 160)*

2. exothermic *(p. 160)*

3. the energy needed to get a chemical reaction started *(p. 160)*

4. A: energy of reactants

   B: energy of products

   C: activation energy *(all p. 160)*

5. exothermic *(pp. 160–61)*

6. Together they must equal the energy originally in the reactants. *(p. 161)*

Speeding Up Reactions

Ask students to suggest ways to affect the speed at which a reaction occurs. They might suggest changing the temperature, stirring, breaking the reactants into smaller pieces, changing the concentration of the reactants, or adding a catalyst.

Reaction Simulation

For a link to a simulation, visit TeacherTools Online.com. Under Additional Resources, click on Chapter 7 Web Links. Then click on *Reactions and Rates*. Students can simulate the effect of varying amounts of reactants and varying temperatures to see the effects on reaction rate and equilibrium.

7D | REACTION RATES AND EQUILIBRIUM

Why do some things burn slowly while others explode?

7.9 REACTION RATES

We see reactions happen all the time. Cars rust, wood burns, dead plants decay, and fireworks explode. Some reactions happen lightning fast, while others occur with almost glacial slowness. Some things explode and others do not. Why is this true?

Collision Model

As we think about why some reactions happen quickly while others are slow, we must consider the conditions needed for a reaction to occur at all. Our current model, the **collision model**, states that we need three things to happen before reactants will react. First, the reactant particles must collide with each other. Then, when they collide, they must be

FACTORS THAT AFFECT REACTION RATES

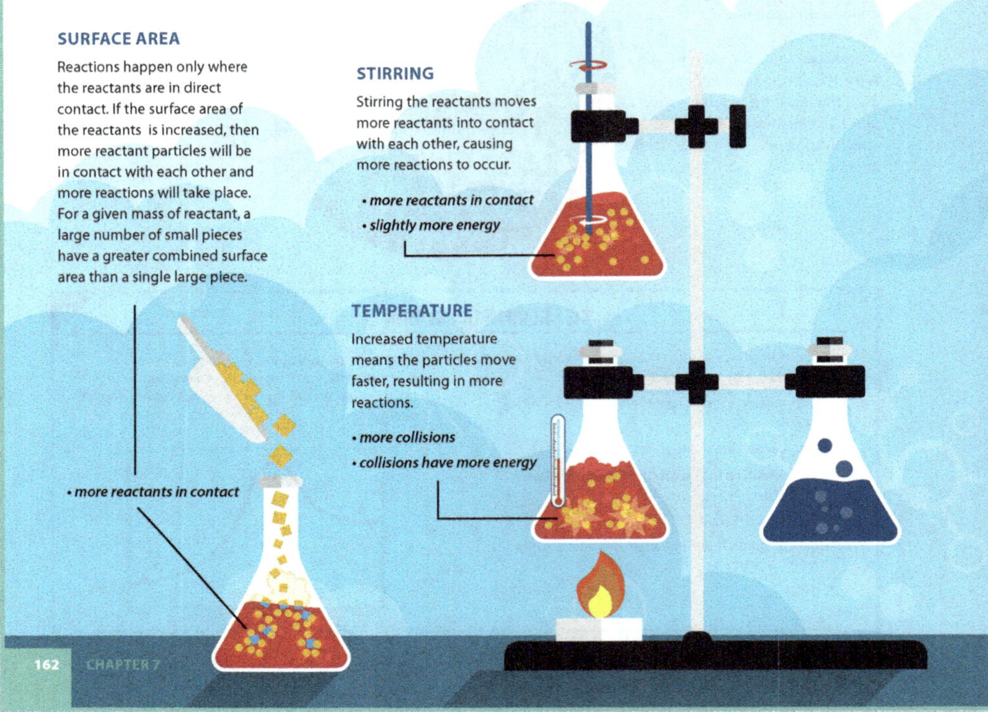

TEACHING THE MATERIAL

ESSENTIAL QUESTION

Why do some things burn slowly while others explode?

OBJECTIVES

- 7D1 Summarize the collision model.
- 7D2 Predict how reaction rate is affected by reaction conditions.
- 7D3 Compare the actions of catalysts, inhibitors, and enzymes.
- 7D4 Summarize the law of chemical equilibrium.
- 7D5 Predict adjustments within a reversible chemical reaction in response to specific changes in conditions by using Le Châtelier's principle.

aligned properly in order to rearrange their bonds. Finally, there must be enough energy. But what does "enough energy" mean? Let's look again at the energy diagrams for endothermic and exothermic reactions.

To move from the reactants to the products on this graph, we have to get all the way to the top of the activation energy "hill." The activation energy is the "enough energy." If the collision occurs, is in the right alignment, and there is the proper amount of energy, then the reaction occurs. How quickly this all happens depends on the likelihood that the reactant particles will collide with the proper alignment and energy.

The speed of a reaction is called its **reaction rate**. This is an indication of how quickly reactants change into products. Many things can cause a chemical reaction to speed up or slow down, that is, change the reaction's rate. These things include temperature, concentration, surface area, and stirring. Let's look at how these things affect the collisions between particles in a reaction.

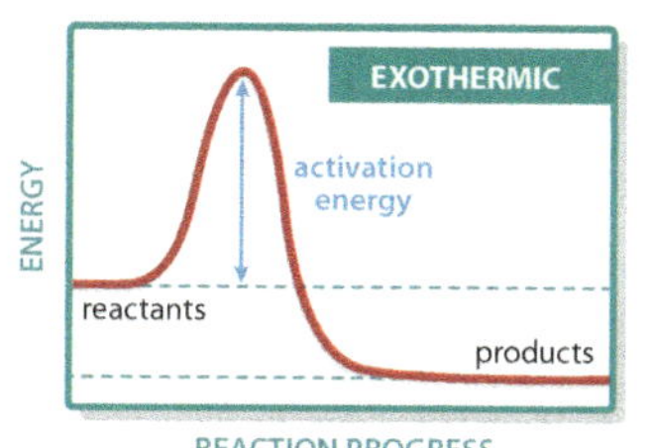

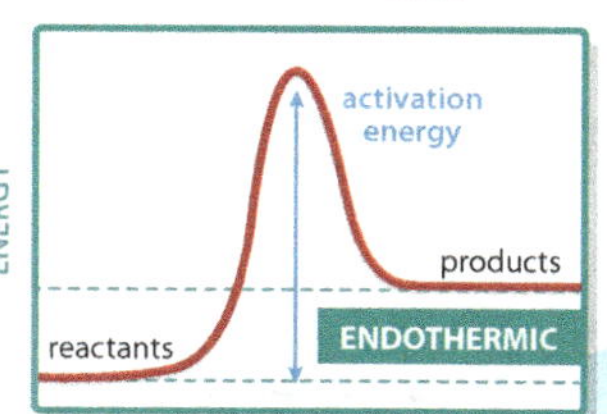

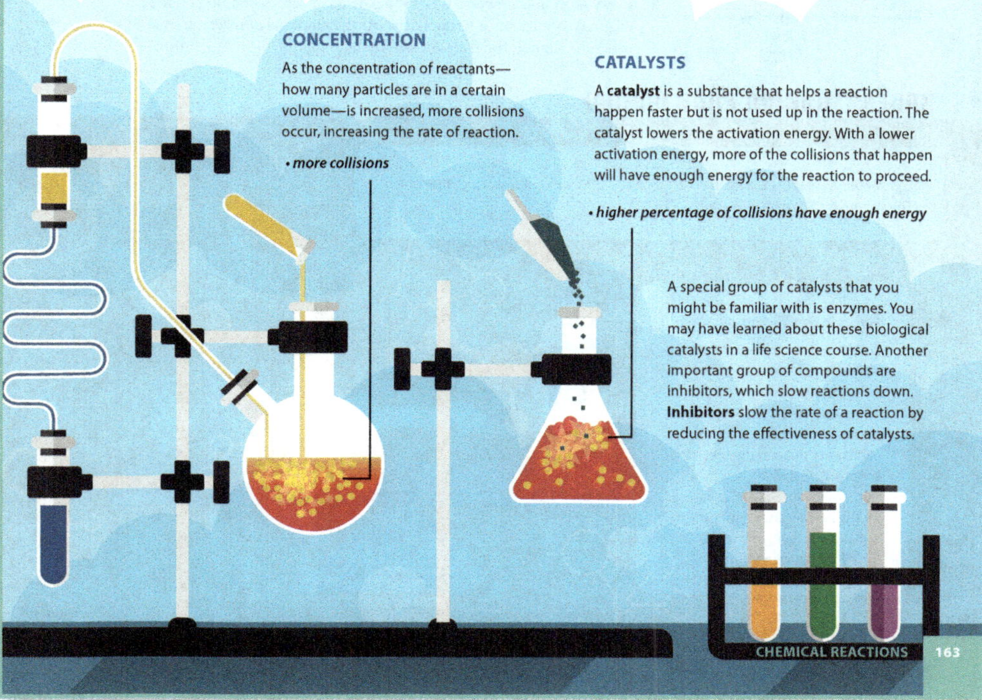

Rates

Students should be familiar with rates and rate problems from mathematics. You may want to review the concept for students that are struggling with it.

Concentration

Some students may not be familiar with the concept of concentration. This will be covered in more depth in Chapter 9. For the purpose of reaction rates, students need to understand that as we increase the concentration, the reactants become closer to each other, increasing the number of collisions.

Fizz or No Fizz

Ask a student volunteer to describe the action of hydrogen peroxide when poured on an open wound. Then pour some household hydrogen peroxide on your hand to show that there is no reaction if the skin is unbroken. Ask, "Why do you think that hydrogen peroxide fizzes on a cut but not on normal, unbroken skin?" *(Blood and cells contain an enzyme called catalase that causes the reaction. If there is no blood or damaged cells, there is no reaction.)*

RESOURCES

Case Studies: Grain Elevators (p. 172), Building Implosion (p. 173)

Reaction Simulation

Lab 7B: *It's in the Bag*—Inquiring into Chemical Reactions

STRATEGIES

Class Opener: Use the Speeding Up Reactions teacher note on page 162 to stimulate students' thinking about reaction rates.

Alternative Class Opener: Use the Fizz or No Fizz teacher note on this page to stimulate students' thinking about why reactions may or may not happen depending on circumstances.

Video: Do a keyword search at TED-Ed for a video on chemical equilibrium.

Ticket Out the Door: Write down three factors that can affect the direction of a reaction that involves substances in a state of chemical equilibrium.

Equilibrium

Some students may wrestle with the concept of equilibrium. For example, if an object is stationary, students tend to think that there are no forces acting on it. But that isn't so, as students will see later in the textbook. In chemistry, the temptation is to assume that when a reversible reaction is in equilibrium, the interplay between reactions has stopped. In reality, both the forward and reverse reactions continue but at the same rate.

7.10 REVERSIBLE REACTIONS AND EQUILIBRIUM

Reversible Reactions

You might think that a reaction continues until all the reactants are turned into products. Think about a burning candle, which is a combustion reaction. If we were to light the candle and leave it, we would expect it to burn until all the fuel (the candle wax) was used. This reaction is an irreversible reaction because it progresses only in one direction. The candle and oxygen form soot and some gases. The soot and gases can't recombine to form the candle again. But many reactions are **reversible reactions**. Not only do the reactants form products, but the products can also react together to re-form the original reactants. The equations for these reactions differ from those you have already seen in that there are yield arrows in both directions. The reaction between hydrogen and iodine to form hydrogen iodide gas is reversible.

$$H_2 + I_2 \rightleftharpoons 2HI$$

For the reverse reaction to occur, the product(s) must remain contained. If this is done for the hydrogen iodide reaction, some of the hydrogen iodide will decompose back into hydrogen and iodine gases.

Equilibrium

Every reversible reaction starts at a particular rate depending on the original conditions in its container. As the amounts of reactants and products change, the reaction rate changes too.

UNDERSTANDING EQUILIBRIUM

Equilibrium represents a state in which multiple influences cancel each other out. The system is said to be balanced.

Forces in Equilibrium

Equilibrium can refer to forces that balance each other. Think of balancing with a friend on a seesaw. The forces that you each apply to the seesaw cancel each other out and the seesaw is balanced, that is, it's in equilibrium. There are still forces acting on the seesaw, but its motion remains unchanged because the forces are balanced.

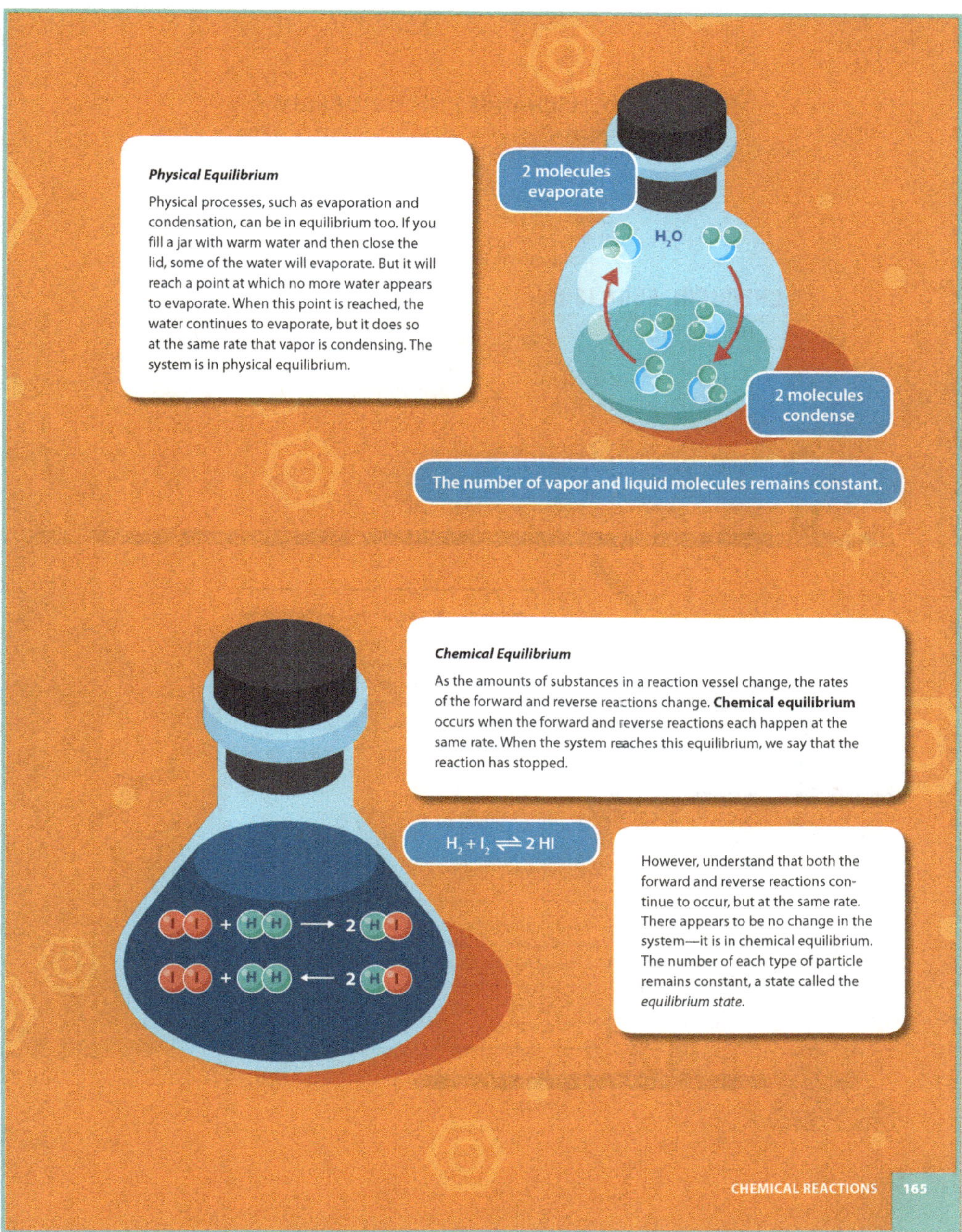

Physical Equilibrium

Physical processes, such as evaporation and condensation, can be in equilibrium too. If you fill a jar with warm water and then close the lid, some of the water will evaporate. But it will reach a point at which no more water appears to evaporate. When this point is reached, the water continues to evaporate, but it does so at the same rate that vapor is condensing. The system is in physical equilibrium.

Chemical Equilibrium

As the amounts of substances in a reaction vessel change, the rates of the forward and reverse reactions change. **Chemical equilibrium** occurs when the forward and reverse reactions each happen at the same rate. When the system reaches this equilibrium, we say that the reaction has stopped.

However, understand that both the forward and reverse reactions continue to occur, but at the same rate. There appears to be no change in the system—it is in chemical equilibrium. The number of each type of particle remains constant, a state called the *equilibrium state*.

Once a system is in equilibrium, changes to the system can alter the equilibrium state. This happens as either the forward or reverse reaction rate increases. Over time the system will reach a new equilibrium state. Changes to temperature, pressure, or concentration can cause these changes. Let's use the following equation for the exothermic reaction of nitrogen and hydrogen gases forming gaseous ammonia to see how changing conditions can affect the equilibrium state.

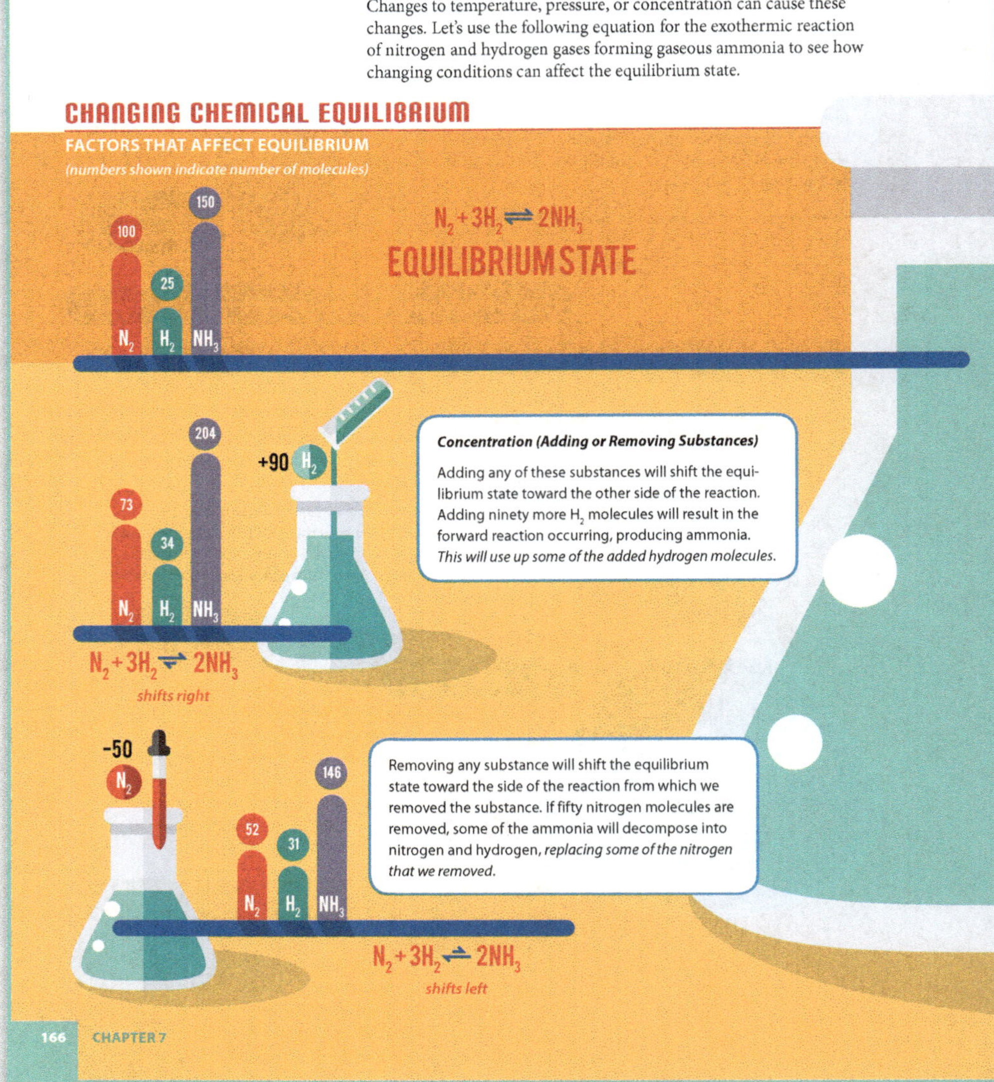

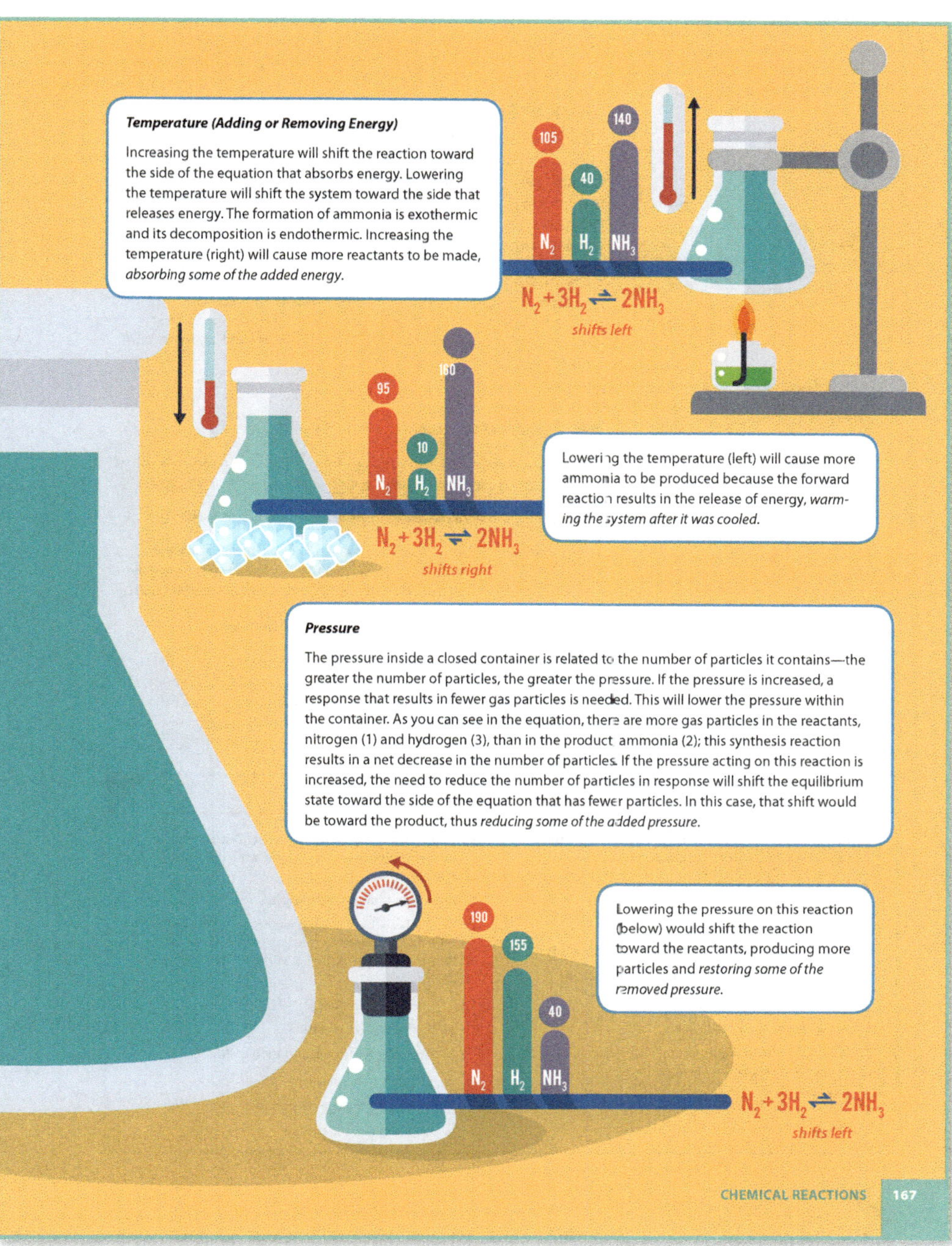

Temperature (Adding or Removing Energy)

Increasing the temperature will shift the reaction toward the side of the equation that absorbs energy. Lowering the temperature will shift the system toward the side that releases energy. The formation of ammonia is exothermic and its decomposition is endothermic. Increasing the temperature (right) will cause more reactants to be made, *absorbing some of the added energy.*

Lowering the temperature (left) will cause more ammonia to be produced because the forward reaction results in the release of energy, *warming the system after it was cooled.*

Pressure

The pressure inside a closed container is related to the number of particles it contains—the greater the number of particles, the greater the pressure. If the pressure is increased, a response that results in fewer gas particles is needed. This will lower the pressure within the container. As you can see in the equation, there are more gas particles in the reactants, nitrogen (1) and hydrogen (3), than in the product ammonia (2); this synthesis reaction results in a net decrease in the number of particles. If the pressure acting on this reaction is increased, the need to reduce the number of particles in response will shift the equilibrium state toward the side of the equation that has fewer particles. In this case, that shift would be toward the product, thus *reducing some of the added pressure.*

Lowering the pressure on this reaction (below) would shift the reaction toward the reactants, producing more particles and *restoring some of the removed pressure.*

Notice the results of the changes that were made to the ammonia synthesis reaction. In each case, the system tries to undo the change. French Chemist Henri Le Châtelier studied the effects of changes to reactions that were in equilibrium. Today **Le Châtelier's principle** states that any chemical system in equilibrium will adjust its equilibrium state in such a way as to reduce the effect of any changes made to the system.

EXAMPLE 7-4:
Predicting the Effect of Changes on Chemical Equilibrium

Consider the exothermic reaction of gaseous sulfur dioxide and oxygen to form the gas sulfur trioxide.

$$2SO_2 + O_2 \rightleftharpoons 2SO_3$$

What will happen if SO_3 is added?

If we add SO_3, the system will try to correct the change, so the reverse reaction will increase, forming additional reactants and decomposing some of the added SO_3.

What will be the effect if the temperature of the system is increased?

Energy is added when the temperature is raised. This will cause the reaction to proceed in the endothermic direction, shifting the equilibrium state toward the left and using up some of the added energy. SO_3 will decompose into SO_2 and O_2.

What will occur if the pressure within the system is increased?

If the pressure is increased, the system will shift to reduce the pressure. This is done by decreasing the number of gas particles. Since the synthesis of SO_3 decreases the number of particles within the system, an increase in pressure will shift the equilibrium to the right, using SO_2 and O_2 to form more SO_3.

7D | REVIEW QUESTIONS

1. According to the collision model, what are the three requirements for a reaction to occur?

2. Predict how the following factors affect reaction rate (if they do at all).

   a. raising the temperature

   b. decreasing the concentration of the reactants

   c. decreasing the surface area of the reactants

   d. stirring

   e. applying a magnetic field

3. Why are there no enzymes that slow down the rates of chemical reactions in living things?

4. Define *chemical equilibrium*.

5. What is the effect of adding a chemical substance to a reversible reaction that is in equilibrium? Give an example.

6. Explain Le Châtelier's principle.

7. Predict how the following factors will shift equilibrium for the following reaction (if they do at all) using Le Châtelier's principle. The forward reaction is exothermic.

$$4Li + O_2 \rightleftharpoons 2Li_2O$$

   a. raising the temperature

   b. decreasing the concentration of lithium

   c. increasing the pressure

   d. increasing the concentration of lithium oxide

   e. applying a magnetic field

7D REVIEW ANSWERS

1. The reactants (1) must collide (2) with the correct alignment and (3) with enough energy (the activation energy). *(pp. 162–63)*

2. a. increases the rate

   b. decreases the rate

   c. decreases the rate

   d. increases the rate

   e. does not affect the rate
   (all pp. 162–63)

3. Enzymes are catalysts, so they can speed up chemical reactions, not slow them down. *(p. 163)*

4. Chemical equilibrium is the state of balance in a reversible reaction in which the forward and reverse reactions are occurring at the same rate. There appears to be no change in the reaction, and the reaction is thought of as being complete. *(p. 165)*

5. Adding any substance will move the equilibrium state toward the other side. For example, if you add reactant, it will move the equilibrium state toward the product(s). *(p. 166)*

6. Le Châtelier's principle states that a system in equilibrium will adjust its equilibrium state to reduce the effect of any changes made to the system. *(p. 168)*

7. a. shifts the reaction toward the reactants

   b. shifts the reaction toward the reactants

   c. shifts the reaction toward the product

   d. shifts the reaction toward the reactants

   e. does not affect equilibrium
   (all pp. 166–68)

👓 **Lab 7B:** *It's in the Bag*

Once students have an understanding of the material in Section 7D, they should be ready to work through Lab 7B.

CHAPTER 7 REVIEW . . . .

7A CHEMICAL CHANGES

- Evidences that a chemical reaction may have occurred include formation of a precipitate or bubbles, the release of energy, temperature or color change, odor, burning, or a change in composition.

- A single evidence by itself may not be sufficient to prove that a chemical reaction has occurred since some physical changes also produce such evidence.

- Chemical equations use chemical formulas and coefficients to show the number and kinds of reactants and products in a chemical reaction.

- The number and kind of atoms shown in chemical equations must be balanced in order to fulfill the requirements of the law of conservation of matter.

- Balanced chemical equations show not only the number of individual particles that enter into a chemical reaction but also the number of moles of each substance.

- The number of moles of a substance can be converted to the mass of that substance using mole-mass conversion factors.

7A Terms

chemical reaction	148
chemical equation	150
reactant	150
product	150
coefficient	150
mole	152
molar mass	153

7B TYPES OF CHEMICAL REACTIONS

- The five main kinds of chemical reactions are synthesis, decomposition, single-replacement, double-replacement, and combustion reactions.

- Combustion occurs when a substance reacts with oxygen.

- In a redox reaction, one reactant is oxidized (loses electrons) while another is reduced (gains electrons).

7B Terms

synthesis reaction	156
decomposition reaction	156
single-replacement reaction	157
double-replacement reaction	157
combustion	157
oxidation	158
reduction	158

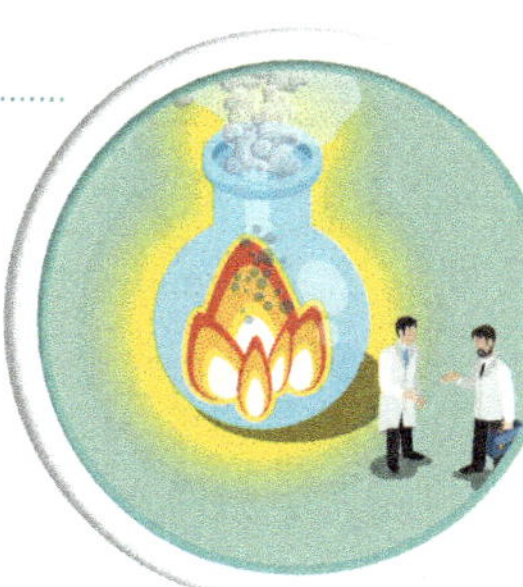

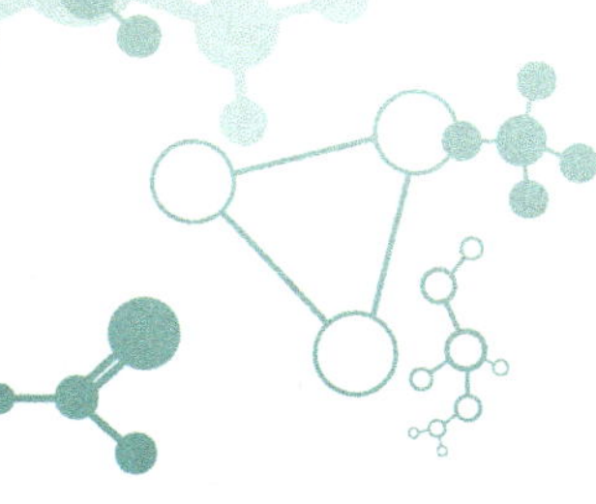

7C   ENERGY IN CHEMICAL REACTIONS

- Exothermic reactions release thermal energy, while endothermic reactions absorb it.
- Chemical reactions require an initial input of energy—the activation energy—in order to proceed.
- The law of conservation of energy states that during a chemical reaction no energy is created or destroyed, but only changed from one form to another.

7C Terms

7D   REACTION RATES AND EQUILIBRIUM

- Reactions occur only when reactants collide with proper alignment and enough energy.
- A reaction rate can be changed by changing the temperature of the reaction, by stirring the reactants, or by changing the concentration or surface area of the reactants.
- Catalysts increase reaction rate by lowering the activation energy. Inhibitors slow the rate by interfering with catalysts. Enzymes are biological catalysts.
- A reversible chemical reaction has reached equilibrium when the forward and reverse reaction rates are equal. The reaction appears to have stopped.
- Le Châtelier's principle states that a system in equilibrium will adjust to reduce the effect of any changes made to the system.

7D Terms

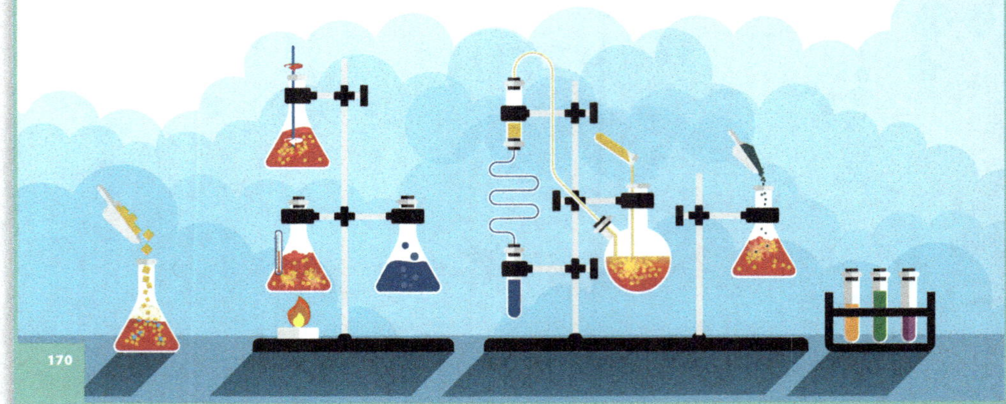

REVIEW

CHAPTER REVIEW QUESTIONS

Recalling Facts

1. Which of the following is never an evidence of chemical change?
   a. formation of bubbles
   b. change in phase from solid to liquid
   c. change in temperature
   d. production of an odor

2. What is a combustion reaction?

3. Why does a reaction in which one substance is oxidized happen only in the presence of another substance that is reduced?

4. Which type of reaction gives off thermal energy? Which kind absorbs thermal energy?

5. How much energy is enough to cause a reaction to occur?

6. List the conditions given in the textbook that can cause a change in the equilibrium state.

7. Why does changing the temperature change the equilibrium state?

Understanding Concepts

In Questions 8–10, write a balanced chemical equation for each reaction.

8. Magnesium and oxygen react to form magnesium oxide.

9. Sodium sulfide decomposes into sodium and sulfur (consider sulfur to exist as individual atoms).

10. Zinc and hydrogen chloride react to form zinc chloride and hydrogen gas (assume +2 oxidation state for zinc).

11. For each of the equations that you balanced in Questions 8–10, how many moles of each reactant and product were involved in the reaction?

Calculate the molar mass for each of the following compounds.

12. K_2O
13. $AgNO_3$
14. C_2H_6
15. MgS

16. If 4 mol of hydrogen molecules are reacted with 4 mol of iodine molecules, how many grams of hydrogen iodide will be produced?

Classify each of the following as either a synthesis, decomposition, single-replacement, or double-replacement reaction.

17. $2AgNO_3 + Zn \rightarrow 2Ag + Zn(NO_3)_2$
18. $2KClO_3 \rightarrow 2KCl + 3O_2$
19. $BaCl_2 + Na_2SO_4 \rightarrow BaSO_4 + 2NaCl$
20. $2K + Cl_2 \rightarrow 2KCl$

21. Compare combustion reactions with redox reactions.

22. Why is it incorrect to say that all oxidation reactions involve oxygen?

11. *Question 8*: 2 mol each of magnesium and magnesium oxide and 1 mol of oxygen

    Question 9: 1 mol of sodium sulfide, 2 mol of sodium, and 1 mol of sulfur

    Question 10: 1 mol of zinc, 2 mol of hydrogen chloride, and 1 mol each of zinc chloride and hydrogen gas
    (all pp. 152–53)

12. 94.20 g/mol (p. 154)

13. 169.88 g/mol (p. 154)

14. 30.08 g/mol (p. 154)

15. 56.37 g/mol (p. 154)

16. 1023.28 g (pp. 153–54) (Students will underestimate the correct value by half if they fail to remember that both hydrogen and iodine are diatomic.)

17. single replacement (pp. 156–57)

18. decomposition (pp. 156–57)

19. double replacement (pp. 156–57)

20. synthesis (pp. 156–57)

21. Combustion reactions require the presence of oxygen. A redox reaction might involve oxygen, but there are substances other than oxygen that can accept electrons in a redox reaction. (pp. 157–58)

22. The concept of oxidation began to be understood when it was realized that metals that reacted with oxygen were losing electrons. The term *oxidation* eventually came to mean any reaction in which electrons are being lost by a reactant whether oxygen is involved or not. (p. 158)

CHAPTER REVIEW ANSWERS

Recalling Facts

1. b. (pp. 148–49)

2. A combustion reaction is one in which a substance reacts with oxygen. (p. 157)

3. The substance being reduced is needed to accept the electrons that are lost by the substance being oxidized. (p. 158)

4. exothermic; endothermic (p. 160)

5. an amount equal to or greater than the activation energy (p. 160)

6. concentration, pressure, and temperature (pp. 166–67)

7. When temperature changes, energy is added or removed from the system. The system will move in the direction of the endothermic process to reduce some of the energy that was added. (p. 167)

Understanding Concepts

8. $2Mg + O_2 \rightarrow 2MgO$ (pp. 150–52)

9. $Na_2S \rightarrow 2Na + S$ (pp. 150–52)

10. $Zn + 2HCl \rightarrow ZnCl_2 + H_2$ (pp. 150–52)

23. **C** *(p. 160)*

24. **B** *(p. 160)*

25. endothermic *(p. 160)*

26. The reaction will not proceed. *(p. 160)*

27. No. This would be a violation of the law of conservation of energy. *(p. 161)*

28. Increasing the temperature makes the particles move faster. This results in more collisions and more energy for those collisions, both of which increase the reaction rate. *(p. 162)*

29. No. For reactants to react, the collision must be properly aligned and have enough energy to cause a reaction. *(pp. 162–63)*

30. All three change the reaction rate. Catalysts cause reactions to happen faster because they lower the activation energy of the reaction. Enzymes are biological catalysts. Inhibitors slow the reaction by interfering with catalysts. *(p. 163)*

31. The equilibrium state will shift left to use some of the added CO_2. This will use some of the H_2O and produce more C_2H_6 and O_2. *(pp. 162, 166)*

32. Since the forward reaction is exothermic, removing energy will shift the equilibrium state toward the right. This will use up some of the C_2H_6 and O_2 and produce more CO_2 and H_2O. *(pp. 163, 167)*

33. Adding CO or O_2, removing CO_2, cooling the system, or increasing the pressure would all cause more CO_2 to be produced. *(pp. 162–63, 167)*

34. Changes in volume change the space available for the chemical substances. If the substances are gases, the pressure would be changed. Increasing the volume would shift the equilibrium state toward the side with more gas molecules. Decreasing volume would shift the equilibrium state toward the side with fewer gas molecules. *(pp. 163, 166–67)*

Use the graph below to answer Questions 23–25.

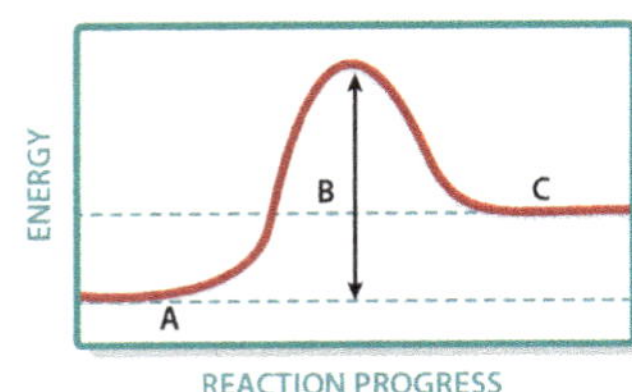

23. Which portion of the graph shows the energy of the products?

24. Which portion of the graph shows the activation energy?

25. Does this graph show an exothermic or endothermic reaction?

26. What will happen if the necessary substances for a reaction are placed together but the activation energy is not available?

27. Can a chemical reaction ever give off more energy than was originally contained in its reactants? Explain.

28. Why does increasing the temperature increase a reaction rate?

29. Do reactants that collide always react together? Explain.

30. Compare catalysts, inhibitors, and enzymes.

For Questions 31–32, consider the following equation for a system in equilibrium. All substances in the reaction are gases and the forward reaction is exothermic.

$$2C_2H_6 + 7O_2 \rightleftharpoons 4CO_2 + 6H_2O$$

31. What would be the effect of adding CO_2 to the system?

32. What would be the effect of cooling the system?

For Questions 33–34, consider the following equation for a system in equilibrium. All substances in the reaction are gases and the forward reaction is exothermic.

$$2CO + O_2 \rightleftharpoons 2CO_2$$

33. How could you cause more CO_2 to be produced?

34. Hypothesize about how changes in volume would affect the equilibrium state.

Critical Thinking

Use the Case Study on the left to answer Questions 35–36.

35. Review the information in Subsection 7.9 on reaction rates. What conditions found in grain storage silos do you think contribute to the danger of possible explosion?

36. What might be done to reduce the risk of explosion in a grain silo?

CASE STUDY: GRAIN ELEVATORS

Grain elevators are large facilities used for storing harvested grain and loading it onto ships or railway cars. On occasion such elevators have exploded, often causing considerable damage and even loss of life. The culprit is the very fine grain dust that circulates within the storage silos at the elevator site.

35. If the silo is located in a warm climate, the increased temperature will increase the number of collisions between grain dust particles and oxygen in the air. The fine dust can circulate on air currents within the silo—a process akin to stirring—adding energy to collisions between the dust and oxygen in the air. The smaller size of the dust particles also increases the total surface area of the grain product, increasing the number of collisions. *(pp. 162–63)*

36. Answers will vary. Since most of the danger is due to the fine particulate nature of the grain dust, preventing or reducing dust accumulation, incorporating dust removal systems into silo designs, and ensuring that such systems operate at maximum efficiency will all reduce the risk of explosion.

37. When sugar ($C_{12}H_{22}O_{11}$) and sulfuric acid (H_2SO_4) are mixed, they form a steaming black solid that grows out of the mixture. Using this information, evaluate whether a chemical change has taken place. Support your claim with evidence.

38. Given the following equation, what could you have changed to produce the results shown in the table? The forward reaction is exothermic and all substances are gases.

$$CO + 2H_2 \rightleftharpoons CH_3OH$$

Equilibrium State	CO	H_2	CH_3OH
Initial	21	18	374
Final	24	24	371

CASE STUDY: BUILDING IMPLOSION

Some pro sports teams play in iconic stadiums, such as Chicago's Wrigley Field. But the old Seattle King-dome, once home to baseball's Mariners and foot-ball's Seahawks, was *not* one of those venues. The Kingdome's vast interior and small baseball crowds once earned it the nickname *The Tomb*. In addition, hit baseballs that crazily bounced off structures like loudspeakers were often still in play, which made for some interesting game outcomes. Not too many tears were shed when the decision was made to demolish the old stadium and build a new one.

Tearing down such a massive structure can be very expensive and time-consuming. One solution: bring the old structure down with explosives, a process called *implosion*.

Use this Case Study to answer Questions 39–40.

39. Think about what a demolition needs to accomplish. Would that best be done by an exothermic or endothermic chemical reaction?

40. Implosions are done with many smaller demolition charges rather than a single large one. Why do you think this is so?

37. A chemical change has taken place: a color change, the production of a gas, and the release of heat are all present. *(pp. 148–49)*

38. There are more reactants and fewer products, indicating a shift to the left. A shift to the left could be caused by adding some product, removing reactants, decreasing pressure, or increasing temperature. Some students may also rule out (1) adding product (would have increased all three) and (2) removing reactants (would have lowered the product and raised only one of the reactants). Thus, only increasing the temperature or decreasing pressure could have caused the data given. *(pp. 164–65, 166–67)*

39. In order to demolish a large building, large amounts of energy must be released. This requires the use of an exothermic chemical reaction. *(pp. 159–60)*

40. Answers will vary. The area over which an explosive scatters debris increases with the size of the charge. Large charges tend to scatter debris over a large area and in this situation have the potential to harm bystanders or damage nearby buildings. A succession of smaller charges is used to weaken a building so that it will collapse under its own weight.

These apples are being irradiated. The process kills any organisms in or on the apples, improving the apples' shelf life. The apples do not become radioactive in the process.

CHAPTER 8
Nuclear Changes

RED, RIPE, AND IRRADIATED?

The world population is quickly heading toward 8 billion people. How can we ensure that all these people have enough food and that the food they have is safe? The food we eat comes from all over the world. But before it reaches the market, some will decay or be eaten by insects. And of the food that does reach us, some is tainted by disease. Scientists are always working to solve these issues.

Scientists have learned that by treating food with *ionizing radiation*, a form of *nuclear radiation*, they can slow decay, kill insects, and destroy the bacteria that cause disease. But is irradiated food safe to eat? Over thirty years of testing has shown us that treated food is safe for humans. Furthermore, the radiation does not affect the nutritional value of food or change its taste. People in the United States eat over 100 million kilograms of treated food each year.

In this chapter, we will learn about nuclear radiation, including the benefits and risks of this amazing aspect of chemistry. So grab a red, ripe, and possibly irradiated apple as we learn about radiation.

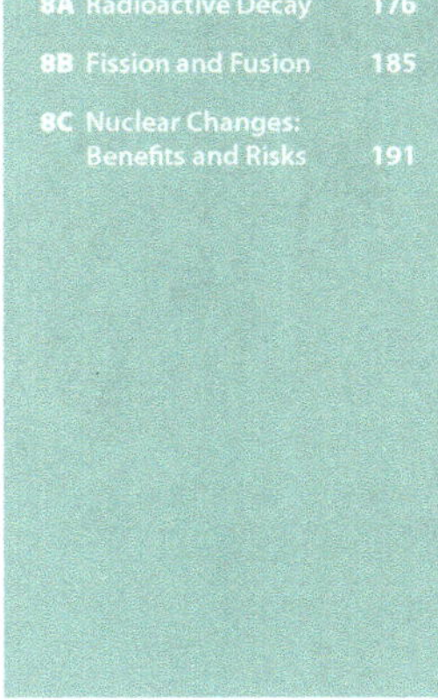

Main image: *The apples shown at left are being treated with ionizing radiation to kill insects and microbes.*

Inset: *The green emblem indicates irradiated food.*

Lab Activities

Lab 8A: *Flipping Out*—In this activity, students use coins to model radioactive decay.

Lab 8B: *Radioactive!*—In this activity, students explore common sources of radiation exposure and how a person's lifestyle affects his exposure.

Superheroes and Monsters

Begin class by asking students to name a monster or superhero that was created by exposure to radiation. (Since this is a common theme in comics and science fiction, there should be no shortage. Examples include Spiderman, the Incredible Hulk, and the Fantastic Four.) Use this to direct their attention to the real history of radiation.

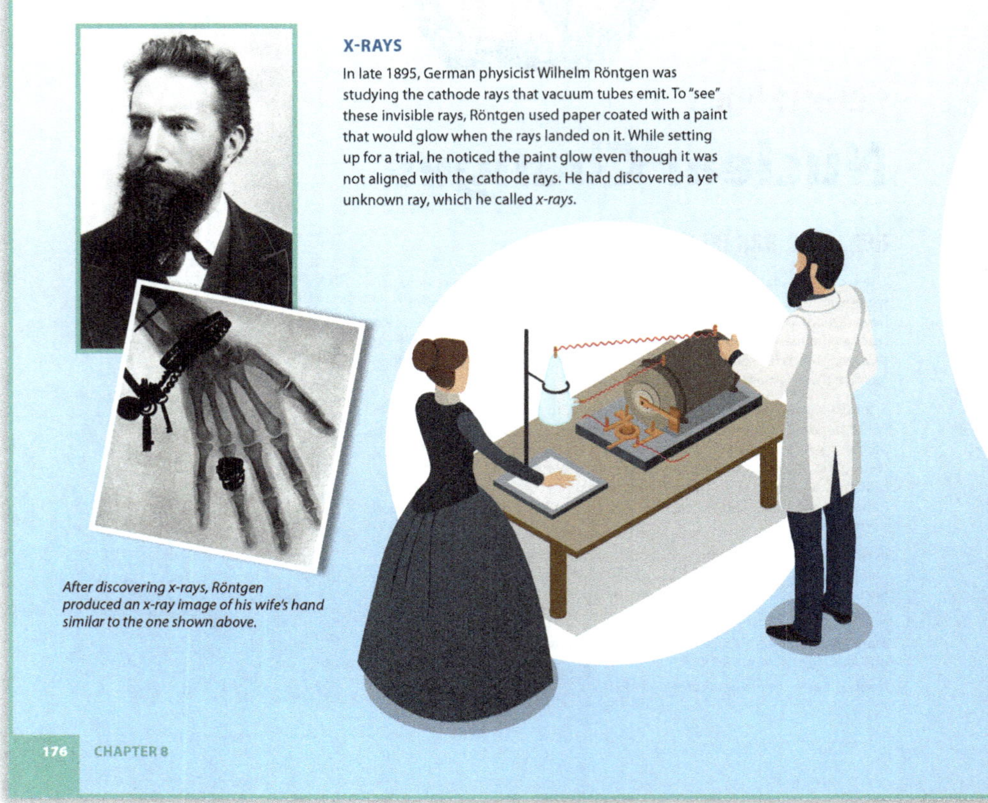

8A Questions

- How was radioactivity discovered?
- Why don't all atoms decay?
- What types of radioactive decay exist?
- Can we know when an atom will decay?
- How can we use decay to determine the age of an object?

8A Terms

strong force, radioactive decay, alpha decay, beta decay, gamma decay, half-life

8A | RADIOACTIVE DECAY

Why do only some isotopes decay?

8.1 RIDDLES IN THE DARK

In Chapter 2 you learned about physical changes—changes that alter the appearance of something but not its makeup. Chapter 2 also mentioned chemical changes, which you learned more about in Chapter 7. A chemical change alters the composition of a material by breaking and forming bonds to rearrange the atoms into new substances. In this chapter, you will look at *nuclear changes*, which change the particles that are in the nucleus, usually resulting in a different element.

DISCOVERING RADIOACTIVITY

X-RAYS

In late 1895, German physicist Wilhelm Röntgen was studying the cathode rays that vacuum tubes emit. To "see" these invisible rays, Röntgen used paper coated with a paint that would glow when the rays landed on it. While setting up for a trial, he noticed the paint glow even though it was not aligned with the cathode rays. He had discovered a yet unknown ray, which he called *x-rays*.

After discovering x-rays, Röntgen produced an x-ray image of his wife's hand similar to the one shown above.

TEACHING THE MATERIAL

ESSENTIAL QUESTION

Why do only some isotopes decay?

OBJECTIVES

- 8A1 Compare physical, chemical, and nuclear changes.
- 8A2 Explain why some isotopes decay and others do not.
- 8A3 Define *radioactive decay*.
- 8A4 Classify nuclear decay by type.
- 8A5 Write balanced nuclear decay equations.
- 8A6 Define *half-life*.
- 8A7 Predict how much of a sample remains after a given amount of time on the basis of its half-life.

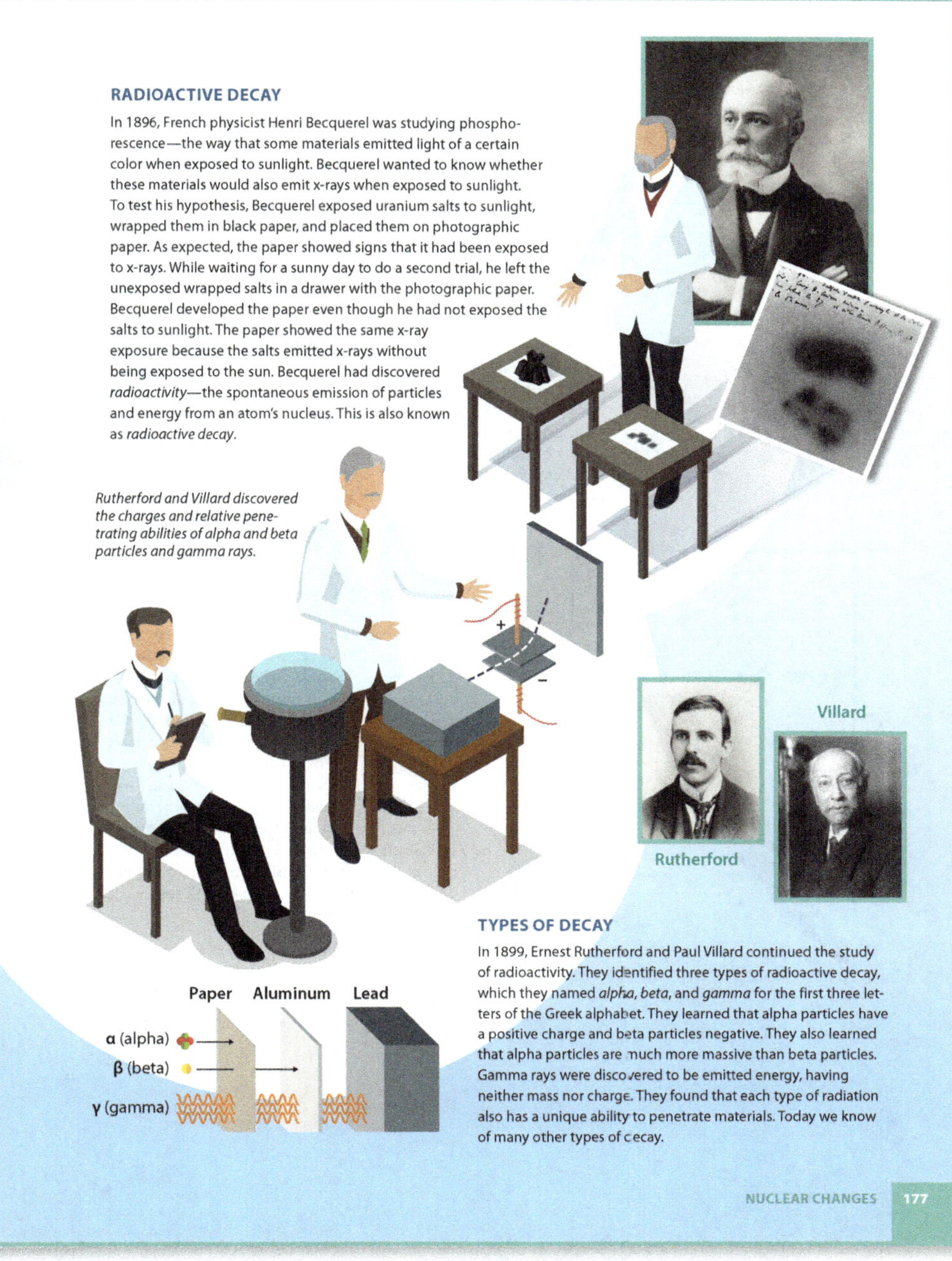

RADIOACTIVE DECAY

In 1896, French physicist Henri Becquerel was studying phosphorescence—the way that some materials emitted light of a certain color when exposed to sunlight. Becquerel wanted to know whether these materials would also emit x-rays when exposed to sunlight. To test his hypothesis, Becquerel exposed uranium salts to sunlight, wrapped them in black paper, and placed them on photographic paper. As expected, the paper showed signs that it had been exposed to x-rays. While waiting for a sunny day to do a second trial, he left the unexposed wrapped salts in a drawer with the photographic paper. Becquerel developed the paper even though he had not exposed the salts to sunlight. The paper showed the same x-ray exposure because the salts emitted x-rays without being exposed to the sun. Becquerel had discovered *radioactivity*—the spontaneous emission of particles and energy from an atom's nucleus. This is also known as *radioactive decay*.

Rutherford and Villard discovered the charges and relative penetrating abilities of alpha and beta particles and gamma rays.

TYPES OF DECAY

In 1899, Ernest Rutherford and Paul Villard continued the study of radioactivity. They identified three types of radioactive decay, which they named *alpha*, *beta*, and *gamma* for the first three letters of the Greek alphabet. They learned that alpha particles have a positive charge and beta particles negative. They also learned that alpha particles are much more massive than beta particles. Gamma rays were discovered to be emitted energy, having neither mass nor charge. They found that each type of radiation also has a unique ability to penetrate materials. Today we know of many other types of decay.

⁉️ *Relative Dangers of Radiation*

Radiation does damage, and students often wonder which type of radiation (alpha, beta, or gamma) is the most damaging. It's a complex issue that requires factoring in the mass of the particles and the ability of the particles to penetrate into the body.

Alpha radiation's massiveness compared with beta particles and gamma rays makes it both the least and the most dangerous form of radiation Alpha radiation can penetrate only a few centimeters of air, and any that strikes a human is almost always stopped by dead skin cells. But if radioactive material is ingested, any that produces alpha particles will be the most dangerous. With nothing to protect the body, the massiveness of alpha particles causes them to do an immense amount of damage to human tissue.

RESOURCES

Case Study: Vikings

Demonstrating Half-Life

How It Works: Smoke Detectors

Lab 8A: *Flipping Out*—Modeling Radioactive Decay

STRATEGIES

Class Opener: Use the Superheroes and Monsters teacher note on page 176 to begin a discussion of radiation.

Formative Assessments: Use the Formative Assessment: Radioactivity and Nuclear Stability teacher note on page 179 to assess students' understanding of these concepts.

Differentiated Instruction: Consider using some of the various techniques mentioned in the teacher notes on pages 180 and 181 to help struggling students.

Formative Assessments: Use the Formative Assessment: Radioactive Decay teacher note on page 181 to check students' understanding of decay.

(continued)

Strong Force Resources

There are several informative video clips available online. Conduct an internet search using the keywords "strong nuclear force video."

Electromagnetic Force

The electromagnetic force mentioned in the Student Edition is the force responsible for protons repelling each other. It is also responsible for like magnetic poles repelling each other.

Some students may notice that the neutron-to-proton ratio for stable nuclei becomes greater as the number of protons increases. This is because the strong nuclear force acts over extremely short distances. Larger nuclei have protons on opposite sides that are not bound by the strong nuclear force but do repel each other. The increased percentage of neutrons increases the distance between these protons, decreasing the effect of the electromagnetic force. The added neutrons also provide for more of the strong force attractions without adding any additional electromagnetic repulsion forces.

Most Atoms Are Stable

Some students may be confused by the statement that the majority of atoms exist as stable isotopes since the graph on this page shows most isotopes as unstable. The reason for this is that stable isotopes account for the overwhelming majority of atoms. Unstable isotopes are relatively rare.

8.2 NUCLEAR STABILITY

So why do some isotopes emit radiation while others don't? It's an issue of nuclear stability. In Chapter 5 you learned about chemical bonding—how atoms bond by sharing or transferring valence electrons to become chemically stable. But radioactivity is related to the stability of the nucleus.

Do you recall the particles found in the nucleus? All nuclei contain protons, and most contain neutrons. Remember that protons have a positive charge and that neutrons are neutral. Since like charges repel, we might expect that nuclei with more than one proton would break apart as the protons push each other away. But there seems to be an attractive force that scientists call the **strong force** that holds protons and neutrons together in nuclei. At very short distances, it is the strongest of four fundamental forces known to science (the others being gravity, the electromagnetic force, and the weak force). The strong force can hold a nucleus together if there is a proper mix of protons and neutrons. In unstable nuclei, the strong force isn't powerful enough to hold the nucleus together. The unstable nucleus changes or is said to decay, resulting in a nucleus with a lower, more stable, energy state.

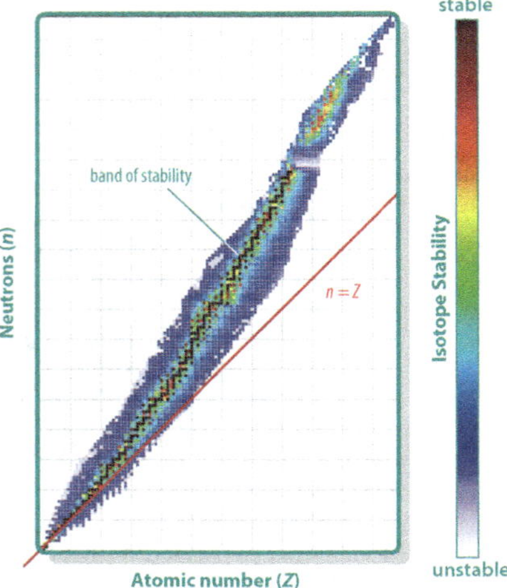

The graph at left shows all the known isotopes for the elements plotted by number of neutrons (n) versus number of protons (Z). Stable isotopes are shown with black data points and form the belt of stability—isotopes with the proper ratio of neutrons to protons that results in stable nuclei. Any isotope outside this belt will experience **radioactive decay**—the naturally occurring change of an unstable isotope to a more stable one as its nucleus emits particles, energy, or both. The red and green data points represent isotopes that are fairly stable, decaying very slowly. The blue data points show isotopes that are very unstable and that decay rapidly. The belt of stability eventually ends because all elements with atomic number 84 and above are unstable. The process of decaying moves isotopes closer to the belt of stability. Some need only one decay event to make them stable, while others need a series of decay events.

The stable isotopes are in black. All the other colors indicate unstable isotopes that undergo some form of radioactive decay. Isotopes farthest from the black isotopes (blues) are the most unstable. While there are many isotopes that are unstable, the majority of atoms exist as stable isotopes.

TEACHING THE MATERIAL

Demonstration: Use the Demonstrating Half-Life teacher note on page 182 to help students understand that half-life is a probabilistic concept.

Case Study: Use the Vikings case study on page 183 to look into the issue of radiometric dating techniques.

Formative Assessments: Use the Formative Assessment: Half-Life teacher note on page 183 to determine how well students understand half-life.

Biblical Worldview Shaping: Use the Radiation Salvation teacher note on page 184 to help students recognize that technology will never solve humanity's ultimate need.

Ticket Out the Door: Write down the symbols for the emissions in the three forms of radioactive decay.

HOW IT WORKS

Smoke Detectors

Have you ever burned something in the oven? You turn the oven off, turn on a fan, and wave your arms around. You are trying to disperse the smoke before it sets off the smoke detector. Too late!

Working smoke detectors don't just notify us that the pizza is burnt. They lower the death rate in house fires by more than 50%. But how do these devices detect fire and sound the alarm? There are two basic designs—photoelectric and ionization devices. Some detectors combine these two technologies.

A photoelectric device contains a light source and sensor, which are set up so that the light is aimed away from the sensor. If smoke enters the device, the light reflects off the smoke particles. When enough light is reflected onto the sensor, the warning tone sounds. This design is most effective for smoky, smoldering fires.

An ionization device contains a small sample of americium-241, which undergoes alpha decay. The sample is in an ionization chamber that has a positive end and a negative end. As the isotope decays, emitted alpha particles collide with oxygen and nitrogen atoms in the chamber, knocking electrons from those atoms. These free electrons are attracted to the positive end of the chamber and the cations to the negative end. These charges produce a current in the detector circuit. If smoke enters the chamber, it blocks some of this current. The alarm sounds when the current drops below a set limit. This design is best at detecting flaming fires.

Smoke detectors save lives, but only if they are installed and working. Keep those batteries fresh!

---

1. What type of radiation was Henri Becquerel experimenting with when he discovered radioactive decay? *(x-rays)*

2. What type of radiation is attracted to a negative electrode? *(alpha particles)*

3. Why would you expect fermium (atomic number 100) to be unstable? *(Elements with an atomic number greater than 83 are unstable.)*

4. Explain what makes a particular isotope nuclear stable. *(Nuclear stability depends on the ratio of neutrons to protons.)*

Differentiated Instruction:
Radioactive

Some students may benefit from seeing demonstrations of nuclear reactions using visuals, whether written equations or 3D atomic models.

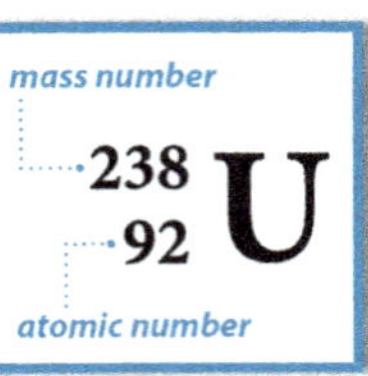

8.3 RADIOACTIVE DECAY

Chemists use equations similar to chemical equations to show the changes that occur during nuclear decay. The only way that these equations differ is that they use isotope notation instead of chemical symbols. Noting the isotope notation for uranium-238 (left) recall that the subscript shows the atomic number, while the superscript shows the mass number. As in chemical equations, the law of conservation of matter still applies to equations for nuclear reactions. Therefore the atomic numbers and mass numbers on each side of the equation must be equal to each other.

TYPES OF RADIOACTIVE DECAY

ALPHA DECAY

Alpha decay results in the emission of an *alpha particle*— a helium-4 nucleus—and some energy. An alpha particle is made of two protons and two neutrons, leaving it with a 2+ charge. The symbol for an alpha particle is $^4_2$He, or α. This decay event lowers the mass number of the remaining nucleus by 4 and the atomic number by 2.

Uranium-238 emits an alpha particle and some energy as it moves toward a more stable element. Uranium-238 has to go through fourteen decay events to become stable lead-206. The first decay event in this *decay chain* is uranium-238 emitting an alpha particle to become thorium-234.

$$^{238}_{92}U \rightarrow\, ^{234}_{90}Th + ^4_2He$$

Notice how the mass numbers, 238 = 234 + 4, and the atomic numbers, 92 = 90 + 2, on each side are equal to each other.

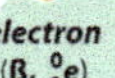

BETA DECAY

Beta decay emits a high-energy electron—a *beta particle*—and some energy from the nucleus. It's produced when a neutron splits into a proton and an electron. The resulting nucleus has the same mass number as before, but its atomic number is greater by 1. The symbol for a beta particle is $^0_{-1}$e, or β.

In the decay chain of uranium-238, the second step is the emission of a beta particle from thorium-234 to form protactinium-234.

$$^{234}_{90}Th \rightarrow\, ^{234}_{91}Pa + ^0_{-1}e$$

Again, the sums of the mass numbers and atomic numbers on each side are equal to each other, 234 = 234 + 0 and 90 = 91 + (−1).

Radon-222 goes through alpha decay to become more stable. Use the law of conservation of matter to write the balanced equation for this decay event.

Write the isotope notation for radon-222.

$$^{222}_{86}\text{Rn}$$

Since this is alpha decay, we will have an alpha particle along with our unknown element on the product side of our equation.

$$^{222}_{86}\text{Rn} \rightarrow ^{4}_{2}\text{He} + ^{A}_{Z}X$$

According to the law of conservation of matter, our mass numbers on each side of the equation must be equal. Therefore,

$$222 = 4 + A$$
$$222 - 4 = A$$
$$A = 218$$

Now we can enter the mass number of our unknown.

$$^{222}_{86}\text{Rn} \rightarrow ^{4}_{2}\text{He} + ^{218}_{Z}X$$

The law of conservation of matter also requires the atomic numbers on each side of the equation to be equal. Therefore,

$$86 = 2 + Z$$
$$86 - 2 = Z$$
$$Z = 84$$

Now we can enter the atomic number of our unknown.

$$^{222}_{86}\text{Rn} \rightarrow ^{4}_{2}\text{He} + ^{218}_{84}X$$

Knowing the atomic number of our element, we look up the symbol on the periodic table. Element number 84 is polonium (Po).

$$^{222}_{86}\text{Rn} \rightarrow ^{4}_{2}\text{He} + ^{218}_{84}\text{Po}$$

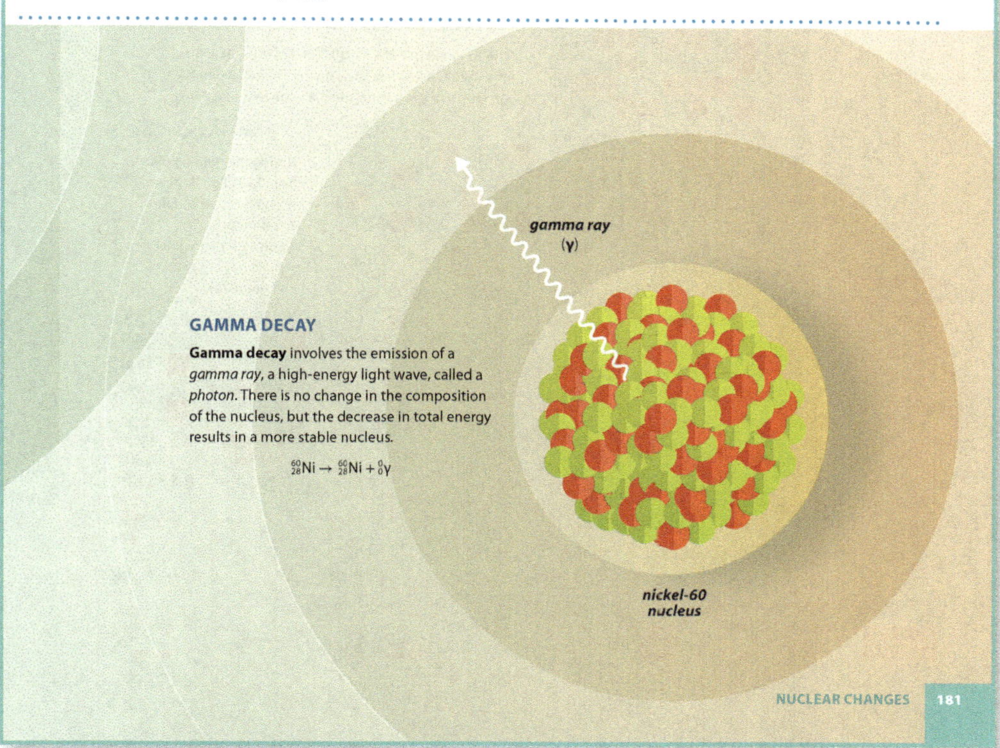

GAMMA DECAY

Gamma decay involves the emission of a *gamma ray*, a high-energy light wave, called a *photon*. There is no change in the composition of the nucleus, but the decrease in total energy results in a more stable nucleus.

$$^{60}_{28}\text{Ni} \rightarrow ^{60}_{28}\text{Ni} + ^{0}_{0}\gamma$$

Differentiated Instruction:
Balancing Nuclear Decay Equations

Some students may struggle with balancing nuclear decay equations and may benefit from extra practice. Alternatively, consider pairing a struggling student with one who excels in the skill.

Formative Assessment:
Radioactive Decay

1. Describe alpha decay. *(Alpha decay is a form of radioactive decay in which an unstable nucleus emits an alpha particle [a helium nucleus] and energy to become more stable.)*

2. What is the charge of an alpha particle? *(2+)*

3. What effect would an electric field have on a gamma ray? *(It would have no effect.)*

4. What form of radioactive decay causes the atom's atomic number to increase? *(beta decay)*

5. What form of radioactive decay doesn't change the identity of the isotope? *(gamma decay)*

6. What form of radioactive decay changes both the mass number and the atomic number? *(alpha decay)*

7. An atom of Mendelevium (atomic number 101) undergoes alpha decay. What is the atomic number of the product of this decay? *(99)*

⁉️ Half-Life

Many students struggle with the concept of half-life. It is really a statistical probability. Each atom has a 50% chance of decaying during one half-life, so approximately half of them will. As with all statistical probabilities, the actual number of atoms that decay will be closer to 50% if the sample is large. Emphasize to students that we cannot predict when any particular atom will decay. Also emphasize that previous results don't change the probability of future events.

A good comparison is flipping coins. Ask, "What is the probability of flipping a head on a standard coin?" *(50%)* Then ask, "What is the probability of flipping a head on the tenth flip after flipping nine heads in a row?" *(50%)*

🔬 Demonstrating Half-Life

Preparation: Provide one penny for each student.

Performance/Discussion:
Tell students that each of them represents an atom of a radioactive isotope. "The penny will determine whether you decay (tails) or not (heads)."

1. Since there are _____ students in the class, how many should we expect to decay the first time we flip the coin? *(should be half the class)*

Record the prediction on the board. Have each student stand by his desk and flip his coin. Those that flip tails should sit down; they have "decayed." (Sitting is more stable than standing.) Write the actual number decayed and the percentage of "decayed atoms" on the board. Repeat until all the students are seated.

2. What did each set of coin flips represent? *(a half-life)*

3. What chance did each student have of "decaying" in each half-life? *(50%)*

4. Did exactly 50% of the "atoms" "decay" each half-life? *(probably not)*

5. Would increasing the size of the class make the results closer to the expected 50% decay rate? *(Yes. Many students may not recognize that statistical probabilities are more accurate when the population is large.)*

Table 8-1		
Half-Life	Fraction Remaining	Fraction Decayed
0	1	0
1	1/2	1/2
2	1/4	3/4
3	1/8	7/8
4	1/16	15/16
5	1/32	31/32
6	1/64	63/64
7	1/128	127/128

8.4 HALF-LIFE

Look back at the graph (page 178) showing the belt of stability. Recall that the isotopes farther from the stable isotopes are the most unstable. Each unstable isotope decays at its own consistent rate. The rate at which an isotope decays is called its **half-life**—the length of time needed for half of the atoms in a sample to decay. After one half-life, 50% of the atoms in a sample will have decayed (changed) to some other element, and the other 50% will remain unchanged. What do you think will happen after a second half-life? Will the rest of the atoms have decayed? Let's look at an example to see what happens.

UNDERSTANDING HALF-LIFE

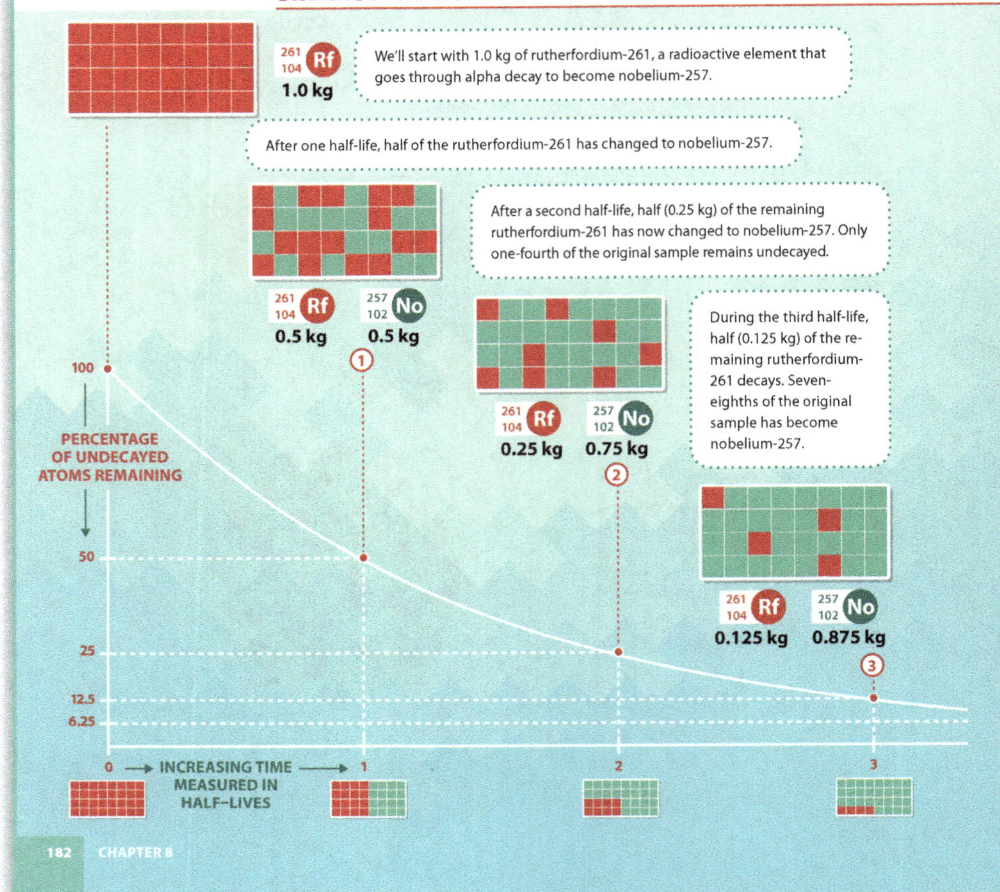

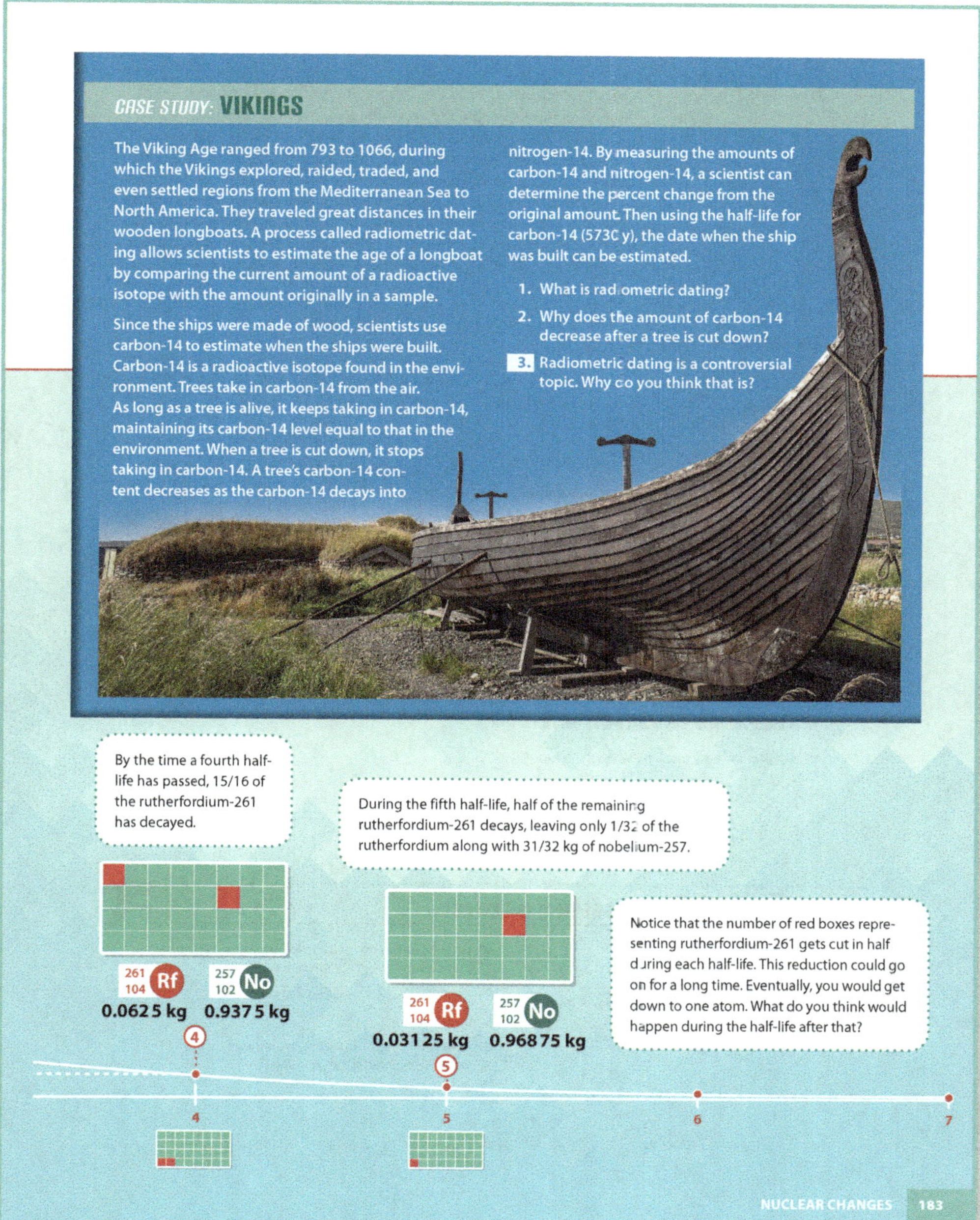

This case study uses the study of Viking materials to explain the theory and uncertainties of radiometric dating.

Answers

1. Radiometric dating is a process that estimates the age of a material by comparing the amount of a radioactive isotope currently in a sample with the amount the sample originally contained.

2. Once a tree stops taking in carbon-14 from the environment, the amount of carbon-14 in the tree decreases as the radioactive isotope decays into nitrogen-14.

3. Answers will vary. Students may recognize the worldview nature of radiometric dating. The interpretation of the results of radiometric dating can be influenced by our worldview. Also, there are natural fluctuations in the amount of carbon-14 in the environment. Therefore, the original amount of carbon-14 is an unknown value, which makes estimating ages accurately extremely difficult.

Formative Assessment: *Half-Life*

1. How many polonium-208 atoms do you expect to remain after two half-lives if the original sample contained 100 atoms? *(25)*

2. How much time must elapse before one-eighth of a radioactive sample with a half-life of 1 h remains? *(3 h)*

3. How many nitrogen-13 atoms would remain after 40 min if its half-life is 10 min and the original sample contained 200 atoms? *(12.5; Students should answer 12 or 13.)*

 *Radiation Salvation*

Radiation has figured prominently in the mindset of the twentieth and early twenty-first centuries. In fiction, many superheroes who were formed as a result of exposure to radiation end up saving humanity. In more serious discussions, radiation has been feared as the destroyer of humanity through a nuclear war or hailed as the savior of humanity by providing an alternative to fossil fuels. But even if fusion can be harnessed to provide cheap, clean energy to the world, it will never save humanity.

Human problems have their root in the Fall. Without redemption from an external source, humans will always use even the best technology for evil purposes.

Oddly enough, the comic book writers weren't entirely wrong. We do need saving by a being far greater than ourselves. But that Savior is not a human becoming a superhuman, but God becoming a man.

 *Lab 8A:* *Flipping Out*

Once students have an understanding of the material in Section 8A, they should be ready to work through Lab 8A.

The Understanding Half-Life infographic on the previous page spread shows an ideal example. In reality, the half-life is the time for half of the sample to *probably* decay. When the samples are large, the statistics predict fairly well what will happen for the entire sample. But we can never predict what will happen to any one particular atom. When you get down to a handful of atoms, about half of the sample will decay in a half-life, but it's not a sure thing. Each atom has a 50% probability of decaying during one half-life, sort of like flipping a coin. So when you get down to a single atom, you have a 50% chance that the atom will decay in one half-life.

EXAMPLE 8-2: Half-Life Problem

Rutherfordium-261 is a radioactive isotope that turns into nobelium-257 through alpha decay. It has a half-life of 78 s. Start with a 3.2 g sample.

$$^{261}_{104}\text{Rf} \rightarrow {}^{4}_{2}\text{He} + {}^{257}_{102}\text{No}$$

How long will it take until 0.40 g of rutherfordium-261 remains?

We start by determining what fraction of rutherfordium-261 remains.

We can see that $\frac{0.4\text{ g}}{3.2\text{ g}} = \frac{1}{8}$ of the original rutherfordium-261 remains. According to Table 8-1, 1/8 of the original substance will remain after three half-lives. Since each half-life is 78 s long, in 234 s (3 × 78 s) there would be 0.40 g of rutherfordium-261 remaining.

How much rutherfordium-261 will remain after 312 s?

First determine how many half-lives 312 s represents.

$$\frac{312\text{ s}}{78\text{ s}} = 4 \text{ half-lives}$$

Table 8-1 shows that after four half-lives, 1/16 of the sample remains.

$$3.2\text{ g} \times \frac{1}{16} = 0.20\text{ g}$$

Therefore 0.20 g of rutherfordium-261 remains after 312 s.

How much of the sample will have changed into nobelium-257 after 390 s?

$$\frac{390\text{ s}}{78\text{ s}} = 5 \text{ half-lives}$$

After five half-lives, 31/32 of the sample will be changed to the new isotope.

$$3.2\text{ g} \times \frac{31}{32} = 3.1\text{ g}$$

Therefore 3.1 g of rutherfordium-261 has changed in 390 s.

8A | REVIEW QUESTIONS

1. Define *radioactive decay*.

2. Identify three types of radioactive decay.

3. Complete the following equation. What type of decay is this?

$$^{227}_{90}\text{Th} \rightarrow {}^{A}_{Z}X + {}^{4}_{2}\text{He}$$

4. Where does the beta particle (electron) come from in the decay event shown below?

$$^{137}_{55}\text{Cs} \rightarrow {}^{137}_{56}\text{Ba} + {}^{0}_{-1}e$$

5. Define *half-life*.

6. How many grams of an 8.4 g sample of carbon-14 would remain after two half-lives?

8A REVIEW ANSWERS

1. Radioactive decay or radioactivity is the naturally occurring change of an unstable isotope to a more stable isotope as its nucleus emits particles, energy, or both. *(p. 178)*

2. alpha, beta, and gamma *(pp. 180–81)*

3. $^{227}_{90}\text{Th} \rightarrow {}^{223}_{88}\text{Ra} + {}^{4}_{2}\text{He}$; alpha decay *(pp. 180–81)*

4. the nucleus of a cesium atom, as a neutron changes into a proton *(p. 180)*

5. Half-life is the amount of time needed for half of the atoms of a radioactive sample to decay into another isotope. *(p. 182)*

6. 2.1 g *(pp. 182–84)*

8B | FISSION AND FUSION

Why is the sun so hot?

In Section 8A, you learned about radioactive decay, which happens when an unstable nucleus spontaneously emits particles, energy, or both to become more stable. As scientists studied these decays, they wondered whether it were possible to trigger such nuclear changes. They learned that by crashing a high-energy particle into the nucleus of an atom, they could cause the nucleus of one element to change into the nucleus of a new one, a process called *artificial transmutation*. Apparently, the alchemists had been on to something (see page 70), though they were trying to accomplish these changes through chemical reactions.

How do nuclear reactions compare with the chemical reactions that you learned about in Chapter 7? Chemical reactions involve only the valence electrons of the atoms. In those reactions, bonds are broken and formed to rearrange the atoms as new substances. Nuclear reactions involve the nucleus, including both the protons and neutrons. Nuclear changes also produce more energy than chemical reactions produce. There are two types of nuclear reactions—fission and fusion.

8B Questions

- How do fission and fusion compare?
- Where does the missing mass go in a nuclear reaction?

8B Terms

fission, chain reaction, critical mass, fusion

The image below shows a section of the Large Hadron Collider (LHC). Notice the man walking through the collider.
Inset: *This image shows the results of a particle collision like those done by the LHC.*

TEACHING THE MATERIAL

ESSENTIAL QUESTION

Why is the sun so hot?

OBJECTIVES

- 8B1 Compare nuclear fission and fusion.
- 8B2 Relate chain reaction and critical mass to nuclear fission.
- 8B3 Explain how nuclear reactions don't violate the law of conservation of matter.

RESOURCES

Case Study: Tsar Bomba

Mini Lab: *Modeling Chain Reactions*

(continued)

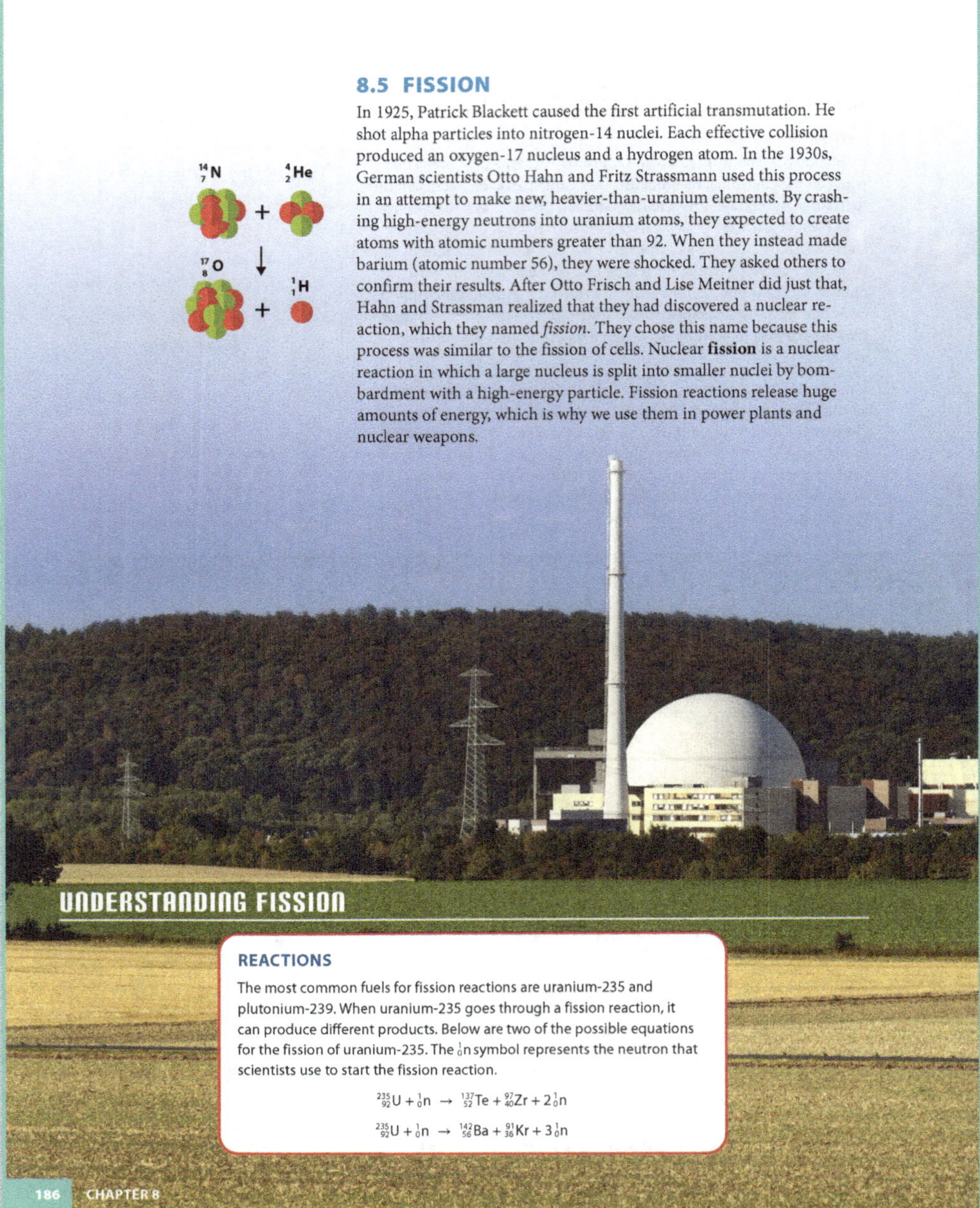

8.5 FISSION

In 1925, Patrick Blackett caused the first artificial transmutation. He shot alpha particles into nitrogen-14 nuclei. Each effective collision produced an oxygen-17 nucleus and a hydrogen atom. In the 1930s, German scientists Otto Hahn and Fritz Strassmann used this process in an attempt to make new, heavier-than-uranium elements. By crashing high-energy neutrons into uranium atoms, they expected to create atoms with atomic numbers greater than 92. When they instead made barium (atomic number 56), they were shocked. They asked others to confirm their results. After Otto Frisch and Lise Meitner did just that, Hahn and Strassman realized that they had discovered a nuclear reaction, which they named *fission*. They chose this name because this process was similar to the fission of cells. Nuclear **fission** is a nuclear reaction in which a large nucleus is split into smaller nuclei by bombardment with a high-energy particle. Fission reactions release huge amounts of energy, which is why we use them in power plants and nuclear weapons.

REACTIONS

The most common fuels for fission reactions are uranium-235 and plutonium-239. When uranium-235 goes through a fission reaction, it can produce different products. Below are two of the possible equations for the fission of uranium-235. The $_0^1n$ symbol represents the neutron that scientists use to start the fission reaction.

$$^{235}_{92}U + {}^1_0n \rightarrow {}^{137}_{52}Te + {}^{97}_{40}Zr + 2{}^1_0n$$

$$^{235}_{92}U + {}^1_0n \rightarrow {}^{142}_{56}Ba + {}^{91}_{36}Kr + 3{}^1_0n$$

TEACHING THE MATERIAL

STRATEGIES

Class Opener: Use the Solar Fusion teacher note on page 185 to begin a discussion of fusion and fission.

Active Learning: Use the Active Learning: Fission Reactions teacher note on page 187 to provide practice with fission reaction equations.

Formative Assessments: Use the Formative Assessment: Fission and Fusion teacher note on page 188 to assess students' understanding of these reactions.

Formative Assessments: Use the Formative Assessment: Nuclear Energy teacher note on page 189 to check students' understanding of the source of energy in nuclear reactions.

Ticket Out the Door: Write down one example each of nuclear fission and fusion.

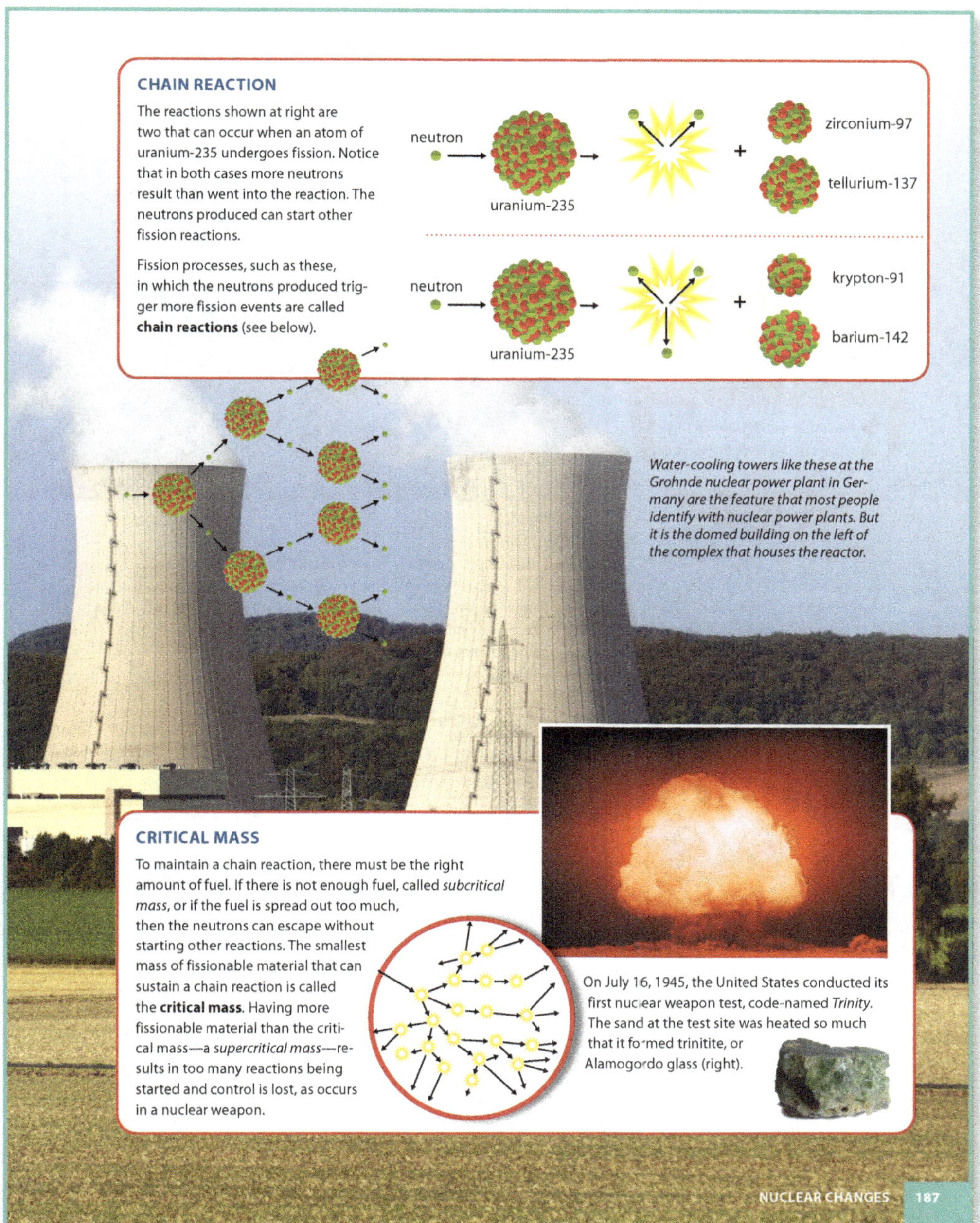

CHAIN REACTION

The reactions shown at right are two that can occur when an atom of uranium-235 undergoes fission. Notice that in both cases more neutrons result than went into the reaction. The neutrons produced can start other fission reactions.

Fission processes, such as these, in which the neutrons produced trigger more fission events are called **chain reactions** (see below).

Water-cooling towers like these at the Grohnde nuclear power plant in Germany are the feature that most people identify with nuclear power plants. But it is the domed building on the left of the complex that houses the reactor.

CRITICAL MASS

To maintain a chain reaction, there must be the right amount of fuel. If there is not enough fuel, called *subcritical mass*, or if the fuel is spread out too much, then the neutrons can escape without starting other reactions. The smallest mass of fissionable material that can sustain a chain reaction is called the **critical mass**. Having more fissionable material than the critical mass—a *supercritical mass*—results in too many reactions being started and control is lost, as occurs in a nuclear weapon.

On July 16, 1945, the United States conducted its first nuclear weapon test, code-named *Trinity*. The sand at the test site was heated so much that it formed trinitite, or Alamogordo glass (right).

Nuclear Chain Reactions

The first manmade chain reaction was created in December 1942 in Chicago by a team led by Enrico Fermi as part of the Manhattan Project that built the first atomic bomb.

Active Learning: *Fission Reactions*

Have students work in groups of four to solve a fission reaction equation together. Don't give the students the equation, but assign each student to represent either a known material or an unknown from the equation. Then have them work together to determine the identity of the unknown. An example is given below.

$$^{235}_{92}\text{U} + {}^{1}_{0}\text{n} \rightarrow {}^{142}_{56}\text{Ba} + {}^{91}_{36}\text{Kr} + 3{}^{1}_{0}\text{n}$$

You could assign three students to be the knowns (uranium-235, barium-142, and four neutrons) and the other student would be the unknown. They would have to work together to figure out that the other student is krypton-91.

Nuclear Power

Although experiments in fusion power are ongoing, the first power plants utilizing fusion are not expected to be operational until around 2050.

Formative Assessment:
Fission and Fusion

1. Define *fission*. *(Fission is a nuclear reaction in which a large nucleus is split into smaller nuclei and is typically initiated by bombardment with a high-energy particle.)*

2. An artificial transmutation results in lithium atoms (atomic number 3) becoming fluorine (atomic number 9). Is this fission or fusion? Explain. *(fusion; Small nuclei combine to form a more massive nucleus.)*

3. The nuclear material in a nuclear power plant forms a subcritical mass. What is likely to happen? *(Nothing. There is not enough matter to maintain a chain reaction.)*

8.6 FUSION

While fission reactions can be helpful for producing energy, life would not exist on Earth without nuclear fusion. We depend on fusion because our sun produces energy by fusion reactions. Every second, our sun emits 3.8×10^{26} J of energy—enough energy to power the United States for 3.7 million years! While only a small portion of this energy reaches us, it provides almost all the energy ultimately used by living things.

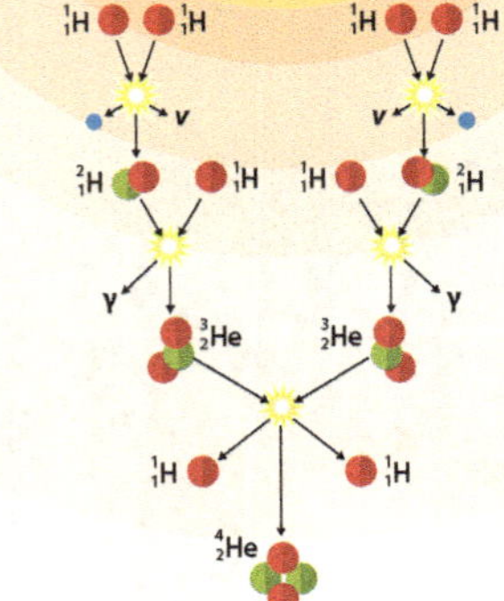

Fusion is a nuclear reaction in which small nuclei combine to form a more massive nucleus. These reactions occur only under extreme pressures and temperatures. Fusion in the sun occurs within the core, where the temperature is about 15.7 million K. Four hydrogen nuclei in our sun fuse together in a three-step process to become a helium nucleus. The equation for this is $4\,{}^{1}_{1}\text{H} \rightarrow {}^{4}_{2}\text{He} + 2\,{}^{0}_{1}\text{e}$ and takes place as a series of three steps:

$$ {}^{1}_{1}\text{H} + {}^{1}_{1}\text{H} \rightarrow {}^{2}_{1}\text{H} + {}^{0}_{1}\text{e} \text{ (occurs twice)} $$

$$ {}^{1}_{1}\text{H} + {}^{2}_{1}\text{H} \rightarrow {}^{3}_{2}\text{He} \text{ (occurs twice)} $$

$$ {}^{3}_{2}\text{He} + {}^{3}_{2}\text{He} \rightarrow {}^{4}_{2}\text{He} + 2\,{}^{1}_{1}\text{H} \text{ (occurs once)} $$

In the above fusion equations, the ${}^{0}_{1}\text{e}$ symbols look like beta particles, but they are actually positrons, which is why the atomic number is positive and not negative. Scientists would love to be able to produce energy through nuclear fusion, and many organizations are working to make it a reality.

8.7 NUCLEAR ENERGY

Every nuclear change involves the release of energy. As isotopes decay, they emit particles and energy to become more stable. As a nucleus splits in a fission reaction, it yields two smaller nuclei and a large amount of energy. Fission of uranium-235 produces 2.7 million times more energy than an equal mass of anthracite coal (the most energy-dense coal). Recall from our discussion about the sun that the fusing of small nuclei releases even larger amounts of energy. Fusion produces three to four times the energy that fission produces. But where does this energy come from?

In each of these cases, the total mass of the products is less than the mass of what went into the reaction. At first glance, this looks like a violation of the law of conservation of matter. Where did the missing mass go? Albert Einstein answered this question with his famous equation, $E = mc^2$. This equation states that mass and energy are interchangeable forms of the same thing. Remember that the law of conservation of matter states that matter cannot be created or destroyed but only transformed. In nuclear decays and reactions, the "missing mass" changes into energy.

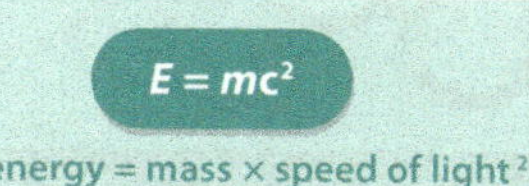

CASE STUDY: TSAR BOMBA

In August of 1945, many people were celebrating around the world. The United States had dropped two fission bombs and brought World War II to an end. The world was at peace—or was it? As soon as the war ended, the Soviet Union decided that it also wanted nuclear weapons. In 1949, the Soviet Union tested its first nuclear weapon and entered into a cold war with its former ally. Each country worked to build larger weapons and stay ahead of the other.

The United States tested many devices on islands in the South Pacific Ocean. On March 1, 1954, the United States tested its largest nuclear weapon in a test code-named *Castle Bravo*. Instead of the planned 5 Mt (five-megaton) yield, the explosive equivalent of 5 million tons of TNT, the blast had an actual yield of 15 Mt. This error convinced scientists in the United States that they didn't need to build any larger weapons.

Castle Bravo inspired the Soviets. The Soviet leader, Nikita Khrushchev, challenged his scientists to show the Americans what the Soviets could do. The result, a fusion weapon code-named *Vanya* but nicknamed *Tsar Bomba* in the West, produced the largest man-made explosion ever. The Soviets designed it as a three-stage, 100 Mt device. Stage 1 was a fission reaction using uranium-238. This reaction would start Stage 2, a fusion reaction of hydrogen. This second stage then started a larger fusion reaction, Stage 3. The test used only the first two stages, which made it a 50 Mt weapon. Even this scaled-down version yielded more than 1500 times the explosive force of either of the two bombs that ended World War II.

1. Why do you think that this period of history was called the *Cold War*?

2. The bombs dropped that ended World War II were fission devices. What does this mean?

3. What enabled the Tsar Bomba to have so much more power than the fission bombs?

1. How does the energy released in a fission reaction compare with burning an equal amount of anthracite coal? *(The fission reaction releases 2.7 million times as much energy.)*

2. How does the energy released in a fusion reaction compare with burning an equal amount of anthracite coal? *(The fusion reaction releases 9–10 million times as much energy.)*

3. How are nuclear reactions able to produce so much energy? *(They transform matter into energy.)*

Tsar Bomba Case Study

This case study explores the use of nuclear changes in weaponry. It highlights the extremes that two nations went to in the arms race by showcasing the fusion weapon that created the largest manmade explosion ever recorded.

Answers

1. A cold war is one in which two sides don't actively fight each other. This war was in weapons development, spheres of influence, and proxy wars.

2. A fission device releases energy through a fission reaction, which splits a large nucleus into two smaller nuclei and some free neutrons. These neutrons trigger additional fission reactions in an uncontrolled explosion that releases huge amounts of energy.

3. Tsar Bomba was a three-stage device, meaning that it had three nuclear explosions, which would make it larger. But Tsar Bomba also included fusion reactions, which release much more energy than fission reactions.

 *Modeling Chain Reactions*

This mini lab uses dominoes to model a nuclear chain reaction.

Answers

1. A chain reaction is one in which products of one reaction start additional reactions.

2. critical mass

3. Each domino represents an atom.

4. an atom undergoing fission

5. Answers will vary. They may not all fall. If they do, the time should be much longer.

6. Yes. The time was the shortest.

7. The falling of each domino triggers the next one or two dominoes to fall. This is similar to the neutrons from a fission reaction starting the next reaction(s).

8. In a nuclear reactor, the number of collisions must be controlled. The ruler provided the control in this reaction.

We don't usually think about how our electricity gets to us or how it is produced. Did you know that about 20% of our power is generated by nuclear power plants? These plants produce energy by fission. The process begins by smashing high-energy neutrons into the fuel, usually uranium-235 or plutonium-239. Each reaction produces two or more high-energy neutrons that can start more fission events. The key to keeping control of the reactions in the power plant is to control the number of new high-energy neutrons that are available.

Essential Question:

How can we model fission reactions with dominoes?

Equipment
dominoes (15)
stopwatch
centimeter ruler

1. What is a chain reaction?

2. What is the term for the smallest amount of matter needed to sustain a fission chain reaction?

Procedure

Ⓐ Standing the dominoes on their narrow end, arrange them in a line spaced 3.0 cm apart.

Ⓑ Time how long it takes for all 15 dominoes to fall when you push over the first domino.

3. What does a domino represent?

4. What does the domino's falling represent?

Ⓒ Repeat Steps Ⓐ and Ⓑ but with the dominoes set 4.0 cm apart.

5. Did all 15 dominoes fall? How did the time change?

Ⓓ Repeat Steps Ⓐ and Ⓑ but with the dominoes arranged so that each domino will fall against two other dominoes.

6. Did all 15 dominoes fall? How did the time change?

Conclusion

7. How does this exercise model a chain reaction?

Going Further

Ⓔ Rearrange your "atoms" once again like Step Ⓓ. Using the ruler, can you interrupt part of the chain reaction?

8. What does Step Ⓔ have to do with nuclear reactors?

TEACHING THE MATERIAL

ESSENTIAL QUESTION

What are the benefits and risks of nuclear changes?

OBJECTIVES

- 8C1 Explain how radioactivity is used in medical technology.
- 8C2 Compare genetic and somatic damage.
- 8C3 Explain how we detect radiation in order to protect people from harmful radiation.
- 8C4 Justify applications of nuclear changes. **BWS**
- 8C5 Evaluate the positions for and against generating energy from nuclear sources. **BWS**

8B | REVIEW QUESTIONS

8B | REVIEW QUESTIONS

1. Define *fission*.
2. Explain what will happen if there is a subcritical mass for a fission reaction.
3. Define *fusion*.
4. What is the nuclear fuel for the fusion within the sun?
5. Complete the equation for the fusion of helium-4 nuclei, shown below.

$$^4_2He + {}^4_2He \rightarrow {}^A_ZX$$

6. What conditions are required for fusion to occur?
7. Where does the missing mass go in a nuclear reaction?

8C | NUCLEAR CHANGES: BENEFITS AND RISKS

What are the benefits and risks of nuclear changes?

You may not know it, but radiation, both natural and manmade, surrounds us all the time. Sources in nature include soil, rocks, food, water, the atmosphere, and even space. Most of this radiation is low level and safe. We are exposed to manmade radiation from nuclear medicine, medical procedures, and electronic devices. These manmade sources produce radiation that is typically more energetic and often more harmful. We need to balance the helpful aspects of using nuclear processes against the harmful effects of being exposed to too much radiation.

8C Questions

- How does radioactivity affect us?
- How can we protect ourselves from radiation?
- What are some benefits of radiation?

8C Terms

radiotracer, somatic damage, genetic damage

8.8 USES OF RADIATION

We've already seen many uses for radiation, such as protecting food supplies, producing power, detecting fires, and estimating the age of historical objects. Radiation also has other uses, such as in medicine and security.

1. Fission is a nuclear reaction in which a large nucleus is split into two smaller nuclei by bombardment with a high-energy particle. *(p. 186)*

2. Subcritical mass means that there is insufficient mass or the mass is too spread out to maintain a chain reaction because the neutrons escape without starting other reactions. *(p. 187)*

3. Fusion is the nuclear reaction by which small nuclei are combined to form a heavier nucleus. *(p. 188)*

4. hydrogen *(p. 188)*

5. ^8_4Be *(p. 188)*

6. extreme pressure and temperature *(p. 188)*

7. The reaction transforms the missing mass into energy. *(p. 189)*

Radiation Uses

Ask students to name a beneficial use of radiation. There are plenty of examples, including x-ray devices, smoke detectors, and nuclear power stations.

RESOURCES

- Ethics: Nuclear Power Generation (p. 199)
- Worldview Sleuthing: Nuclear Waste
- Lab 8B: *Radioactive!*—Exploring Radiation Dose

STRATEGIES

Class Opener: Use the Radiation Uses teacher note on this page to begin a discussion on the practical uses of radioactive material.

Peer Teaching: Use the Peer Teaching Radiation Uses teacher note on page 192 to provide students the opportunity to learn from each other.

Formative Assessments: Use the Formative Assessment: Using Radiation teacher note on page 193 to assess students' understanding of how we use radiation.

Biblical Worldview Shaping: Use the Using Radiation Wisely teacher note on page 194 to help students understand that using technology wisely is even more important than developing that technology.

(continued)

Why Do We Still Use X-Rays?

Although CT scans improve on x-ray scans, most patients who need imaging still receive x-rays. An x-ray scan is much cheaper and exposes the patient to much less radiation than a CT scan, so doctors are more likely to request an x-ray unless the benefits of the CT scan's better imaging outweigh its disadvantages.

Where's the MRI?

Some students might wonder why MRIs are not mentioned in this section. Magnetic resonance imaging uses strong magnetic fields rather than ionizing radiation to produce its images.

 ### Peer Teaching Radiation Uses

Consider assigning student groups one of the radiation technologies. Allow them time to research the topic and report back either in small groups or to the whole class.

 ### Using Radiation Videos

There are many video sources about using radiation. Consider showing a couple such videos. Do an internet search using the keywords "CT scan video" or "airport security scans video."

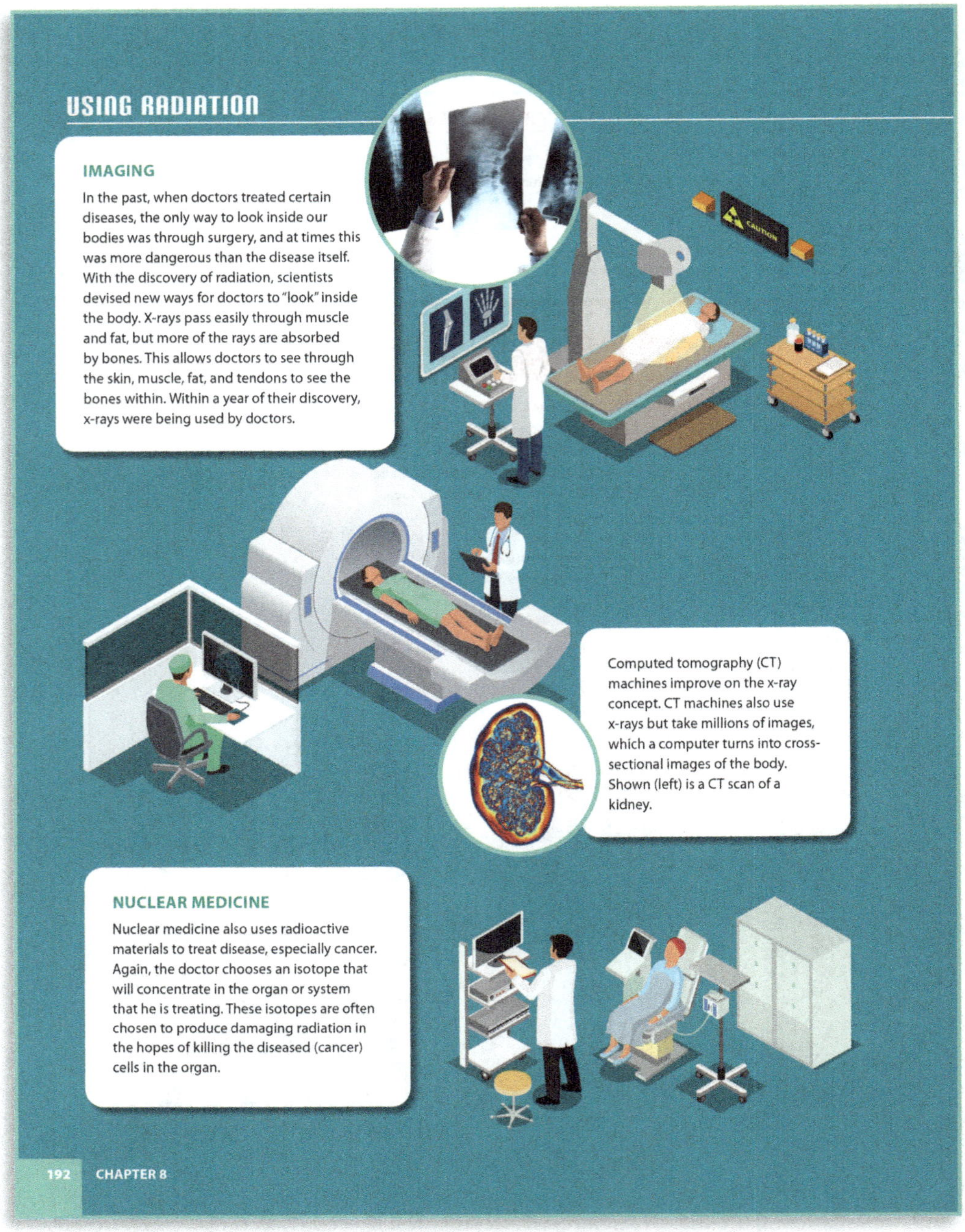

TEACHING THE MATERIAL

Formative Assessments: Use the Formative Assessment: Radiation Safety teacher note on page 195 to check students' progress in learning about nuclear safety.

Biblical Worldview Shaping: Use the Nuclear Waste Worldview Sleuthing box on page 196 to have students research how to properly deal with nuclear wastes.

Biblical Worldview Shaping: Use the Nuclear Power Generation ethics box on page 199 to have students think through the benefits and risks of nuclear power generation.

Ticket Out the Door: Name one of the devices used to detect radiation.

RADIOTRACERS

Radioactive isotopes help doctors diagnose many illnesses. Doctors inject these isotopes, called **radiotracers**, to see how they move through or collect in a certain organ or system. The PET scan (right) shows how fluorine-18 has collected in a brain. The red shows large amounts of the fluorine. When selecting radiotracers, doctors must consider the part of the body they are studying. They also need to select isotopes with fairly short half-lives, long enough to do the testing but without exposing the patient to more radiation than needed. The doctor's goal is to treat the patient while minimizing the harmful effects.

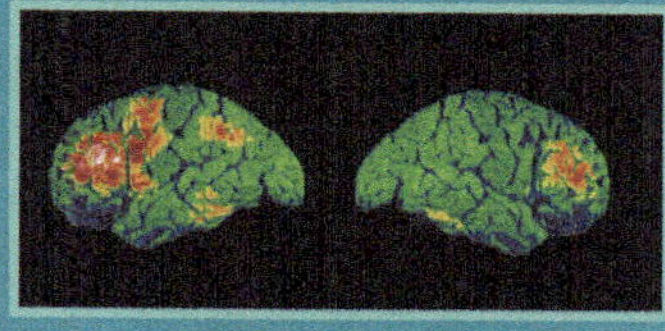

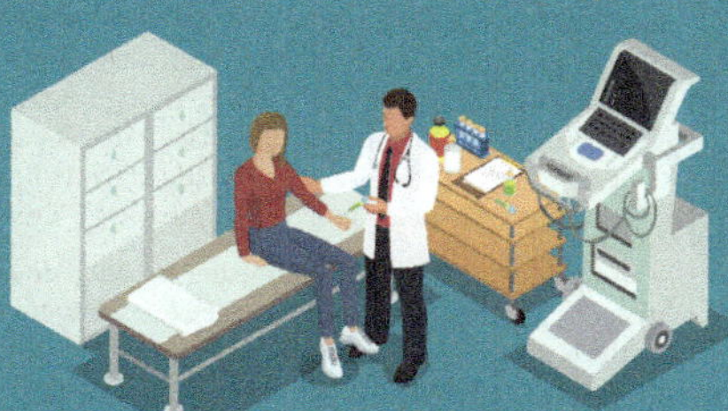

Medical Tracers		
Isotope	**Half-Life**	**Target**
carbon-11	20 min	brain, thyroid
carbon-14	5730 y	pancreas
cobalt-57	272 d	intestines
fluorine-18	110 min	bone
iodine-123	13 h	thyroid
krypton-81m	13 s	lungs
oxygen-15	122 s	brain, heart
technetium-99m	6.0 h	numerous

SECURITY

Before air passengers enter secure areas of an airport, security personnel use x-ray machines to scan them and their luggage. They are able to detect weapons, bombs, and other dangerous objects. Advanced systems use the same CT technology that is used in hospitals. Law enforcement agencies also use different x-ray systems to scan for weapons, drugs, and even people being smuggled into the country. Below, backscatter x-rays show thirty-seven people hidden inside a truck full of bananas.

Formative Assessment:
Using Radiation

1. What is the difference between an x-ray scan and a CT scan? *(A CT scan consists of multiple x-ray scans that a computer uses to form a cross-sectional image.)*

2. How does a doctor justify using radioactive isotopes to treat a cancer patient even when they know it will harm healthy cells? *(Doctors understand that in the end the treatment may destroy the cancer and provide many healthy years to the patient.)*

3. Why would krypton-85, with a half-life of 10.8 years, not be a good radiotracer? *(Its half-life is too long, which means it would emit few particles that could be traced. Also, with such a long half-life, it might remain in the patient too long and expose the patient to too much radiation.)*

Other ways that we benefit from radiation include promoting food safety, as noted in the Chapter opener, and in scientific and industrial applications, which you can study in the worldview sleuthing activity on page 196. These benefits arose because scientists did the pure, investigative science needed to learn new things about God's creation.

8.9 EFFECTS OF RADIATION

The discovery of x-rays in November 1895 had doctors excited about the possibilities. Everyone was experimenting to see what could be done with them. Then in February 1896, Vanderbilt University researchers reported the first *injuries* from the new rays. Other reports flooded in, and doctors realized that while radiation can be useful, many forms of radiation were also harmful.

The degree to which radiation can damage cells depends on how it affects them. As some radiation enters the cells, it can knock electrons out of the atoms and molecules, creating ions. This type of radiation is called *ionizing radiation*. Other forms of radiation, such as radio waves, microwaves, and visible light, do not have enough energy to turn atoms into ions and are less dangerous. We must be careful not to expose people to too much ionizing radiation while using it to help improve their lives.

The amount of damage that ionizing radiation does depends on certain factors. How deep does radiation go into the person? How massive is the particle? How much energy does the radiation have? How much radiation is the person exposed to? How long was he exposed?

Radiation can do two types of damage. Damage to cells that are not involved in reproduction is called **somatic damage**. This type of damage harms the organism but can't be passed on to offspring. Somatic damage causes injury and illness, such as cancer. Damage to the DNA is called **genetic damage**. This damage will affect the reproduction and growth of new cells. It may also be passed to offspring if the damaged DNA is in a reproductive cell. To limit the extent of damage, we must be able to detect radiation and then take steps to protect people from exposure.

8.10 DETECTING RADIATION

Detection of radiation allows us to know when we need to protect ourselves from that radiation. There are a number of devices for detecting radiation. Some tell us only whether there is radiation present, while others tell us whether there is much or little present. Still other devices specifically indicate the amount of radiation present.

DETECTING RADIATION

GEIGER COUNTER

One device used for detecting radiation is the *Geiger counter*. This device consists of a closed tube containing inert gas molecules that connects to an indicating circuit and a counter. As ionizing radiation enters the tube, it knocks electrons from the atoms of gas in the tube. The electrons are attracted to a positively charged wire that runs the length of the tube. As these electrons enter the wire, they produce a current, which produces a click that can be heard and is displayed on the counter. The current produced indicates the amount of radiation present. Geiger counters are limited—they can't identify the type of radiation, and they also work only for low levels of radiation.

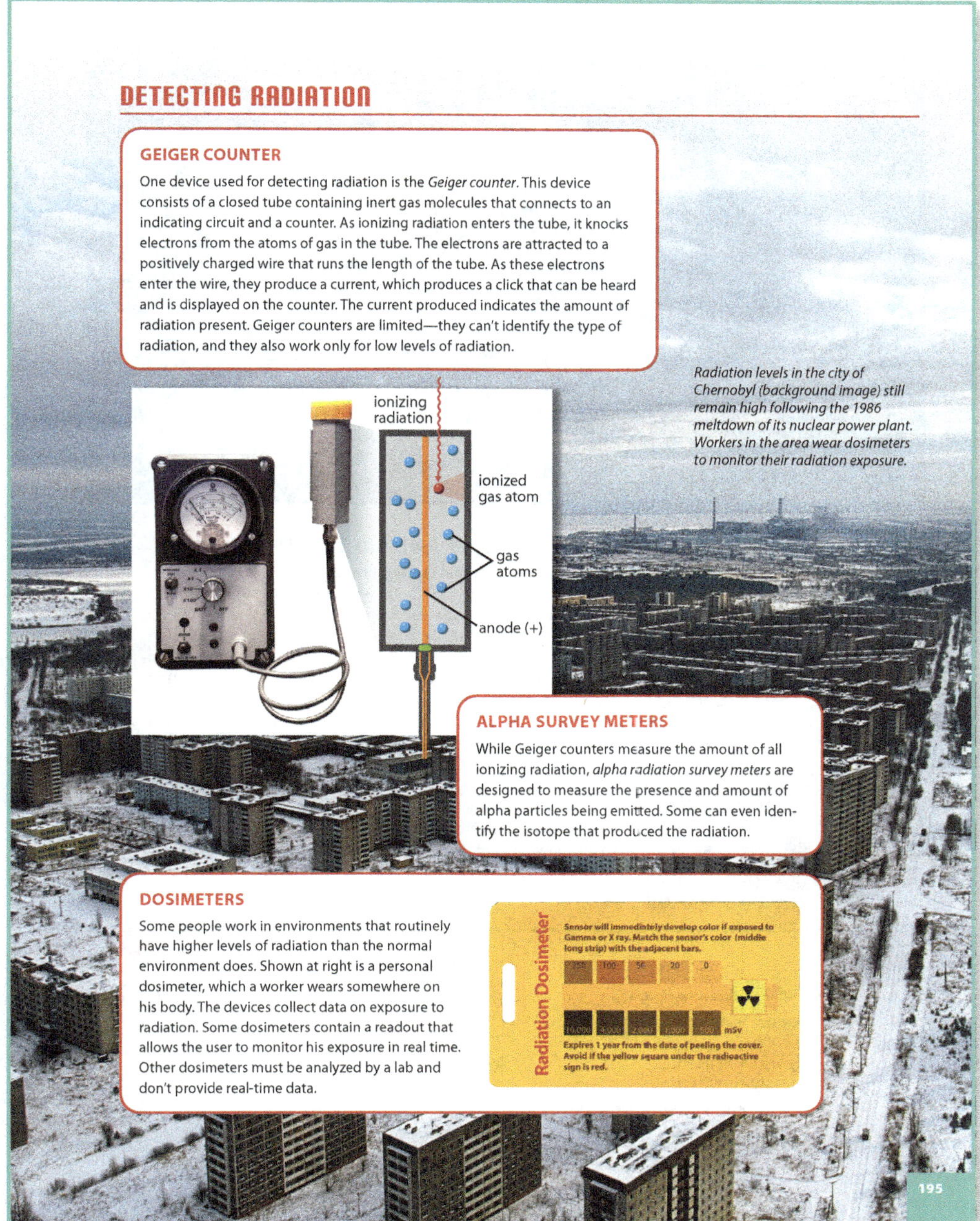

Radiation levels in the city of Chernobyl (background image) still remain high following the 1986 meltdown of its nuclear power plant. Workers in the area wear dosimeters to monitor their radiation exposure.

ALPHA SURVEY METERS

While Geiger counters measure the amount of all ionizing radiation, *alpha radiation survey meters* are designed to measure the presence and amount of alpha particles being emitted. Some can even identify the isotope that produced the radiation.

DOSIMETERS

Some people work in environments that routinely have higher levels of radiation than the normal environment does. Shown at right is a personal dosimeter, which a worker wears somewhere on his body. The devices collect data on exposure to radiation. Some dosimeters contain a readout that allows the user to monitor his exposure in real time. Other dosimeters must be analyzed by a lab and don't provide real-time data.

Formative Assessment:
Radiation Safety

1. Name two factors that determine the amount of damage done by ionizing radiation. *(Accept any two: radiation penetration depth, particle mass, radiation energy, amount of radiation, length of exposure.,*

2. Why do workers in a nuclear power plant wear dosimeters? *(Dosimeters maintain a record of radiation exposure.)*

Worldview Sleuthing Rubric

Appendix G includes a reproducible rubric to assess the Nuclear Waste worldview sleuthing activity.

Lab 8B: *Radioactive!*

Once students have an understanding of the material in Section 8C, they should be ready to work through Lab 8B.

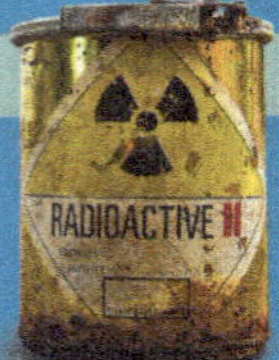

WORLDVIEW SLEUTHING: NUCLEAR WASTE

There is a give-and-take related to many things in our lives. The use of radioactive materials is one of those things. While there are many beneficial uses of radiation, we must limit the harm to people and the environment. Most applications of radioactive materials generate waste that remains radioactive. What do we do with this material?

TASK

You are a member of the city planning board. A company has proposed building a new facility in the city to process, store, and dispose of nuclear waste. This facility will create many jobs and boost the local economy. But the company will be transporting, processing, storing, and disposing of large amounts of radioactive waste. You have volunteered to report on the issues related to radioactive wastes and have agreed to give a five-minute presentation at the next board meeting before the vote regarding this proposal.

PROCEDURE

1. Research the issue by doing a keyword search for "radioactive waste management," "radioactive waste storage," "radioactive waste hazards," and "nuclear waste disposal."

2. Plan your presentation and collect any required materials. Remember to cite your sources.

3. Show your presentation to a classmate or friend for feedback.

4. Present your findings to your classmates.

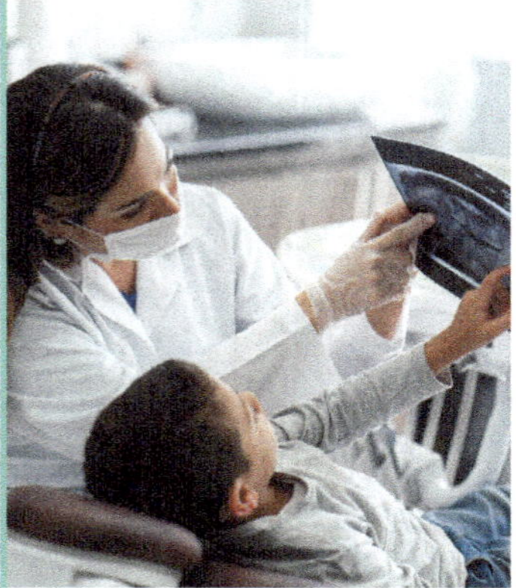

8.11 RADIATION AND WORLDVIEW

The study of nuclear chemistry began as a pursuit of pure science, seeking scientific knowledge for its own sake. Often people think that science glorifies God only when it helps others, but even pure science glorifies God. Many of the technologies that help people were made possible because of discoveries in pure science. In most cases, we protect ourselves from radiation by avoiding exposure to it. In Chapter 3 you had the opportunity to investigate radon-222 exposure in homes. Homeowners that identify issues with radon in their home can install a system to remove the radon. By doing this, they avoid exposure. Think about getting x-rays at the dentist. As a patient, you sit there because it's the only way to get images of your teeth, but your dental hygienist leaves the room. For a patient, the once- or twice-a-year exposure to x-rays during dental exams is acceptable, but a hygienist would be exposed to too much radiation if she stayed in the room for every x-ray. Both our study of and use of radioactivity fulfill the Creation Mandate by glorifying God and helping others.

8C | REVIEW QUESTIONS

1. Name three natural sources of radiation.

2. Briefly describe two applications of radiation in medical technology.

3. List three ways outside of the medical field that radiation is used to benefit people.

4. What form of radiation is especially dangerous to living cells?

5. What type of radiation damage has the potential to be passed on to an organism's offspring?

6. Why is it beneficial to detect radiation?

7. Why are dental patients covered with a heavy, lead-lined blanket when getting x-rays?

8C REVIEW ANSWERS

1. Accept any three: soil, rocks, food, water, the atmosphere, space. *(p. 191)*

2. Accept any two: Doctors use radiation to "look" inside the body without surgery. X-rays pass easily through flesh but are absorbed by bones. This allows doctors to produce a picture of the bones within the body. CT machines produce cross-sectional images of the body. Doctors also use radiotracers when making PET scans. Isotopes are used to treat cancer. *(pp. 192–93)*

3. The Student Edition mentions power production, food safety, smoke detectors, and airport security. Other answers are possible. (Accept any three.) *(pp. 175, 179, 186–87, 193)*

4. ionizing radiation *(p. 194)*

5. genetic damage *(p. 194)*

6. In most cases we want to avoid radiation, so the ability to detect radiation allows us greater opportunity to avoid exposure. Also, some career fields require people to work near radiation sources. To protect these people we need to be able to measure how much radiation they are exposed to. *(pp. 194–95)*

7. The blanket protects patients' vital organs from radiation exposure. *(p. 196)*

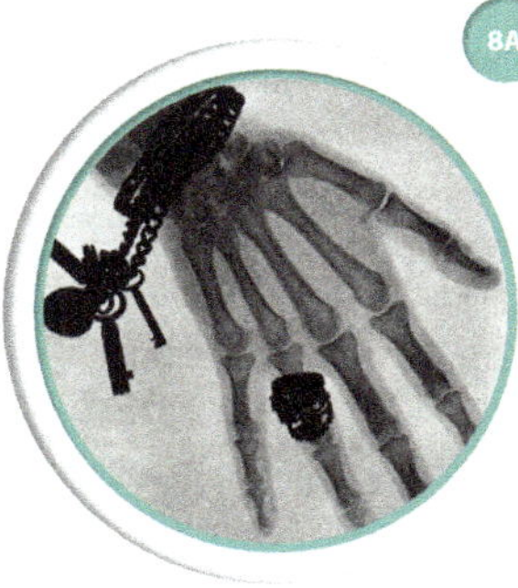

8A RADIOACTIVE DECAY

- Radioactivity was accidentally discovered when uranium salts were observed to emit radiation spontaneously.

- The ratio of neutrons to protons in the nucleus determines whether isotopes are stable or unstable.

- Unstable isotopes move toward greater stability by emitting particles, energy, or both, a change known as radioactive decay.

- Three forms of radioactive decay are alpha, beta, and gamma.

- Each isotope decays at a specific rate. The time needed for half of a sample of radioactive material to decay is called its half-life.

- Scientists estimate the age of objects by using the half-life of radioactive elements.

8A Terms

strong force	178
radioactive decay	178
alpha decay	180
beta decay	180
gamma decay	181
half-life	182

8B FISSION AND FUSION

- Scientists can trigger changes within the nucleus through artificial transmutations.

- Fission is a nuclear reaction in which a large nucleus splits into two or more smaller nuclei.

- Nuclear power plants rely on products of one reaction to start additional reactions, a process known as a chain reaction.

- A chain reaction needs a minimum amount of radioactive fuel—the critical mass—to sustain the reaction.

- Fusion is a nuclear reaction in which two small nuclei combine to form a larger nucleus.

- The energy in a nuclear reaction comes from converting some of the mass into energy.

- Fission produces over a million times the energy of fossil fuels. Fusion produces three to four times the energy of fission reactions.

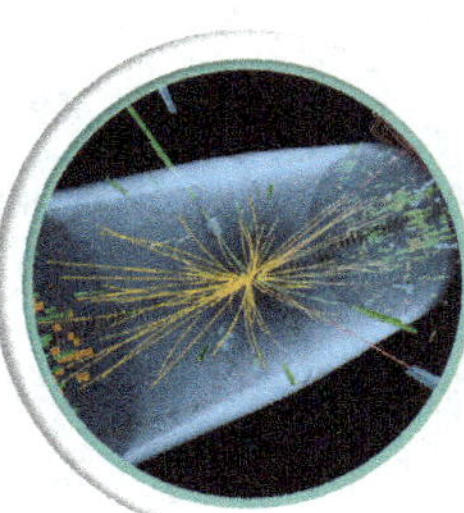

8B Terms

fission	186
chain reaction	187
critical mass	187
fusion	188

8C NUCLEAR CHANGES: BENEFITS AND RISKS

- Scientists have discovered a number of beneficial applications for radiation, including medical, food safety, security, and power production uses.

- Radiation can cause damage to cells, including both genetic and somatic damage.

- We must balance the benefits of using radiation with the risks of overexposure.

- Studying radiation has led to many technologies that have benefited people.

8C Terms

radiotracer	193
somatic damage	194
genetic damage	194

Recalling Facts

1. Wilhelm Röntgen began the study of radiation when he discovered x-rays in 1895. Henri Becquerel discovered radioactivity as he was studying x-rays. He realized that uranium salts were emitting radiation spontaneously as the uranium decayed. Ernest Rutherford and Paul Villard observed three forms of decay—alpha, beta, and gamma. *(pp. 176–77)*

2. An atom spontaneously emits particles, energy, or both from its nucleus in order to become more stable. *(p. 178)*

3. The strong force holds protons and neutrons together in the nucleus. For a nucleus to be stable, it has to have the proper ratio of neutrons to protons; some isotopes lack this. Unstable nuclei will decay, resulting in a nucleus with a lower energy state, which is more stable. *(p. 178)*

4. Gamma decay causes a change in energy in the nucleus but doesn't change its composition. *(p. 181)*

5. In one half-life, approximately half the nuclei will decay into more stable nuclei. *(p. 182)*

6. They were trying to form heavier-than-uranium elements, and barium is significantly lighter. *(p. 186)*

7. A chain reaction is one in which products, typically neutrons, from one reaction can start additional reactions. *(p. 187)*

8. Supercritical mass would be the goal in a fission weapon. The mass would allow the reaction to proceed uncontrolled, producing a lot of energy in a short period of time. *(p. 187)*

9. There isn't enough mass or the mass is too spread out to maintain a chain reaction. The neutrons that are produced escape before they can start other reactions. *(p. 187)*

10. Fusion is essential to life on Earth. Fusion is the nuclear reaction that produces the energy of the sun. That energy is emitted into space and provides almost all the food energy for living organisms on the planet. *(p. 188)*

Recalling Facts

1. Summarize the discoveries that led to the discovery of radioactivity.
2. What happens to an atom during radioactive decay?
3. Why are some isotopes stable while other isotopes are unstable and experience radioactive decay?
4. Which type of radioactive decay does not result in a new element?
5. What happens to a sample of an isotope during one half-life?
6. Why were scientists surprised that they produced barium after crashing a neutron into a uranium atom?
7. What is a chain reaction?
8. In what application of fission would supercritical mass be desired? Explain.
9. What does it mean for a fissionable material to have subcritical mass?
10. How does fusion impact our lives on Earth?
11. Name a manmade source of radiation.
12. How does CT technology differ from x-rays?
13. How are radiotracers used in medicine?

Understanding Concepts

14. Compare physical, chemical, and nuclear changes.
15. Why is it surprising that protons remain close together in the nucleus? What force is believed to hold these protons close to each other?
16. Complete the following equation of a decay event.

$$^{210}_{84}\text{Po} \;\rightarrow\; ^{A}_{Z}X + ^{4}_{2}\text{He}$$

17. Using a graphic organizer, summarize what we know about alpha, beta, and gamma decay. Include the symbol, charge, mass, and penetrating power.
18. Can we predict when a single atom of a radioactive isotope will decay? Explain.
19. A 179.2 g sample of silicon-31, which decays into phosphorus-31 (half-life: 2.6 h), has decayed, leaving 5.6 g of silicon-31. How much time has passed?
20. How much of 11.2 g of iodine-135 (half-life: 6.6 h) would remain after 19.8 hours?
21. How do alpha and beta decay differ from artificial transmutations?
22. Complete the equation for the fission of plutonium-239.

$$^{239}_{94}\text{Pu} + ^{1}_{0}\text{n} \;\rightarrow\; ^{134}_{54}\text{Xe} + ^{A}_{Z}X + 3\,^{1}_{0}\text{n}$$

23. Compare subcritical, critical, and supercritical mass.
24. Compare nuclear fission and fusion.
25. Create a concept map using the terms *radioactive decay, stable nucleus, unstable nucleus, fusion, fission, critical mass,* and *chain reaction.*
26. Explain why nuclear reactions don't violate the law of conservation of matter.
27. Compare somatic and genetic damage.
28. Respond to the following statement: "The purely scientific study of radiation is not a valuable use of time because it doesn't help others."

11. Accept any one: nuclear medicine, medical procedures, electronic devices, and food irradiation devices. *(p. 191)*

12. CT combines multiple x-ray images into a cross-sectional image of the body. *(p. 192)*

13. Radiotracers are radioactive isotopes that affect a specific organ or system within the body. Doctors inject the radiotracer and then, using imaging technology, study the organ or system for diagnosing certain conditions. *(p. 193)*

Understanding Concepts

14. *Physical changes* alter the appearance of a substance without changing its chemical composition. *Chemical changes* alter the composition of compounds by breaking and forming chemical bonds to rearrange atoms into new substances. *Nuclear changes* convert one isotope into another by changing the composition of the nucleus. *(pp. 41, 176)*

15. Protons have a positive charge, and like charges repel. But the strong force holds the protons and neutrons together in the nucleus. *(p. 178)*

16. $^{206}_{82}\text{Pb}$ *(p. 180)*

REVIEW

Critical Thinking

29. Would you expect an unstable isotope that was above atomic number 83 to emit an alpha particle ($^4_2$He) or a beta particle ($^0_{-1}$e) to move toward the belt of stability?

30. Would you expect an isotope that undergoes beta decay to be above or below the belt of stability?

31. A 10.0 g sample of radioactive material has decayed to 2.0 g after 6 h. Estimate the half-life of the material.

32. If a neutron collided with a plutonium-239 nucleus and produced cerium-140 and krypton-97, how many neutrons would also be produced? Write the nuclear equation that supports your answer.

33. Why does a dental hygienist ask her female patient whether she is pregnant before taking x-rays?

ETHiCS — NUCLEAR POWER GENERATION

THE ISSUE: CLEAN ENERGY?

Nuclear reactions can produce large amounts of energy without generating any air pollution. Some people feel that nuclear power can meet our growing energy needs and reduce air pollution, including greenhouse gases. Others fear that the possibility of a mishap, like at Fukushima in Japan or Chernobyl in Ukraine, makes the use of nuclear power too risky a proposal. A third group likes the reduction in air pollution but wonders what we will do with the leftover radioactive material.

34. What information can you find about this issue?

35. What does the Bible say about this issue?

36. What are some of the acceptable and unacceptable options?

37. What might be the consequences of adopting the acceptable and unacceptable options that you have listed?

38. What are the motivations for the acceptable options?

39. Write a one-page position paper indicating whether you believe that a Christian should support an initiative to develop more energy from nuclear sources.

Use the rubric below to help you determine whether you have completed all the necessary tasks for writing your response.

Self-Assessment Rubric

	Task Completed	Task Partially Completed	Task Not Completed
Additional Information	two or more sources	just one source	none
Biblical Principles	two or more passages	just one passage	none
Acceptable and Unacceptable Options	both	only one	neither
Consequences	yes	—	no
Biblical Outcomes	all three	only one or two	none
Biblical Motivations	all three	only one or two	none
Opinion Statement	yes	—	no

17. See table below (pp. 180–81)

Decay	Symbol	Charge	Mass (amu)	Penetrating Power (stopped by)
Alpha	$^4_2$He, α	2+	4	low (paper, skin, few centimeters of air)
Beta	$^0_{-1}$e, β	1–	0	moderate (few millimeters of aluminum or wood, 1 m of air)
Gamma	γ	none	0	high (thick lead, steel, concrete)

18. No. Radioactive decay is a random process that occurs spontaneously. While half-life gives the time for half of a sample to decay, it does not allow us to predict when a decay event will take place. (p. 184)

19. 13 h (Since 5.6 g/179.2 g = 0.031 25, or 1/25, five half-lives have passed.) (p. 184)

20. 1.4 g (19.8 h /6.6 h per half-life = 3 half-lives) (p. 184)

21. Alpha and beta decay occur spontaneously, while artificial transmutations are manmade. (p. 185)

22. $^{103}_{40}$Zr (p. 186)

23. Subcritical mass refers to an insufficient amount of material available to maintain a fission chain reaction. A critical mass is sufficient to produce a controlled chain reaction. A supercritical mass will produce an uncontrolled reaction. (p. 187)

24. Fission and fusion are both nuclear reactions. Fission is the splitting of a large nucleus into two or more smaller nuclei. Fusion is the combining, or fusing, of two smaller nuclei into a larger nucleus. Fusion releases more energy than fission does. (pp. 186–88)

25. See concept map below. (pp. 177, 178, 185–88)

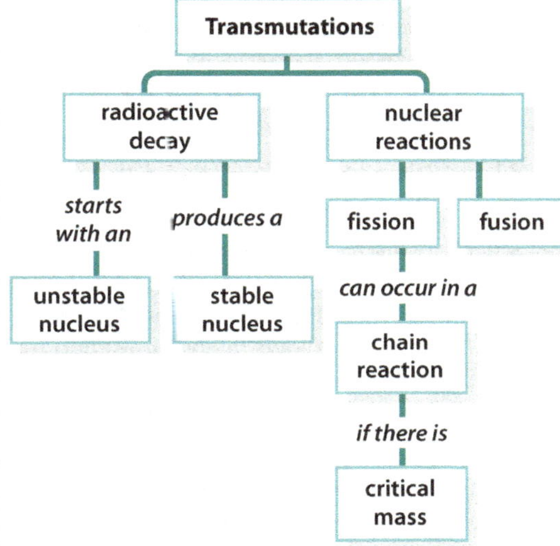

26. During nuclear processes, mass is converted into energy, not lost. Einstein's equation $E = mc^2$ tells us that energy and mass are equivalent; mass is not destroyed, only transformed to a new form. (pp. 188–89)

27. Ionizing radiation can cause both types of damage. Somatic damage occurs in nonreproductive cells and can result in injury and illness to the organism itself. Genetic damage results in damage to the DNA of the organism. This can affect growth and development of the organism and may also be passed on to offspring. (p. 194)

28. Pure science is valuable. We are called to have dominion over God's creation. We glorify Him when we learn more about the creation. Also, history has shown that the pure scientific study of radiation has greatly benefited people through the different applications of that knowledge. (p. 196)

29. The isotope should undergo alpha decay. Beta decay causes the atomic number to go up instead of down. *(pp. 178, 180)*

30. This isotope would be above the belt of stability and so has too many neutrons. During beta decay, a neutron splits into an electron (beta particle) and a proton, lowering the neutron-to-proton ratio. *(pp. 178, 180)*

31. After one half-life, 5 g would remain, after two, 2.5 g, but after three half-lives, 1.25 g would remain. Since 2.0 g of the material in question remains, between two and three half-lives have passed. Since 2.0 g is closer to 2.5 then it is to 1.25, the half-life is closer to 2. A good estimate is that 2.4 half-lives have passed. Therefore, the half-life is estimated as 6 h/2.4 = 2.5 h. The actual value would be 2.58 h. *(pp. 182–84)*

32. 3; $^{239}_{94}\text{Pu} + ^{1}_{0}\text{n} \rightarrow ^{140}_{58}\text{Ce} + ^{97}_{36}\text{Kr} + 3^{1}_{0}\text{n}$
(p. 186)

33. Exposure to radiation can have a harmful effect on people. The effect is even greater for a developing baby. Asking the question allows the hygienist and patient to assess the benefits and risks of taking x-rays. *(pp. 194, 196)*

34. Compared with other power sources, nuclear power is both the cleanest and safest per unit of energy produced. But even with the environmental and safety record mentioned above, there is much controversy. Many people are concerned about a major mishap, such as took place in Fukushima or Chernobyl.

35. The Bible doesn't address nuclear power directly. But it does teach us to study the world in which we live and to use that knowledge to care for people and the rest of creation. Part of good and wise dominion is pressing creation to its maximum usefulness to people, while considering risks and benefits. If we could safely use nuclear power, that would fall within the Creation Mandate.

36. *Acceptable:* Work to find ways to safely use nuclear power to benefit people and protect the world in which we live. *Unacceptable:* Use nuclear power in a way that endangers people and the environment. Ignore a potentially clean and safe energy source without studying the issue sufficiently.

37. *Acceptable:* Meet the energy needs of the growing world population. Help reduce air and water pollution by decreasing our dependence on fossil fuels. *Unacceptable:* We could harm people and the environment by pursuing nuclear power in a rash manner. We could limit development or harm the environment by ignoring a potentially clean and safe power source.

38. We glorify God by studying the world that He created. We fulfill the Creation Mandate by meeting the needs (energy, safety, and clean environment) of God's image bearers. We protect God's creation by safely using clean energy sources.

39. Use the rubric located in Appendix H to assess students' essays.

REVIEW

29. Would you expect an unstable isotope that was above atomic number 83 to emit an alpha particle ($^{4}_{2}\text{He}$) or a beta particle ($^{0}_{-1}\text{e}$) to move toward the belt of stability?

30. Would you expect an isotope that undergoes beta decay to be above or below the belt of stability?

31. A 10.0 g sample of radioactive material has decayed to 2.0 g after 6 h. Estimate the half-life of the material.

32. If a neutron collided with a plutonium-239 nucleus and produced cerium-140 and krypton-97, how many neutrons would also be produced? Write the nuclear equation that supports your answer.

33. Why does a dental hygienist ask her female patient whether she is pregnant before taking x-rays?

ETHiCS — NUCLEAR POWER GENERATION

THE ISSUE: CLEAN ENERGY?

Nuclear reactions can produce large amounts of energy without generating any air pollution. Some people feel that nuclear power can meet our growing energy needs and reduce air pollution, including greenhouse gases. Others fear that the possibility of a mishap, like at Fukushima in Japan or Chernobyl in Ukraine, makes the use of nuclear power too risky a proposal. A third group likes the reduction in air pollution but wonders what we will do with the leftover radioactive material.

34. What information can you find about this issue?

35. What does the Bible say about this issue?

36. What are some of the acceptable and unacceptable options?

37. What might be the consequences of adopting the acceptable and unacceptable options that you have listed?

38. What are the motivations for the acceptable options?

39. Write a one-page position paper indicating whether you believe that a Christian should support an initiative to develop more energy from nuclear sources.

Use the rubric below to help you determine whether you have completed all the necessary tasks for writing your response.

Self-Assessment Rubric

	Task Completed	Task Partially Completed	Task Not Completed
Additional Information	two or more sources	just one source	none
Biblical Principles	two or more passages	just one passage	none
Acceptable and Unacceptable Options	both	only one	neither
Consequences	yes	—	no
Biblical Outcomes	all three	only one or two	none
Biblical Motivations	all three	only one or two	none
Opinion Statement	yes	—	no

When certain candies are dropped into a soda bottle, the rapid release of carbon dioxide from solution results in a great deal of liquid being shot into the air. Diet soda is often used to avoid the stickiness caused by the sugar or corn syrup contained in regular soda.

CHAPTER 9
Solutions

FROTHY FOUNTAIN OF FOAM

You reach into the refrigerator and grab an ice-cold bottle of soda. As you unscrew the top, you hear the hiss and see bubbles forming inside the bottle. But every so often something goes wrong: the hiss is louder and the bubbles form rapidly, spraying soda everywhere. Why did that happen? Most of us have seen someone add Mentos™ to soda, creating spraying soda to the extreme.

Soda is a mixture of water, sweetener, and carbon dioxide. Normally when we open a bottle, the carbon dioxide slowly comes out of solution. In the occasional mishap or in the Mentos and soda demonstration, the gas comes out of solution rapidly. We will learn through this chapter the what and how of solutions. You'll see how mixtures and solutions are a vital part of understanding and using the world that God has given us.

<table>
<tr><td>9A Mixtures and Solutions</td><td>202</td></tr>
<tr><td>9B Solution Concentration</td><td>212</td></tr>
</table>

CHAPTER 9 OBJECTIVES

- Compare kinds of mixtures.
- Model the dissolving process.
- Evaluate how colligative properties can be used to help people.

Overview

Chapter 9 is a foundational chapter. An understanding of mixtures and solutions is required for understanding acid and base chemistry in addition to being useful in many everyday situations.

Lab Activities

Lab 9A: *All Mixed Up*—In this inquiry lab activity, students devise and execute a procedure to separate and recover the components of a mixture.

Lab 9B: *That's Cold!*—In this lab activity, students investigate the factors that affect freezing point depression.

Demonstrating Mentos™ and Diet Soda

As a great start to the chapter, consider doing the Mentos and Diet Soda demonstration or view one of the many videos available on the internet. Do a keyword search for "Mentos and Diet Soda."

Misunderstanding Mentos and Diet Soda

Many people believe that putting Mentos in soda causes a chemical reaction that results in the production of gas. The process is actually a physical process, a faster version of the slow bubbling that occurs when you open a soda bottle under normal conditions.

In this process, the carbon dioxide is coming out of solution. It happens rapidly because of the properties of the Mentos. The candy has a very rough surface, and each tiny imperfection on its surface is a site where CO_2 can come out of solution. There are so many sites that the CO_2 escapes rapidly. The gas gets caught behind the soda, pressure builds, and a geyser like Old Faithful erupts.

🔷 *How Should We Classify Milk?*

Students often think that science is clear-cut. While scientific modeling is useful, it is almost always oversimplified. The divisions of suspension, colloid, and solution are helpful, but some substances do not fit neatly into one of these categories. Milk is one of those substances.

Have students research the classification of milk. Is it a solution, a colloid, a suspension, or some combination of the three? They will find that a lot depends on whether they are dealing with skim milk, homogenized milk, or raw milk. Students should keep in mind that the difference between a suspension and a colloid is the size of the particles.

Emulsion

Ask students, "What do you know about oil and water?" Most will know that they do not mix easily. From their study of organic chemistry in Chapter 6, many may recognize that oils are nonpolar and so will not dissolve in polar water. Then ask them, "What would you say if I told you that mayonnaise is mostly oil and water? What must have happened so that the oil and water in mayonnaise don't separate?" Many will recognize that another substance must be allowing them to mix together. Such a substance is called an *emulsifier*. In mayonnaise, egg yolk acts as an emulsifier and keeps the solute particles mixed in the solvent.

9A | MIXTURES AND SOLUTIONS

How do things dissolve?

In Chapter 2 we learned about matter. We learned about pure substances, which can be elements or compounds, and mixtures, which are either heterogeneous or homogeneous. In this chapter we will look more closely at mixtures, especially homogeneous mixtures. Recall that a mixture is two or more substances that are physically combined in a changeable ratio. Mixtures can form from substances that are in the same state of matter (e.g., liquid in liquid or gas in gas) or in different states (e.g., gas in liquid or solid in liquid). Some common examples of mixtures include trail mix and tossed salad (heterogeneous) as well as salt water and tea (homogeneous).

9.1 HETEROGENEOUS MIXTURES

Recall from Chapter 2 that heterogeneous mixtures do not have a uniform appearance since the substances are unevenly distributed—the materials are not evenly spread out. In a trail mix, you can point to areas in the mixture that are only raisins or only peanuts. Italian salad dressing is another great example since the seasonings can be seen floating around in the mixture. Scientists classify heterogeneous mixtures as suspensions or colloids according to particle size.

TYPES OF HETEROGENEOUS MIXTURES

SUSPENSIONS

A **suspension** is a heterogeneous mixture of a fluid and large particles that will *settle out*—separate naturally—over time. The particles are typically larger than 1 μm in diameter. Human hair has a diameter of 10 to 200 μm.

TEACHING THE MATERIAL

ESSENTIAL QUESTION

How do things dissolve?

OBJECTIVES

• 9A1 Compare suspensions, colloids, and solutions.

• 9A2 Demonstrate the process by which solutions are made.

• 9A3 Explain energy changes during dissolving.

• 9A4 Compare different methods for separating mixtures.

RESOURCES

🌐 Worldview Sleuthing: Sports Drinks

How It Works: Hot and Cold Packs

🔬 Demonstrating Mentos and Diet Soda, Demonstrating the Tyndall Effect, Demonstrating Mixtures, Demonstrating Energy in Solution, Demonstrating Rates of Dissolving, Demonstrating Boiling

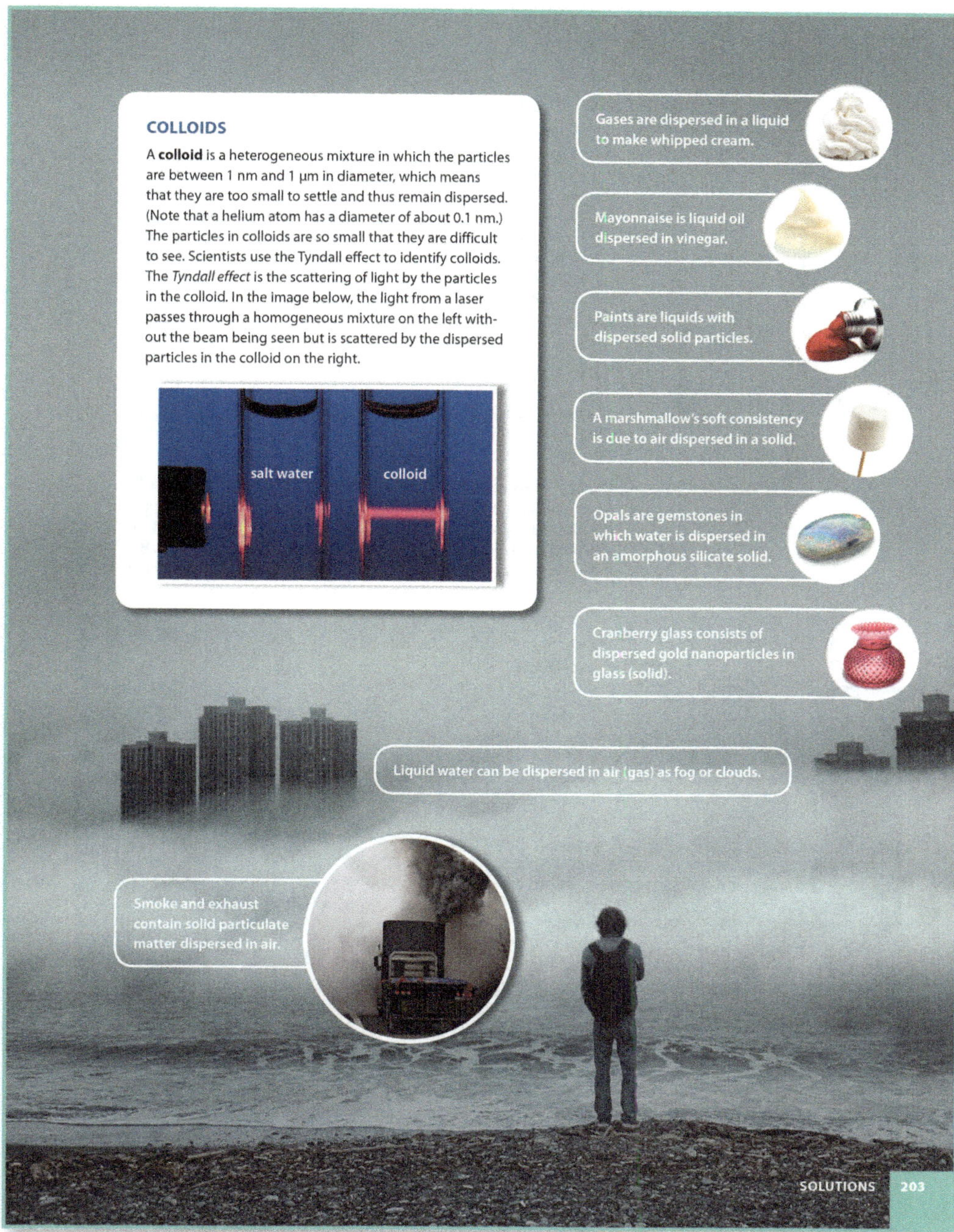

⊟ Mentos and Diet Soda, Dissolving Rate Applets

☁ Lab 9A: *All Mixed Up*—Inquiring into Separating Mixtures

STRATEGIES

Class Opener: Use the Demonstrating Mentos and Diet Soda teacher note on page 201 to begin a discussion on mixtures.

Differentiated Instruction: Consider using the How Should We Classify Milk? teacher note on page 202 to help challenge advanced students.

Discussion: Use the Demonstrating the Tyndall Effect teacher note on this page to help students understand how the Tyndall effect can be used to distinguish colloids and solutions.

Formative Assessment: Use the Formative Assessment: Heterogeneous Mixtures teacher note on this page to assess students' understanding of these mixtures.

Discussion: Use the Demonstrating Mixtures teacher note on page 204 to help students distinguish between suspensions, colloids, and solutions.

Biblical Worldview Shaping: The Worldview Sleuthing: Sports Drinks box on page 204 can help students examine claims about foods and drinks to better equip them to glorify God with their bodies.

(continued)

🔬 *Demonstrating the Tyndall Effect*

Preparation: Prepare a clear solution and a clear colloid in separate 50 mL beakers. A solution can be made by dissolving a small amount of sugar in water. A colloid can be made by mixing a few drops of milk in water. You will also need a laser pointer.

Note: If you choose to do this demonstration, be cautious with the laser. A laser can be dangerous if the beam is directed into the eyes. Pay particular attention if you use a green laser—green lasers are more powerful than red ones.

Performance: Shine the laser through the solution and the colloid. Students should view the effect from a direction perpendicular to the direction of the beam. ***Never look directly at a laser beam.***

Discussion:

1. What did you notice about how the laser light interacted with the two mixtures? (*The light was scattered by one of the mixtures but not by the other.*)

2. On the basis of your observations, which beaker do you think contains a colloid and which contains a solution? (*The beaker in which the light was scattered contains the colloid.*)

3. Why did the colloid scatter the light while the solution didn't? (*The particles dispersed in the colloid are larger than the particles dissolved in the solution.*)

🔎 *Formative Assessment:*
Heterogeneous Mixtures

1. What is the identity of a mixture containing particles that are 500 nm in diameter? (*a colloid*)

2. Why might letting an unknown mixture sit on a shelf for a week help identify it? (*Suspension particles eventually settle out, so if it is a suspension, letting the mixture sit might give it time to begin settling out.*)

3. Name three common colloids. (*Answers will vary. Examples: whipped cream, mayonnaise, paint, marshmallows, opals, smoke, fog*)

Preparation: Prepare one or two samples each of suspensions, colloids, and solutions. Consider selecting fairly easily distinguished mixtures such as Italian dressing for the suspension and sugar water for the solution. A few drops of milk in water makes an excellent example of a colloid, and a small amount of chalk in water is an excellent suspension. Feel free to identify the mixtures.

Performance: Ask students to distinguish between the mixtures. They should be able to identify the suspension fairly easily. They may ask to use the Tyndall effect to distinguish between the colloid and the solution. If so, it is advisable that you perform the test to avoid any potential injuries from the laser.

Discussion:

1. How did you distinguish the suspension? *(Answers will vary. In more easily identified suspensions, the two substances have begun separating.)*

2. How did you distinguish the colloid? *(Answers will vary. In clear colloids, the Tyndall effect is an effective indicator.)*

3. How did you distinguish the solution? *(Answers will vary. It may be by process of elimination.)*

Worldview Sleuthing Rubric

Appendix G includes a reproducible rubric to assess this worldview sleuthing activity.

9.2 HOMOGENEOUS MIXTURES

Recall from Chapter 2 that homogeneous mixtures, also known as **solutions**, are mixtures with a uniform appearance throughout. This uniform appearance occurs because the dissolved particles are smaller than the particles in colloids and are uniformly distributed throughout the mixture. The particles are so small that we can't see them and they don't scatter light. The particles are said to have dissolved into the mixture. The material that dissolves in a solution is called the **solute**. The substance into which the solute dissolves is the **solvent**. In some cases these definitions don't work well, such as mixtures of two gases, two liquids, or two solids. In these cases the solvent is the material in greater amount, while the solute is the material in lesser amount.

Salt water is a familiar example of a solution. We make it by dissolving salt (the solute) in water (the solvent) to form the saltwater solution. They are physically combined but not chemically. The sodium and chloride ions are uniformly and completely distributed among the water molecules.

When people think about solutions, they often think of things like salt water and sugar water. These are considered liquid solutions because the solvent is liquid. Solutions can also have either gases or solids as the solvent. Earth's atmosphere, for example, is a solution. Nitrogen gas is the solvent, and other gases such as oxygen and argon are dissolved in it.

Watch a sporting event and you will see sports drinks. They are on the sidelines. They are in advertisements. You can find shelves and shelves of these drinks in the supermarket. Manufacturers tell you that they will help you before, during, and after the event.

TASK

You are the newly elected student body president at your school. A group of students has asked you to convince the administration that having sports drinks available throughout the school day will help students. They tell you that they will help improve the performance of student athletes. They also tell you that the drinks will help students pay better attention in class and therefore do better in their studies. Your principal tells you to do some research and come back with a one-page justification for offering sports drinks in school. She also said that you have to be able to defend your findings.

PROCEDURE

1. Research the issue by doing keyword searches for "sports drinks," "do you really need sports drinks," and "what do sports drinks do."

2. Plan, organize, and write your paper. Support each claim in your paper with specific evidence. Remember to cite your sources.

3. Show your paper to a classmate or friend for feedback.

4. Turn in your paper.

CONCLUSION

Do sports drinks do what they claim? Do they benefit all people or only certain groups? When in a leadership position, we want to meet the needs of those we represent. The challenge can be in recognizing true needs, not mere wants.

TEACHING THE MATERIAL

Formative Assessment: Use the Formative Assessment: Solutions teacher note on page 205 to check students' understanding of homogeneous mixtures.

Demonstration: Use the Demonstrating Energy in Solution teacher note on page 207 to help students recognize that making solutions can be endothermic or exothermic.

Formative Assessment: Use the Formative Assessment: Forming Solutions teacher note on page 208 to see whether students understand how solutions form.

Demonstration: Use the Demonstrating Rates of Dissolving teacher note on page 209 to help students learn the factors that influence dissolving rates.

Demonstration: Use the Demonstrating Boiling teacher note on page 210 to help students see an example of separating a solute from a solvent.

Formative Assessment: Use the Formative Assessment: Dissolving Rates and Separating Mixtures teacher note on page 210 to determine whether students understand these concepts.

Ticket Out the Door: Write one of the processes used to separate mixtures.

Alloys are another very common type of solution. An **alloy** is a solid solution made of two or more elements, at least one of which is a metal. Bronze (tin dissolved in copper) and steel (carbon dissolved in iron) are common alloys. Let's look at some other examples of solutions.

SOLUTIONS

Solutions are homogeneous mixtures with particles that are tiny and evenly distributed so that they have a uniform appearance throughout. Solutions can be made with solutes and solvents in any state of matter.

Pewter is a solution of tin with other metals (e.g., copper, lead, antimony) dissolved in it.

Palladium hydride is a solid solution of solid palladium with metallic hydrogen dissolved in it.

Amalgam is a liquid solute (mercury) dissolved in a solid solvent (silver) and has been used in dentistry applications.

Air is a solution in which the solute (oxygen) and the solvent (nitrogen) are both gases.

Sports drinks are made by dissolving solids (salts) in liquid (water).

Soda is carbonated because it is a mixture of liquid (soda) with a gas (carbon dioxide).

Rubbing alcohol is a homogeneous mixture of two liquids: isopropyl alcohol and water.

Solvents and Solutes in Alloys

Why do we say that bronze is tin dissolved in copper and steel is carbon dissolved in iron? In each case, the material in greater amount (copper and iron) is considered the solvent and the material in lesser amount (tin and carbon) is the solute.

Formative Assessment: Solutions

1. What is the solute in sugar water? What is the solvent? *(sugar; water)*

2. What is the solvent in coffee? *(water)*

3. Some 18 karat white gold is made from 75% gold and 25% palladium. Which metal is the solute and which is the solvent? Explain. *(Gold is the solvent and palladium is the solute because gold is present in greater abundance.)*

Each of the different types of solution (solid, liquid, and gaseous) forms by different methods. We will look at the most familiar solution form, a solid dissolving in a liquid. Think about making salt water. We know that salt dissolves in water, but what happens to allow that?

SOLVATION

The process of dissolving a solute into a liquid solvent is called *solvation*. The solvation process involves attractive forces between the solvent particles and the solute particles, similar to but weaker than those that cause bonding. The result is that the solute particles become surrounded by particles of the solvent. Consider the dissolving of table salt in water.

This whole process involves energy. Energy is needed to separate the solvent particles from each other ($E_{solvent}$). More energy is needed to move the solute particles away from each other (E_{solute}). The process also releases some energy as the solvent and solute particles attract each other ($E_{solution}$).

$$E_{solvation} = E_{solute} + E_{solvent} - E_{solution}$$

If $E_{solvation} > 0$, then the process is endothermic.

If $E_{solvation} < 0$, then the process is exothermic.

Demonstrating Energy in Solution

Preparation: Prepare two thermometers, two stirring rods, two beakers with 100 mL of room-temperature water in each, and two medicine cups. In one cup place 6 g of potassium chloride, and in the other place 25 g of calcium chloride. Be sure that everyone wears goggles.

1. Describe the two chemicals. (*Both are white solids. One is similar to table salt and the other has larger granules.*)

2. What do you think will happen to the temperature of each mixture as the solid dissolves? (*Answers will vary.*)

Performance: Place the thermometers in the beakers and pour the salts into separate beakers. Begin stirring while students observe the thermometers. Record the value of the temperature after the greatest change.

Discussion:

3. How did potassium chloride dissolving affect the temperature of the solution? (*The temperature dropped.*)

4. How did calcium chloride dissolving affect the temperature of the solution? (*The temperature rose.*)

5. Did a chemical reaction occur in either of these beakers? (*No*)

6. What are the terms for a process that releases thermal energy? for one that absorbs thermal energy? (*exothermic; endothermic*)

7. How do you account for the temperature change in the endothermic process? (*More energy was absorbed to separate the solvent and solute particles than was released when the solvent and solute moved together.*)

1. What is solvation? *(the process of dissolving a solute into a liquid solvent)*

2. What occurs when ions are solvated? *(Their ions are surrounded by solvent particles.)*

3. Why does sugar not dissociate in solution? (Dissociation occurs to ionic solutes. Sugar is a covalent solute.)*

4. According to the graph on page 208, which solute is the most soluble at 20 °C? *(KI [potassium iodide])*

5. Is soda more likely to go flat in the refrigerator or at room temperature? Explain. *(Soda will likely go flat at room temperature because gas becomes less soluble at higher temperatures.)*

6. Does butter dissolve in water? Explain. *(Water is a polar solvent and butter is a nonpolar solute. Since like dissolves like, butter does not dissolve in water.)*

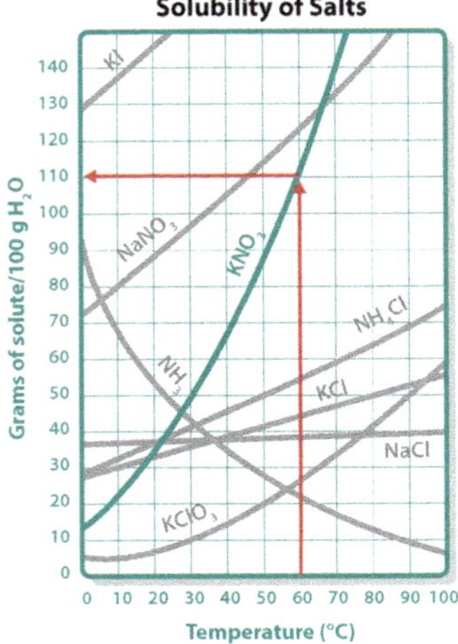

Solubility

The explanation of solvation describes the dissolving of an ionic compound in water. Making solutions with molecular solutes works in a similar way. But does every solute dissolve in every solvent? You actually already know the answer to that question. Think about any ocean beach around the world. While the sea is a saltwater solution made of different salts dissolved in water, sand does not dissolve in water. Why do some substances dissolve in a particular solvent while others don't? It's because of the amount of attraction between the solute and solvent particles as well as the energy relationship outlined on the previous page. We measure the degree of dissolving with **solubility**—the maximum amount of solute that can dissolve in a given amount of solvent at a certain temperature.

The graph at left shows the solubility curves for a number of substances. Each curved line represents the maximum amount of solute that can be dissolved in 100 g of water at temperatures from 0 °C to 100 °C. For example, if you wanted to know the solubility of KNO_3 at 60 °C, you would enter the graph at 60 °C, move vertically until reaching the KNO_3 curve, then move left until reaching the vertical axis. Now you can read the solubility of KNO_3 at 60 °C, approximately 113 g.

HOW IT WORKS

Hot and Cold Packs

If you enjoy outdoor activities, you may have needed either an ice pack or hot pack. Many people have ice packs in their freezer and can soak in a hot bath at home. But these items are not available to you on a hike, at the ball field, or at the skating rink. Chemists have provided these in a form that gets cold or hot only when needed, and both operate according to the properties of solutions.

Instant Cold Pack. Many injuries cause pain and swelling. A common first-aid practice is to use an ice pack to ease the pain and reduce the swelling. Instant cold packs remain at room temperature until they are needed. The pack contains a solid—such as ammonium nitrate, calcium ammonium nitrate, or urea—and water. The two compounds are in separate chambers until needed. When cold is needed, you break the divider between the two compounds, usually by twisting the package, allowing the solid to dissolve in the water. This process is endothermic, so it absorbs heat from the environment, making it feel cold.

Chemical Heat Pack. Whether you participate in or watch winter sports, you probably don't want an ice pack, but a hand warmer would be nice. Chemical hand warmers can work either by chemical reactions or by the action of chemical solutions. The hand warmer above contains a *supersaturated* solution of sodium acetate. Supersaturated means that there are excess solute particles dissolved in the water (see Subsection 9.6). When the metal disk is bent, the solute begins to come out of solution. This is an exothermic process, so heat is released. These warmers can reach about 55 °C and maintain that temperature for about an hour.

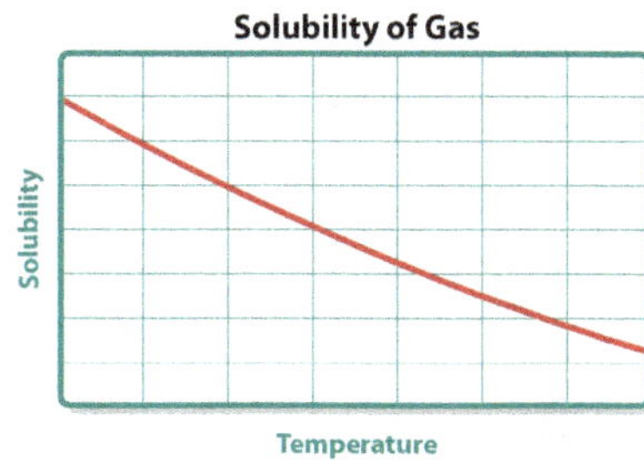

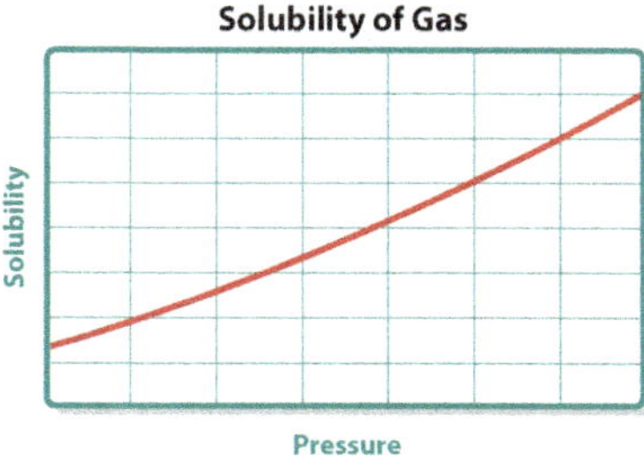

The graphs above show the effect of temperature and pressure on the solubility of a gas in a liquid.

Solubility depends on the chemical and physical properties of the substances involved. The attraction between the solvent and solute has to be great enough to overcome the attractive forces between solvent particles and between solute particles. In general terms, polar and ionic solutes will dissolve in polar solvents while nonpolar solvents are needed to dissolve nonpolar solutes. Chemists remember this by saying "like dissolves like." The solubility of a solute in a certain solvent changes on the basis of solvent conditions, such as temperature and pressure (gaseous solutes only). For example, you can dissolve more sugar in hot tea than in iced tea.

9.4 RATE OF DISSOLVING

On a hot summer day, you may like a nice tall glass of iced tea, and perhaps you like your tea sweetened. So you add sugar cubes and wait, and wait, and wait. Are there ways to speed up the process of dissolving? What you learned in Chapter 7 can help because many of the things that affect reaction rates also affect dissolving rates.

As temperature increases, not only does solubility increase for most solid solutes, but the rate of dissolving increases also. This is because the solvent particles are moving faster, which in turn allows them to come in contact with the solute particles more frequently. Dissolving the sugar in the tea while it is warm and then adding the ice afterward makes the process faster.

The process of stirring also speeds up the dissolving process. By stirring, you are physically moving more solvent particles into contact with the solute. So stir that tea but not too vigorously because you don't want to work up a sweat on an already hot day!

Finally, you could crush those sugar cubes. By crushing them, you expose more of the sugar to the water in the tea. You are providing more surface area for the dissolving process to occur.

🧪 *Demonstrating Rates of Dissolving*

Preparation: Prepare four 100 mL beakers, a hot plate, a stirring rod, a sugar cube, and three medicine cups each with an amount of granulated sugar with the same mass as the sugar cube. You will also need four timers. Consider assigning a group of students to each beaker.

Performance: Pour 40 mL of water into each beaker. Set one on the hot plate and heat on medium-low. Arrange your four groups. One should have the beaker on the hot plate, one should have the sugar cube, and one should have the stirring rod. At the same time, each group should add the sugar to their beaker. The group with the stirring rod should begin stirring their beaker. Continue until the sugar in one of the beakers has completely dissolved.

Discussion:

1. In which beaker did the sugar dissolve most quickly? *(Answers will vary. It was probably either the stirred or the heated beaker.)*

2. Which beaker took the longest time to dissolve? *(the one with the sugar cube)*

3. What combination of conditions would make sugar dissolve most quickly? *(granulated sugar in stirred hot water)*

💻 *Dissolving Rate Applets*

There are numerous applets available online that demonstrate rates of dissolving. Do an internet search using the keywords "solubility applets."

Preparation: Prepare about 10 mL of salt water in a 50 mL beaker. You will also need a hot plate.

Performance: Wearing goggles, turn on the hot plate to medium-low and place the beaker on it. Continue heating until most of the water has boiled off. Turn off the hot plate and allow the remaining heat to evaporate the rest of the water.

Note: This would be a great time to stress that a hot plate should never be left on and unattended.

Discussion:

1. What means of separating mixtures was used to separate the salt from the water? *(boiling and evaporation)*

2. Would this work to separate mulch and gravel? Explain. *(No. Neither boils under normal conditions.)*

3. How might a desert nation near an ocean use this process? *(By evaporating and recapturing the water, drinkable water could be obtained from seawater.)*

Formative Assessment: *Dissolving Rates and Separating Mixtures*

1. Why is water for iced tea heated? *(to speed the dissolving of the tea and the sugar)*

2. What is the difference between evaporation and distillation? *(Evaporation is typically used to separate a solid solute from a liquid solvent. Distillation is used to separate two liquids. Evaporation is typically slower, while distillation involves boiling.)*

3. What determines whether a solute is removed by filtration? *(particle size)*

4. Is magnetism a good choice for separating alloyed gold and silver? Explain. *(No. Neither is magnetic, so a magnet has no effect on either.)*

9.5 SEPARATING MIXTURES

So far we have looked at how to make a mixture, but what if we want to separate a mixture into its components? Since mixtures are physical combinations that are formed by physical processes, we use physical processes to separate them. We choose a method of separation on the basis of the physical properties of the substances in the mixture. While there are many methods of separation, we will look at just a few.

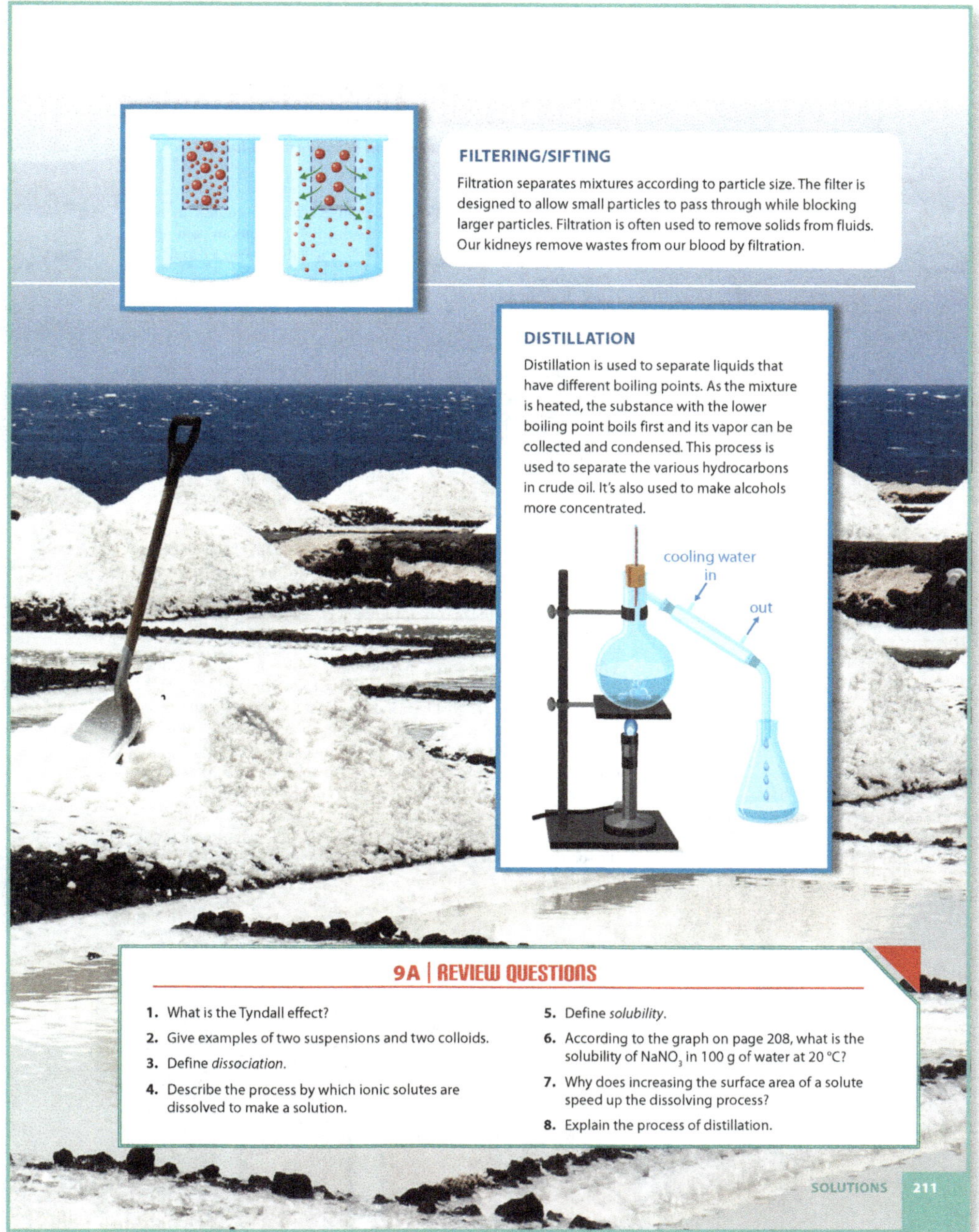

Lab 9A: *All Mixed Up*

Once students have an understanding of the material in Section 9A, they should be ready to work through Lab 9A.

9A | REVIEW QUESTIONS

1. What is the Tyndall effect?
2. Give examples of two suspensions and two colloids.
3. Define *dissociation*.
4. Describe the process by which ionic solutes are dissolved to make a solution.
5. Define *solubility*.
6. According to the graph on page 208, what is the solubility of $NaNO_3$ in 100 g of water at 20 °C?
7. Why does increasing the surface area of a solute speed up the dissolving process?
8. Explain the process of distillation.

9A REVIEW ANSWERS

1. The Tyndall effect is the scattering of light by the particles in a colloid. *(p. 203)*

2. *suspensions*: hairspray, mud; *colloids*: clouds, paint (Other answers are possible.) *(pp. 202–3)*

3. Dissociation is the separating of the ions in an ionic solute to make a solution. *(p. 207)*

4. Solutions are made by dissociating the ions in the ionic solid and dispersing them in the solvent. Each pole of the solvent particles attracts the oppositely charged ions in the ionic solute, pulling them away from each other. Once they are separated enough, they become solvated. *(pp. 206–7)*

5. Solubility is the maximum amount of solute that can be dissolved in a certain amount of solvent at a given temperature. *(p. 208)*

6. approximately 88 g *(p. 208)*

7. By increasing the surface area, more solute particles are exposed to solvent particles. *(p. 209)*

8. Distillation is a process for separating mixtures that is based on the boiling points of the solute and solvent. As a mixture is heated, the lower of the two boiling points will be reached first. This substance will boil off and can be collected by condensation. *(p. 211)*

Ask students how many of them enjoy swimming. Then ask whether they know how pools are kept safe. Some may know that most pools use either chlorine or a chloride salt. Then ask them what happens when the disinfectant substance becomes too concentrated or too weak. Lead them to understand that a solution that is too concentrated is harmful to swimmers. On the other hand, if too dilute, it does not kill pathogens, also making the water dangerous to swimmers.

Formative Assessment: *Saturation*

1. If 50 g of KCl has been dissolved in 100 g of water, and the solution is now at 20 °C, is the solution unsaturated, saturated, or supersaturated? Use the graph on page 208. *(supersaturated)*

2. How would this solution be made? Use the graph on page 208. *(The salt would have to be dissolved after the water had been heated to 80 °C. The solution would then be carefully cooled to 20 °C.)*

3. How much KNO_3 must be dissolved in 100 g of water at 30 °C to make a saturated solution? *(approximately 49 g)*

4. What type of solution (saturated, unsaturated, or supersaturated) would 87 g of $NaNO_3$ make at 40 °C? *(unsaturated)*

9B Questions

- What does "juice from concentrate" mean?
- How can we describe the amount of solute in a solution?
- How can solutions be helpful?

9B Terms

concentration, saturated solution, unsaturated solution, supersaturated solution, molarity, colligative property

Concentration can vary from dilute (left) to concentrated (right), even to the point of being supersaturated.

9B | SOLUTION CONCENTRATION

How can we describe the amount of solute in a solution?

9.6 SATURATION

Have you ever tried to see how much salt or sugar you could dissolve in water? Just as there is a limit to how much water you can pour into a glass, there is a limit to how much solute will dissolve in a certain amount of solvent. Recall that solubility is the *maximum* amount of solute that can dissolve in a given amount of solvent at a certain temperature. We use the term **concentration** to describe the actual amount of solute dissolved in a given amount of solution.

We can use descriptive terms for how concentrated a solution is. When we have a solution with a relatively large amount of solute, we call it a *concentrated* solution. When the amount of solute is relatively low, we say that it is *dilute*. For example, frozen orange juice is a concentrated solution. To prepare the juice for drinking, we dilute it back into orange juice by adding water. We can also use the terms *saturated, unsaturated,* or *supersaturated* as descriptive ways to compare the amount of dissolved solute with the solubility of that solute.

A **saturated solution** contains the maximum amount of solute that the solution can hold at that temperature. If we were to dilute that solution, it would then be an **unsaturated solution**, meaning that the solution contains less than the maximum amount of solute that it could hold at that temperature.

In the case of a **supersaturated solution**, the solution contains more than the maximum amount of solute that it can normally hold at that temperature. How can that be? Under normal conditions, adding more solute to a saturated solution will result in that added solute just settling to the bottom since no more solute can dissolve. The solution is in equilibrium, as some solute continues to dissolve while some dissolved particles return to their undissolved form. But remember that solubility usually increases as temperature increases. So if we make a saturated solution of some solutes at a high temperature and then carefully cool the solution, we can make a supersaturated solution.

TEACHING THE MATERIAL

ESSENTIAL QUESTION

How can we describe the amount of solute in a solution?

OBJECTIVES

- 9B1 Compare unsaturated, saturated, and supersaturated solutions.

- 9B2 Explain measures of concentration, including percent by mass, percent by volume, and molarity.

- 9B3 Relate colligative properties to measures of concentration.

- 9B4 Assess the hidden costs of the usage of salt or a brine solution to prevent roads from freezing. **BWS**

- 9B5 Formulate a position on pollution mitigation from a Christian perspective. **BWS**

Chemists are typically more interested in quantitative descriptions of concentration, and there are many ways to express concentration numerically. We will look at percent by mass, percent by volume, and molarity.

9.7 CALCULATING CONCENTRATION

Percent by Mass

Percent by mass is a comparison of the mass of the solute with the mass of the entire solution. For example, chlorine bleach is typically a 5% by mass solution of NaClO dissolved in water. That means that for every 100 g of bleach solution, there are 5 g of sodium hypochlorite. The formula for percent by mass is

$$\%_m = \left(\frac{m_{solute}}{m_{solution}}\right)100\%,$$

where $\%_m$ is the percent by mass, m_{solute} is the mass of the solute, and $m_{solution}$ is the mass of the entire solution (the mass of the solute plus the mass of the solvent).

EXAMPLE 9-1: Using Percent by Mass

How many grams of salt are needed to make 175 g of a 13.1% salt solution?

Write what you know.

$$\%_m = 13.1\%$$

$$m_{solution} = 175 \text{ g}$$

$$m_{solute} = ?$$

Write the formula and solve for the unknown.

$$\%_m = \left(\frac{m_{solute}}{m_{solution}}\right)100\%$$

$$\frac{\%_m}{100\%} = \left(\frac{m_{solute}}{m_{solution}}\right)\frac{100\%}{100\%}$$

$$\left(\frac{\%_m}{100\%}\right)m_{solution} = \left(\frac{m_{solute}}{m_{solution}}\right)m_{solution}$$

$$m_{solute} = \left(\frac{\%_m}{100\%}\right)m_{solution}$$

Plug in known values and evaluate.

$$m_{solute} = \left(\frac{13.1\%}{100\%}\right)175 \text{ g}$$

$$= 22.9 \text{ g}$$

Differentiated Instruction:
Remediation and Peer Tutoring

Some students may struggle with concentration calculations. Giving the students extra practice problems and pairing struggling students with those more advanced for some of the problems may be helpful. To find extra practice problems, conduct an internet search using the keywords "solution concentration worksheet."

RESOURCES

Ethics: Pollution (p. 221)

Case Studies: Road Salt, Maple Syrup (p. 221)

Mini Lab: *Mass and Volume in Solutions;* Lab 9B: *That's Cold!*—Investigating Freezing Point Depression

STRATEGIES

Class Opener: Use the Pool Concentration teacher note on page 212 to begin a discussion on the importance of measuring solution concentration.

Formative Assessment: Use the Formative Assessment: Saturation teacher note on page 212 to check students' understanding of the terms *saturated*, *unsaturated*, and *supersaturated*.

Differentiated Instruction: Consider using the Differentiated Instruction: Remediation and Peer Tutoring teacher note on this page to help struggling students.

Discussion: Use the Discussion: 70% Isopropyl Solution teacher note on page 214 to help students recognize that the solute and the solvent of a solution are sometimes interchangeable.

(continued)

 **Discussion:** *70% Isopropyl Solution*

Preparation: Bring a bottle of rubbing alcohol to class.

Performance: Pass around the unopened bottle to allow students to examine the label.

Discussion: Begin a discussion using the following questions.

1. What is the percent by volume concentration of isopropyl? *(70%)*

2. What is the solvent in this solution? Explain. *(The isopropyl is the solvent because it makes up more of the solution than the water.)*

3. What is the solvent in a 30% solution of isopropyl? *(water)*

4. We call a solution of salt dissolved in water a salt solution and a solution of sugar dissolved in water a sugar solution. Why do we call 30% water dissolved in 70% isopropyl alcohol an isopropyl alcohol solution? *(While typically the name of the solution identifies the solute, the name actually identifies the substance that makes the solution unique. In the case of 70% isopropyl alcohol, it is the alcohol that makes the solution unique, even though the water is the solute.)*

Percent by Volume

Percent by volume is used when both the solute and solvent are liquids. It is a comparison of the volume of the solute with the volume of the entire solution. You may have cleaned a cut with isopropyl alcohol—typically a 70% by volume solution. This mixture contains 70 mL of isopropyl alcohol in every 100 mL of solution. The formula for percent by volume is

$$\%_V = \left(\frac{V_{solute}}{V_{solution}} \right) 100\% ,$$

where $\%_V$ is the percent by volume, V_{solute} is the volume of the solute, and $V_{solution}$ is the volume of the entire solution (the volume of the solute plus the volume of the solvent).

EXAMPLE 9-2: Using Percent by Volume

What is the volume of solution if 24.3 mL of ethanol is used to make an 11.4% solution?

Write what you know.

$$\%_V = 11.4\%$$

$$V_{solute} = 24.3 \text{ mL}$$

$$V_{solution} = ?$$

Write the formula and solve for the unknown.

$$\%_V = \left(\frac{V_{solute}}{V_{solution}} \right) 100\%$$

$$\%_V V_{solution} = \left(\frac{V_{solute}}{V_{solution}} \right) V_{solution} 100\%$$

$$\frac{\%_V V_{solution}}{\%_V} = \frac{V_{solute} \, 100\%}{\%_V}$$

$$V_{solution} = \frac{V_{solute} \, 100\%}{\%_V}$$

Plug in known values and evaluate.

$$V_{solution} = \frac{(24.3 \text{ mL})(100\%)}{11.4\%}$$

$$= 213 \text{ mL}$$

TEACHING THE MATERIAL

Formative Assessment: Use the Formative Assessment: Calculating Concentration teacher note on page 215 to see whether students understand these calculations.

Formative Assessment: Use the Formative Assessment: Colligative Properties teacher note on page 217 to assess students' understanding of how solutes can alter physical properties.

Case Study: Use the Road Salt case study on page 218 to help students assess the costs and benefits of treating roads with salt.

Case Study: Use the Maple Syrup case study on page 221 to help students understand how the concepts that they have learned apply to making maple syrup.

Biblical Worldview Shaping: Use the Pollution ethics box on page 221 to help students think biblically about the subject of air pollution.

Ticket Out the Door: Write the formula for calculating a solution's percent by mass.

Molarity

Probably the most common measure of concentration used by chemists is molarity. **Molarity** (symbol M) is the number of moles of solute per liter of solution. For example, in a lab activity, you may use 2 M HCl, which we would read as "two molar hydrochloric acid." This solution would contain 2 mol of hydrogen chloride in 1 L of HCl solution. The formula for molarity is

$$M = \frac{mol_{solute}}{V_{solution}},$$

where M is the molarity in mol/L, mol_{solute} is the number of moles of solute, and $V_{solution}$ is the total volume of the solution in liters.

EXAMPLE 9-3: Using Molarity

You need 2.75 mol of NaOH to make a batch of soap. How many liters of 8.4 M NaOH solution would you need to have the required moles of NaOH?

Write what you know.

$$M = 8.4 \text{ mol NaOH/L}$$

$$mol_{solute} = 2.75 \text{ mol NaOH}$$

$$V_{solution} = \,?$$

Write the formula and solve for the unknown.

$$M = \frac{mol_{solute}}{V_{solution}}$$

$$M V_{solution} = \left(\frac{mol_{solute}}{V_{solution}}\right) V_{solution}$$

$$\frac{M V_{solution}}{M} = \frac{mol_{solute}}{M}$$

$$V_{solution} = \frac{mol_{solute}}{M}$$

Plug in known values and evaluate.

$$V_{solution} = \frac{2.75 \text{ mol NaOH}}{8.4 \frac{\text{mol NaOH}}{\text{L}}}$$

$$= 0.33 \text{ L}$$

Formative Assessment:
Calculating Concentration

1. What is the percent by mass of a solution made by mixing 90 g of water with 10 g of sugar? *(10%)*

2. A sample of vinegar is a 5% by volume solution of acetic acid. How much acetic acid is in 1 L of vinegar? *(0.05 L, or 50 mL)*

3. What is the molarity of 1 L of solution containing 2 mol of solute? *(2 M)*

Mass and Volume in Solutions

In this mini lab, students test how the mass and volume will change as a solute (sugar) is added to a solvent (water).

Answers

1. Mass is the amount of matter in an object. Volume is the amount of space it takes up.

2. Answers will vary. Most students will predict that the masses will be added.

3. Answers will vary. Some students may predict that the volume of the water will increase by the volume of the sugar. Others may predict that the final volume of the solution will be equal to the volume of the water because the sugar will not be visible. The correct answer is that the volume will increase but by less than the volume of the sugar. Accept any reasonable hypothesis.

4. Answers will vary. Make sure that students connect their results back to their prediction.

5. Since mass is the amount of matter in a substance, masses add up.

6. The dissolved sugar takes up some space but not as much as the crystalline structure of the solid sugar. Therefore the volume of the solution is greater than the volume of the water but less than the sum of the volumes of water and dry sugar.

7. Since the mass of the solution is the sum of the constituent masses but the volume of the solution is less than the sum of the constituent volumes, we should expect that the density of the solution will be greater than the density of the water alone.

MASS AND VOLUME IN SOLUTIONS

Essential Question:

How do mass and volume change as a solution is made?

Equipment

laboratory balance

weighing boat or paper

graduated cylinder, 10 mL

graduated cylinder, 25 mL

stirring rod

sugar, 5.0 g

We have seen throughout this textbook that chemical and physical properties often change when we make changes to a substance. This activity will allow you to investigate how making a solution affects the mass and volume of the substances.

1. What is the difference between mass and volume?

PROCEDURE

A. In the weighing boat, measure out exactly 5.0 g of sugar.

B. Using the dry 10 mL graduated cylinder, measure the volume of the sugar. Record your data on a piece of paper.

C. Put exactly 15 mL of water in the 25 mL graduated cylinder.

D. Measure the mass of the 25 mL graduated cylinder and water. Record your data on a piece of paper.

2. Predict the change in mass when you add the sugar to the 25 mL graduated cylinder.

3. Predict the change in the volume of the water when you add the sugar.

E. Add the sugar to the 25 mL graduated cylinder and stir until the sugar completely dissolves.

F. Measure the mass and volume of the 25 mL graduated cylinder and sugar solution. Record your data on a piece of paper.

CONCLUSION

4. Did the mass and volume change as expected?

5. Why did the mass change as it did?

6. Why did the volume change as it did?

GOING FURTHER

7. How do you think the density of the solution changed compared with the density of water alone?

9.8 COLLIGATIVE PROPERTIES

As you saw in the in-text lab activity, a solution has different properties compared with its parent solvent. Some of the properties that change are called **colligative properties**, meaning that they are caused by a collection of particles. Colligative properties depend only on the concentration of dissolved particles and not on the identity or properties of the solute. We will look specifically at two colligative properties: *boiling point elevation* and *freezing point depression*.

Boiling Point Elevation

As the relative amount of solute increases, the boiling point of the solution becomes higher than that of the pure solvent. To understand why this happens we have to recall how liquids boil. As we warm a liquid, its vapor pressure increases. Boiling occurs at the temperature where the liquid's vapor pressure becomes equal to or greater than the atmospheric pressure. As solute dissolves in a solvent, the vapor pressure of the solvent becomes lower. Therefore we have to warm the solution to a higher temperature to get the liquid's vapor pressure to equal or exceed the atmospheric pressure.

In the past, chemists used boiling point elevation to measure both the molar mass of the solute and the degree to which ionic solutes dissociate in a solution. Today, other techniques have replaced these uses. But a common use of boiling point elevation is in candy making. As a cook boils a candy solution, it becomes more concentrated, and the boiling point increases. On the candy thermometer (right), you will notice the markings "thread," "soft ball," and "hard ball." The candy maker has to heat each type of candy to a particular temperature and uses the target temperatures on the thermometer for the type of candy that he is making. If you were making fudge, you would heat the mixture to 250 °F ("softball").

Formative Assessment:
Colligative Properties

1. What is a colligative property? *(A colligative property is a property of a substance that changes when a solute is dissolved in a solvent dependent only on the number of particles, not on their identity.)*

2. Does a 2 M solution of KI or a 2 M solution of LiCl have a higher boiling point? *(They have the same boiling point.)*

3. Does a 2 M solution of NaCl or a 2 M solution of $MgCl_2$ have a higher boiling point? *(The $MgCl_2$ has a higher boiling point because there are more dissolved particles.)*

Challenge Question

4. Does a 2 M solution of NaCl or a 1 M solution of $MgCl_2$ have a higher boiling point? Explain. *(NaCl has a higher boiling point because even though there are more particles in each $MgCl_2$, the higher concentration of NaCl produces more total particles.)*

Freezing Point Depression

A common misconception is that since the boiling point increases when a solute is added to a solvent, the freezing point must increase too. The opposite is true—the freezing point of a solution is actually lower than that of the solvent alone. This phenomenon, called *freezing point depression*, also happens because of the lowering of the vapor pressure in the solution.

As they do with boiling point, chemists can use freezing point depression to measure dissociation in a solution and the molar mass of the solute. Road crews in northern climates use freezing point depression to help keep roadways clear of ice as they apply solid salt or a salt solution to roads. The salt lowers the freezing point of the water on the roads, keeping the water from freezing, even though the water temperature is below 0 °C. Freezing point depression is also key to getting the very cold temperatures for making home-made ice cream.

As the boiling point rises and the freezing point decreases, the temperature range of the liquid phase increases. This again is very useful in our everyday lives. We use a solution of propylene glycol and water in the radiators of our cars. The solute raises the boiling point, which makes the solution useful as a coolant to prevent the engine from overheating in hot weather. The same solute also lowers the freezing point, enabling the same solution to act as an antifreeze in winter. Chemistry is both amazing and useful!

CASE STUDY: ROAD SALT

In many climates, winter means snow, and along with it fleets of snowplows. But snowplows alone can't keep the roads clear, so many locations have those same trucks apply salt or brine (salt water solution) to the roads. The application of salt to the roads melts the snow and ice and also prevents new snow from freezing on road surfaces. The most common salt used is sodium chloride, which can lower the freezing point of water to about −10 °C. Other salts are also used, such as potassium chloride, calcium chloride, and potassium acetate.

Lab 9B: *That's Cold*

Once students have an understanding of the material in Section 9B, they should be ready to work through Lab 9B.

9B | REVIEW QUESTIONS

1. Define *concentration*.

2. While making a salt solution, you notice that after stirring the mixture for a long period, there is still salt on the bottom of the glass and it won't dissolve. Explain why this occurs.

3. How much sugar is in 275 g of a sugar solution that is 10.5% sugar by mass?

4. What is the molarity of a solution made by dissolving 0.35 mol of $MgCl_2$ in enough water to make 250 mL of solution?

5. How are colligative properties related to measures of concentration? Give an example.

Use the Case Study above to answer Questions 6–8.

6. Calcium chloride ($CaCl_2$) lowers the freezing point of a solution more than sodium chloride (NaCl) does. Why do you think this is true?

7. Why does salt or brine melt the snow and keep it from freezing even when the temperature is below the freezing point of water?

8. What are some of the hidden costs of using road salt?

9B REVIEW ANSWERS

1. Concentration is the actual amount of solute dissolved in a given amount of solution. *(p. 212)*

2. The solution has reached saturation and the rates of dissolving and precipitating (solidifying) are equal, so no more solute appears to dissolve. *(p. 212)*

3. What we know: $m_{solution} = 275$ g, $\%_m = 10.5\%$, $m_{solute} = ?$

Write the formula and solve for the unknown:

$$\%_m = \left(\frac{m_{solute}}{m_{solution}}\right) 100\%$$

$$\frac{\%_m}{100\%} = \left(\frac{m_{solute}}{m_{solution}}\right) \frac{\cancel{100\%}}{\cancel{100\%}}$$

$$\left(\frac{\%_m}{100\%}\right) m_{solution} = \left(\frac{m_{solute}}{\cancel{m_{solution}}}\right) \cancel{m_{solution}}$$

$$m_{solute} = \left(\frac{\%_m}{100\%}\right) m_{solution}$$

Plug in known values and evaluate:

$$m_{solute} = \left(\frac{10.5\cancel{\%}}{100\cancel{\%}}\right) 275 \text{ g}$$

$$= 28.9 \text{ g}$$

(p. 213)

4. What we know:
$V_{solution} = 250$ mL $= 0.250$ L, $mol_{solute} = 0.35$ mol, $M = ?$

Write the formula:

$$M = \frac{mol_{solute}}{V_{solution}}$$

Plug in known values and evaluate:

$$M = \frac{0.35 \text{ mol}}{0.250 \text{ L}}$$

$$= 1.4 \text{ M}$$

(p. 215)

9A MIXTURES AND SOLUTIONS

- Heterogeneous mixtures are classified on the basis of solute particle size. Two examples are suspensions and colloids.

- Solutions are homogeneous mixtures made by dissolving a solute in a solvent.

- Solutes and solvents can be in any of the states of matter.

- Solutions form due to the electrostatic attractions between the solute and solvent particles. This process can be either exothermic or endothermic.

- Solubility is the maximum amount of solute that can be dissolved in a given amount of solvent at a particular temperature. Solubility of solids in liquids generally increases with increased temperature. Solubility of gases in liquids generally decreases as temperature increases.

- The rate of dissolving can be altered by changing the temperature of the solution, increasing the surface area of the solute, or by stirring the mixture.

- Because mixtures are physical combinations, they are separated by physical means. The technique needed to separate a particular mixture depends on the properties of the substances in the mixture.

9A Terms

suspension	202
colloid	203
solution	204
solute	204
solvent	204
alloy	205
dissociation	207
solubility	208

9B SOLUTION CONCENTRATION

- We can describe solutions as unsaturated, saturated, or supersaturated, depending on the amount of dissolved solute compared with its solubility.

- Concentration can be measured in many ways, including percent by mass, percent by volume, and molarity.

- Colligative properties depend on the number of dissolved particles in a solution, not their identity or properties.

- Boiling point elevation is a colligative property in which the boiling point of a solution is increased above the normal boiling point of the pure solvent.

- Freezing point depression, another colligative property, is the lowering of the freezing point of a solution below the normal freezing point of the pure solvent.

9B Terms

concentration	212
saturated solution	212
unsaturated solution	212
supersaturated solution	212
molarity	215
colligative property	217

---

5. Colligative properties are properties that change according to the number of dissolved particles in a solution. As concentration changes, the colligative properties change. For example, as the concentration of a solution increases, its boiling point also increases. *(p. 217)*

6. The more dissolved particles, the greater the freezing point depression. Calcium chloride dissociates into three particles, while sodium chloride dissociates into only two. *(pp. 217–18)*

7. The dissolved salt depresses the freezing point of the salt water solution below the normal freezing point of water. *(p. 218)*

8. Answers will vary. The salts can be corrosive to car bodies, road surfaces, and bridges. Chemical runoff from some of these salts can be harmful to plants. Runoff from others can affect the oxygen levels in nearby water, possibly affecting aquatic organisms. Salt can also end up in the groundwater system and have a health impact on people.

Recalling Facts

1. Solutes are dissolved in a solvent to make a solution. *(p. 204)*

2. **a.** homogeneous

   b. heterogeneous

   c. heterogeneous

   d. homogeneous *(all pp. 202–5)*

3. The solute particles are surrounded by solvent particles. *(p. 207)*

4. Solubility changes with temperature; therefore, a solubility value is valid only at a particular temperature. *(p. 208)*

5. The solubility of a gas dissolved in a liquid decreases as temperature increases. At high temperatures, less of the gas can stay in solution. *(p. 209)*

6. For solutions generally, polar solvents will dissolve polar solutes and nonpolar solvents will dissolve nonpolar solutes. So if the polar nature of the solute and solvent are alike, the solute will typically be soluble. *(p. 209)*

7. The dissolving process will occur faster and more sugar can be dissolved. *(p. 209)*

8. **a.** magnetism

   b. filtration

   c. distillation or evaporation. *(all pp. 210–11)*

9. A dilute solution is one that has a relatively small amount of solute particles dissolved in it. *(p. 212)*

10. Molarity is a measure of concentration that indicates the moles of solute in 1 L of solution. *(p. 215)*

11. As solute is added, the vapor pressure is lowered. Therefore, to raise the liquid's vapor pressure to atmospheric pressure, its boiling point increases. *(p. 217)*

Understanding Concepts

12. The cutoff between a colloid and a suspension is 1 μm; therefore, this would be considered a suspension. *(pp. 202–5)*

13. See hierarchy chart below. *(pp. 202–5)*

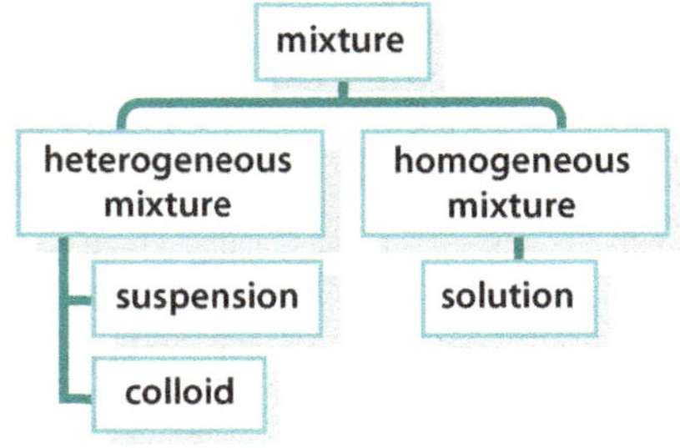

CHAPTER REVIEW QUESTIONS

Recalling Facts

1. Relate the terms *solute*, *solvent*, and *solution* in a sentence.

2. Classify the following as heterogeneous or homogeneous.

   a. salt water   **b.** smog   **c.** chalk in water   **d.** brass alloy

3. During the dissolving process, the solute becomes solvated. Explain.

4. Why does the definition of solubility include the temperature?

5. Why does a warm soda fizz more than a cold soda?

6. What does "like dissolves like" mean?

7. Why do people add sugar to the tea while it is hot when making sweetened iced tea?

8. Which method would work to separate

   a. iron filings from sand?

   b. pollen from the air?

   c. table salt from water?

9. What is a dilute solution?

10. Define *molarity*.

11. Why does the boiling point of a solution rise as solute is added?

Understanding Concepts

12. Would a mixture of a liquid with a solid with particle size 2 μm be considered a suspension, colloid, or solution? Explain.

13. Create a hierarchy chart relating the terms *heterogeneous mixture*, *homogeneous mixture*, *colloid*, *suspension*, and *solution*.

14. Dissolving sodium hydroxide (NaOH) is a highly exothermic process. Where does this energy come from?

15. According to the graph (left), what is the solubility of $NaNO_3$ in 100 g of water at 50°C?

16. According to the graph (left), what happens to the solubility of NH_3 as the temperature increases?

17. Refer to the graphs on pages 208 and 209 to create a table summarizing the solubility trends with changes to temperature and pressure. Include both solid and gaseous solutes.

18. Evaluate the following statement. "I am going to separate this mixture with a chemical reaction."

19. Decanting is a technique used to separate mixtures in which a scientist pours off a less dense material from atop a more dense material. Give an example of a mixture that could be separated by decanting.

20. Explain how to calculate percent by mass.

21. Calculate the mass of solution prepared when 51.8 g of solute is used to make a 14.7% solution.

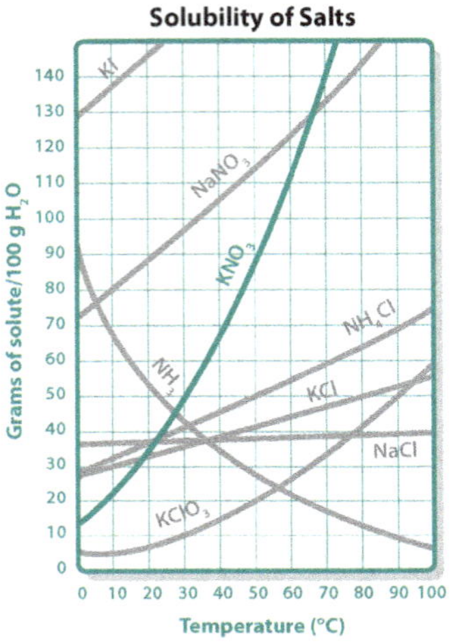

Solubility of Salts

Grams of solute/100 g H_2O vs. Temperature (°C)

(curves labeled: KI, NaNO₃, KNO₃, NH₄Cl, NH₃, KCl, NaCl, KClO₃)

---

14. As solute is dissolved, the solute particles have to be separated from each other and the solvent particles have to be separated from each other. Both processes are endothermic. The process of the solute and solvent particles attracting each other is exothermic. When dissolving NaOH, the energy released in the third process must be significantly higher than the energy needed for the first two. *(p. 207)*

15. approximately 115 g *(p. 208)*

16. The solubility of NH_3 decreases with increased temperature. *(p. 208)*

17. See table at right. *(pp. 208–9)*

18. This statement is false. Mixtures are physical combinations and therefore must be separated by physical means. *(p. 210)*

GENERAL TRENDS IN SOLUBILITY IN LIQUID SOLVENTS		
	Solute	**Solubility**
Increasing Temperature	solid	increases
	gas	decreases
Decreasing Temperature	solid	decreases
	gas	increases
Increasing Pressure	gas	increases
Decreasing Pressure	gas	decreases

19. Answers will vary. *Examples*: pouring grease off the top of a gravy; pouring water off sand and silt that has settled to the bottom of a container. (Other answers are possible.) *(pp. 210–11)*

REVIEW

22. What is the percent by volume of a solution made by mixing 345 mL of ethanol with enough water to form 1325 mL of solution?

23. Create a concept map using the terms *solute, solvent, solution, solubility, concentration, concentrated, dilute, unsaturated solution, saturated solution, supersaturated solution, percent by mass, percent by volume,* and *molarity.*

Critical Thinking

24. Where would saturated, unsaturated, and supersaturated solutions be on the graph on page 220?

25. Calculate the percent by mass of 155 g of a solution that contains 142 g of solvent.

26. How many grams of salt should be added to 100 g of water to make a 14.5% salt solution?

27. Show how you could convert a percent by mass to a percent by volume. (*Hint:* Remember that mass and volume are related by density [$d = m/V$].)

28. Some classmates tell you that if they mix 100 mL of a 3.5 M sugar solution with 100 mL of a 5.0 M sugar solution, they will have 200 mL of an 8.5 M solution. Are they correct? Explain.

Use the Case Study at right to answer Questions 29–32.

29. Which material is the solvent and which is the solute in the sap?

30. What process is being used to separate the water from the sucrose in the sap?

31. What is the percent by mass of sucrose when done?

32. Why does the boiling point rise while the water is boiling off?

Use the Ethics Box below to answer Question 33.

33. Using the strategy presented in Chapter 3, write a one-page essay about how Christians should approach this issue.

ETHiCS — POLLUTION

The Issue: Air Pollution

Air pollution is an issue in many areas around the world. The compounds that cause air pollution come from sources that are both natural (volcanoes, plants and animals, and forest fires) and manmade (fossil fuel power plants, motor vehicles, and aerosols). While most people think of air pollution being outdoors in major urban areas, air pollution can be indoors also. The impact of this pollution is widespread. Air pollution can damage both the natural environment and manmade structures. The health impact is enormous, increasing the number of cases of asthma and allergies as well as lung and heart disease. The World Health Organization estimates that air pollution causes about 7 million deaths each year.

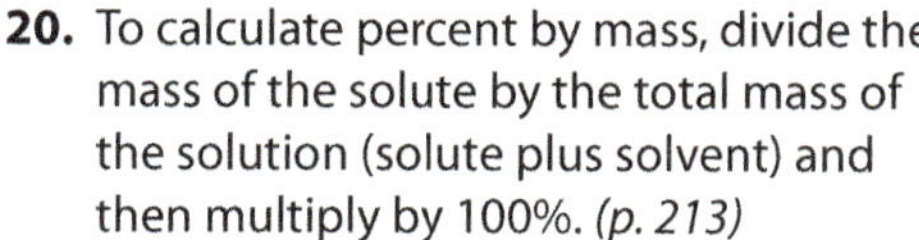

CASE STUDY: MAPLE SYRUP

Many people love the unique taste of maple syrup, which is a mixture made of primarily sucrose and water. The syrup is made from the sap of the sugar maple tree. The sap is collected in late winter and early spring. All of the collected sap is then boiled, removing much of the water. The process is complete when the boiling point has risen to 104.1 °C. The final syrup contains about 60 g of sucrose for every 100 g of syrup.

Equation Solving Techniques (Question 26)

Throughout the Student Edition, equations will normally be solved algebraically and then known values will be substituted so that the answer can be evaluated. In this case, the numerous similar variables and scripts would make the solution difficult to follow. We have therefore opted to substitute known values and solve for the unknown.

Ethics Rubric

A rubric to grade this essay is available in Appendix H.

23. See concept map below. (*pp. 204–15*)

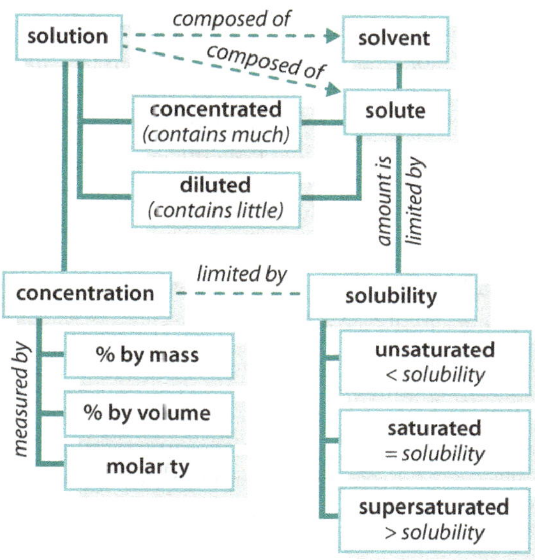

Critical Thinking

24. The curves on the graph represent saturated solutions for each temperature. The area above each curve would represent supersaturated solutions; below each curve, unsaturated solutions. (*pp. 208, 212*)

25. What we know: $m_{solution} = 155$ g, $m_{solvent} = 142$ g, $\%_m = ?$

   Write the formula:

   $$\%_m = \left(\frac{m_{solute}}{m_{solution}}\right) 100\%$$

   Plug in known values and evaluate:

   $$\%_m = \left(\frac{155\ g - 142\ g}{155\ g}\right) 100\%$$

   $$= \left(\frac{13\ g}{155\ g}\right) 100\%$$

   $$= 8.4\%$$

   (*p. 213*)

---

20. To calculate percent by mass, divide the mass of the solute by the total mass of the solution (solute plus solvent) and then multiply by 100%. (*p. 213*)

21. What we know: $m_{solute} = 51.8$ g, $\%_m = 14.7\%$, $m_{solution} = ?$

   Write the formula and solve for the unknown:

   $$\%_m = \left(\frac{m_{solute}}{m_{solution}}\right) 100\%$$

   $$\%_m\, m_{solution} = \left(\frac{m_{solute}}{m_{solution}}\right) (100\%)\, m_{solution}$$

   $$\frac{\%_m\, m_{solution}}{\%_m} = \left(\frac{m_{solute}}{\%_m}\right) 100\%$$

   $$m_{solution} = \left(\frac{m_{solute}}{\%_m}\right) 100\%$$

   Plug in known values and evaluate:

   $$m_{solution} = \left(\frac{51.8\ g}{14.7\%}\right) 100\%$$

   $$= 352\ g \quad (p.\ 213)$$

22. What we know: $V_{solute} = 345$ mL, $V_{solution} = 1325$ mL, $\%_v = ?$

   Write the formula:

   $$\%_m = \left(\frac{V_{solute}}{V_{solution}}\right) 100\%$$

   Plug in known values and evaluate:

   $$\%_m = \left(\frac{345\ mL}{1325\ mL}\right) 100\%$$

   $$= 26.0\%$$

   (*p. 214*)

26. See solution on facing page.

27. $\% \, m = \left(\dfrac{m_{solute}}{m_{solution}} \right) 100\%$

(pp. 213–14)

We can convert masses to volumes by dividing mass by density.

$$\%_{volume} = \left(\dfrac{m_{solute} \div d_{solute}}{m_{solution} \div d_{solution}} \right) 100\%$$

$$= \dfrac{m_{solute} \div \dfrac{m_{solute}}{V_{solute}}}{m_{solution} \div \dfrac{m_{solution}}{V_{solution}}} (100\%)$$

$$= \dfrac{m_{solute} \left(\dfrac{V_{solute}}{m_{solute}} \right)}{m_{solution} \left(\dfrac{V_{solution}}{m_{solution}} \right)} (100\%)$$

$$\%_{volume} = \left(\dfrac{V_{solute}}{V_{solution}} \right) 100\%$$

28. They are incorrect. While the moles of solute and volumes of solutions can add together, the molarities would not. The final molarity would end up between the two starting molarities as the more concentrated solution was diluted and the more dilute solution was made more concentrated. *(p. 215)*

29. Water is the solvent and sugar is the solute. *(p. 204)*

30. distillation *(p. 211)*

31. $\%_m = 60\%$ *(p. 213)*

32. As the water boils, the concentration of the solute (sugar) increases. The increased concentration causes boiling point elevation, a colligative property. *(p. 217)*

33. Biblical principles include stewardship of the world that God provided as well as caring for God's image bearers. Reducing pollution addresses both of these principles.

Biblical outcomes include protecting lives and improving quality of life. Pollution is a challenging issue because much of the pollution is caused by technological advancements. But many of those advancements have improved our quality of life.

REVIEW

22. What is the percent by volume of a solution made by mixing 345 mL of ethanol with enough water to form 1325 mL of solution?

23. Create a concept map using the terms *solute, solvent, solution, solubility, concentration, concentrated, dilute, unsaturated solution, saturated solution, supersaturated solution, percent by mass, percent by volume,* and *molarity.*

Critical Thinking

24. Where would saturated, unsaturated, and supersaturated solutions be on the graph on page 220?

25. Calculate the percent by mass of 155 g of a solution that contains 142 g of solvent.

26. How many grams of salt should be added to 100 g of water to make a 14.5% salt solution?

27. Show how you could convert a percent by mass to a percent by volume. (*Hint:* Remember that mass and volume are related by density [$d = m/V$].)

28. Some classmates tell you that if they mix 100 mL of a 3.5 M sugar solution with 100 mL of a 5.0 M sugar solution, they will have 200 mL of an 8.5 M solution. Are they correct? Explain.

Use the Case Study at right to answer Questions 29–32.

29. Which material is the solvent and which is the solute in the sap?

30. What process is being used to separate the water from the sucrose in the sap?

31. What is the percent by mass of sucrose when done?

32. Why does the boiling point rise while the water is boiling off?

Use the Ethics Box below to answer Question 33.

33. Using the strategy presented in Chapter 3, write a one-page essay about how Christians should approach this issue.

CASE STUDY: **MAPLE SYRUP**

Many people love the unique taste of maple syrup, which is a mixture made of primarily sucrose and water. The syrup is made from the sap of the sugar maple tree. The sap is collected in late winter and early spring. All of the collected sap is then boiled, removing much of the water. The process is complete when the boiling point has risen to 104.1 °C. The final syrup contains about 60 g of sucrose for every 100 g of syrup.

ETHiCS — POLLUTION

The Issue: Air Pollution

Air pollution is an issue in many areas around the world. The compounds that cause air pollution come from sources that are both natural (volcanoes, plants and animals, and forest fires) and manmade (fossil fuel power plants, motor vehicles, and aerosols). While most people think of air pollution being outdoors in major urban areas, air pollution can be indoors also. The impact of this pollution is widespread. Air pollution can damage both the natural environment and manmade structures. The health impact is enormous, increasing the number of cases of asthma and allergies as well as lung and heart disease. The World Health Organization estimates that air pollution causes about 7 million deaths each year.

Biblical motivations include honoring and glorifying God by caring for His creation. We also honor God by protecting His image bearers. When trying to mitigate pollution, we also learn to hope in God's promises of His providence over the creation. We must make wise decisions to learn how to love others in mitigating pollution while making people's lives better.

Student opinions should incorporate biblical principles, outcomes, and motivations.

26. $\%_m = \left(\dfrac{m_{solute}}{m_{solution}}\right)100\%$

Remember that

$m_{solution} = m_{solvent} + m_{solute}$

Therefore,

$\%_m = \dfrac{m_{solute}}{(m_{solution} + m_{solvent})}100\%$

Substitute what we know and solve:

$$14.5\% = \left(\frac{m_{solute}}{m_{solute} + 100\text{ g}}\right)100\%$$

$$\frac{14.5\%}{100\%} = \frac{\left(\dfrac{m_{solute}}{m_{solute} + 100\text{ g}}\right)\cancel{100\%}}{\cancel{100\%}}$$

$$0.145(m_{solute} + 100\text{ g}) = \frac{m_{solute}}{\cancel{(m_{solute} + 100\text{ g})}}\cancel{(m_{solute} + 100\text{ g})}$$

$$\cancel{0.145\,m_{solute}} + 14.5\text{ g} - \cancel{0.145\,m_{solute}} = m_{solute} - 0.145\,m_{solute}$$

$$\frac{14.5\text{ g}}{0.855} = \frac{\cancel{0.855}\,m_{solute}}{\cancel{0.855}}$$

$$m_{solute} = 17.0\text{ g}$$

(p. 213)

The caustic salt water of Lake Natron is colored red by cyanobacteria. The white spots are precipitated salt deposits.

CHAPTER 10

Acids, Bases, and Salts

NOT YOUR IDEAL SWIMMING HOLE

Were it not for a hint of blue sky and clouds, you might think that the image on the left is from another planet—but it isn't. Lake Natron is located in Tanzania in East Africa. It's an example of a *soda lake*, a lake rich in sodium carbonate. The lake has no outflow. Rivers carry dissolved minerals into the lake, and evaporation increases their concentration. Between the high concentration of minerals and water temperatures that can reach 40° C, you might think that nothing could live here. But Lake Natron's red color is caused by a type of blue-green algae, whose color, despite the name, is due to a red pigment that helps with photosynthesis. The algae forms the base of a food chain that supports a large population of flamingoes. Even in such harsh conditions, some of God's creatures can find what they need to thrive.

 **More About Lake Natron**

For video clips about Lake Natron, conduct an internet search. Some full-length documentaries are also available commercially.

Overview

Chapter 10 is a key chapter that explains the important concepts of acid and base chemistry, pH, and salts.

 **Lab Activities**

Lab 10A: *pH pHun*—In this lab activity, students measure the pH of different solutions using an indicator and a pH meter.

Lab 10B: *Feeling the Burn*—In this lab activity, students determine the concentration of a sample of hydrochloric acid by performing a titration.

Acids

Ask students to name some acidic foods. Common examples include lemonade, Italian salad dressing, and mustard. Ask, "What taste do these foods have in common?" *(They are all sour.)* Use this question to guide the discussion into characteristics of acids. You might also take this opportunity to stress the importance of never tasting substances in the laboratory.

Definition of Acids

The definition for acids given here is one of several different ways to define acids on the basis of our modeling the ways that acids work. Other definitions are covered in BJU Press *Chemistry*.

Word Origin

Our English word *acid* is derived from the Latin word *acidus*, which means "sour."

Demonstrating an Acid Reacting with a Metal

Preparation: Prepare approximately 50 mL *of* 3 M hydrochloric acid in a 250 mL Erlenmeyer flask and approximately 10 g of zinc (not powder). Since this demonstration involves HCl, be sure to wear goggles and nitrile gloves. Hydrogen gas is produced, so make sure that there are no open flames.

Performance: Carefully place the zinc in the Erlenmeyer flask. Allow students to watch—without getting too close—the hydrogen bubbles forming on the surface of the zinc.

Discussion:

1. What is this reaction demonstrating? *(an acid's reactivity with metals)*

2. What type of reaction is occurring? *(a single-replacement reaction)*

3. What substances are being produced? *(Some students may be able to deduce that zinc chloride and hydrogen are being produced.)*

Once the reaction has stopped, the zinc chloride solution may be carefully poured down the drain with copious amounts of water if the drain connects to a water treatment plant. As always, follow all local regulations for the disposal of chemicals.

• What are the characteristics of acids and bases?

10A Terms

acid, aqueous solution, hydronium ion, indicator, base, hydroxide ion

10A | ACIDS AND BASES

What's the difference between an acid and a base?

10.1 ACIDS

There's a good chance that you might start salivating when you see the lemon on the next page. You might even know that citric acid is the substance that gives lemons, oranges, and grapefruit their tart flavors. But what exactly is citric acid or any acid for that matter? Let's find out!

PROTON DONORS

Most **acids** are substances that produce hydrogen ions (H⁺) in a solution with water. Since a hydrogen atom has just one proton and one electron, a hydrogen ion is a proton alone. Acids are therefore also considered proton donors. Water-based solutions are called **aqueous solutions**. In water the hydrogen ions combine with water molecules to form **hydronium ions** (H_3O^+). The more hydronium ions an acid produces in solution, the more *acidic* we say that it is.

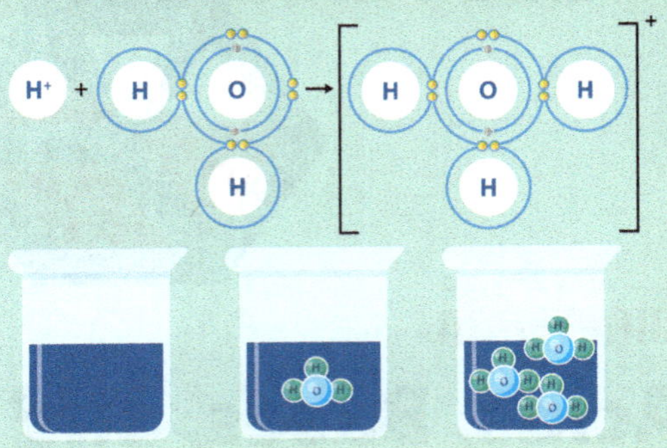

REACTIVITY WITH METALS

Acids can chemically react with some metals in a single-replacement reaction that also releases hydrogen gas. A common use of this property is an industrial process called *etching*. Etching can be used to create delicate metal structures, such as the patterns found on electronic circuit boards.

224   CHAPTER 10

TEACHING THE MATERIAL

ESSENTIAL QUESTION

What's the difference between an acid and a base?

OBJECTIVES

• 10A1 Define *acid* and *base* and give common characteristics of each.

• 10A2 Identify a substance as an acid or base on the basis of its characteristics.

RESOURCES

Case Study: The King of Chemicals

Demonstrating an Acid Reacting with a Metal, Demonstrating Conductivity, Demonstrating Indicators

CONDUCTIVITY

Have you ever wondered why you should never plug in your hair dryer near the bathtub? Pure water doesn't conduct electricity, but any ions dissolved in the water will. Since acids dissolved in water produce ions, an acid solution *will* conduct an electric current. Virtually all household water supplies have some dissolved ions in them, though not necessarily hydronium ions, so you should always remain alert for shock hazards.

COLOR CHANGE WITH INDICATORS

An **indicator** is a substance that will change color in the presence of an acid or base. A very common indicator used in chemistry classes is litmus, which is a kind of dye. Litmus-saturated strips of paper are easily dipped into solutions. Acidic solutions will turn a strip of blue litmus paper red.

SOUR TASTE

Many foods owe their sour taste to the presence of acids. In addition to the citric acid mentioned earlier, common acids found in foods include acetic acid, which is found in vinegar, and malic acid, the acid that produces the tart taste of apples and berries. Strong acids can cause chemical burns, so chemicals in a chemistry lab should *never* be tasted to see whether they are acids!

Demonstrating Conductivity

If you have a conductivity tester, consider giving a demonstration of the conductivity of an acid. For example, dip the tester in white vinegar to show that the acid conducts electricity. After the demonstration, be sure to rinse the conductivity tester with water.

Demonstrating Indicators

If you have access to other indicators, consider giving a demonstration of them. For more information on the kinds available, do an internet search for pH indicators.

Indicators

There are actually many different kinds of indicators besides litmus paper, which today is considered old-fashioned. Indicators are available as both paper strips and solutions. *Narrow range* pH indicators, as the name suggests, can indicate a change in pH over only a narrow range of pH values. The turmeric solution used in this chapter's lab activity is one such indicator. Different narrow range indicators can be combined to form a *wide-range* or *universal* indicator that can indicate pH changes over a broader spectrum of values. One such commonly available product is simply known as *pH paper*. Today, inexpensive pH meters or probes are widely used.

STRATEGIES

Class Opener: Use the Acids teacher note on page 224 to begin discussing the characteristics of acids and bases.

Video: Do an internet search for a video on Lake Natron.

Demonstration: Use the Demonstrating an Acid Reacting with a Metal teacher note on page 224 to show students one of the acid characteristics listed.

Demonstration: Use the Demonstrating Conductivity teacher note on this page to show students one of the acid characteristics listed.

Demonstration: Use the Demonstrating Indicators teacher note on this page to show students one of the acid characteristics listed.

Peer Teaching: Use the Collaboration: Acids and Bases teacher note on page 226 to help students learn the characteristics of acids and bases.

Think, Pair, Share: Use the Think, Pair, Share: Detecting Acids and Bases teacher note on page 226 to help students recognize the value of knowing the characteristics of acids and bases.

(continued)

Collaboration: *Acids and Bases*

Have half of the students think of a characteristic of an acid while the other half think of a characteristic of a base. Pair students from the two groups and have them share with one another their characteristics. Then ask for pairs to share their characteristics with the class.

Water Producers

More observant students may notice that a hydrogen ion and a hydroxide ion together form a water molecule. Additionally, a hydronium ion and a hydroxide ion form two water molecules. This observation is quite accurate and these reactions are part of neutralization reactions, which are covered in Section 10C.

More on Coffee

Caffeine and several other compounds found in coffee beans are bases, but the roasting and brewing of coffee produces chemical changes in other coffee compounds that result in an acidic drink. A typical cup of coffee has a pH value around 5.

Think, Pair, Share:
Detecting Acids and Bases

Write on the board, "How can you safely tell whether a clear liquid sample is deionized water, hydrochloric acid, or sodium hydroxide in only one or two steps?" Have each student think of an answer and then share with a neighbor. Once students have had sufficient time, ask for a few examples of their ideas.

10.2 BASES

As you'll see, some characteristics of bases are similar to those of acids, but others are quite different. Take a look at some identifying features of bases.

BASES

PROTON ACCEPTORS

Many **bases** are substances that produce **hydroxide ions** (OH^-) in an aqueous solution. Such solutions are described as being *basic*, or *alkaline*. The more hydroxide ions that a base can produce in solution, the more basic it is. Hydroxide ions act as proton acceptors. Other molecules or ions that can accept protons are also considered bases. Hydroxide ions can react with hydrogen ions to form water.

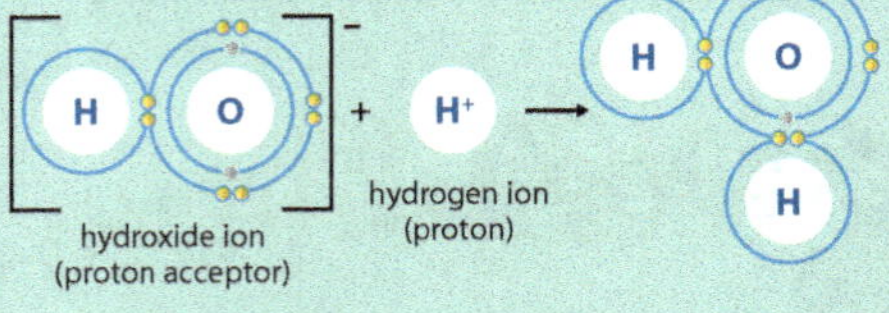

SLIPPERY FEEL

Some strong bases in solution can react with the oil on your skin in much the same way that soap is made. Such bases are thus slippery to the touch, just as soap is slippery. Like acids, strong bases can cause chemical burns, so there is no safe "touch and taste" test for unknown chemicals!

CONDUCTIVITY

The ions produced by bases in solution conduct electrical current just as those in acidic solutions do.

TEACHING THE MATERIAL

Case Study: Use The King of Chemicals case study on page 227 to discuss sulfuric acid and its importance in industry and economics.

Formative Assessment: Use the Formative Assessment: Acids and Bases teacher note on page 227 to assess students' understanding of acids and bases.

Differentiated Instruction: Consider using the Differentiated Instruction: Note on Chemical Equations teacher note on page 228 to differentiate for advanced students or those who are interested in a career in chemistry.

Ticket Out the Door: Write two characteristics of an acid and two characteristics of a base.

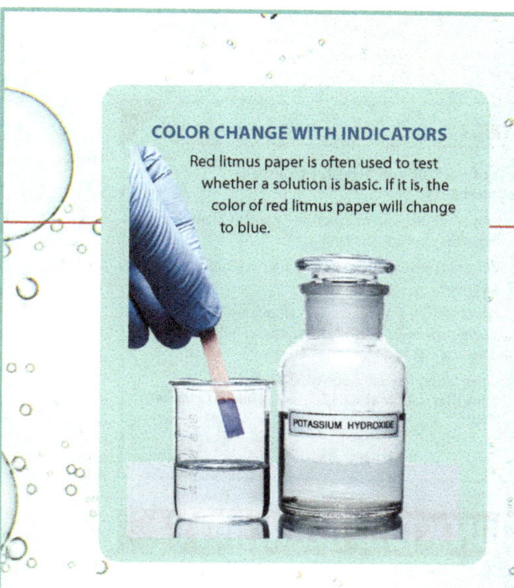

COLOR CHANGE WITH INDICATORS

Red litmus paper is often used to test whether a solution is basic. If it is, the color of red litmus paper will change to blue.

BITTER TASTE

Bases tend to taste bitter. The caffeine in coffee beans and a chemical called *theobromine* in unsweetened chocolate are both proton acceptors and thus bases.

Chemical companies in the United States produce more sulfuric acid (H_2SO_4) than any other single chemical. This acid is used in almost every industrial process. In fact, economists consider sulfuric acid so essential to industry that they sometimes measure the economic condition of a country by how much sulfuric acid it uses. Generally, when a nation's usage of sulfuric acid drops, its whole economy is headed for a downturn. You can easily understand why sulfuric acid is often referred to as the king of chemicals.

A dense, oily substance with a high boiling point, concentrated sulfuric acid is highly corrosive and can dissolve many metals in a matter of minutes! Most importantly, sulfuric acid can react with numerous other chemicals to produce thousands of useful products. More than 60% of the sulfuric acid used worldwide goes into the manufacture of fertilizers and detergents. In addition, chemists use sulfuric acid to produce paints, dyes, plastics, fibers, and a vast array of other products.

The chemical formulas for acids and bases often hint at their acid or base nature. Look at the formulas for some common acids and bases in the table below and see whether you can spot the sources of hydrogen or hydroxide ions.

Common Acids and Bases

Source	Common Name	Formula
vinegar	acetic acid	$HC_2H_3O_2$
fruits	ascorbic acid	$H_2C_6H_6O_6$
soft drinks	carbonic acid	H_2CO_3
citrus fruit	citric acid	$H_3C_6H_5O_7$
insects	formic acid	$HCHO_2$
sour milk	lactic acid	$H_3C_3H_4O_3$
rhubarb, spinach	oxalic acid	$H_2C_2O_4$
household ammonia	ammonium hydroxide	NH_4OH
milk of magnesia	magnesium hydroxide	$Mg(OH)_2$
lye	sodium hydroxide	$NaOH$

Ⓐ Formative Assessment:
Acids and Bases

This activity can be done by placing signs labeled "acid" and "base" at opposite sides of the room and having students point to one of them, or it can be done with multiple forms of technology.

Does this description describe acids, bases, or both?

1. hydrogen ion donor *(acid)*
2. electricity conductor *(both)*
3. proton donor *(acid)*
4. turns blue litmus paper red *(acid)*
5. turns red litmus paper blue *(base)*
6. proton acceptor *(base)*
7. produces ions *(both)*

⁉️ Strange Formulas?

Students may notice that some of the chemical formulas in the Common Acids and Bases table do not have all the atoms of hydrogen grouped together. This is simply an alternative convention for writing formulas—with the H or OH isolated—that allows chemists to quickly identify the ions formed by an acid.

10A REVIEW ANSWERS

1. a solution in which the solvent is water *(p. 224)*

2. hydrogen ions (H^+) *(p. 224)*

3. Accept any three: They have a sour taste, react with metals, conduct electric current in solution, and turn blue litmus paper red. Some students may say that they are proton donors. *(pp. 224–25)*

4. The solution could be tested to see whether it would react with a metal or could be tested with an indicator. *(pp. 224–25)*

5. hydroxide ions (OH^-) *(p. 226)*

6. Accept any three: They taste bitter, feel slippery, conduct electric current in solution, and turn red litmus paper blue. Some students may say that they are proton acceptors. *(pp. 226–27)*

7. Any aqueous solution with ions in it, whether from acids, bases, or both, will conduct electric current. *(pp. 225–26)*

8. Sulfuric acid is used in a wide variety of industrial processes. The more sulfuric acid that is produced, the healthier other industries and the overall economy are assumed to be.

9. hydrogen (H^+) and sulfate ($SO_4{}^{2-}$)

10. $3H_2SO_4 + 2Al \rightarrow Al_2(SO_4)_3 + 3H_2$

11. The prefix *di-* means "two," so diprotic must refer to the two protons (H^+) that sulfuric acid can donate in solution. *(p. 224)*

⟨⟩ Note on Chemical Equations (Question 10)

Throughout the Student Edition, the writers have chosen to omit the parenthetical indicators used in chemical equations to show the states of matter of the various substances. Such formatting is typically introduced in a standard chemistry course but adds an unnecessary dimension of complexity at this level. You may want to include a description of it for advanced students.

10A | REVIEW QUESTIONS

1. What is an aqueous solution?

2. What ions do many acids produce in aqueous solutions?

3. List three characteristics of acids.

4. Suppose you find a beaker in the laboratory that contains an unidentified aqueous solution. Describe one *safe* way to test whether the solution might be an acid.

5. What ions do many bases produce in aqueous solutions?

6. List three characteristics of bases.

7. Why would testing the solution to see whether it will conduct electrical current not be an acceptable answer for Question 4?

Use the Case Study on page 227 to answer Questions 8–11.

8. Why is sulfuric acid production looked at as an indicator of economic conditions?

9. What ions would you expect to find in a sulfuric acid solution?

10. Dilute sulfuric acid reacts with aluminum in a single-replacement reaction. Write the balanced chemical equation for this reaction.

11. Considering the definition of an acid, why do you think that sulfuric acid (H_2SO_4) is considered a *diprotic* acid?

10B | ACIDITY AND ALKALINITY

Why are strong acids and bases dangerous?

10.3 ACID AND BASE STRENGTHS

It might be confusing to hear your science teacher talk about the dangers of acids in the laboratory but then learn that there are acids in our foods. Why is sulfuric acid dangerous but not vinegar? The answer is found in an understanding of two concepts: *strength* and *concentration*. Let's look first at what determines the strength of an acid or base.

Ionization of Acids

You learned in Subsection 10.1 that acids produce hydrogen ions in aqueous solutions. But not all acids give up their hydrogen ions easily. As in other covalently bonded compounds, the bonds between the atoms in some acids are stronger than in others. This means that some acids ionize in water more easily than others. If all or most of the molecules in an acid will ionize in water, then that acid will produce many hydrogen ions in solution. Such acids are called **strong acids**. In contrast, most of the molecules in **weak acids** do not ionize in water, so fewer hydrogen ions are produced.

The more ions an acid produces in solution, the more reactive it is. Vinegar is not particularly dangerous because acetic acid, the acid in vinegar, is a weak acid. Sulfuric acid is far more hazardous because it is a strong acid. Below, see a list of strong and weak acids.

10B Questions

- What makes an acid or base strong or weak?
- What is pH?
- How does pH relate to acid and base strength?
- Is acid and base strength related to concentration?
- How do buffers work?

10B Terms

strong acid, weak acid, strong base, weak base, pH, buffer

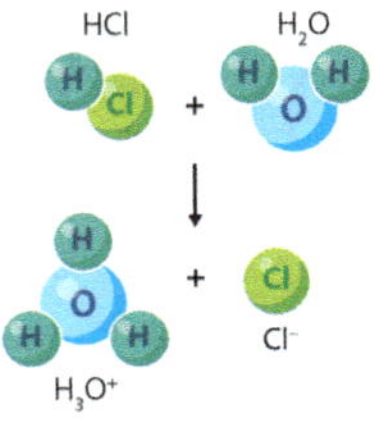

Strong Acids	Weak Acids
perchloric acid ($HClO_4$)	hydrosulfuric acid (H_2S)
sulfuric acid (H_2SO_4)	carbonic acid (H_2CO_3)
hydrochloric acid (HCl)	acetic acid ($HC_2H_3O_2$)
nitric acid (HNO_3)	nitrous acid (HNO_2)

TEACHING THE MATERIAL

ESSENTIAL QUESTION

Why are strong acids and bases dangerous?

OBJECTIVES

- 10B1 Explain the relationship between acid and base strengths.

- 10B2 Relate the terms *concentration* and *strength*.

- 10B3 Explain the pH scale.

- 10B4 Relate pH values with acid or base strength and concentration.

RESOURCES

- Demonstrating Concentration and pH, Demonstrating a Buffered System

- Lab 10A: *pH pHun*—Determining pH; Lab 10B: *Feeling the Burn*—Comparing the Concentrations of Basic Solutions

Dissociation of Bases

Bases vary in their strength, just as acids do. Some, like sodium hydroxide, are very hazardous. Others, like baking soda (sodium bicarbonate), pose far less danger. Many common bases are ionic compounds. As you learned in Chapter 9, ionic compounds dissociate in water. **Strong bases** include those that readily dissociate, producing large numbers of hydroxide ions, which in turn act as proton acceptors. The metal hydroxides shown in the table below are easily identifiable as bases because of the presence of the OH in their chemical formulas.

Ionization of Bases

As you can see in the table below, not all bases contain a readily identifiable OH^- ion. Covalent compounds such as ammonia (NH_3) do not dissociate in water. Instead, these **weak bases** produce hydroxide ions by ionizing in a reaction with water. In this reaction, a hydroxide anion is formed when a water molecule donates one of its protons (H^+) to a weak base. The weak base that accepts the proton becomes a cation. Weak bases do not ionize completely and thus produce fewer hydroxide ions in aqueous solutions than strong bases. The ionization of ammonia in water, which is a reversible reaction, is shown in the following equation.

$$NH_3 + H_2O \rightleftharpoons NH_4^+ + OH^-$$

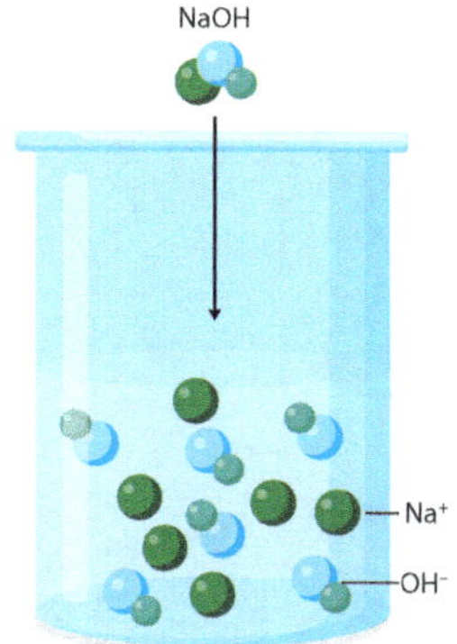

Strong Bases	Weak Bases
lithium hydroxide (LiOH)	ammonia (NH_3)
sodium hydroxide (NaOH)	methanamine (CH_3NH_2)
potassium hydroxide (KOH)	aniline ($C_6H_5NH_2$)
rubidium hydroxide (RbOH)	pyridine (C_5H_5N)

Concentration

Acetic acid is a weak acid, so chugging large amounts of your favorite balsamic vinegar should be okay, right? Not exactly. Vinegar is relatively safe only because it's not pure acetic acid. Vinegar is actually a solution of acetic acid in water. It's usually only about 5% acetic acid by mass. That's good, because pure acetic acid is rather hazardous and can damage your lungs, eyes, and skin.

Like all solutions, acid and base solutions can vary in their concentration—the amount of solute per unit of solution. So if we're talking about a concentrated solution of acetic acid, we mean that it has a lot of acetic acid dissolved in it. Similarly, sodium hydroxide is a very hazardous strong base, but if it is made into a very dilute solution—one with a small amount of solute in a large amount of solvent—then it will pose less risk than a more concentrated solution. Your favorite balsamic vinegar is thus doubly safe because it is both a weak acid *and* a dilute solution.

STRATEGIES

Class Opener: Consider following the suggestions in the pH teacher note on this page to gauge students' awareness of the pH scale and conduct the class period accordingly.

Biblical Worldview Shaping: Use the Water as Acid and Base teacher note on this page to help students recognize the amazing design of water molecules.

Formative Assessment: Use the Formative Assessment: Acid and Base Strengths teacher note on this page to assess students' understanding of acid and base strength.

Demonstration: Use the Demonstrating Concentration and pH teacher note on page 230 to help students understand the connection between an acid's concentration and its pH.

Formative Assessment: Use the Formative Assessment: pH teacher note on page 230 to assess students' understanding of pH.

Demonstration: Use the Demonstrating a Buffered System teacher note on page 231 to help students understand the action of a buffer.

Ticket Out the Door: Write the pH ranges of acids, neutral substances, and bases.

 ### pH

The basic concepts of pH are often covered in elementary grades. You might ask for a show of hands to see how many students are familiar with the concept of pH. Consider asking a few questions about the pH scale, and if students seem to have a good grasp of the content, it might be best to spend most of class time on acid and base strength, concentration, and buffers.

 ### Water as Acid and Base

Ask students whether they think water acts like an acid or a base. The answer is that it behaves as both. When a water molecule accepts a proton to become a hydronium ion, it is acting like a base (proton acceptor). When it loses a proton to a weak base to become a hydroxide ion, it is acting like an acid (proton donor). Even by itself, it acts as both. Substances like water that can donate or accept a proton are called *amphiprotic*. Substances that can act as an acid or a base are called *amphoteric*.

Ask students to explain how water's amphiprotic nature shows God's marvelous design of the world.

Ammonium

More inquisitive students might ask about the identity of the NH_4^+ ion ammonium, a weak acid that exists in equilibrium with ammonia in water.

 ### Formative Assessment:
Acid and Base Strengths

1. Predict the number of hydrogen ions in a 1 M solution of hydrochloric acid. *(1 mol)*

2. If 2 mol of citric acid (a weak acid) are dissolved in enough water to make 1 L of solution, why would you not expect the solution to contain 2 mol of hydrogen ions? *(Citric acid is a weak acid, so many of its molecules will not ionize.)*

3. What happens when methanamine is dissolved in water? *(The water molecule donates a proton and becomes a hydroxide ion.)*

4. Which acid solution is the least concentrated: a concentrated strong acid, a concentrated weak acid, a dilute strong acid, or a dilute weak acid? *(a dilute weak acid)*

Preparation: Prepare three beakers, water, white vinegar, and pH test paper, universal indicator solution, or a pH meter.

Performance: Pour 20 mL of vinegar in Beaker 1. Mix 2 mL of vinegar with 18 mL of water in Beaker 2, and pour 20 mL of water in Beaker 3. Test the pH of each beaker of water.

Discussion: Lead a discussion using the following questions.

1. What is the pH of pure vinegar? *(Answers will vary: usually around 2)*

2. What is the pH of pure water? *(7; Impurities may cause this to vary slightly.)*

3. What is the pH of the dilute vinegar solution? *(Answers will vary: usually around 4)*

4. What effect does concentration have on the pH of an acid? *(Reduced concentration results in a higher pH.)*

5. How does concentration cause this effect? *(Reducing the concentration of acid reduces the concentration of hydrogen ions, which is measured by pH.)*

6. On the basis of this demonstration, what effect would you expect diluting a base to have on its pH? *(Diluting a base would lower its pH.)*

Formative Assessment: *pH*

1. If the hydronium ion concentration in an acidic solution with a pH of 2.0 is reduced by a factor of ten, what would its new pH value be? *(3.0)*

2. Arrange the following substances from most acidic to most basic: water (pH 7), cranberry juice (pH 2.4), and orange juice (pH 3.8). *(cranberry juice, orange juice, water)*

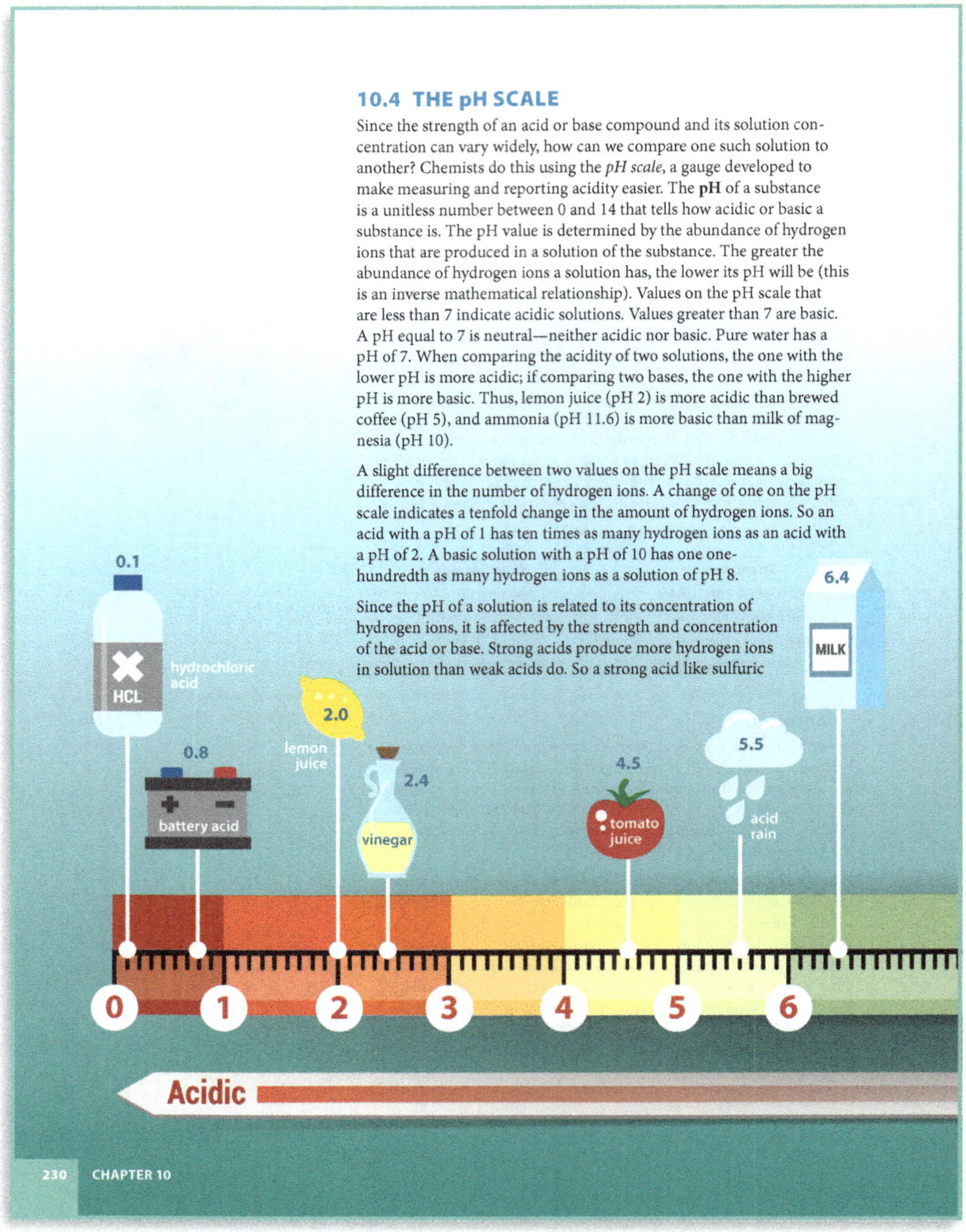

10.4 THE pH SCALE

Since the strength of an acid or base compound and its solution concentration can vary widely, how can we compare one such solution to another? Chemists do this using the *pH scale*, a gauge developed to make measuring and reporting acidity easier. The **pH** of a substance is a unitless number between 0 and 14 that tells how acidic or basic a substance is. The pH value is determined by the abundance of hydrogen ions that are produced in a solution of the substance. The greater the abundance of hydrogen ions a solution has, the lower its pH will be (this is an inverse mathematical relationship). Values on the pH scale that are less than 7 indicate acidic solutions. Values greater than 7 are basic. A pH equal to 7 is neutral—neither acidic nor basic. Pure water has a pH of 7. When comparing the acidity of two solutions, the one with the lower pH is more acidic; if comparing two bases, the one with the higher pH is more basic. Thus, lemon juice (pH 2) is more acidic than brewed coffee (pH 5), and ammonia (pH 11.6) is more basic than milk of magnesia (pH 10).

A slight difference between two values on the pH scale means a big difference in the number of hydrogen ions. A change of one on the pH scale indicates a tenfold change in the amount of hydrogen ions. So an acid with a pH of 1 has ten times as many hydrogen ions as an acid with a pH of 2. A basic solution with a pH of 10 has one one-hundredth as many hydrogen ions as a solution of pH 8.

Since the pH of a solution is related to its concentration of hydrogen ions, it is affected by the strength and concentration of the acid or base. Strong acids produce more hydrogen ions in solution than weak acids do. So a strong acid like sulfuric acid will have a lower pH (more acidic) in solution than a weak acid in a solution of the same concentration. Concentrated solutions (i.e., those with greater amounts of solute) contain more ion-producing particles than dilute solutions do. This means that a highly concentrated solution of sulfuric acid will be more acidic than a dilute solution and will thus have a lower pH.

10.5 BUFFERS

Human blood is a chemical solution with a pH of about 7.4. Our health depends on our blood remaining at or very near this particular pH value (7.35–7.45). Blood pH levels that are too low (acidemia) or too high (alkalemia) can cause some very serious medical issues. Happily for us, our blood is an example of a *buffered system*. The **buffers** in a buffered system are weak acids or bases in solution. Buffers respond to changes in pH in order to maintain the pH level at a certain value. If excess acid is added to the system, it is neutralized by the weak base. If there is too much base in the system, the weak acid neutralizes it. In human blood, the primary weak acid is carbonic acid (H_2CO_3), and the primary weak base is bicarbonate ion (HCO_3^-). Similar buffering systems work to maintain a stable pH inside your cells.

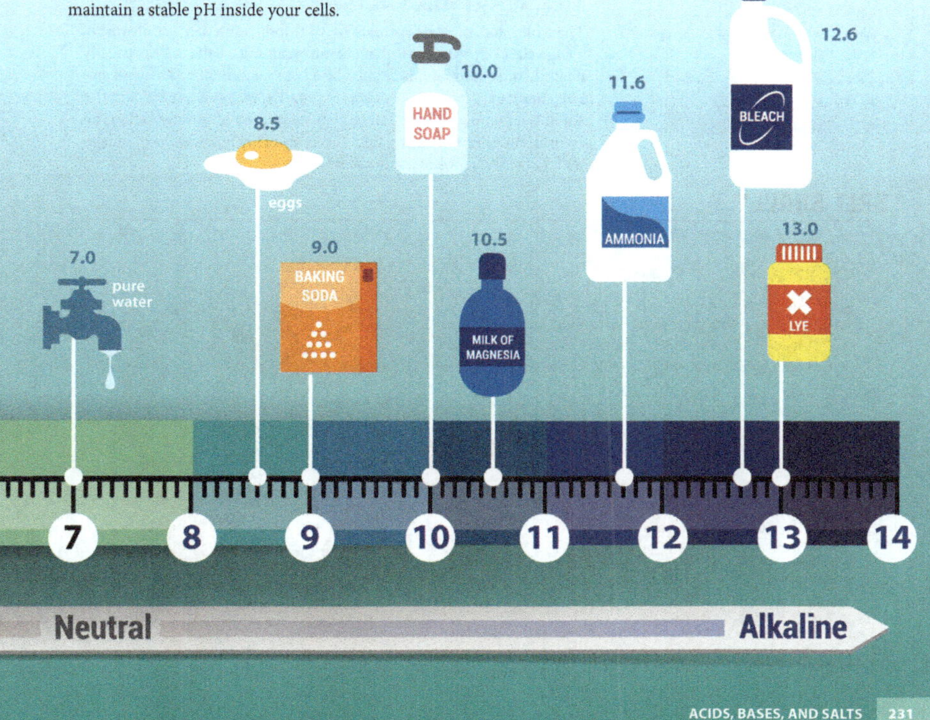

What About the Concentration of Hydroxide?

Some students may puzzle over how the pH of basic solutions can be measured if pH is based only on the abundance of hydrogen ions. Point out to them that because water can act as an acid or a base, even basic solutions contain some hydrogen ions and that this concentration of hydrogen ions decreases as the concentration of hydroxide ions increases. Chemists sometimes use a similar scale called the *pOH scale* to indicate the concentration of hydroxide ions in a solution.

Le Chatelier's Principle

A buffered solution is a great example of Le Chatelier's principle. The ionized and un-ionized forms of the weak acid or base used as a buffer exist in equilibrium. When an acid is added to the solution, the equilibrium changes, resulting in the basic form of the buffer reacting with the acid to neutralize it. The opposite occurs when a base is added.

Demonstrating a Buffered System

Preparation: Prepare two Erlenmeyer flasks, each containing 100 mL of water and a few drops of universal indicator solution. Add some buffer solution to one of the flasks. Prepare a few milliliters each of hydrochloric acid and sodium hydroxide of equal molarity in separate containers. Place a small pipette in the acid and one in the base.

Performance: Place a drop of the acid in each flask. The universal indicator solution in the unbuffered solution should indicate a low pH; add more drops if needed. The buffered solution should be unchanged. Add twice the number of drops of sodium hydroxide to each flask. The color of the unbuffered solution should indicate a high pH, while the buffered solution's pH remains unchanged.

 Lab 10A: *pH pHun*

Lab 10B: *Feeling the Burn*

Once students have an understanding of the material in Section 10B, they should be ready to work through Labs 10A and 10B.

10B REVIEW ANSWERS

1. Acid or base strength is based on how completely a substance ionizes in solution. *(p. 228)*

2. A strong acid is one in which acid ionizes easily and produces a lot of ions in solution. A concentrated solution of that acid is one that contains a lot of solute. It is possible to have a concentrated weak acid since the term simply indicates that such a solution has a large amount of the solute. *(pp. 228–29)*

3. Strength, which determines how readily an acid or base forms ions, and concentration, which describes how much solute is in a solution, both have an effect on the number of ions produced in solution. The stronger the acid or base and the more concentrated the solution formed from it, the more ions that are produced. More ions means greater chemical reactivity. *(pp. 228–29)*

4. Your friend is incorrect. The acidity of a solution increases as pH values decrease, so a pH of 4 indicates that a solution is more acidic than one of pH 6. *(p. 230)*

5. The nitric acid with a pH of 1 has 100 times more hydrogen ions than the nitric acid with a pH of 3. *(p. 230)*

6. The strong base will produce a solution with a higher pH since it will produce more hydroxide ions in solution. *(pp. 230–31)*

7. Buffered systems need to be able to respond to changes in the pH of the system toward either a more acidic or more basic value. *(p. 231)*

Bonneville Salt Flats

Most people think of salt as sodium chloride, but there are many other salts. The Bonneville Salt Flats in Utah contain many salts other than sodium chloride. This location has been used to set many land speed records, including one that broke the sound barrier. To find videos on this and other land speed records at Bonneville Salt Flats, conduct an internet search using the keywords "Bonneville Salt Flats land speed record."

10B | REVIEW QUESTIONS

1. What determines the strength of an acid or base?

2. What is the difference between a *strong acid* and a *concentrated solution* of that acid? Is it possible to have a concentrated weak acid? Explain.

3. Are the chemical properties of an acid or base solution based on the strength of the acid or base, the concentration of the acid or base, or both? Explain.

4. Your friend says that a solution with a pH of 6 is more acidic than one with a pH of 4. Is your friend correct? Explain.

5. How does the amount of hydrogen ions in a nitric acid solution with a pH of 1 compare with the amount of hydrogen ions in a nitric acid solution with a pH of 3?

6. If a strong base and a weak base are each used to make a solution of equal concentration, which is more likely to have a higher pH value? Explain.

7. Why do buffered systems require the presence of both acids and bases?

10C Questions

- How is an acid or base neutralized?
- How can I identify the ions in an acid or base?
- What are the products of a neutralization reaction?

10C Terms

salt, neutralization

10C | SALTS

What happens when acids and bases mix?

10.6 WHAT ARE SALTS?

The white stuff that you sprinkle on your food, table salt, is only one of a large class of compounds that chemists call *salts*. **Salts** are electrically neutral, ionic compounds formed by a combination of cations and anions. In many salts, known as *binary salts*, the cation is a metal and the anion is a nonmetal. Table salt (NaCl), composed of sodium cations and chloride anions, is one such salt. There are many other kinds of salts as well. Let's take a look at some examples.

SALT SURVEY

TEACHING THE MATERIAL

ESSENTIAL QUESTION

What happens when acids and bases mix?

OBJECTIVES

- 10C1 Explain how a neutralization reaction occurs.
- 10C2 Identify the cation of a base and the anion of an acid.
- 10C3 Predict the salt compound that will be formed in a neutralization reaction.
- 10C4 Relate the properties of buffers with how they can benefit people. **BWS**

RESOURCES

- Ethics: Antacids
- Demonstrating Neutralization
- Mini Lab: *Basic Problem*

AMMONIUM DICHROMATE: $(NH_4)_2Cr_2O_7$

Early photographers used ammonium dichromate in a process called *gum bichromate photography*. Using this method, they were able to produce photographs with painting-like qualities.

CALCIUM CARBONATE: $CaCO_3$

Plain, white calcium carbonate crystals aren't much to look at, but calcium carbonate is the main component of pearls and seashells. Animals aren't the only ones who use calcium carbonate to make stuff either. In the form of limestone, it's used by humans to build stone structures and to make concrete.

MALACHITE GREEN: $C_{23}H_{25}ClN_2$

Malachite green is a member of a large group of salts called *triarylmethane dyes*. While malachite green obviously produces a green color, other members of the group yield shades of blue, red, or violet. Malachite green solutions are also used to treat fungal diseases in aquarium fish.

STRATEGIES

Class Opener: Use the Bonneville Salt Flats teacher note on page 232 to introduce the wide variety of salts.

Video: Do a keyword search on "Flinn Scientific neutralization reaction of an antacid" to find a video to demonstrate a neutralization reaction of an antacid and a strong acid.

Demonstration: Use the Demonstrating Neutralization teacher note on page 236 as an alternative to a video to demonstrate the neutralization reaction of an antacid and a strong acid.

Formative Assessment: Use the Formative Assessment: Neutralization teacher note on page 236 to assess students' understanding of neutralization reactions.

Formative Assessment: Use the Formative Assessment: Predicting Salt Compounds teacher note on page 237 to check students' abilities to predict the salt compounds formed during neutralization.

Biblical Worldview Shaping: Use the Antacids ethics box on page 238 to begin a discussion of the benefits and disadvantages of antacids.

(continued)

Basic Problem

Use 1/4 tsp of turmeric powder per 1/4 c of isopropyl alcohol.

Possible basic solutions to test include soap or detergent in water, window cleaner, household ammonia, and baking soda in water.

Be sure to test the pH of your solutions in advance of the activity. You'll need this information to grade students' responses.

For an added element of fun, you may have students test unknown solutions and then guess what substance they think each is.

Answers

1. Answers will vary.

2. orange-yellow

3. Answers will vary. *Example*: Ammonia will be more basic than soapy water.

4. No

5. Yes; a dark red

6. Turmeric indicator shows that a substance is an acid if the indicator does not change color. A color change indicates that the substance is a base.

More on Turmeric Indicator

Strictly speaking, turmeric indicator starts to change color somewhere between pH 7.4 and 8.6. It is therefore not entirely trustworthy for distinguishing between acids and bases since it may not change color in the presence of some mild bases.

Turmeric is an orange-yellow spice that is used in Asian cuisine. Spicy yellow curries get their color from it. Turmeric is also an indicator. In this activity, you'll use turmeric indicator to test some common household products.

Essential Question:

Is there a simple way to determine which of two solutions is more basic?

Equipment

goggles

pipette

small bowls (4)

stirring rod or spoon

turmeric indicator

lemon juice

basic solutions to test

PROCEDURE

A Use the small bowls to obtain some turmeric indicator, lemon juice, and samples of two basic solutions from your teacher. You'll need to either use a separate stirring rod for each solution or rinse the stirring rod in between testing each solution.

  1. Which two basic solutions will you be testing?

  2. What color is the turmeric indicator?

  3. Which of the two basic solutions do you think is more basic? Write your answer in the form of a hypothesis.

B Add one drop of turmeric indicator to the lemon juice. Gently stir the indicator and juice together.

  4. Does the turmeric indicator change color in lemon juice?

C Add one drop of turmeric indicator to one of the bases. Gently stir the indicator and base together.

  5. Does the turmeric indicator change color in the base? If so, what color is produced?

  6. How does turmeric indicator show whether a substance is an acid or base?

TEACHING THE MATERIAL

Biblical Worldview Shaping: Use the Not Always Their Choice teacher note on page 238 to help students recognize that health problems such as heartburn are not always the result of poor decisions on the part of the sufferer.

Ticket Out the Door: Write the products of a neutralization reaction involving HCl and LiOH.

D Now add drops of lemon juice to the base, stirring gently after each drop. Keep count of how many drops are added. Stop when the color changes back to a shade that matches that of the indicator in lemon juice alone.

7. What is happening to the basic solution as you add drops of lemon juice?

8. How many drops did you add to obtain the desired color change?

E Repeat Steps **C** and **D** for the second base.

9. How many drops did you add before obtaining the desired color change in the second basic solution?

ANALYSIS

10. Which of the two basic solutions is more basic? How do you know?

11. Did this result support your hypothesis?

CONCLUSION

The technique that you used to compare the basicity of two solutions is called *titration*. It is a common laboratory procedure. You'll see it again in a standard chemistry course.

GOING FURTHER

12. The two images below show hydrangea flowers, which act as natural indicators. What would explain their two colors?

13. What would a gardener need to do to if she desired the same plant to have blue flowers one year and pink the next?

Answers (continued)

7. The acid in the lemon juice is neutralizing the base in the basic solution.

8. Answers will vary.

9. Answers will vary.

10. Answers will vary. The solution that requires more drops of lemon juice has more base ions to neutralize, so it is the more basic solution.

11. Answers will vary. Check to make sure that students' answers are consistent with their hypotheses.

12. Answers will vary. Since they are natural indicators, these two hydrangeas must be growing in soils of different pH. *(The blue is growing in acidic soil and the pink in alkaline soil.)*

13. She would need to amend the soil so that its pH is that which produces the desired color.

 ### *Don't Try This in Class!*

It may be tempting to demonstrate the neutralization of hydrochloric acid with sodium hydroxide since the products are harmless table salt and water. But the reaction also produces a great deal of heat, so it should be performed only with very dilute reagents. Because this makes the reaction less visually interesting and because of the potential safety hazard, we do not recommend demonstrating this reaction.

  ### *Demonstrating Neutralization*

For a video showing a very visible neutralization reaction, conduct an internet search using the keywords "Flinn Scientific neutralization reaction of an antacid." You may choose to perform this demonstration for your students or just have them view the clip.

This demonstration also connects with the ethics box on page 238.

 ### *Formative Assessment:*
Neutralization

Show each of the following reactions and have students use thumbs up or down to vote on whether each is a neutralization reaction.

1. $HCl + KOH \rightarrow KCl + H_2O$ *(Yes)*

2. $2H_2 + O_2 \rightarrow 2H_2O$ *(No. A salt is not produced.)*

3. $NaHCO_3 + C_6H_8O_7 \rightarrow$
   $NaC_6H_7O_7 + CO_2 + H_2O$ *(Yes)*

4. $C_6H_{12}O_6 + C_6H_{12}O_6 \rightarrow C_{12}H_{22}O_{11} + H_2O$

   (No. Sucrose is a covalent compound, not a salt.)

Questions 3 and 4 may be difficult for some students, so consider allowing them to work on these questions in groups of two or three.

10.7 NEUTRALIZATION

Salts can be produced by the chemical reaction of an acid with a base, called a *neutralization reaction*. **Neutralization** is a double-replacement reaction that generally produces a salt and water. The salt consists of the cation from a base and the anion from an acid. For example, sodium chloride contains a sodium cation (Na^+) and a chloride anion (Cl^-). The sodium cations for a sodium chloride–forming neutralization reaction come from a sodium base, such as sodium hydroxide. The chloride anions come from a chloride acid, such as hydrochloric acid. Below is the neutralization reaction that results when sodium hydroxide and hydrochloric acid are combined.

$$HCl + NaOH \rightarrow NaCl + H_2O$$

Since these compounds completely ionize or dissociate in water, this equation could be written the following way.

$$H^+ + Cl^- + Na^+ + OH^- \rightarrow Na^+ + Cl^- + H_2O$$

The double-replacement reaction produces water, which mingles with the water in the original solution, and the soluble table salt. The table salt remains dissociated as long as it is in solution, but it can be crystallized if the water is removed by evaporation.

We can use the previous example of neutralization to create a general formula for such reactions.

$$acid + base \rightarrow salt + water$$

This general formula can be used to predict the products of a neutralization reaction.

10.8 PREDICTING SALT COMPOUNDS

We've already seen that sodium hydroxide, a strong base, reacts with a strong acid like hydrochloric acid. What happens when sodium hydroxide ($NaOH$) reacts with a weak acid such as carbonic acid (H_2CO_3)? Let's use what we know about the general formula for neutralization reactions, along with the ionization of acids and dissociation of bases, to predict the products formed by a reaction of those two compounds. The following process will work well for neutralization reactions involving strong bases.

EXAMPLE 10-1: Predicting a Salt

Begin writing out a chemical equation by writing down the reactants on the left side of the equation. Identify which compound is the acid and which is the base.

$$H_2CO_3 + NaOH \rightarrow salt + water$$

The OH^- group in sodium hydroxide alerts us to its being the base in this reaction. It will serve as the proton acceptor.

Remember, water will be formed by the anion from the base (OH^-) and a hydrogen ion (a single H^+ ion, that is, a proton) from the acid.

$$H_2CO_3 + NaOH \rightarrow salt + H_2O$$

The salt will be formed from the leftover cation from the base (Na^+ in this example) combined with the anion from the acid (H_2CO_3 less one H^+ ion leaves HCO_3^-). The finished equation thus looks like this:

$$H_2CO_3 + NaOH \rightarrow NaHCO_3 + H_2O$$

Remember that list of common polyatomic ions back in Chapter 5? HCO_3^- is bicarbonate, so the salt formed by this reaction is called *sodium bicarbonate*.

10.9 PUTTING NEUTRALIZATION TO WORK

Have you ever eaten too much spicy food at one sitting? Perhaps afterward you felt a burning sensation in your chest and esophagus—acid indigestion. It's caused by the excess digestive juices that your stomach produces in response to a large meal. These juices are highly acidic and have a pH value around 1. How can the symptoms associated with too much stomach acid be treated? One way is to take an antacid, a base compound used to relieve the symptoms of indigestion. Common bases found in antacids are calcium carbonate, sodium bicarbonate, magnesium hydroxide, and aluminum hydroxide. These bases work to neutralize stomach acid. Many people find great relief through these kinds of products. But as with any medicine, antacids must be taken in limited doses.

Formative Assessment:
Predicting Salt Compounds

Students will likely not know the names of some of these compounds, but they should be able to create the formulas.

1. HNO_3 + KOH *(KNO_3, potassium nitrate)*

2. H_2SO_4 + 2NaOH *(Na_2SO_4, sodium sulfate)*

3. H_2CO_3 + $2NH_4OH$ *[$(NH_4)_2CO_3$, ammonium carbonate]*

For Questions 4 and 5, show the salt and water formulas and ask students to predict the acid and base used to form the products.

4. KBr + H_2O *(HBr + KOH)*

5. $FeCl_2$ + $2H_2O$ *[2HCl + Fe(OH)$_2$]*

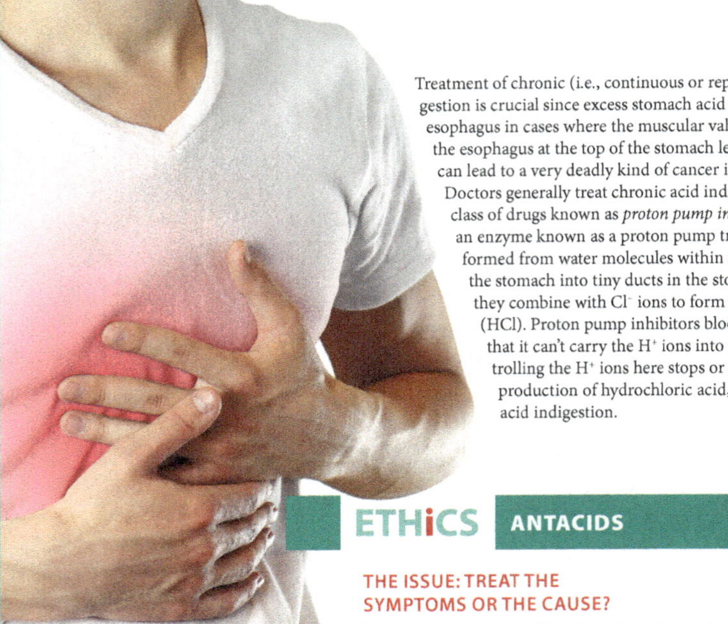

Treatment of chronic (i.e., continuous or repeated) acid indigestion is crucial since excess stomach acid can damage the esophagus in cases where the muscular valve that shuts off the esophagus at the top of the stomach leaks. This damage can lead to a very deadly kind of cancer in the esophagus. Doctors generally treat chronic acid indigestion with a class of drugs known as *proton pump inhibitors*. Normally, an enzyme known as a proton pump transports H^+ ions formed from water molecules within the cells that line the stomach into tiny ducts in the stomach wall. There they combine with Cl^- ions to form hydrochloric acid (HCl). Proton pump inhibitors block the enzyme so that it can't carry the H^+ ions into the ducts. Controlling the H^+ ions here stops or greatly reduces the production of hydrochloric acid, preventing chronic acid indigestion.

ETHiCS — ANTACIDS

THE ISSUE: TREAT THE SYMPTOMS OR THE CAUSE?

Americans consume one of the richest diets of any nation on Earth. Such a diet can lead to a variety of problems, one of which is acid indigestion, also called heartburn. In 2017 alone, Americans spent over $2.6 billion on over-the-counter (OTC) treatments for acid indigestion. But the overuse or long-term use of antacids can cause its own set of problems. Some issues, such as kidney stones, are serious.

10C | REVIEW QUESTIONS

1. Of what two parts is a salt composed?
2. What two kinds of substances react in a neutralization reaction?
3. What are the products of a neutralization reaction?

Use the following neutralization reaction to answer Questions 4–7.

$$Mg(OH)_2 + 2HF \;\rightarrow\; ? + ?$$

4. Which of the reactants is an acid?
5. What is the name of the cation that is contained in the base? What is the name of the anion that is part of the acid?
6. Write out the complete balanced equation.
7. What is the name of the salt produced by this reaction?

Use the Ethics box above to answer Question 8.

8. Using the strategy presented in Chapter 3, write a one-page essay about how Christians should approach the usage of OTC antacids.

 **Not Always Their Choice**

The figure mentioned in the ethics box does not include the amount spent on prescription remedies for more serious digestive issues, such as gastroesophageal reflux. We need to be careful to distinguish between health issues that are linked to poor dietary choices and those that may have underlying physiological causes, including genetics. Individuals with the latter may have little or no control over such health issues. The focus should be on eating choices, not on the use of antacids per se.

Ethics Rubric

A rubric to grade the essay for Question 6 is available in Appendix H.

10C REVIEW ANSWERS

1. a cation and an anion *(p. 232)*
2. an acid and a base *(p. 236)*
3. a salt and water *(p. 236)*
4. HF (hydrofluoric acid) *(p. 236)*
5. magnesium; fluoride *(p. 236)*
6. $Mg(OH)_2 + 2HF \rightarrow MgF_2 + 2H_2O$ *(pp. 236–37)*
7. magnesium fluoride *(pp. 236–37)*
8. Biblical principles include stewardship of the bodies that God has given us.

The Bible is not primarily a book about healthy eating, of course, but in the New Testament we do read some principles that in a broader context can be applied to eating. Paul wrote in 1 Corinthians 6:19–20, for example, in response to sexual immorality, but the principle of our bodies not belonging to us and the requirement to honor God with our bodies can be applied to making wise eating choices.

Biblical outcomes include protecting lives and improving quality of life. Although antacids can provide a convenient and welcome relief after the occasional bout of improper eating, frequent use of them may indicate that a change in eating habits is in order.

One of the ways that we show love for God is by taking care of the bodies that God has given us. A healthy body not only glorifies God but also gives a person greater opportunities to be used by God.

Student opinions should incorporate biblical principles, outcomes, and motivations.

CHAPTER 10 REVIEW

10A ACIDS AND BASES

- Most acids produce hydrogen ions in aqueous solutions. Most bases produce hydroxide ions.
- Acids are proton donors; bases are proton acceptors.
- Acids taste sour, react with metals, and turn blue litmus paper red.
- Bases taste bitter, feel slippery, and turn red litmus paper blue.
- Both acids and bases produce ions that will conduct electricity in water.

10A Terms

acid	224
aqueous solution	224
hydronium ion	224
indicator	225
base	226
hydroxide ion	226

10B ACIDITY AND ALKALINITY

- Strong acids or bases are those that ionize or dissociate completely in water. Weak acids and bases do not ionize or dissociate completely.
- The pH scale assigns values between 0 and 14 that indicate how acidic or basic a solution of a substance is. A value of 7 is neutral, while values less than 7 are acidic and more than 7 are basic.
- The pH of a solution depends both on the strength of its acid or base solute and on its concentration.
- A buffered system contains both a weak acid and a weak base that work to maintain a particular pH value.

10B Terms

strong acid	228
weak acid	228
strong base	229
weak base	229
pH	230
buffer	231

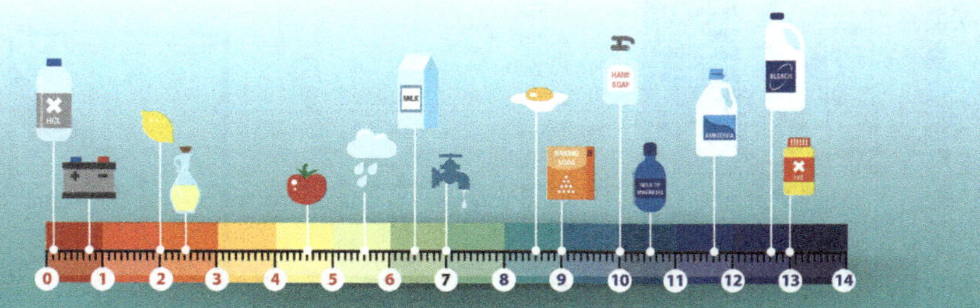

10C SALTS

- Acids and bases neutralize each other in a double-replacement reaction that produces a salt and water.
- Salts are electrically neutral compounds formed by a combination of cations and anions.
- Antacids contain bases that neutralize excess stomach acid.

10C Terms

salt	232
neutralization	236

Recalling Facts

1. Hydronium forms when the hydrogen ion from an acid combines with a water molecule. *(p. 224)*

2. a substance that will change color in the presence of an acid or base *(p. 225)*

3. See concept maps below.

(pp. 224–27)

```
        substance that
        produces hydrogen
        ions in solution

acetic acid,         proton
hydrochloric acid,   ACID   donor
citric acid

        tastes sour,
        reacts with metals,
        conducts electricity,
        turns blue litmus paper red
```

```
        substance that
        produces hydroxide
        ions in solution

sodium hydroxide,    proton
ammonia,    BASE     acceptor
aniline

        tastes bitter,
        feels slippery,
        conducts electricity,
        turns red litmus paper blue
```

4. He is incorrect. Both acids and bases can cause chemical burns. *(pp. 225–26)*

5. A strong acid or base is one that dissociates easily or produces many ions. *(pp. 228–29)*

6. decreases *(p. 230)*

7. There are 1000 (10^3) times more ions. *(p. 230)*

8. Buffers are solutions that contain weak acids and bases that respond to changes in pH in order to maintain the pH level at a certain value. *(p. 231)*

9. Salts in general are ionic compounds containing cations and anions. In binary salts, the cation is a metal and the anion is a nonmetal. *(p. 232)*

10. a double-replacement reaction *(p. 236)*

Understanding Concepts

11. Both are correct. Acids produce hydrogen ions in solution, which then react with water molecules to produce hydronium ions. *(p. 224)*

12. It is most likely an acid. The bubbles indicate that the metal (zinc) is reacting with the liquid. Metals react readily with acids but not with bases. *(pp. 224, 226–27)*

13.
  a. base
  b. acid
  c. base
  d. base
  e. acid *(all pp. 224–27)*

14. The substance being tested is a base because an acid would turn the blue litmus paper red. *(p. 227)*

15. Since it is a strong base, sodium hydroxide dissociates in water rather than ionizing. *(p. 229)*

16. The unknown base must be a weak base. It does not appear to be producing large numbers of ions in solution and is thus probably only partially dissociating. *(p. 229)*

17. Both have an effect. A substance's pH is determined by the number of ions produced in solution, which is determined by both the ease with which particles are formed (strength) and by the amount of particle-producing solute in a solution (concentration). *(pp. 230–31)*

CHAPTER 10

CHAPTER REVIEW

Recalling Facts

1. How are hydronium ions formed when an acid is dissolved in water?

2. What is an indicator?

3. Create concept definition maps for the terms *acid* and *base*.

4. A classmate tells you that acids are dangerous but bases are not. Is he correct? Explain.

5. What does the word *strong* mean when describing an acid or base?

6. If the pH of a solution increases, does its acidity increase or decrease?

7. A change from pH 5 to pH 2 represents what change in the amount of hydrogen ions?

8. What are buffers?

9. What is the difference between salts in general and binary salts?

10. What type of chemical reaction is a neutralization reaction?

Understanding Concepts

11. One of your classmates says that acids produce hydronium ions in solution. Another says that acids produce hydrogen ions. Which is correct? Explain.

12. A chemist accidentally drops a small piece of metal into an unmarked test tube of clear liquid (left). Small bubbles begin to form on the metal. Is the liquid in the test tube most likely an acid or a base? Explain.

13. Classify each of the following substances as either an acid or a base.

  a. KOH
  b. H_3BO_3
  c. NH_4OH
  d. $Zn(OH)_2$
  e. $HClO_2$

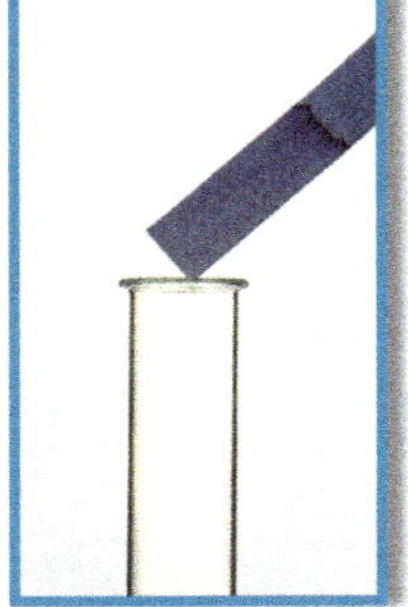

14. Is the substance being tested at right an acid or a base? Explain.

15. Why is it incorrect to say that sodium hydroxide (NaOH) ionizes in an aqueous solution?

16. A chemist attempts to make a basic solution by adding an unknown base to a liter of distilled water. A relatively large amount of solute shifts the pH of the solution only from 7.0 to 7.4. What can the chemist conclude about the strength of the unknown base?

17. Is pH affected by the strength of an acid or base, the concentration of the acid or base, or both? Explain.

18. Two acid solutions of equal concentration have different pH values. How is this possible?

19. What would happen if human blood contained bicarbonate as a buffer but not carbonic acid?

240 CHAPTER 10

REVIEW

20. Nitric acid (HNO_3) forms hydrogen cations and nitrate anions in solution. Why is it not considered a salt?

21. Predict the salt that will form during a neutralization reaction between cesium hydroxide (CsOH) and hydrobromic acid (HBr). Write the balanced chemical equation for the reaction.

Critical Thinking

22. Water is an example of an *amphiprotic* substance, one that can either donate or accept a proton. Use information and examples from the textbook to explain how this is possible.

23. Your study buddy tells you that 1 M hydrochloric acid is weaker than 2 M hydrochloric acid. What is the flaw in your partner's statement? What would be a better way to compare the two solutions?

24. Do you agree or disagree with the statement, "Vinegar tastes strong, so it must be a strong acid"? Defend your choice.

25. Scientists use conductivity meters to measure how well a solution can conduct an electrical current. The higher the conductivity measured by the meter, the better the solution is at conducting electricity. If one basic solution conducts electrical current better than another basic solution of equal concentration, what can be inferred about the strengths of the two bases? Explain.

26. Carbonic acid is a weak acid that forms when carbon dioxide dissolves in water. Why does this make it nearly impossible to maintain samples of pure water?

27. Zinc reacts with hydrochloric acid (HCl) as shown in the following equation. State at least two reasons why the reaction is not considered a neutralization reaction.

$$Zn + 2HCl \rightarrow ZnCl_2 + H_2$$

28. How is the presence of blood buffers an indication of God's creative design?

---

18. The two solutions are made from acids of different strengths. *(pp. 230–31)*

19. The body wouldn't be able to correct the pH of the blood if it became too basic. *(p. 231)*

20. Nitric acid is a covalent compound, not an ionic one, so it does not meet the definition of a salt. The ions are formed through ionization and not dissociation of an ionic compound. *(p. 232)*

21. cesium bromide

$$CsOH + HBr \rightarrow CsBr + H_2O$$

(pp. 236–37)

Critical Thinking

22. Water can act as a proton acceptor, as when it forms hydronium ions with the hydrogen ions that result from the ionization of acids. When water reacts with a weak base, such as ammonia, it acts as a proton donor. *(pp. 224, 229)*

23. Your friend is describing molarity, a property that is based on the concentration of the acid, not on its strength. He should have described the 1 M solution as more dilute than the 2 M solution. Since both solutions contain hydrochloric acid, they are both made from a strong acid. *(p. 229)*

24. A strong taste refers to a subjective sensation. But a strong acid is one that meets a particular requirement, that is, it completely or nearly completely ionizes. Acetic acid does not meet that requirement and is thus classified as a weak acid. *(p. 228)*

25. The base that conducts electricity better is a stronger base because it is producing more current-conducting ions in solution. *Note*: In some solutions, once a certain concentration is reached, the conductivity will actually decrease. *(p. 229)*

26. Atmospheric carbon dioxide dissolves into pure water, rendering it impure.

27. It is not a double-replacement reaction nor does it produce both a salt and water as products. *(p. 236)*

28. God designed our blood to function best at a particular pH level. The buffers show that not only did God design our blood, He also included safeguards in the design that work to keep our blood at the optimal pH level. Such a providential buffer system is important, especially in a fallen world.

APPENDIX A

UNDERSTANDING SCIENTIFIC TERMS

You may find science a little intimidating because of all the long, unfamiliar words that scientists use. However, you can often unravel the meaning of these words by breaking them down into simple parts that *do* have meaning to you. When you see a difficult scientific word, look at the entries in this appendix to help you understand that term. These word parts may come at the beginning, at the end, or in the middle of the term, depending on their meaning.

For example, if you ran into the word *magnetohydrodynamics*, you could separate it into three parts—*magneto*, *hydro*, and *dynamics*. *Magneto* means "magnetic," *hydro* means "water" or "fluid," and *dynamics* means "study of power." So magnetohydrodyamics has to do with the study of the magnetic properties of moving fluids.

Changes in the sun's surface are due to the movement of the charged particles that make up the sun.

a, an (Gk.)—not, without

ab (L.)—away from

ac, **ad**, **ag** (L.)—to, toward

acou (Gk.)—hearing

aer, **aero** (Gk.)—air

alter (L.)—change

amal (Gk.)—soft

amphi, **ampho** (Gk.)—on both sides

ant, **anti** (Gk.)—opposite, against

ante (L.)—before

aqua (L.)—water

audio (L.)—hear

aut, **auto** (Gk.)—self

bar, **baro** (Gk.)—weight, pressure

bi (L.)—two, twice, double

bio, **bios**, **biot** (Gk.)—life

calc, **calci** (L.)—calcium

calor (L.)—heat

centi (L.)—a hundred

centr, **centri**, **centro** (Gk.)—center

chem, **chemi**, **chemo** (Gk.)—chemistry

chrom, **chromo** (Gk.)—color

chron, **chrono** (Gk.)—time

cline (Gk.)—sloping

co, **com**, **con** (L.)—with, together

cupr (Gk.)—copper

cycl, **cyclo** (Gk.)—circle, wheel

de (L.)—loss, removal

deci (L.)—tenth

di (Gk.)—two

div (Gk.)—apart

duce, **duct** (L.)—to lead

dyna (Gk.)—power

eco (Gk.)—house

electro (L.)—electricity

en, **end**, **endo** (Gk.)—within, inner

equ, **equa**, **equi** (L.)—equal

ex, **exo** (Gk.)—out, outside, without

extra (L.)—outside, more, beyond, besides

fissi (L.)—split, divide

flam (L.)—fire

fund (L.)—basis

fusi (L.)—to join together

grad (L.)—step, walk, slope

graph, **grapho**, **graphy** (Gk.)—to write

grav (L.)—heavy

gyro (Gk.)—spinning

halo (Gk.)—salt

hemi (Gk.)—half

hetero (Gk.)—other, different

homeo, **homio**, **homo** (Gk.)—
like, same, resembling

hydr, **hydra**, **hydro** (Gk.)—water, fluid

hyper (Gk.)—over, beyond

hypo (Gk.)—under, beneath

ic (Gk.)—of, relating to

inter (L.)—between

is, **iso** (Gk.)—equal

ism (Gk.)—belief, process of

kine, **kinema**, **kinemato**, **kines**,
kinesi, **kinet**, **kineto** (Gk.)—
move, moving, movement

log, **logo**, **logus**, **logy**
(Gk.)—word, study of

macr, **macro** (Gk.)—large

magneto (Gk.)—magnetic

medi, **media**, **medio** (L.)—middle

met, **meta** (Gk.)—between,
with, after, change

meter, **metry** (Gk.)—measure

micro (Gk.)—small

mill, **mille**, **milli**, **millo** (L.)—
one thousand

mit (L.)—to send

mono (Gk.)—one

morph, **morpha**, **morpho**
(Gk.)—form, shape

multi (L.)—many

nan, **nani**, **nano**, **nanus** (Gk.)—
dwarf, one billionth

nomy (Gk.)—the science of

nuc, **nucle**, **nucleo** (L.)—central part

ocul, **oculi**, **oculo**, **oculus** (L.)—eye

opt, **opti**, **opto** (Gk.)—eye, vision

organ (Gk.)—living

orth, **ortho** (Gk.)—upright,
perpendicular

ox, **oxy** (Gk.)—oxygen

par, **para** (Gk.)—beside

pause (Gk.)—to stop

pend (L.)—hanging

peri (Gk.)—around, near

phon, **phono** (Gk.)—sound

phos, **phot**, **photo** (Gk.)—light

phyt, **phyto**, **phytum** (Gk.)—plant

poly (Gk.)—many

post (L.)—after

pre (L.), **pro** (Gk.)—before, in front of

prot, **prote**, **proto** (Gk.)—first, original

pyro (Gk.)—fire

radi, **radia**, **radio** (L.)—spoke, ray

retro (L.)—backward

sal (L.)—salt

scient (L.)—knowledge

scope, **scopy** (Gk.)—to see, watch

sect (L.)—to cut

seism (Gk.)—earthquake

semi (L.)—half

sol (L.)—sun

son (L.)—sound

spec (L.)—to see, look at

sphere (Gk.)—ball, globe

stasis (Gk.)—to stand still

sub (L.)—below, under

super (L.)—above, over

syn (Gk.)—together

tele (Gk.)—distant

terra (L.)—earth

tetr, **tetra** (Gk.)—four

therm, **thermo** (Gk.)—heat

top, **topo** (Gk.)—place

tran, **trans** (L.)—across, through

trop, **tropae**, **trope**, **tropo**
(Gk.)—turn, change

uni (L.)—single, one

vacu (L.)—empty

vari, **vario** (L.)—difference

vect (L.)—to carry

volu (L.)—bulk, amount

APPENDIX B

MATH HELPS

This appendix provides you the necessary helps for solving the many kinds of physical science problems found in this textbook. It's not intended to be all-inclusive. We assume that you understand or are learning the methods for solving single-variable algebraic equations, including using the order of operations and solving equations for an unknown quantity.

Rounding Rules

We round numbers all the time. Sometimes it's convenient to round, for example, the number of residents in your town to the nearest 100 people or the distance around the earth to the nearest 1000 kilometers. Rounding is useful for several reasons. It helps us grasp large numbers when the specific number is not needed but the approximate size is. Rounding also allows us to report measured scientific data correctly.

For the purposes of measuring scientific data and performing calculations, scientists have agreed on certain rounding rules. These ensure that calculation solutions are properly rounded to show the needed precision of numerical quantities. We suggest that you use the following rules for problems in this textbook to closely imitate those used by scientists.

1. Identify the place value that you are going to round to. This is the rounded place value.

2. If the digit to the right of the rounded place value is 0–4, the digit in the rounded place value remains unchanged ("rounded down"). If the digit to the right is 5–9, the digit in the rounded place value is increased by 1 ("rounded up").

   a. If the rounded place value is to the *right* of the decimal point, all digits to the *right* of the rounded place value will be dropped after rounding.

   Examples: 105.639 g → 105.6 g

   82.47 mm → 82.5 mm

   32.95 cm → 33.0 cm

   b. If the rounded place value is to the *left* of the decimal point (or assumed decimal point), all digits between the rounded place value and the decimal point become zeros and the decimal point (if one is shown) is omitted.

   Examples: 1694 m → 1700 m

   19.6 °C → 20 °C

3. Notice in two of the examples above that rounding to a particular place value may affect other digits to the left of the rounding place value. When a 9 is rounded up, the 9 becomes 0 and 1 is added to the digit to the left.

Mathematics and Measurement

When measured quantities appear in equations as terms or factors, follow the same arithmetic principles that you have learned for solving any equation. But acting as a scientist, you have to think about some extra things before arriving at a solution to a problem.

Measured quantities include *units*. Units, such as meters (m), seconds (s), and meters per second (m/s), are treated as parts of a measurement. So any arithmetic operation applies to both the number and the unit factors in a measurement.

Measured quantities also have *precision* (see Subsection 1.9). Precision is the exactness of a measurement. Scientists exert a lot of effort to ensure that their data properly represents the exactness that their instruments can read. Scientists alert other scientists to the precision of their measurements by the number of *significant figures* (SF) they include in their reported data.

On the following pages you will find some rules for doing basic arithmetic operations using measurements.

Determining Significant Figures

When we are reading measured data collected by someone else, we have to be able to identify the significant figures. (Significant figures do not apply to counts or definitions—only to measured data.) Scientists follow an established set of rules to determine which digits in a measurement are significant.

>Rule 1: All nonzero digits are significant.

>Rule 2: All zeros between significant figures are significant.

>Rule 3: All ending zeros to the right of the decimal point are significant.

Examples:

>**a.** 1.694 mL → 4 SF (all four digits are nonzero—Rule 1)

>**b.** 51.021 m → 5 SF (four nonzero digits and a zero between SF—Rule 2)

>**c.** 22.7500 g → 6 SF (four nonzero digits and two ending zeros to the right of the decimal point—Rule 3)

>**d.** 0.097 N → 2 SF (the zeros are not ending zeros—Rule 3)

The fourth example can be confusing, but it is much easier to understand if you change the number to scientific notation.

$$0.097 \text{ N} = 9.7 \times 10^{-2} \Rightarrow \text{only two SF}$$

One other case in which determining SF can be difficult is when ending zeros occur to the left of the decimal point.

Example:

How many SF does the measurement 7600 m have? You know that two of the digits must be significant, but you as the reader cannot determine which, if either, of the zeros are significant. Does this number have two, three, or four SF? Again, the person reporting the value can use scientific notation to avoid confusion.

>$7600 \rightarrow 7.6 \times 10^3$ (two SF)

>$7600 \rightarrow 7.60 \times 10^3$ (three SF)

>$7600 \rightarrow 7.600 \times 10^3$ (four SF)

Adding and Subtracting Physical Science Data

Math Rule 1: Added or subtracted data must have the same units.
Even if the kinds of data are the same, if their units are not the same, then you are adding apples and oranges. For example, you can't add the lengths 3.1 m and 45 cm together without first converting one of the measurements to the other's units.

Math Rule 2: The result of adding or subtracting data can't be more precise than the least precisely measured data used in the calculation. After adding or subtracting, round the result to the decimal place of the estimated digit in the least precise measurement.

EXAMPLE 1: Adding Data: Length

Add 3.1 m and 45 cm.

Unit Conversion:

$$45\ \cancel{cm}\left(\frac{1\ m}{100\ \cancel{cm}}\right) = 0.45\ m$$

Add data in column:

$$\begin{array}{r} 3.\underline{1}\ m \\ +0.4\underline{5}\ m \\ \hline 3.\underline{55}\ m = 3.6\ m \end{array}$$

The estimated digits in each piece of data and the sum are underlined. The sum contains two underlined digits that result from the summing operation. These must be rounded to the place having the least precise estimated value, in this case the 0.1 m place.

Multiplying and Dividing Physical Science Data

Math Rule 3: The result of multiplying or dividing data can't have more SF than the data with the fewest SF used in the calculation. After multiplying or dividing, round the results to the same number of significant figures as the data with the fewest SF.

EXAMPLE 2: Dividing Data: Density

Calculate the density of a sample of quartz with a mass of 27.55 g and a volume of 10.4 cm^3.

Given data: $m = 27.55$ g (four SF), $V = 10.4\ cm^3$ (three SF)

Calculation:

$$d = \frac{m}{V}$$

$$= \frac{27.55\ g}{10.4\ cm^3}\left(\frac{\text{four SF}}{\text{three SF}}\right)$$

$$= 2.6\underline{4}9038462\frac{g}{cm^3}$$

$$= 2.65\frac{g}{cm^3}\ \text{(three SF allowed)}$$

Since 10.4 cm^3 has only three SF, only three SF are allowed in the quotient.

Math Rule 4: The result of multiplying or dividing a measured quantity and a pure number has the same number of decimal places, or the same precision, as the measured quantity used in the calculation.

Examples:

    a. $(7)(2.35 \text{ cm}) = 16.45 \text{ cm}$ (not 16.5 cm)

    b. $2.63 \text{ cm} \div 5 = 0.53 \text{ cm}$ (not 0.526 cm)

Notice that the number of SF in the measured data do not determine the number of SF in the result when multiplying or dividing by pure numbers. Math Rule 4 ensures that you preserve the precision of the original measurement, even if you change the number by multiplying or dividing by a pure number.

Proportionalities in Physical Science

As you study physical science, you will discover that many measurable quantities change or vary in a reliable, predictable way with other quantities. For example, the gravitational potential energy of a ball increases with its height above the ground. If its height is doubled, then its potential energy is doubled as well. Similarly, if you double your running speed around a cross-country track, you halve the time it takes to run the course. These kinds of relationships are called *proportionalities*.

Proportionality

A minute has 60 seconds, right? How many seconds are in 2 minutes? Correct, 120 seconds. The number of seconds is *directly proportional* to the number of minutes. The word *proportional* means "having a constant ratio." In the case of seconds and minutes, the ratio is always

$$\frac{60 \text{ s}}{1 \text{ min}} \quad \text{or} \quad \frac{1 \text{ min}}{60 \text{ s}}$$

—they are directly proportional.

If two numbers, A and B, are directly proportional, then they can be set equal to each other by including a proportionality constant factor—k. The equation is

$$A = kB,$$

which means that

$$\frac{A}{B} = k,$$

and k is a constant for all ratios of these two quantities. For our example, suppose A is time in minutes and B is time in seconds. The proportionality constant k is equal to 1 min/60 s, as shown below.

$$A = \left(\frac{1 \text{ min}}{60 \text{ s}}\right)B$$

You'll work with many quantities that are directly proportional. You will study direct proportions that involve momentum, force, acceleration, mechanical advantage, and other concepts.

You'll also find that other pairs of quantities are *inversely proportional*. If the value of one increases, the other decreases in proportion.

The equations for these proportions set up as shown below.

$$A = \frac{k}{B}$$

If you multiply both sides of the equation by B, you end up with

$$AB = k.$$

So if k is a constant, as A increases, B must decrease in the same proportion to maintain the equality.

Examples of inverse proportions include the pressure and volume of a confined gas (Boyle's law), electric current and resistance in Ohm's law, frequency and period, and wavelength and frequency of a wave.

Lastly, there are other kinds of proportionalities that are neither entirely direct nor inverse relationships. For example, the kinetic energy of an object increases much faster than expected when its speed increases. If a baseball's speed is doubled, it will have four times as much kinetic energy than it did at the slower speed. Similarly, the force of gravity and the electric force decrease much faster with distance than would be the case with a simple inverse proportion. Double the distance between two massive objects and the gravitation force is only one-fourth its value at the closer distance. These kinds of proportionalities include an extra math operation in the direct or inverse relationship. Scientists give the latter case a special name—the *inverse square law.*

GRAPHING

Scientists often like to compare two or more groups of numbers visually to see whether they are related in some way. When scientists plot data to compare different quantities, they make a graph. The simplest graphs compare two changing quantities, called *variables.* Usually, one of the variables is determined by the scientist or else changes in a regular way (e.g., time). This is called an *independent variable.* Its value doesn't depend on anything in the data. The other variable is expected to change in some way related to the independent variable. Its value depends on the first variable, so it is called the *dependent variable.* That makes sense! The values of the independent and dependent variables are called the *coordinates of the data.* To plot the data, an ordered pair of coordinates is used. You have graphed ordered pairs in the form (x, y) in a math class.

Scientists usually plot the independent variable on the horizontal axis (x-axis), with increasing values to the right. The dependent variable is typically plotted on the vertical axis (y-axis), increasing upward. These are not hard and fast rules, and many graphs are arranged differently to more clearly see the relationships in the plotted data. Useful graphs include a title describing the graph's purpose and labels identifying the quantities and units used on each axis. Graphs are often depicted on a scaled grid with the x- and y-axis scales shown so that estimates of the values of the plotted variables can be made.

Scatterplots

Graphs come in different forms. A simple plot of points on a graph is called a *scatterplot*. This is the starting point for many graphs. Scientists like to detect trends in the data and then create a mathematical equation (model) that describes the trend. They draw a *best-fit curve* through the pattern of dots. The kind of curve drawn depends on how the data changes. We still call it a best-fit curve even if it doesn't actually curve. Curves that are straight lines are called *linear graphs*. These are fairly rare in nature. Most trends in nature range from slightly curved to really wavy! These graphs, logically enough, are called *nonlinear graphs*.

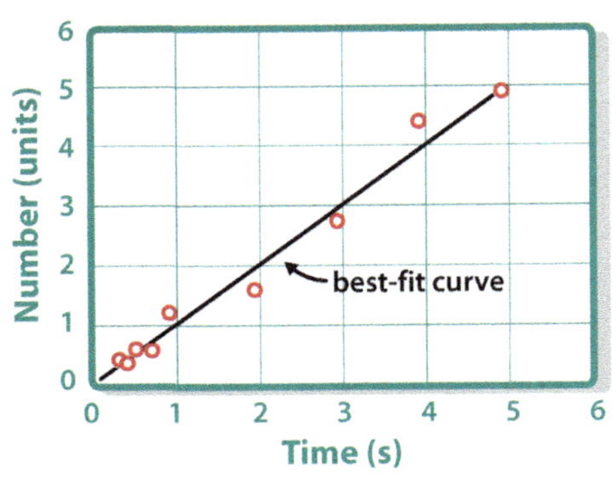

A scatterplot with a direct proportion

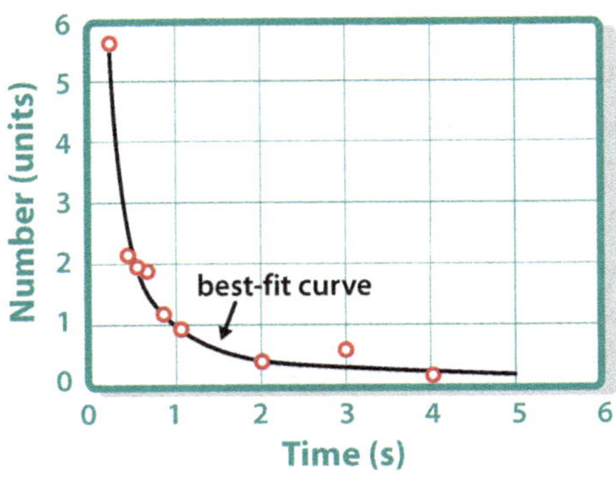

A scatterplot with an inverse proportion

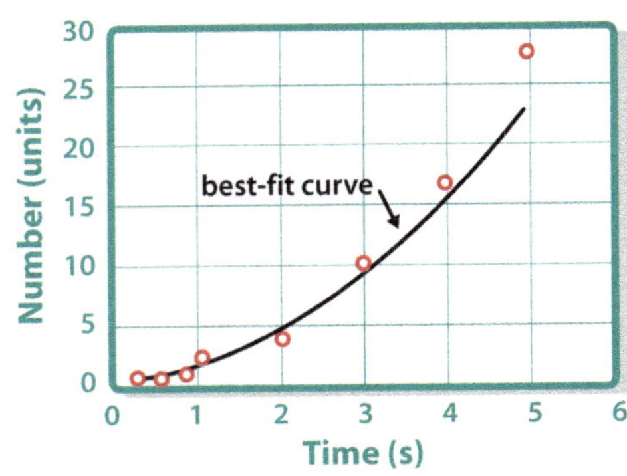

A scatterplot with a square proportion

Slope

One of the most important things to be learned from a graph is the rate at which the dependent variable changes in comparison to the independent variable. This rate is indicated by the *slope* of the graph. A graph with a steep slope shows that things are changing quickly. One with a horizontal slope shows that the dependent variable is not changing at all. A graph that rises from left to right has a positive slope; a dropping curve has a negative slope.

To calculate the slope (m) of a curve, use the slope formula,

$$m = \frac{\Delta y}{\Delta x} = \frac{y_2 - y_1}{x_2 - x_1},$$

where m is the slope, Δy is the change in the y variable, and Δx is the change in the x variable.

What is the slope of a line between (–4, 8) and (5, 2)?

Given data: $x_1 = -4$, $y_1 = 8$, $x_2 = 5$, $y_2 = 2$; $m = ?$

Calculation:

$$m = \frac{\Delta y}{\Delta x} = \frac{y_2 - y_1}{x_2 - x_1}$$

$$m = \frac{2 - 8}{5 - (-4)} = \frac{-6}{9} = -\frac{2}{3}$$

The graph slopes downward, losing two units vertically for every three units moved horizontally.

Estimating Data

You can sometimes use scatterplots and trend lines to obtain information that is not in the measured data set. If you follow the trend line between two data points, these values are estimated—not measured. Obtaining unmeasured data this way is called *interpolation*. Scientists also often try to predict the values of the dependent variable beyond the range of the measured independent variable data points. This method of analysis is called *extrapolation*. Extrapolating data assumes that the trend will continue in the same fashion as observed within the measured data.

Other Types of Graphs

In a bar graph, dependent data is plotted as vertical bars at each independent data value. Area graphs fill in the area of a graph. When several dependent variables are plotted on a single graph, the areas can be compared to show their relationships visually. Pie charts are another type of area graph. They are especially useful for showing percentages of a whole.

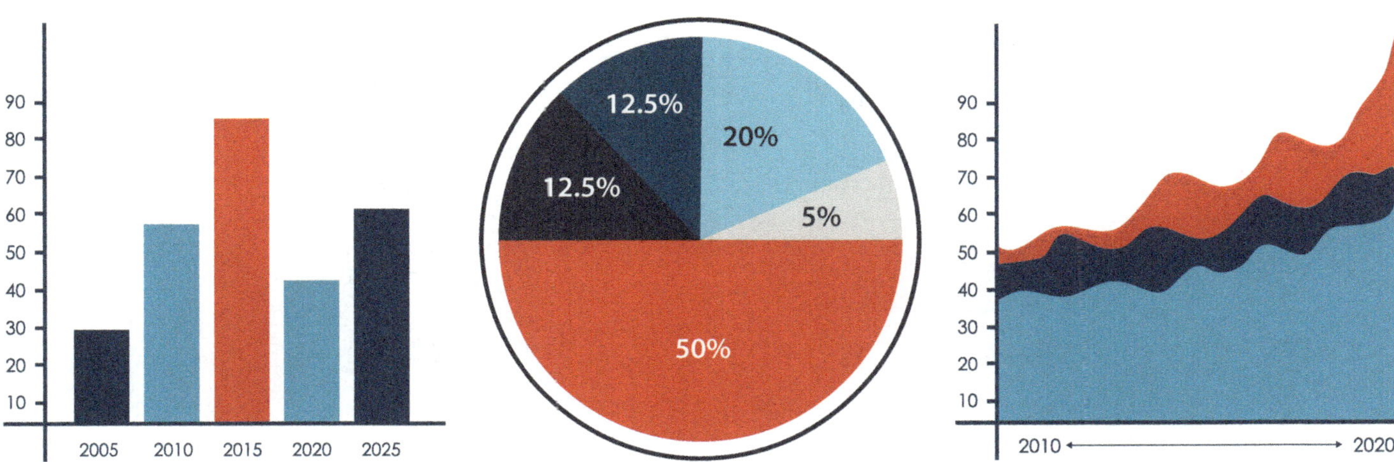

A bar graph (left), pie chart (center), and an area graph (right)

APPENDIX C

FUNDAMENTAL AND DERIVED UNITS OF THE SI

Fundamental Units

Dimension	Name	Symbol	Definition
length	meter	m	The meter is the length of the path traveled by light in a vacuum during a time interval of 1/299 792 458 of a second.
mass	kilogram	kg	The kilogram is the unit of mass. It is defined on the basis of the Planck constant ($6.626\ 070\ 15 \times 10^{-34}$ J·s), the meter, and the second.
time	second	s	The second is the duration of 9 192 631 770 cycles of the radiation associated with a specific transition of the cesium-133 atom.
electrical current	ampere	A	The ampere is the current required for one coulomb of charge to pass through a conductor every second. The coulomb is defined on the basis of the elementary charge, the electric charge carried by a single proton.
temperature	kelvin	K	The kelvin, the unit of thermodynamic temperature, is defined on the basis of the Boltzmann constant (1.380649×10^{-23} J/K), the kilogram, the meter, and the second.
amount of substance	mole	mol	A mole is the amount of material that contains $6.022\ 140\ 76 \times 10^{23}$ entities. This value is the Avogadro constant.
luminous intensity	candela	cd	The candela is the luminous intensity, in a given direction, of a source that emits monochromatic radiation of frequency 540×10^{12} hertz and that has a radiant intensity in that direction of 1/683 watt per steradian.

Note: The unit definitions are a contribution of the National Institute of Standards and Technology and come directly from the NIST website.

Commonly Used Derived Units

Dimension	Name	Symbol	Definition
force	newton	N	$\dfrac{kg \cdot m}{s^2}$
energy, work, heat	joule	J	$N \cdot m = \dfrac{kg \cdot m^2}{s^2}$
power	watt	W	$\dfrac{J}{s} = \dfrac{kg \cdot m^2}{s^3}$
pressure	pascal	Pa	$\dfrac{N}{m^2} = \dfrac{kg}{m \cdot s^2}$
electrical charge	coulomb	C	$A \cdot s$
voltage	volt	V	$\dfrac{J}{C} = \dfrac{kg \cdot m^2}{A \cdot s^3}$
electrical resistance	ohm	Ω	$\dfrac{V}{A} = \dfrac{kg \cdot m^2}{A^2 \cdot s^3}$
frequency	hertz	Hz	$\dfrac{cycles}{s} = s^{-1}$
particle mass	atomic mass unit	u	1/12 of a carbon-12 nuclide

METRIC PREFIXES

Prefix	Meaning (Origin)	Symbol	Factor	Example	Application
exa-	six (Gk.)	E	10^{18}	exabyte (EB)	amount of data created every second
peta-	five (Gk.)	P	10^{15}	petahertz (PHz)	frequency of ultraviolet radiation
tera-	monster (Gk.)	T	10^{12}	terawatt (TW)	average annual power generation in the United States
giga-	giant (Gk.)	G	10^{9}	gigabyte (GB)	hard-drive storage capacity
mega-	great (Gk.)	M	10^{6}	megajoule (MJ)	energy to heat 10 L of water from 0 °C to 100 °C
kilo-	thousand (Gk.)	k	10^{3}	kilometer (km)	distances on Earth
hecto-	hundred (Gk.)	h	10^{2}	hectometer (hm)	wavelength of radio waves
deca-	ten (Gk.)	da	10^{1}	decapascal (daPa)	pressure of sound
(base)			10^{0}		
deci-	tenth (L.)	d	10^{-1}	decibel (dB)	sound loudness
centi-	hundredth (L.)	c	10^{-2}	centimeter (cm)	distances in a laboratory
milli-	thousandth (L.)	m	10^{-3}	millivolt (mV)	EKG signal from heart at skin
micro-	small (Gk.)	μ	10^{-6}	microsecond (μs)	time for light to travel 1 km
nano-	dwarf (Gk.)	n	10^{-9}	nanometer (nm)	size of atoms
pico-	small (Sp.)	p	10^{-12}	picometer (pm)	long gamma-ray wavelength
femto-	fifteen (Dan.)	f	10^{-15}	femtogram (fg)	mass of a virus
atto-	eighteen (Dan.)	a	10^{-18}	attosecond (as)	period required to image an orbiting electron

APPENDIX E

COMMON ABBREVIATIONS AND SYMBOLS

Abbreviations

Unit	Abbreviation	Dimension
ampere	A	current
atmosphere	atm	pressure
atomic mass unit	u, Da	mass
bar	bar	pressure
candela	cd	light intensity
coulomb	C	electrical charge
day	d	time
decibel	dB	relative power
degree	°	temperature or angle
degree (Celsius)	°C	temperature
degree (Fahrenheit)	°F	temperature
gram	g	mass
hertz	Hz, s^{-1}	frequency
hour	h	time
joule	J	energy or work
kelvin	K	temperature
kilogram	kg	mass
kilowatt-hour	kW·h	energy
liter	L, ℓ	volume
meter	m	length/distance
millibar	mbar	pressure
minute	min	time
molarity	M	solution concentration
mole	mol	quantity (particles)
newton	N	force
ohm	Ω	electrical resistance
pascal	Pa	pressure
second	s	time
volt	V	electrical potential difference
watt	W	power
year	y	time
–	×	magnification power
–	α, $_2^4\text{He}$	alpha particle

Unit	Abbreviation	Dimension
–	AC	alternating current
–	AM	amplitude modulation
–	AMA	actual mechanical advantage
–	(aq)	chemical state: aqueous
–	β, $_{-1}^0 e$	beta particle
–	CAT/CT	computed axial tomography
–	DC	direct current
–	e^-	electron
–	EM	electromagnetic
–	EPE	elastic potential energy
–	FM	frequency modulation
–	γ	gamma ray
–	(g)	chemical state: gas
–	GFI	ground-fault interrupter
–	GPE	gravitational potential energy
–	IMA	ideal mechanical advantage
–	IR	infrared
–	IUPAC	International Union of Pure and Applied Chemistry
–	KE	kinetic energy
–	(l)	chemical state: liquid
–	LED	light-emitting diode
–	MA	mechanical advantage
–	n	neutron
–	NIST	National Institute of Standards and Technology
–	p^+	proton
–	PE	potential energy
–	pH	acidity/alkalinity
–	(s)	chemical state: solid
–	SF	significant figure(s)
–	sp gr	specific gravity
–	SI	Système International d'Unités
–	STP	standard temperature and pressure
–	UV	ultraviolet

Formula Symbols and Quantities

Symbol	Quantity
A	area; mass number
$\mathbf{a}$	acceleration
c	speed of light
c_{sp}	specific heat
d	density, distance
$\Delta\mathbf{x}$	displacement
$\mathbf{F}$	force
$\mathbf{F}_g$	gravitational force, weight
$\mathbf{F}_{in}$	input force
$\mathbf{F}_{out}$	output force
f	frequency
G	gravitational constant
$\mathbf{g}$	gravitational acceleration
$\mathbf{h}$	height
h	Planck's constant
h_i	image height
h_o	object height
I	electrical current
k	constant (unspecified)
λ	wavelength
L	length
ℓ	lever arm; liter

Symbol	Quantity
ℓ_{in}	input lever arm
ℓ_{out}	output lever arm
L_f	heat of fusion
L_v	heat of vaporization
m	mass
N_{in}	number of turns, input
N_{out}	number of turns, output
P	power; pressure
p	linear momentum
Q	charge; heat
R	electrical resistance
r	radius
s	speed
T	period
T_C	Celsius temperature
T_F	Fahrenheit temperature
T_K	Kelvin temperature
Δt	time interval
V	volume; voltage, electrical potential difference
$\mathbf{v}$	velocity
W	work
Z	atomic number

APPENDIX F

CREATING GRAPHIC ORGANIZERS

Several times in this textbook you are asked to demonstrate that you understand a concept or a group of concepts by creating a *graphic organizer*—a visual way to represent data. Use this appendix as a guide to help you create different kinds of graphic organizers to show how concepts in physical science compare, how they are structured, or how they move through a process.

COMPARING CLASSES OF LEVERS

	First-Class Lever	Second-Class Lever	Third-Class Lever
Order of parts	F_{out} – fulcrum– F_{in}	fulcrum– F_{out} – F_{in}	fulcrum– F_{in} – F_{out}
Comparing lengths of input arm (ℓ_{in}) and output arm (ℓ_{in})	any	$\ell_{in} > \ell_{out}$	$\ell_{in} < \ell_{out}$
Mechanical advantage	any	> 1	< 1

Table/T-Chart

A table is often used in science to organize data, but it can also be used to organize descriptions and other information. A table can be a helpful way to compare two or more concepts in physical science. For example, the three classes of levers are compared in the table at left. If you are comparing only two concepts, this kind of table with only two columns can be called a *T-chart* because of its shape.

When you create a table, place the concepts that you are comparing in the top row. The characteristics under consideration should go in the leftmost column. Fill in the cells, considering how each characteristic or category differs for the concepts that you are comparing.

Hierarchy Chart

A *hierarchy chart* is a diagram that shows the relationships between several concepts. For example, a company has an organizational chart that does the same thing. Hierarchy charts are especially useful in physical science when studying classification.

When you create a hierarchy chart, think of the different categories that a larger category includes. The example below shows how all the objects in the solar system can be classified as planets, small solar system bodies, dwarf planets, the sun, or moons.

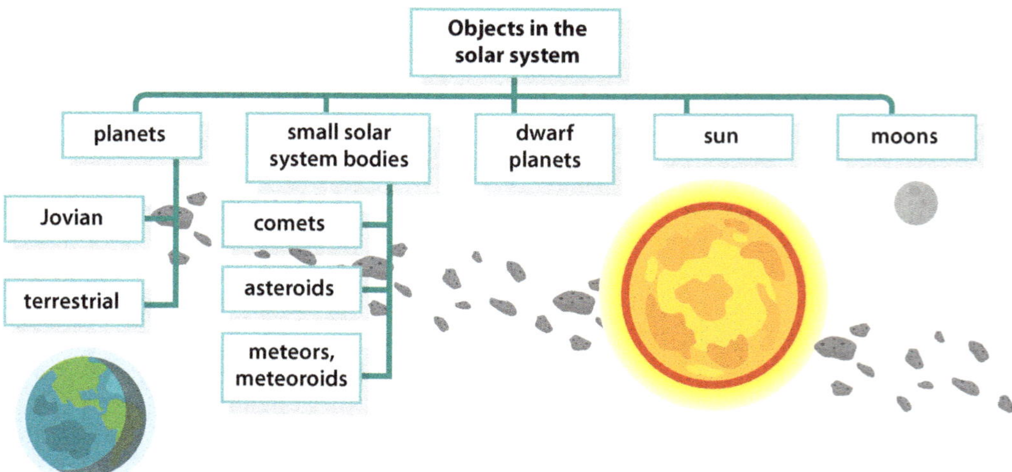

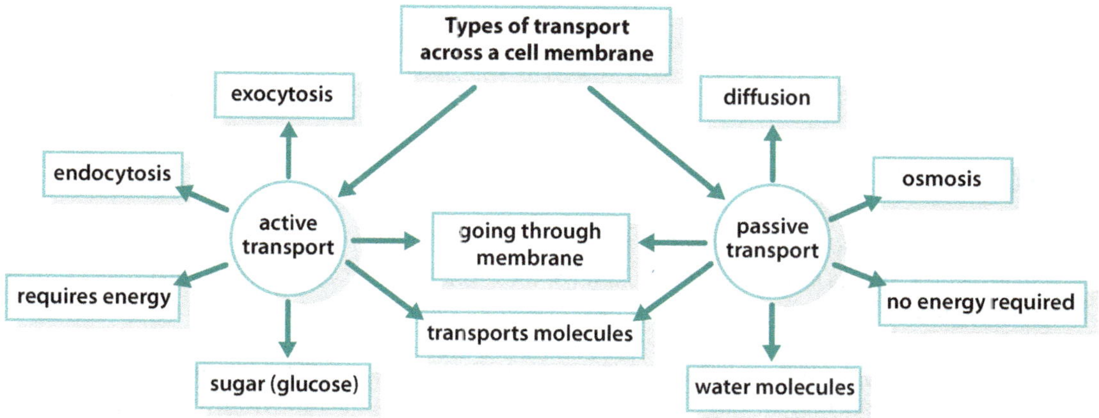

Concept Map

A *concept map* (above) is a freeform graphic organizer that relates concepts, sub-concepts, and their related characteristics. It's similar to an outline in visual form. Because concept maps are so flexible, they can be more difficult to create.

To make a concept map, start with the biggest concept and put it in a circle. Use lines or arrows to connect this main concept with smaller concepts. You can use linking words to help a viewer see the connection between two concepts and read the concept map like a sentence. The further you get from the main idea, the more specific the information should become.

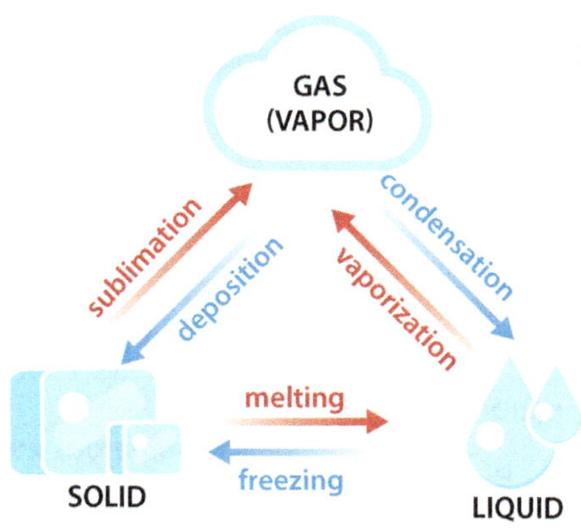

Process Map/Flowchart

A *process map* (left) illustrates a specific chain of events from the first in a physical science process. Because of its purpose, it almost always includes arrows to indicate the direction of the process.

When you create a process map, think of the process that you are illustrating and the different steps that are part of this process. Make sure that you put them in order! You can show a process in a line of events such as a timeline, or, if the process repeats itself, you can show it in a circle as a cyclical process that doesn't have a beginning or an end.

Concept Definition Map

The last type of graphic organizer is a *concept definition map*. The graphic organizer on the right takes a concept and considers four questions: What is it? What is it not? What are its properties? What are some examples?

Think through these four questions for a concept when you create a concept definition map. Then put the concept in the middle of the map and organize these four questions in boxes labeled with these questions around the concept.

APPENDIX G

WORLDVIEW SLEUTHING GRADING RUBRICS

Chapter 2 Report: *Bulletproof! (Student Edition p. 44)*

Use the standards below to assess the criteria listed in the left column of the rubric. Specific values are left blank to allow teachers to assign those values. Use the boxes in the rubric to take notes and provide feedback.

Excellent—Student demonstrates <u>thorough</u> understanding of the concept/completion of the task. Work may include minor conceptual errors that don't affect communication of understanding/completion of the task.

Good—Student demonstrates <u>good</u> understanding of the concept/completion of the task. Work may have some conceptual errors. Work may not communicate effectively.

Fair—Student demonstrates <u>some</u> understanding of the concept/completion of the task. Work may have major conceptual errors.

Poor—Student demonstrates <u>little or no</u> understanding of the concept/completion of the task.

	Excellent (___ pts)	Good (___ pts)	Fair (___ pts)	Poor (___ pts)
Conceptual				
Research — Sources and Information	__ sources	__ sources	__ sources	__ sources
Reference Citations	clear and complete	complete but with __ minor errors	__ significant errors	not cited
Product				
Understanding of the Issues				
Grammar, Usage, Mechanics, and Spelling	__ minor errors	__ minor errors or __ major errors	__ minor errors or __ major errors	numerous errors throughout
Appearance — Neatness and Organization				
Quality				
Timeliness	on time			late
Worldview				
Biblical Worldview				

Use the standards below to assess the criteria listed in the left column of the rubric. Specific values are left blank to allow teachers to assign those values. Use the boxes in the rubric to take notes and provide feedback.

Excellent—Student demonstrates <u>thorough</u> understanding of the concept/completion of the task. Work may include minor conceptual errors that don't affect communication of understanding/completion of the task.

Good—Student demonstrates <u>good</u> understanding of the concept/completion of the task. Work may have some conceptual errors. Work may not communicate effectively.

Fair—Student demonstrates <u>some</u> understanding of the concept/completion of the task. Work may have major conceptual errors.

Poor—Student demonstrates <u>little or no</u> understanding of the concept/completion of the task.

	Excellent (___ pts)	Good (___ pts)	Fair (___ pts)	Poor (___ pts)
Conceptual				
Information — Research and Understanding Both Sides				
Point — Focus, Statement, and Support				
Counterpoint — On topic and Challenging				
Organization				
Delivery				
Language — Grammar, Pronunciation, Usage	__ minor errors	__ errors	__ errors	__ errors
Poise — Use of Voice, Gesticulation, and Composure				
Persuasion	convinces completely	partially convinces		fails to convince
Time Usage	__% of time allotted	__% of time allotted	__% of time allotted	__% of time allotted
Result	won	lost		disqualified
Worldview				
Biblical Worldview				

Use the standards below to assess the criteria listed in the left column of the rubric. Specific values are left blank to allow teachers to assign those values. Use the boxes in the rubric to take notes and provide feedback.

Excellent—Student demonstrates <u>thorough</u> understanding of the concept/completion of the task. Work may include minor conceptual errors that don't affect communication of understanding/completion of the task.

Good—Student demonstrates <u>good</u> understanding of the concept/completion of the task. Work may have some conceptual errors. Work may not communicate effectively.

Fair—Student demonstrates <u>some</u> understanding of the concept/completion of the task. Work may have major conceptual errors.

Poor—Student demonstrates <u>little or no</u> understanding of the concept/completion of the task.

	Excellent (___ pts)	Good (___ pts)	Fair (___ pts)	Poor (___ pts)
Conceptual				
Research — Sources and Information	__ sources	__ sources	__ sources	__ sources
Reference Citations	clear and complete	complete but with __ minor errors	__ significant errors	not cited
Product				
Content — Understanding of the Issue				
Graphics — Visual Presentations				
Speaking — Audience Awareness, Projection, Enunciation, and Posture				
Grammar, Usage, Mechanics, and Spelling	__ minor errors __ mispronunciations	__ minor errors __ mispronunciations	__ minor errors __ mispronunciations	__ minor errors __ mispronunciations
Quality				
Timeliness	on time			late
Worldview				
Biblical Worldview				

Use the standards below to assess the criteria listed in the left column of the rubric. Specific values are left blank to allow teachers to assign those values. Use the boxes in the rubric to take notes and provide feedback.

Excellent—Student demonstrates <u>thorough</u> understanding of the concept/completion of the task. Work may include minor conceptual errors that don't affect communication of understanding/completion of the task.

Good—Student demonstrates <u>good</u> understanding of the concept/completion of the task. Work may have some conceptual errors. Work may not communicate effectively.

Fair—Student demonstrates <u>some</u> understanding of the concept/completion of the task. Work may have major conceptual errors.

Poor—Student demonstrates <u>little or no</u> understanding of the concept/completion of the task.

	Excellent (___ pts)	Good (___ pts)	Fair (___ pts)	Poor (___ pts)
Conceptual				
Research Sources and Information	__ sources	__ sources	__ sources	__ sources
Reference Citations	clear and complete	complete but with __ minor errors	__ significant errors	not cited
Product				
Understanding of the Issues				
Grammar, Usage, Mechanics, and Spelling	__ minor errors	__ minor errors or __ major errors	__ minor errors or __ major errors	numerous errors throughout
Appearance Neatness and Organization				
Quality				
Timeliness	on time			late
Worldview				
Biblical Worldview				

Use the standards below to assess the criteria listed in the left column of the rubric. Specific values are left blank to allow teachers to assign those values. Use the boxes in the rubric to take notes and provide feedback.

Excellent—Student demonstrates <u>thorough</u> understanding of the concept/completion of the task. Work may include minor conceptual errors that don't affect communication of understanding/completion of the task.

Good—Student demonstrates <u>good</u> understanding of the concept/completion of the task. Work may have some conceptual errors. Work may not communicate effectively.

Fair—Student demonstrates <u>some</u> understanding of the concept/completion of the task. Work may have major conceptual errors.

Poor—Student demonstrates <u>little or no</u> understanding of the concept/completion of the task.

	Excellent (___ pts)	Good (___ pts)	Fair (___ pts)	Poor (___ pts)
Conceptual				
Research — Sources and Information	__ sources	__ sources	__ sources	__ sources
Reference Citations	clear and complete	complete but with __ minor errors	__ significant errors	not cited
Product				
Content — Understanding of the Issue				
Graphics — Visual Presentations				
Speaking — Audience Awareness, Projection, Enunciation, and Posture				
Grammar, Usage, Mechanics, and Spelling	__ minor errors __ mispronunciations	__ minor errors __ mispronunciations	__ minor errors __ mispronunciations	__ minor errors __ mispronunciations
Quality				
Timeliness	on time			late
Worldview				
Biblical Worldview				

Use the standards below to assess the criteria listed in the left column of the rubric. Specific values are left blank to allow teachers to assign those values. Use the boxes in the rubric to take notes and provide feedback.

Excellent—Student demonstrates <u>thorough</u> understanding of the concept/completion of the task. Work may include minor conceptual errors that don't affect communication of understanding/completion of the task.

Good—Student demonstrates <u>good</u> understanding of the concept/completion of the task. Work may have some conceptual errors. Work may not communicate effectively.

Fair—Student demonstrates <u>some</u> understanding of the concept/completion of the task. Work may have major conceptual errors.

Poor—Student demonstrates <u>little or no</u> understanding of the concept/completion of the task.

	Excellent (___ pts)	Good (___ pts)	Fair (___ pts)	Poor (___ pts)
Conceptual				
Research Sources and Information	__ sources	__ sources	__ sources	__ sources
Reference Citations	clear and complete	complete but with __ minor errors	__ significant errors	not cited
Product				
Understanding of the Issues				
Grammar, Usage, Mechanics, and Spelling	__ minor errors	__ minor errors or __ major errors	__ minor errors or __ major errors	numerous errors throughout
Appearance Neatness and Organization				
Quality				
Timeliness	on time			late
Worldview				
Biblical Worldview				

Use the standards below to assess the criteria listed in the left column of the rubric. Specific values are left blank to allow teachers to assign those values. Use the boxes in the rubric to take notes and provide feedback.

Excellent—Student demonstrates <u>thorough</u> understanding of the concept/completion of the task. Work may include minor conceptual errors that don't affect communication of understanding/completion of the task.

Good—Student demonstrates <u>good</u> understanding of the concept/completion of the task. Work may have some conceptual errors. Work may not communicate effectively.

Fair—Student demonstrates <u>some</u> understanding of the concept/completion of the task. Work may have major conceptual errors.

Poor—Student demonstrates <u>little or no</u> understanding of the concept/completion of the task.

	Excellent (___ pts)	Good (___ pts)	Fair (___ pts)	Poor (___ pts)
Conceptual				
Research — Sources and Information	__ sources	__ sources	__ sources	__ sources
Reference Citations	clear and complete	complete but with __ minor errors	__ significant errors	not cited
Product				
Content — Understanding of the Issues				
Grammar, Usage, Mechanics, and Spelling	__ minor errors	__ minor errors or __ major errors	__ minor errors or __ major errors	numerous errors throughout
Appearance — Neatness and Organization				
Quality				
Timeliness	on time			late
Worldview				
Biblical Worldview				

Use the standards below to assess the criteria listed in the left column of the rubric. Specific values are left blank to allow teachers to assign those values. Use the boxes in the rubric to take notes and provide feedback.

Excellent—Student demonstrates <u>thorough</u> understanding of the concept/completion of the task. Work may include minor conceptual errors that don't affect communication of understanding/completion of the task.

Good—Student demonstrates <u>good</u> understanding of the concept/completion of the task. Work may have some conceptual errors. Work may not communicate effectively.

Fair—Student demonstrates <u>some</u> understanding of the concept/completion of the task. Work may have major conceptual errors.

Poor—Student demonstrates <u>little or no</u> understanding of the concept/completion of the task.

	Excellent (___ pts)	Good (___ pts)	Fair (___ pts)	Poor (___ pts)
Conceptual				
Research — Sources and Information	__ sources	__ sources	__ sources	__ sources
Reference Citations	clear and complete	complete but with __ minor errors	__ significant errors	not cited
Product				
Content — Understanding of the Issues				
Graphics — Visual Presentations				
Speaking — Audience Awareness, Projection, Enunciation, and Posture				
Grammar, Usage, Mechanics, and Spelling	__ minor errors __ mispronunciations	__ minor errors __ mispronunciations	__ minor errors __ mispronunciations	numerous errors throughout
Quality				
Timeliness	on time			late
Worldview				
Biblical Worldview				

Use the standards below to assess the criteria listed in the left column of the rubric. Specific values are left blank to allow teachers to assign those values. Use the boxes in the rubric to take notes and provide feedback.

Excellent—Student demonstrates <u>thorough</u> understanding of the concept/completion of the task. Work may include minor conceptual errors that don't affect communication of understanding/completion of the task.

Good—Student demonstrates <u>good</u> understanding of the concept/completion of the task. Work may have some conceptual errors. Work may not communicate effectively.

Fair—Student demonstrates <u>some</u> understanding of the concept/completion of the task. Work may have major conceptual errors.

Poor—Student demonstrates <u>little or no</u> understanding of the concept/completion of the task.

	Excellent (___ pts)	Good (___ pts)	Fair (___ pts)	Poor (___ pts)
Conceptual				
Research Sources and Information	__ sources	__ sources	__ sources	__ sources
Reference Citations	clear and complete	complete but with __ minor errors	__ significant errors	not cited
Product				
Content Understanding of the Issues				
Grammar, Usage, Mechanics, and Spelling	__ minor errors	__ minor errors or __ major errors	__ minor errors or __ major errors	numerous errors throughout
Appearance Neatness and Organization				
Quality				
Timeliness	on time			late
Worldview				
Biblical Worldview				

ETHICS ESSAY GRADING RUBRIC

Use the standards below to assess the criteria listed in the left column of the rubric. Specific values are left blank to allow teachers to assign those values. Use the boxes in the rubric to take notes and provide feedback.

Excellent—Student demonstrates <u>thorough</u> understanding of the concept/completion of the task. Work may include minor conceptual errors that don't affect communication of understanding/completion of the task.

Good—Student demonstrates <u>good</u> understanding of the concept/completion of the task. Work may have some conceptual errors. Work may not communicate effectively.

Fair—Student demonstrates <u>some</u> understanding of the concept/completion of the task. Work may have major conceptual errors.

Poor—Student demonstrates <u>little or no</u> understanding of the concept/completion of the task.

	Excellent (___ pts)	Good (___ pts)	Fair (___ pts)	Poor (___ pts)
Conceptual				
Research — Sources and Information	__ sources	__ sources	__ sources	__ sources
Reference Citations	clear and complete	complete but with __ minor errors	__ significant errors	not cited
Scriptural Principles				
Acceptable and Unacceptable Options	considered all of both	considered some of both	considered few of both	considered none or only either acceptable or unacceptable options
Consequences	all	at least one for each acceptable option	at least one for some acceptable options	none
Biblical Outcomes — Human Flourishing, Thriving Creation, and God's Glory	all three	only two	only one	none
Biblical Motivation — Faith in God, Hope in God's Promises, and Love for God and Others	all three	only two	only one	none
Action — Urged a feasible action				
Product				
Grammar, Usage, Mechanics, and Spelling	__ minor errors	__ minor errors or __ major errors	__ minor errors or __ major errors	numerous errors throughout
Appearance — Neatness and Organization				
Quality				
Timeliness	on time			late

APPENDIX I

ELEMENT CARDS

Chapter 4 Mini Lab, Organizing Elements (SE p. 73)

Student handout

<table>
<tr><td>Radius</td><td>Mass</td></tr>
<tr><td>Charge</td><td>State</td></tr>
<tr><td colspan="2">Electronegativity
symbol</td></tr>
</table>

example

3.54　　2.01 +1　　Solid 0.9 **Cc**	2.71　　7.63 −2　　Gas 3.5 **Cw**	5.81　　31.75 +1　　Solid 0.7 **Lt**	7.53　　47.97 +2　　Solid 0.9 **Dw**
2.81　　4.81 +2　　Solid 1.1 **Eg**	7.61　　48.12 +1　　Solid 0.7 **Lk**	5.81　　56.11 −1　　Solid 2.8 **Dq**	2.64　　24.33 −1　　Gas 3.9 **Ma**
3.77　　14.97 +2　　Solid 1.0 **Ph**	3.17　　19.21 −2　　Solid 2.7 **Rc**	3.91　　39.63 −2　　Solid 2.4 **Rv**	2.11　　8.92 −1　　Gas 4.1 **Rs**
3.36　　44.33 −1　　Liquid 3.0 **Sl**		4.92　　12.11 +1　　Solid 0.8 **Te**	6.42　　53.81 −2　　Solid 2.0 **Tt**

3.54 2.01 +1 Solid 0.9 **Cc**	2.81 4.81 +2 Solid 1.1 **Eg**	2.71 7.63 −2 Gas 3.5 **Cw**	2.11 8.92 −1 Gas 4.1 **Rs**
4.92 12.11 +1 Solid 0.8 **Te**	3.77 14.97 +2 Solid 1.0 **Ph**	3.17 19.21 −2 Solid 2.7 **Rc**	2.64 24.33 −1 Gas 3.9 **Ma**
5.81 31.75 +1 Solid 0.7 **Lt**		3.91 39.63 −2 Solid 2.4 **Rv**	3.36 44.33 −1 Liquid 3.0 **Sl**
7.61 48.12 +1 Solid 0.7 **Lk**	7.53 47.97 +2 Solid 0.9 **Dw**	6.42 53.81 −2 Solid 2.0 **Tt**	5.81 56.11 −1 Solid 2.8 **Dq**

MINI LAB EQUIPMENT AND MATERIALS LIST (ALPHABETICAL)

Amount required is specified for each lab group.

Item	Chapter	Size	# Req.	Remarks
apple juice	16	10 mL		Choose five or six of these substances for the demonstration.
balance, laboratory	2, 3, 9		1	
ball, foam	5	5 cm	1	
		2.5 cm	4	
	6	5 cm	6	
		2.5 cm	14	
balloon	18, 19		1	
basic solutions	10			Examples: soap solution, household ammonia, bleach, and baking soda
bowl	10	small	4	
can opener	18		1	
colored candies	7		96	
computer paper	11		1 pc.	
converging lens	22		1	
corn oil	16	10 mL		
corn syrup	16	10 mL		
cube, metal	15		1	
cube, plastic	15		1	
cube, wood	15		1	
cup	7	small	6	
diverging lens	22		1	
dominoes	8		15	
element cards	4		1 set	Element cards can be found in Appendix I.
empty tin can	18		1	
fine sand	14			
glue	18			
glycerol	16	10 mL		
goggles	10		1 pair	
graduated cylinder	2		1	plastic
	9	10 mL	1	
	9	25 mL	1	
	16	100 mL	1	
graph paper	11		2 pcs.	

Item	Chapter	Size	# Req.	Remarks
honey	16	10 mL		
iron filings	20			
isopropyl alcohol	10		1/4 c	used to make tumeric indicator
lemon juice	10			
light ray box	22		1	
magnet	20		3	Each magnet should have a different shape.
marbles	3		12	to make eggogen atoms
mass	13	100 g	6	
	14			various
metal object	2		1	irregularly shaped
meter stick	13, 14		1	
mineral oil	16	10 mL		
molasses	16	10 mL		
motor oil	16	10 mL		
pan, large & shallow	14		1	
paper	12			one or two pieces
	20		1 pc.	
pencil	12, 13		1	
petri dish	21		3	
piece of mirrored plastic	18	1 cm^2 or smaller	1	
pipette	10		1	substitute: eyedropper
plastic eggs	3		10	to make eggogen atoms
prism, rectangular	1		1	
protractor	5		1	
red laser pointer	18		1	
rubber eraser	16	small	1	Can cut an eraser into pieces small enough to fit into graduated cylinder.
rubbing alcohol	16	10 mL		

Item	Chapter	Size	# Req.	Remarks
ruler, metric	1, 7, 8, 11, 14		1	
ruler, US customary	1		1	
scissors	18		1 pair	
Slinky	17		1	substitute: long spring
stirring rod	9		1	
	10		1	substitute: spoon
stopwatch	8, 11		1	
sugar	9	5.0 g		
sunscreen	21			
tape	18			
tape, clear plastic	19			
tape, masking	11			
tongs	15		2	
toothpicks	5		4	
	6		19	
triangular prism	22		2	
turmeric	10		1/4 tsp	used to make turmeric indicator
ultraviolet detection beads	21		18	
water	2			
	16	10 mL		
weighing boat	9		1	substitute: weighing paper
white cardboard	22	small piece	1	

MINI LAB EQUIPMENT AND MATERIALS LIST (BY LAB)

Amount required is specified for each lab group.

Chapter	Item	Size	# Req.	Remarks
1	prism, rectangular		1	
	ruler, metric		1	
	ruler, US customary		1	
2	balance, laboratory		1	
	graduated cylinder		1	plastic
	metal object		1	irregularly shaped
	water			
3	balance, laboratory		1	
	marbles		12	to make eggogen atoms
	plastic eggs		10	to make eggogen atoms
4	element cards		1 set	Element cards can be found in Appendix I.
5	ball, foam	5 cm	1	
	ball, foam	2.5 cm	4	
	protractor		1	
	tooth picks		4	
6	ball, foam	5 cm	6	
	ball, foam	2.5 cm	14	
	toothpicks		19	
7	colored candies		96	
	cups	small	6	
	ruler, metric		1	
8	dominoes		15	
	ruler, metric		1	
	stopwatch		1	
9	balance, laboratory		1	
	graduated cylinder	10 mL	1	
	graduated cylinder	25 mL	1	

Chapter	Item	Size	# Req.	Remarks
9	stirring rod		1	
	sugar	5.0 g		
	weighing boat		1	substitute: weighing paper
10	basic solutions			examples: soap solution, household ammonia, bleach, and baking soda
	bowl	small	4	
	isopropyl alcohol		1/4 c.	used to make tumeric indicator
	goggles		1 pair	
	lemon juice			
	pipette		1	substitute: eyedropper
	stirring rod		1	substitute: spoon
	turmeric		1/4 tsp	used to make turmeric indicator
11	computer paper		1 pc.	
	graph paper		2 pcs.	
	ruler, metric		1	
	stopwatch		1	
	tape, masking			
12	paper			one or two pieces
	pencil		1	
13	mass	100 g	6	
	meter stick		1	
	pencil		1	
14	fine sand			
	masses			various
	meter stick		1	
	pan, large & shallow		1	
	ruler, metric		1	
15	cube, metal		1	
	cube, plastic		1	
	cube, wood		1	
	tongs		2	

Chapter	Item	Size	# Req.	Remarks
16	graduated cylinder	100 mL	1	
	apple juice	10 mL		Choose five or six of these substances for the demonstration.
	corn oil	10 mL		
	corn syrup	10 mL		
	glycerol	10 mL		
	honey	10 mL		
	mineral oil	10 mL		
	molasses	10 mL		
	motor oil	10 mL		
	rubbing alcohol	10 mL		
	water	10 mL		
	rubber eraser	small	1	Can cut an eraser into pieces small enough to fit into graduated cylinder.
17	Slinky		1	substitute: long spring
18	balloon		1	
	can opener		1	
	empty tin can		1	
	glue			
	piece of mirrored plastic	1 cm² or smaller	1	
	red laser pointer		1	
	scissors		1 pair	
	tape			
19	balloon		1	
	tape, clear plastic			
20	iron filings			
	magnets		3	Each magnet should have a different shape.
	paper		1	
21	petri dishes		3	
	sunscreen			
	ultraviolet detection beads		18	

Chapter	Item	Size	# Req.	Remarks
22	converging lens		1	
	diverging lens		1	
	light ray box		1	
	triangular prisms		2	
	white cardboard	small piece	1	

APPENDIX K

EXPLAINING THE GOSPEL

One of the greatest desires of Christian teachers is to see their students repent and believe in Christ. Relying on the Holy Spirit, you should take advantage of the opportunities that arise for presenting the gospel. You may find the following outline helpful, especially when dealing individually with a young person.

1. The Lord God is King over all His creation (Rev. 4:11).

- God created everything that is (Gen. 1–2).
- God created the world with laws about how His world is to work and the way that people are to live.

2. All have sinned—including me (Rom. 3:23).

- Since Adam first sinned, all people are born rebels against the rule of God (Rom. 5:12).
- God has made me in His own image so that I might declare His glory by being like Him (Gen. 1:26–27).
- But I am a sinner. I disobey God's Word. The Bible teaches that I am to love God more than anything or anyone (Mark 12:30). It also teaches that I should love other people at all times (Mark 12:31). But sometimes I don't enjoy doing what God wants me to do, I don't always delight in obeying my parents, and I don't always like being kind to other people.
- God will punish me for my rebellion and sin (Rom. 6:23). God hates sin, and there is nothing I can do to get rid of my sin. I can try to change my behavior, but I can never change my heart.

3. Jesus died and rose again for me (Rom. 5:8).

- God loves me even though I am a sinner.
- He sent His Son, Jesus Christ, to live a perfect life and to die on the cross, suffering the punishment for my sin.
- Three days later, God raised Jesus from the dead and made Him the ruler of His eternal kingdom. Jesus is alive today. This is the gospel of Jesus Christ: He died on the cross and was raised up to be God's appointed King (1 Cor. 15:1–4).
- God desires to restore me to bear His image fully by making me like His Son, the perfect image bearer of God (Rom. 8:29).
- God gives a new heart—new loves and desires—to every one of His children (Ezek. 36:26–27).

4. I need to put my trust in Jesus (Rom. 10:9–10).

- I must repent of, or turn away from, my sin and let Jesus control my life. I must also believe what God has done through Jesus (Mark 1:15).
- If I repent and believe in what Jesus has done, I am putting my trust in Jesus.
- Everyone who has trusted in Jesus is forgiven of sin.
- Everyone who has trusted in Jesus submits to having Him as his King.
- Everyone who has been saved by God will continue to trust and obey Jesus for the rest of his life (Heb. 3:14).

When talking individually with a young person, ask questions to discern sincerity or any misunderstanding. What is sin? Are you a sinner? What is the gospel? What does it mean to repent? Read the verses from your Bible. If a student shows genuine sorrow for sin and a sufficiently accurate understanding of the basics of the gospel, encourage him to call on the Lord as you listen. Perhaps he will pray something like the following:

"God, I know that You hate sin. But I also know that You love me. I believe that Jesus died for me and rose from the dead for me. I now turn away from my sin, and I am trusting in Jesus to forgive me and to be my King forever. I want to follow Him wherever He leads me."

Show the student the Bible's command that believers unite together in regular fellowship (Heb. 10:24–25) and encourage him to get involved right away in a gospel-preaching church. Tell him that whenever he sins, he will be forgiven as he confesses his sins to God (1 John 1:9).

GLOSSARY

absolute zero   The coldest temperature possible.

acceleration (a)   The rate of change in velocity.

accuracy   The comparison of a measurement to an accepted or expected value.

acid   A substance that produces hydrogen ions in an aqueous solution.

acoustic amplification   The process of making a sound louder.

acoustic spectrum   The continuum of all possible sound waves.

activation energy   The minimum energy needed for a chemical reaction to occur.

active sonar   An underwater device that produces short pulses of sound that echo back to the sending object; used to find the bearing and range of submerged objects.

actual mechanical advantage (*AMA*)   The mechanical advantage that accounts for all mechanical losses sustained through using a machine.

additive color   A color that is produced by combining the wavelengths of different colors.

alchemists   Scientists who were interested in turning low-value materials such as lead into high-value substances like gold.

alcohol   A substituted hydrocarbon in which a hydroxyl group (OH) replaces a hydrogen atom.

aldehyde   A substituted hydrocarbon in which the replacement by an oxygen atom of a pair of hydrogen atoms at the end of a hydrocarbon chain forms a carbonyl group (C=O).

alkali metal   An element in Group 1 of the periodic table, having one valence electron that it can easily lose to form a 1+ cation, making it extremely reactive; the most reactive of all the metals.

alkaline earth metal   An element in Group 2 of the periodic table, having two valence electrons that it tends to lose easily to become a 2+ cation, making it very reactive.

alkane   A saturated hydrocarbon that has only single bonds between its carbon atoms.

alkene   An unsaturated hydrocarbon with at least one double bond between carbon atoms.

alkyne   An unsaturated hydrocarbon with at least one triple bond between carbon atoms.

allotrope   One of multiple different forms of the same element in the same state.

alloy   A solid solution made of two or more elements, including at least one metal.

alpha decay   A nuclear decay that results in the emission of an alpha particle.

alpha particle   A helium nucleus (α or ^4_2He) that is emitted from a nucleus when a radioactive isotope experiences alpha decay.

alternating current (AC)   Electric current in which the charge carriers change direction periodically.

amino acid   A class of organic compounds that serve as the building blocks of proteins.

ammeter   A meter used to measure electric current through a circuit.

amorphous solid   A solid that consists of a mass of particles with no discernible pattern.

ampere (A)   The fundamental SI unit of electric current.

amplitude   The maximum displacement from the equilibrium position during periodic motion.

amplitude modulation (AM)   The process of putting information into a radio wave by changing, or modulating, the amplitude of the wave.

anion   A negatively charged ion.

antenna   A device that converts electrical energy to radio waves and vice versa.

antinode   The position on a standing wave that has maximum displacement.

anvil   One of the three bones, along with the hammer and stirrup, of the middle ear that transmit energy from the outer ear to the inner ear.

aqueous solution   A water-based solution.

Archimedes's principle   The principle that states that an immersed object displaces an amount of fluid equal to its volume and that the weight of the displaced fluid is equal to the buoyant force acting on the object.

aromatic hydrocarbon   An unsaturated hydrocarbon that contains at least one benzene structure.

artificial transmutation   A manmade nuclear change.

atom   The building block of all matter; consists of protons, electrons, and (usually) neutrons.

atomic mass   The mass of an atom expressed in atomic mass units; the weighted average of the masses of all the naturally occurring isotopes of an element.

atomic number   The unique number of protons found in every atom of a particular element.

atomic radius   The distance from the center of an atom's nucleus to its outermost energy level.

balanced forces   Simultaneous forces whose pushes and pulls cancel each other out.

base   A substance that produces hydroxide ions in an aqueous solution.

battery   A power source for DC electrical systems consisting of two or more electrochemical cells.

benzene   A six-carbon unsaturated hydrocarbon ring with the electrons from the C–C bonds equally distributed among the carbon atoms. It is the key feature in aromatic hydrocarbons.

Bernoulli's principle   The principle that states that the fluid pressure of a flowing fluid decreases as its speed increases.

beta decay   A nuclear decay that results in the emission of a beta particle.

beta particle   An electron (β or $_{-1}^{\ 0}e$) emitted from a nucleus when a radioactive isotope experiences beta decay.

binary compound   A compound made from only two elements.

biochemistry   The chemistry of living organisms and their processes.

biomass energy   Chemical potential energy obtained from renewable organic materials.

block and tackle   A system of fixed and movable pulleys connected by ropes.

Bohr model   The atomic model developed by Niels Bohr in which electrons travel in distinct spherical regions called *energy levels* at fixed distances from the nucleus.

boiling   The relatively fast form of vaporization in which the energy within a liquid creates higher pressure within the liquid than the air pressure outside the liquid.

boiling point   The temperature at which a liquid starts to boil.

boiling point elevation   The increase in the boiling point of a solvent due to the presence of dissolved solute.

bonding pair   Two shared electrons in a covalent bond.

Boyle's law   A gas law stating that the pressure and volume of a sample of gas at a constant temperature are inversely proportional.

branched chain   An organic molecule with carbon atoms connected to each other in such a way as to create more than one chain of carbons.

Brownian motion   The random movements of microscopic particles due to collisions.

buffer   A weak acid or base in solution that resists changes in pH when a moderate amount of either an acid or base is added.

buoyancy   The tendency of an object to float when immersed in a fluid.

buoyant force   The upward force caused by the displacement of a fluid.

caloric theory   The now-obsolete theory that stated that heat was an invisible self-repelling fluid.

calorimeter   A device that enables scientists to measure the thermal energy transferred in reactions and between systems.

capacitor   A device that stores electric charge.

carbohydrate   An organic compound comprised of carbon, hydrogen, and oxygen atoms that is the basic energy source for living organisms; includes sugars and starches.

carbonyl group   A functional group comprised of an oxygen atom double bonded to a carbon atom (C=O); a substituent in an aldehyde or ketone.

carboxide   One of a group of compounds made of carbon and oxygen.

catalyst   A substance that helps a reaction happen faster but is not used up in the reaction.

cation   A positively charged ion.

centripetal acceleration   Acceleration that causes an object to move along a circular path.

centripetal force   A force that accelerates an object toward the center of a circular path.

chain reaction   A self-sustaining nuclear fission process in which neutrons produced in one fission reaction trigger more fission events.

charging by conduction   The process by which one charged object can produce a second charged object by the two objects being placed in contact with each other and the excess charge being shared.

charging by friction   The process by which an object can gain excess charge while being rubbed by another object.

charging by induction   The process by which one charged object can produce a second charged object by allowing the electric force to move excess charge onto the second object while the two objects are not in contact with each other.

Charles's law   A gas law stating that the volume and absolute (Kelvin) temperature of a sample of gas at constant pressure are directly proportional.

chemical bond   An electrostatic attraction that forms between atoms when they share or transfer valence electrons.

chemical change   A change that alters the chemical composition of a substance.

chemical equation A combination of chemical formulas and symbols that models a chemical reaction.

chemical equilibrium The state that occurs when forward and reverse reactions each happen at the same rate.

chemical formula A shorthand way of identifying a chemical compound consisting of the element symbols and subscripts that represent the number of atoms of each element.

chemical property A property of a substance that describes how its chemical identity changes in the presence of another substance or under certain conditions.

chemical reaction A process that rearranges the atoms in one or more substances into one or more new substances.

chemiluminescence The emission of light energy from chemical reactions.

chemistry The study of the composition, structure, and properties of matter, and the changes that take place in matter.

circuit breaker An electrical safety device consisting of an automatic switch that opens when there is too much current in a circuit.

circular motion Movement along a circular path.

classical elements Earth, air, water, fire, and aether; considered by the Greeks to be the basic elements of which all materials were formed.

classical mechanics The study of how and why large objects move; also called *Newtonian mechanics*.

closed circuit A complete electric circuit that allows charge to flow.

cochlea The inner-ear organ that converts kinetic energy to electrical impulses.

coefficient A number placed in front of a chemical formula within a chemical equation that shows how many units of a reactant or product are needed to balance the chemical equation.

cohesion The attraction of like particles to each other.

colligative property A physical property of solutions that depends only on the concentration of dissolved particles and not on the identity or properties of the solute.

collision model The model that states that for a reaction to occur, the reactant particles must collide with each other, with the proper alignment, and with enough energy.

colloid A heterogeneous mixture in which the dispersed particles are between 1 nm and 1 μm in diameter and will not settle out.

combined gas law A gas law that states that the ratio of the product of a gas sample's pressure and volume to its absolute (Kelvin) temperature is a constant.

combustion An exothermic chemical reaction in which a fuel reacts with oxygen.

commutator A structure in DC motors and generators that converts current between DC and AC.

compound A pure substance consisting of atoms of two or more elements that are chemically combined in a fixed ratio.

compound machine A machine that combines two or more simple machines.

compressibility The property of a substance indicating how easily its particles can be pushed closer together.

compression A region of high density and pressure in a longitudinal (compression) wave.

compression force Any force that pushes two objects together.

compression wave See *longitudinal wave*.

computed tomography Medical imaging technology that uses x-rays to take millions of images that are combined into cross-sectional images of the body.

concave lens See *diverging lens*.

concave mirror A curved mirror with the reflective side on the inside of the curve.

concentration The actual amount of solute dissolved in a given amount of solution.

condensation The change of state from a gas (vapor) to a liquid.

conduction Movement of electric charge or thermal energy through an object or from object to object through direct contact.

conductivity The physical property indicating how easily a material transfers thermal energy or electric charge.

constructive interference An adding of overlapping waves that creates a wave with a larger amplitude.

contact force A force that acts only when one object touches another.

convection Movement of thermal energy as fluids move.

convection current The cycle of warmer fluid rising while cooler fluid sinks.

conventional current The direction in a DC circuit that positive charges would flow; decided by agreement as the standard current direction.

converging lens A lens that causes light rays to come together (converge); also known as a *convex lens*.

converging rays Rays of light that come together after being reflected or refracted.

conversion factor Two quantities with different units that are equivalent to each other and that are written as a fraction.

convex lens See *converging lens*.

convex mirror A curved mirror with the reflective side on the outside of the curve.

coulomb (C) The derived SI unit for electric charge; $1\ C = 1\ A{\cdot}s$

covalent bond A chemical bond formed as a result of two atoms sharing valence electrons.

Creation Mandate God's command that directs us to exercise wise and good dominion over His creation to the glory of God and for the benefit of fellow humans.

crest The highest point of a wave.

critical mass The smallest mass of fissionable material that can sustain a chain reaction.

crystal lattice An extensive three-dimensional structure of atoms or ions built up by repeating subunits, such as ions in ionic compounds, or individual atoms or molecules, as in metals or covalent solids.

crystalline solid A solid with particles arranged in regular, repeating patterns.

current electricity Electricity involving moving electric charges.

D

damped oscillator A system that is designed to oscillate but includes a force to reduce the amplitude.

damping Reducing the amplitude of periodic motion by applying a force that works against the motion.

data The collection of observations made by scientists during investigations.

decibel (dB) The unit for measuring relative sound intensity.

decomposition reaction A chemical reaction in which a single reactant breaks down into two or more products.

density The mass per unit volume of an object.

deoxyribonucleic acid See *DNA*.

deposition The change in state directly from a gas (vapor) to a solid without condensing first.

derived unit A unit that is a mathematical combination of fundamental SI units.

destructive interference An adding of overlapping waves that creates a wave with a smaller amplitude.

diaphragm A large muscle that forms the floor of the chest cavity and controls the volume of air that passes over the vocal cords.

diatomic element An element found in nature as a molecule consisting of two atoms of that element.

diatomic molecule A molecule made of two atoms.

diffraction The bending of waves around an obstacle or through an opening.

diffuse reflection The reflection off a rough or uneven surface that reflects light rays in all directions.

diffusion The process of spreading out and mixing due to random particle motion.

direct current (DC) Electric current in which electric charges move in only one direction.

displacement (Δx) A vector quantity that describes a change in position.

dissociation The physical separation of the ions in a solid ionic compound by the action of a solvent.

distance (d) How far an object moves during a time interval.

distillation A process for separating mixtures that uses the different boiling points of the materials.

diverging lens A lens that causes light rays to separate (diverge); also known as a *concave lens*.

diverging rays Rays of light that move apart after being reflected or refracted.

DNA The nucleic acid responsible for most cellular reproduction, growth, and development; abbreviation for *deoxyribonucleic acid*.

Doppler effect The perceived change in the frequency of a wave due to the motion of the wave's source or of the receiver.

dosimeter A device used to measure exposure to radiation.

double (covalent) bond A covalent bond formed by two pairs of shared valence electrons.

double-replacement reaction A chemical reaction in which two compounds swap cations or anions with each other; also known as a *double-displacement reaction*.

driven oscillator A system that is designed to oscillate at a constant amplitude by including a force to overcome frictional losses.

ductility The physical property indicating how easily a material can be pulled into a wire.

dynamics The branch of physics that studies forces and how they can change an object's motion.

E

echolocation The process that uses the time interval and direction of an echo to determine the position of an object.

efficiency A comparison of the amount of usable energy remaining after a process with the original amount of energy that went into the process.

electric circuit The loop through which current electricity can flow.

electric current (I) The movement of electric charge through a complete loop.

electric field A three-dimensional region around a charged object that will apply a force on other charged objects within that region.

electric force The field force between two charged objects.

electric potential difference See *voltage*.

electric potential energy Energy that is stored by a charged object in an electric field. The quantity depends on the magnitude of the charges and the object's position in the field.

electrical conductor A material through which electric charge moves easily.

electrical insulator A material through which electric charge does not move easily.

electrical load An electrical device in a circuit that consumes electrical energy.

electrical power The work per second done or produced by an electrical system.

electrical resistance (*R*) An object's opposition to the movement of charge carriers.

electromagnet The combination of a strong magnet and an electrified wire coil.

electromagnetic force The force produced by static and moving charges.

electromagnetic induction The principle that voltage is produced in a conductor whenever the conductor and a magnetic field are changed relative to each other.

electromagnetic radiation See *electromagnetic wave*.

electromagnetic spectrum The entire range of all electromagnetic waves.

electromagnetic wave A disruption in an electromagnetic field that carries energy, even through the vacuum of space.

electromagnetism The theory that electricity and magnetism are parts of the same fundamental force, called the *electromagnetic force*.

electron The smallest of the main subatomic particles, located outside the nucleus, having a negative charge and a mass that is 1/1836th that of a proton.

electron dot notation A representation of an atom consisting of its chemical symbol with surrounding dots representing its valence electrons.

electronegativity A measure of an element's ability to attract and hold electrons when bonded to other atoms.

electroscope An instrument that detects the presence and relative magnitude of static charge.

element A pure substance that consists of atoms with the same atomic number.

EM wave See *electromagnetic wave*.

endothermic reaction A chemical reaction that absorbs more thermal energy than it releases.

energy The ability to do work.

energy level In the Bohr model of the atom, the regions located at fixed distances from the nucleus of an atom in which electrons are found.

ethics A system of moral values; a theory of proper conduct.

eustachian tube The canal that connects the middle-ear cavity with the throat to allow the equalization of pressure on both sides of the eardrum.

evaporation The relatively slow form of vaporization in which liquid particles obtain sufficient energy to change to the gaseous state through the random collisions of particles.

exothermic reaction A chemical reaction that releases more thermal energy than it absorbs.

F

family A column of elements in the periodic table having similar valence electron arrangement, resulting in similar chemical properties; also known as a *group*.

fat A lipid comprised of carbon, hydrogen, and oxygen atoms; serves as long-term energy storage for plants, animals, and humans.

ferromagnetism The physical property in which a material's magnetic domains spontaneously align to an external magnetic field.

field force A force that acts between objects that are not touching; also called *force at a distance*.

filtration A process for separating solids from fluids by using a barrier that allows only the fluid to pass through.

first-class lever A lever in which the fulcrum is between the input and output forces.

first law of thermodynamics The law that states that energy cannot be created or destroyed but only transferred between objects or transformed; also known as the *law of conservation of energy*.

fission A nuclear reaction in which a large nucleus splits into smaller nuclei.

flammability The chemical property indicating how well a material burns in the presence of oxygen.

fluid Any substance that can flow.

fluorescence The emission of light after an object absorbs high-energy electromagnetic waves.

focal length The distance from the center of a lens or mirror to its focal point.

focal point The point on an optical axis at which all reflected or refracted light rays from incident rays that are parallel to the optical axis converge.

force (F)   A push or pull on an object.

force diagram   See *free-body diagram*.

forensic science   A kind of scientific investigation that attempts to reconstruct the scenario of a crime.

formula unit   The smallest whole-number ratio of ions within an ionic compound.

fossil fuel   A fuel that has formed from the remains of plants and animals that lived in the past, including coal, petroleum (oil), and natural gas.

frame of reference   A coordinate system used to describe the motion of an object.

free-body diagram   A sketch that shows an object and the forces acting on it; also called a *force diagram*.

free fall   The motion of an object that falls due to gravity alone, with no other forces acting on it.

freezing   The change of state from a liquid to a solid.

freezing point depression   The lowering of the freezing point of a solvent due to the presence of a dissolved solute.

frequency   The number of waves or cycles that occurs per second.

frequency modulation (FM)   The process of putting information into a radio wave by changing, or modulating, the frequency of the wave.

friction   A contact force that works against the motion of objects trying to move past each other.

fulcrum   The point about which a lever pivots or rotates.

functional group   An atom or group of atoms that replaces a hydrogen atom to form a substituted hydrocarbon; also called a *substituent*.

fundamental force   Any one of the four forces that appear to underlie all the other known forces: gravity, strong force, weak force, and electromagnetic force.

fundamental tone   The longest (lowest frequency) standing wave produced by a vibration of a structure.

fundamental units   The seven units (meter, kilogram, second, ampere, Kelvin, mole, and candela) in the SI from which all other units are derived.

fuse   An electrical safety device that opens the circuit by melting when an overheated condition occurs due to excessive current.

fusion   A nuclear reaction in which small nuclei combine to form a more massive nucleus.

G

gamma decay   A nuclear decay that results in the emission of gamma rays.

gamma ray   A high-energy photon (γ) that is emitted from a nucleus when a radioactive isotope experiences gamma decay.

gas   The state of matter in which particles are far apart, move rapidly, and have little interaction with each other.

Gay-Lussac's law   A gas law stating that the pressure and absolute (Kelvin) temperature of a sample of gas at a constant volume are directly proportional.

gear   A simple machine that consists of a wheel with teeth on its perimeter that mesh with similar teeth of other gears to do work.

Geiger counter   A device designed to measure ionizing radiation.

generator   A device that converts mechanical energy into electrical energy by electromagnetic induction.

genetic damage   Any damage done to DNA in cells that can affect growth and reproduction of the cells; can be passed to offspring if it occurs in reproductive cells.

geocentric model   An obsolete model of the universe with an unmoving Earth at its center and with all planets, the sun, and other celestial objects moving about it.

geographic North Pole   The "top" of the earth at the axis of Earth's spin, where the lines of longitude intersect.

geographic South Pole   The "bottom" of the earth at the axis of Earth's spin, where the lines of longitude intersect.

geothermal energy   Thermal energy that originates deep within the earth's interior.

glycogen   A polymer made of glucose monomers that animals and people use for short-term energy storage.

gravitational potential energy (*GPE*)   Potential energy due to an object's position within a gravitational field.

gravity   A field force that acts between the masses of any two objects.

ground-fault interrupter   A safety device consisting of an outlet with a built-in circuit breaker.

grounding   The act of providing a path for electrical charge to move into the earth.

group   See *family*.

H

half-life   The time in which half the atoms of a radioactive sample will probably decay.

halogen   An element in Group 17 of the periodic table having seven valence electrons. It easily gains an electron, forming a 1– anion, which causes it to be highly reactive.

hammer   One of the three bones, along with the anvil and stirrup, of the middle ear that transmit energy from the outer ear to the inner ear.

harmonics   The fundamental tone and its overtones.

heat   Movement of thermal energy from an area of higher temperature to one of lower temperature.

heat of fusion   The heat required to melt a quantity of a solid.

heat of vaporization   The heat required to vaporize a quantity of liquid.

heliocentric model   The modern model of the solar system with the sun as the gravitational center.

hertz (Hz)   The derived SI unit of frequency;

$$1 \text{ Hz} = 1 \, \frac{\text{cycle}}{\text{s}} = \frac{1}{\text{s}} = \text{s}^{-1}$$

heterogeneous mixture   A mixture that does not have a uniform appearance since the combined substances are unevenly distributed.

homogeneous mixture   A mixture that has a uniform appearance throughout; also known as a *solution*, especially of a solid dissolved in a liquid.

hydraulics   The branch of physics concerned with the forces within and work done by liquids.

hydrocarbon   An organic compound consisting of only hydrogen and carbon atoms.

hydroelectric energy   Electrical energy generated by the movement of water.

hydronium ion   An ion (H_3O^+) formed by the combination of a water molecule and a hydrogen ion.

hydroxide ion   An ion (OH^-) formed by oxygen and hydrogen.

hydroxyl group   A functional group made of an oxygen atom covalently bonded to a hydrogen atom (OH).

hypothesis   An initial, testable explanation of a phenomenon that stimulates and guides scientific investigation.

I

ideal mechanical advantage (*IMA*)   The mechanical advantage for an ideal machine (100% efficient) with no friction.

illuminated object   An object that is visible because it reflects light from an external source.

incandescence   The emission of light due to an object's high temperature.

incident ray   A light ray approaching a surface; an incoming ray.

inclined plane   A simple machine that consists of a plane whose opposite ends are at different heights.

indicator   A substance that changes color in the presence of an acid or base.

inertia   The tendency of matter to resist changes in its motion.

inert substance   A substance that has almost no chemical reactivity except under extreme conditions.

infrared wave (IR)   An electromagnetic wave that is just below (less energetic than) red visible light; used in astronomy, medical imaging, and wireless devices.

infrasonic sound   Sound having frequencies below the range of human hearing.

inhibitor   A substance that slows the rate of a reaction by reducing the effectiveness of catalysts.

inner transition metal   An element from either of two rows usually placed below the periodic table; a member of either the lanthanide or actinide series. It typically has two valence electrons.

instantaneous speed   Speed at a particular point in time.

intensity   A measure of the power contained in a wave; often refers to sound or electromagnetic waves.

interference   The combining of waves where they overlap.

International System of Units   See *SI*.

International Union of Pure and Applied Chemistry (IUPAC)   The international organization responsible for standardization in chemistry.

ion   A charged atom or group of atoms that has gained or lost electrons, producing an unequal number of protons and electrons.

ionic bond   A chemical bond formed when atoms transfer valence electrons.

ionization energy   The minimum amount of energy required to remove the first electron from the outermost shell of a neutral atom in its gaseous state.

ionizing radiation   Radiation that is energetic enough to knock electrons out of atoms or molecules. This radiation is the most damaging to living organisms.

isomer   Any of a group of compounds that have the same molecular formula but different structures.

isotope   An atom of an element that has a different number of neutrons compared with other atoms of that element, resulting in a different mass number.

isotope notation   A symbol that distinguishes between different isotopes. It is written in the form $^{A}_{Z}X$ where A is the mass number, Z is the atomic number, and X is the element symbol.

IUPAC   See *International Union of Pure and Applied Chemistry*.

J

joule (J)   The derived SI unit of work or energy;

$$1 \text{ J} = 1 \text{ N·m} = 1 \, \frac{\text{kg·m}^2}{\text{s}^2}$$

K

ketone A substituted hydrocarbon in which a pair of hydrogen atoms bonded to a carbon atom other than those at the end of the carbon chain are replaced by an oxygen atom, forming a carbonyl group (C=O).

kinematics The study of how things move.

kinetic energy (*KE*) Energy that an object possesses due to its motion.

kinetic-molecular model The model that says that all particles are in constant motion and that thermal energy is the sum of the kinetic energies of these particles.

L

larynx The box-like structure located at the top of the trachea that supports the vocal cords.

law A model, often expressed as a mathematical equation, that describes phenomena under certain conditions.

law of acceleration The law that states that the acceleration of an object is directly proportional to the net force acting on the object and is inversely proportional to its mass; also called *Newton's second law of motion*.

law of action-reaction The law that states that for every action force, there is an equal and opposite reaction force; also called *Newton's third law of motion*.

law of cause and effect The basic law that states that every effect has a specific, identifiable cause, and for every cause there is a definite and predictable effect.

law of conservation of charge The principle that states that charge cannot be created or destroyed but only transferred between objects.

law of conservation of energy The law that states that energy cannot be created or destroyed but only transferred between objects or transformed; also known as the *first law of thermodynamics*.

law of conservation of matter The law that states that matter can neither be created nor destroyed, but can only change form.

law of conservation of momentum The law that states that within a closed system, the total momentum remains constant.

law of definite proportions The law that states that the masses of chemical substances combine in definite, characteristic integer ratios when forming compounds.

law of electrostatic charges The law that states that opposite electrical charges attract each other, while like charges repel each other.

law of inertia The law that states that objects at rest remain at rest and objects in motion continue in a straight line at a constant velocity unless acted on by a net external force; also called *Newton's first law of motion*.

law of octaves The principle published by chemist John Newlands that stated that the properties of the forty-nine then-known elements repeated every eighth element, as in a musical octave.

law of reflection The law that states that when a light ray reflects off a surface, the angle of the incident ray equals the angle of the reflected ray.

law of torques The law that states that in a lever system in rotational equilibrium, the sum of the torques must be zero.

law of universal gravitation The law that states that the strength of gravity varies in direct proportion to the masses of the objects involved and inversely to the square of the distance between their centers of mass.

Le Châtelier's principle The principle that states that a chemical system in equilibrium will adjust its equilibrium position in such a way as to reduce the effect of any changes made to the system.

lever A simple machine that consists of a rigid bar that turns about a pivot point (the fulcrum).

Lewis structure A system for modeling the covalent bonds between atoms in a molecule and any unbonded electrons in the molecule.

light Electromagnetic energy in the visible light range.

lipid Organic compounds that provide long-term energy storage in living organisms; includes fats, oils, waxes, and steroids.

liquid The state of matter in which particles are close together but able to move around.

longitudinal wave A wave in which the disruptions are parallel to the direction of wave travel.

loudness Human perception of the intensity of a sound wave.

luminous object An object that is producing visible light.

luster The physical property indicating how well a substance reflects light.

M

macromolecule A very large molecule.

magnet Any material or object that can produce a magnetic field.

magnetic declination The angular difference between the directions to the geographic and magnetic North Poles.

magnetic domain A group of atoms whose individual magnetic fields are aligned.

magnetic field   The region surrounding a magnet or current-containing wire that can apply a magnetic force on magnetic materials.

magnetic North Pole   The magnetic pole in the Northern Hemisphere. It is currently located beneath the Arctic Ocean and is moving toward Siberia. This pole is actually the south pole of Earth's magnetic field.

magnetic pole   One of the two regions of concentrated magnetic fields in a magnet; traditionally called the *north* and *south poles*.

magnetic South Pole   The magnetic pole in the Southern Hemisphere. It is currently located beneath the Antarctic Ocean south of Australia. This pole is actually the north pole of Earth's magnetic field.

magnetism   The properties of magnets and magnetic fields and the phenomena that they produce.

malleability   The physical property indicating how easily a material can be hammered or pressed into sheets.

mass   The measure of the amount of matter in an object.

mass number (A)   The total number of particles found in the nucleus of a particular isotope of an element.

materials science   An interdisciplinary science that studies the properties of materials to discover new materials, new uses for existing materials, and changes to existing materials to meet the needs of people.

matter   Anything that has mass and occupies space.

measurement   Data that is based on numbers or quantities; includes a number and a unit; also known as *quantitative data*.

mechanical advantage (MA)   The amount by which a simple machine multiplies an input force to produce an output force.

mechanical energy   The sum of the kinetic energy and all the forms of potential energy in a system.

mechanical wave   A wave that carries energy through a physical medium.

mechanics   The study of motion.

melting   The change of state from a solid to a liquid.

melting point   The temperature at which a solid turns to a liquid.

metal   An element that is typically dense, solid, ductile, malleable, highly conductive, and chemically reactive, especially in the presence of nonmetal elements. Metals are located on the left end of the periodic table.

metallic bond   The attraction between metal atoms and their collectively shared valence electrons.

metalloid   An element with characteristics between those of metals and nonmetals; also called *semiconductors*; located between metals and nonmetals on the periodic table.

metric prefixes   A set of prefixes representing powers of ten that are attached to metric units to create smaller or larger units.

microwave   A wave from the upper end of the radio wave band; used for navigation, communication, and cooking.

mixed group   Any of Groups 13–16 in the periodic table; so named because they contain metals, nonmetals, and metalloids. These groups are often named for the first element in the family.

mixture   A physical combination of two or more substances in a changeable ratio.

model   A workable explanation or description of a phenomenon.

molarity (M)   A measure of solution concentration; the number of moles of solute per liter of solution.

molar mass   The mass of one mole of a substance.

mole (mol)   SI fundamental unit for the quantity of matter in a substance; a count equal to 6.022×10^{23}.

molecular formula   A formula that shows the exact composition of a molecule. It is comprised of element symbols along with subscripts for the number of atoms of each element.

molecule   A particle consisting of two or more atoms covalently bonded together.

momentum (p)   A property of a moving system that is equal to its velocity times its mass.

monomer   A simple molecule that can link with other monomers to form large molecules called *polymers*. For example, glucose monomers combine to form the polymer starch.

N

natural science   The study of objects and phenomena of the physical world.

net force   The vector sum of all the forces acting on an object.

neutral buoyancy   The condition when an object submerged in a fluid neither ascends nor descends because its density is equal to that of the fluid.

neutralization   A double-replacement reaction between an acid and a base that produces a salt and water.

neutron   A subatomic particle found in the nucleus of most atoms that has no electrical charge and a mass slightly greater than that of a proton.

newton (N)   The derived SI unit of force; $1\ \text{N} = 1\ \frac{\text{kg} \cdot \text{m}}{\text{s}^2}$

Newton's first law of motion   See *law of inertia*.

Newton's second law of motion   See *law of acceleration*.

Newton's third law of motion   See *law of action-reaction*.

noble gas   An element in Group 18 on the periodic table having eight valence electrons that fill the outer energy level. (Helium is an exception with only two.) With a full outer energy level, it is inert (i.e., nonreactive).

node   The position on a standing wave that has no displacement.

nonmetal   An element that typically has four or more valence electrons and that does not exhibit the general properties of metals; located on the right side of the periodic table.

nonpolar bond   A covalent bond in which electrons are shared equally.

nonrenewable energy resource An energy resource that is not replaced naturally.

normal line   A line that is perpendicular to a surface (e.g., the direction of the normal force or the line from which a light ray's angle of incidence is measured).

normal force   The force that acts in a direction that is perpendicular to the surface where two objects make contact.

north magnetic pole   The end of a magnet from which magnetic field lines leave.

nuclear change   Any change that alters the composition of the nucleus in an atom.

nuclear decay   See *radioactive decay*.

nuclear energy   Energy produced by nuclear reactions.

nuclear model   The atomic model developed by Ernest Rutherford in which an atom is made up of a tiny, dense, positively charged central nucleus surrounded by negatively charged electrons.

nuclear stability   The likelihood of an isotope to not undergo radioactive decay.

nuclear waste   Any waste that is radioactive.

nucleic acid   A biochemical polymer that encodes, stores, and provides instructions for cellular processes.

nucleotide   Any of a group of biochemical molecules that act as the monomers to make nucleic acids. Each consists of a sugar, a nitrogen base, and a phosphate group.

O

octet rule   The principle that states that atoms generally are most stable when they have eight electrons in their valence energy level.

ohm (Ω)   The derived SI unit for electrical resistance;
$$1\ \Omega = 1\ \frac{V}{A} = 1\ \frac{kg \cdot m^2}{s^3 A^2}$$

Ohm's law   The law that states that the current in a circuit is directly related to the voltage and inversely related to the resistance.

opaque material   A material through which visible light cannot pass.

open circuit   An incomplete electric circuit that prevents the movement of charge.

orbit   The circular path of an electron around the nucleus in the Bohr model.

orbital   The region within an atom where an electron will most probably be located; part of the quantum-mechanical model of the atom.

organic chemistry   The study of the composition, structure, and properties of carbon-containing compounds.

organic compound   A covalently bonded compound containing carbon.

overtone   A shorter, faster vibration (higher pitch) in addition to the fundamental tone produced by a vibrating structure.

oxidation   A loss of electrons in a chemical reaction.

oxidation-reduction reaction   See *redox reaction*.

oxidation state   A positive or negative number showing the electric charge on an element when it forms a compound.

oxide   A compound that contains oxygen, often formed in an oxidation reaction.

P

parallel circuit   A circuit with multiple paths that electric current can take.

particle model of matter   A model that states that all physical matter exists in the form of particles.

pascal (Pa)   The derived SI unit of pressure;
$$1\ Pa = 1\ \frac{N}{m^2} = 1\ \frac{kg}{m \cdot s^2}$$

Pascal's principle   The principle that states that pressure applied to a fluid is transmitted throughout the fluid.

passive sonar   A system of underwater microphones that can only receive, not produce, underwater sounds in order to detect a submerged object.

peer review   The assessment process for scientific research in which other scientists review and respond to research.

period   (1) A row in the periodic table of the elements; also called a *series*. (2) The time interval (T) for one complete cycle of periodic motion.

periodicity   The idea that properties of elements repeat in regular patterns in relation to some basic characteristic such as atomic mass or atomic number.

periodic law   The law that states that the properties of the elements vary with their atomic numbers in a regular, repeated pattern.

periodic motion   Motion that repeats in equal time intervals.

periodic table of the elements   A table of the chemical elements arranged to display their periodic properties in relation to their atomic numbers.

periodic trend   The change of a particular property along rows or columns of the periodic table.

pH   A unitless number between 0 and 14 that tells how acidic or basic a substance is.

phase diagram   A diagram summarizing the temperature and pressure conditions under which a pure substance exists as a solid, liquid, or gas.

phenomenon   An observable or measureable event, object, process, or property.

phosphorescence   A slow emission of absorbed electromagnetic energy, similar to fluorescence.

photoelectric effect   The dislodging of electrons from metals caused by light of sufficiently high frequency, supporting the particle model of light.

photon   A wave bundle, or particle, of electromagnetic energy.

physical change   Any change in matter that does not alter its chemical or nuclear composition.

physical property   Anything about a substance that can be observed or measured without altering its chemical composition.

physical science   The study of nonliving matter and energy.

physics   The study of matter and energy and the interactions between them.

pigment   A substance that absorbs some wavelengths of light and reflects others; used to create paints, inks, and dyes.

pitch   How high or low an audible tone sounds to the human ear; related to the concept of wave frequency.

plane mirror   A flat mirror.

plasma   A gas-like state of matter, formed at very high temperatures, that consists of high-energy ions and free electrons.

plum pudding model   The atomic model developed by J. J. Thomson that views atoms as spheres of positively charged material with embedded electrons.

polar covalent bond   A covalent bond in which electrons are shared unequally.

polarity   The unequal distribution of electric charge in a covalent bond.

polar molecule   A molecule with negatively charged and positively charged regions as a result of polar covalent bonds and molecular geometry.

polyatomic ion   A group of covalently bonded atoms that together have gained or lost electrons and act as a single ionized particle.

polyatomic molecule   A molecule formed by two or more atoms, including diatomic molecules.

polymer   A large molecule formed by linking smaller molecules, called *monomers*. For example, nucleic acid, a polymer, is formed by linking nucleotide monomers.

potential energy (*PE*)   Stored energy that can be used later.

power   The rate of doing work.

precipitate   A solid that forms when two solutions react.

precision   The degree of exactness of a measurement; can indicate the closeness or repeatability of measurements.

pressure (*P*)   A measurement of the amount of force acting upon a unit area.

primary color   One of the three colors (red, blue, and green) of visible light that the human eye can sense; mixed to produce other colors.

prime mover   The agent that provides the mechanical energy to turn the rotor of a generator.

principle of uniformity   The basic law that says that nature acts the same today as it did yesterday and that we can fully expect it to act the same way tomorrow.

product   A substance that is formed during a chemical reaction.

projectile motion   The two-dimensional motion of any flying object whose path is determined by the influence of an external force only, such as gravity.

protein   A biochemical polymer made of amino acids. Proteins are the building blocks for muscle, blood, skin, and hair in humans and animals.

proton   A subatomic particle found in the nucleus of an atom and having a positive charge and a mass slightly less than that of a neutron.

pulley   A simple machine that consists of a wheel and axle system with a groove around the perimeter of the wheel in which a rope, cable, or belt moves with the wheel as it rotates.

pure substance   A material made of only one kind of element or compound.

Q

qualitative data   Data that is based on qualities of an object or phenomenon.

quantitative data   Data that is based on measureable dimensions of an object or phenomenon; also known as *measurement*.

quantum-mechanical model   The currently accepted atomic model in which electrons are found in orbitals that are positioned around a nucleus that contains protons and (usually) neutrons.

quantum mechanics   The branch of physics that explores the behavior of matter and energy at the atomic and subatomic level.

R

radiation   Movement of energy in the form of electromagnetic waves.

radioactive decay   The naturally occurring, spontaneous change of an unstable isotope to a more stable one by emitting particles or energy or both; also known as *nuclear decay* or *radioactivity*.

radioactivity   See *radioactive decay*.

radiometric dating   A method for mathematically estimating the age of an object by measuring the present amount of a radioactive isotope and comparing that with the starting amount of that isotope in the object.

radiotracer   A radioactive isotope used in nuclear medicine to study how an isotope moves through or collects in a certain organ or system.

radio wave   The longest and lowest energy type of electromagnetic wave; used for navigation and communication.

rarefaction   A region of lower density and pressure in a longitudinal (compression) wave.

ray   (1) A half-line. (2) A model of light in which the endpoint represents the light's source and the light travels in one direction.

reactant   A substance that enters into a chemical reaction.

reaction rate   The speed of a reaction.

reactivity   The chemical property indicating to what degree a material reacts with another substance.

real image   An image that forms at the point where converging rays of light intersect.

redox reaction   A chemical reaction in which electrons are transferred; oxidation and reduction occurs.

reduction   A gain of electrons in a chemical reaction.

reflected ray   The ray that bounces off a surface.

reflection   The bouncing of waves off a surface.

refraction   A change in wave direction due to a change in a wave's speed as it enters a new medium.

regular reflection   Reflection that occurs when a smooth surface reflects light rays in mostly the same direction; also known as a *specular reflection*. A perfectly regular reflection is mirrorlike, reflecting all rays and wavelengths uniformly.

relativistic mechanics   The study of the motion of objects whose speeds are near the speed of light.

renewable energy resource   An energy resource that is easily replaced by natural methods.

resistor   An electrical device that slows down electric current by converting electrical energy into other forms, such as thermal energy.

resonance   An increase in the amplitude of a vibration due to additional wave input.

retrograde motion   The apparent motion of some planets in which they appear to slow down, stop, reverse their direction, and then resume their normal motion.

reversible reaction   A reaction in which the products can react together to re-form the original reactants.

ribonucleic acid   See *RNA*.

right-hand rule   A mnemonic device used to remember the direction of the magnetic field induced by an electric current moving through a wire.

ring   An organic molecule made by connecting the two ends of a carbon chain.

RNA   A nucleic acid that plays a critical role in protein synthesis; abbreviation for *ribonucleic acid*.

rolling friction   The frictional force between two objects when one is rolling relative to the other.

rotor   The rotating portion of a generator.

S

salt   An ionic compound formed by a combination of cations and anions.

saturated hydrocarbon   A hydrocarbon that has only single bonds between its carbon atoms.

saturated solution   A solution that contains the maximum amount of solute that it can hold at a given temperature.

scalar   A measurable quantity that consists of magnitude (size) only.

science   The systematic study of the universe that produces observations, inferences, and models, including the products that it creates through this systematic study.

scientific inquiry   An ongoing, orderly, cyclical approach used to investigate the world.

screw   A simple machine that consists of an inclined plane wrapped around a cone or cylinder in a spiral pattern.

secondary color   A color of light that is produced when primary colors are mixed.

second-class lever   A lever in which the output force is between the input force and the fulcrum.

second law of thermodynamics   The law that states that energy can flow from a colder object to a warmer object *only* if something does work.

semiconductor   A material with conductivity between those of conductors and insulators.

series circuit   A circuit with only one path that electric current can take.

short circuit   An unintended path for an electric current.

SI   A standardized system of measurement units used for science. SI stands for *International System of Units* (from the French *Système International d'Unités*).

simple harmonic motion   Periodic motion that is caused by a restoring force that is proportional to the displacement of the system.

simple machine   A basic mechanical device that changes the magnitude, direction, or distance traveled of the force used when doing work.

single (covalent) bond   A covalent bond formed by one pair of shared valence electrons.

single-replacement reaction   A chemical reaction in which one element in a compound is replaced by another element; also known as a *single-displacement reaction*.

sliding friction   The frictional force between two objects when one is sliding past the other; also called *kinetic friction*.

soda lake   A lake with a high pH (basic) due to the presence of carbonates.

solar energy   Energy produced by the sun.

solenoid   A coil of current-carrying wire used as a magnet.

solid   The state of matter in which particles vibrate in fixed positions, giving a substance a fixed shape and volume.

solubility   The maximum amount of solute that can dissolve in a given amount of solvent at a given temperature.

solute   In a solution, the dissolved material or the material in lesser abundance.

solution   A mixture with a uniform appearance throughout; also called a *homogeneous mixture*.

solvation   The process of dissolving a solute into a liquid solvent.

solvent   In a solution, the dissolving material or the material in greater abundance.

somatic damage   Any damage to cells that are not involved in reproduction, thus harming the organism but not any future offspring.

sonar   A device for detecting submerged objects by using sound; an acronym for <u>so</u>und <u>na</u>vigation and <u>ra</u>nging.

sonography   Technology that uses ultrasound to create images of objects found inside other objects.

sound energy   A type of mechanical wave energy that can be detected by the human ear.

south magnetic pole   The end of a magnet into which magnetic field lines return.

specific heat (c_{sp})   The energy required to raise the temperature of 1 g of a substance 1 °C.

specular reflection   See *regular reflection*.

speed   A scalar quantity indicating the rate at which an object moves.

speed of light   The speed at which all electromagnetic waves travel in a vacuum (299 792 458 m/s).

standing wave   A wave that is moving even though the locations of the crests and troughs appear to be stationary.

state of matter   A physical form of matter determined by the forces between and energy of its particles. The three most common states of matter on Earth are solid, liquid, and gas.

static electricity   The accumulated electric charge on an object.

static friction   The frictional force between two objects that are touching but not moving relative to each other.

stator   The stationary portion of a generator.

step-down transformer   A transformer that is designed to decrease an output voltage.

step-up transformer   A transformer that is designed to increase an output voltage.

stirrup   One of the three bones, along with the anvil and hammer, of the middle ear that transmit energy from the outer ear to the inner ear.

straight chain   An organic molecule consisting of a single continuous series of any number of carbon atoms bonded to each other.

strong acid   An acid in which all or most of the molecules produce hydrogen ions in an aqueous solution.

strong base   A base that readily produces large numbers of hydroxide ions in an aqueous solution.

strong force   An attractive force that holds protons and neutrons together in a nucleus.

structural formula   A drawing depicting the composition and arrangement of atoms in a molecule.

sublimation   The change in state directly from a solid to a gas (vapor) without melting first.

substituent   See *functional group*.

substituted hydrocarbon   A hydrocarbon in which at least one of the hydrogen atoms has been replaced with another atom or group of atoms. The replacement atom or group of atoms is called a *functional group*.

subtractive color   A color that forms because of pigments absorbing some wavelengths of light that strike an object.

superconductor   A material with zero resistance.

supersaturated solution   A solution that contains more than the maximum amount of solute that it can normally hold at a given temperature.

surface wave   A wave that occurs along the interface between two media.

suspension   A heterogeneous mixture of a fluid and large particles (< 1 µm) that will settle out over time.

synthesis reaction   A chemical reaction in which two or more reactants combine into a single, more complex product.

system   A portion of a larger motion that we are interested in studying.

T

temperature (*T*) The measure of the hotness or coldness of a substance; proportional to the average kinetic energy of the particles within the substance.

tension A pulling force that is transmitted through a rope, chain, or similar object.

theory A model that explains a related set of phenomena; can be used to predict unobserved aspects of the phenomena.

thermal conductor A material through which thermal energy moves easily.

thermal energy The sum of the kinetic energies of all the particles within an object.

thermal expansion The property of many materials to increase in volume when heated and contract when cooled.

thermal insulator A material through which thermal energy does not easily move.

thermodynamics The study of thermal energy and heat and how they relate to work and other forms of energy.

thermometric property Any property that changes predictably with changes in temperature.

third-class lever A lever in which the input force is between the output force and the fulcrum.

third law of thermodynamics The law that states that entropy would be at its minimum value at absolute zero. Therefore, absolute zero can never be achieved.

tidal energy Mechanical energy in rising and falling tides.

timbre The distinctive sound of an instrument; also called *quality*.

torque A force that tends to cause a rotation about a pivot point.

toxicology The scientific study of poisons and their negative effects on organisms.

traction The frictional force between a vehicle's tires and the road; responsible for accelerating the vehicle.

trajectory The curved path of a projectile.

transformer A device that uses electromagnetic induction to increase or decrease the voltage of an AC current.

transition metal Any elements in Groups 3–12 of the periodic table, typically having one or two valence electrons, which it easily loses, resulting in cations with charges of 1+ or 2+.

translucent material A material through which light passes but the light is scattered and does not transmit a clear image.

transmutation Any process that converts one isotope into another by changing the number of protons or neutrons.

transparent material A material through which light passes without scattering, transmitting a clear image.

transverse wave A wave in which the disruptions move perpendicular to the direction of wave travel.

triads A model of periodicity developed by Johann Döbereiner that is based on groups of three elements with similar properties.

triple (covalent) bond A covalent bond formed by three pairs of shared valence electrons.

triple point The temperature and pressure at which the solid, liquid, and vapor states of a substance exist in equilibrium.

trough The lowest point of a wave.

tympanic membrane The thin, flexible membrane that separates the outer ear from the middle ear and converts acoustic energy to kinetic energy.

Tyndall effect The scattering of light by the particles in a colloid.

U

ultrasonic sound Sound having frequencies above the range of human hearing.

ultraviolet wave (UV) An electromagnetic wave that is just beyond (more energetic than) violet visible light; used for medical treatment, dentistry, and killing bacteria.

unbalanced forces A collection of forces on an object that don't cancel out and thus cause an acceleration.

unsaturated hydrocarbon A hydrocarbon that has at least one double or triple bond between its carbon atoms.

unsaturated solution A solution that contains less than the maximum amount of solute that it could hold at a given temperature.

urban heat island Term used to describe the difference in temperature of an urban area compared with that of the surrounding region.

V

valence electron Any electron in the outermost energy level of a neutral atom. Unpaired valence electrons are usually involved in chemical bonding.

vaporization The change of state from a liquid to a gas (vapor).

vector A measurable quantity with both magnitude and direction.

vector addition The mathematical process of combining vectors that takes into account both the magnitude and direction of the vectors.

velocity (v) A vector quantity indicating the rate at which an object's position changes.

virtual image An image produced by diverging rays. The image is formed at the point from which the diverging rays would have originated. Because the rays don't actually intersect, virtual images cannot be projected onto a screen.

viscosity A measure of a fluid's resistance to flowing.

visible light Electromagnetic waves that humans can see.

vocal cords Folds of tissue in the throat that when vibrated produce the sound waves that humans use to communicate.

volt (V) The derived SI unit for electric potential difference; $1\ V = 1\ A \cdot \Omega = 1\ \dfrac{kg \cdot m^2}{s^3\ A}$

voltage The "force" that moves electric charge carriers through an electrical circuit; also called *electric potential difference*.

voltmeter A meter used to measure voltage.

volume (V) The space enclosed or occupied by an object.

W

water displacement A method for determining the volumes of irregularly shaped objects on the basis of Archimedes's principle.

watt (W) The derived SI unit of power; $1\ W = 1\ \dfrac{J}{s} = 1\ \dfrac{kg \cdot m^2}{s^3}$

wave A disruption that carries energy from one location to another.

wave height The vertical distance between the trough and crest of a wave.

wavelength (λ) The distance between two identical points on successive waves.

wave-particle duality The currently accepted model of light indicating that light has both wave and particle characteristics.

wave pulse A single wave disruption.

wave train A series of wave pulses.

weak acid An acid that produces few hydrogen ions in an aqueous solution.

weak base A base that produces few hydroxide ions in an aqueous solution.

weak force The extreme short-distance force that holds the subatomic particles together inside protons and neutrons.

wedge A simple machine consisting of two inclined planes attached at an acute angle and used to spread a material apart as it is forced into the material.

weight The force of gravity acting on the matter in an object.

wheel and axle A simple machine consisting of a wheel with a rod running through its axis that acts as the pivot point.

wind energy Energy from the wind; can be used to turn turbine blades that are attached to generators.

wind shear A significant change in wind speed, direction, or both in a small geographic area; a significant hazard for aircraft.

work The energy transferred to a system by an external force when it acts on the system to move it.

workability The basis upon which a model is assessed, taking into account how well it explains or describes a set of observations and how well the model makes predictions.

worldview How a person views the world; a set of basic beliefs, assumptions, and values that arises from an overarching narrative about the world and that produces individual and group action.

X

x-ray (1) A form of electromagnetic energy with wavelengths between 10 pm (picometers, 10^{-12}) and 10 nm that is used for medical imagery, transportation security, and nondestructive inspection. (2) A photograph taken with x-rays.

Y

Young double-slit experiment An early experiment that demonstrated that electromagnetic waves interfere with each other, supporting the wave model of light.

Key: (t) top; (c) center; (b) bottom; (l) left; (r) right; (i) inset, (bg) background

COVER

Picsfive/Shutterstock.com

FRONT MATTER

i Picsfive/Shutterstock.com; **ii–iii** alexeys/iStock/Getty Images Plus/Getty Images; **iv** Pattadis Walarput/iStock zz/Getty Images Plus/Getty Images; **v** Mr. Klein/Shutterstock.com; **vi** DRogatnev/Shutterstock.com; **vii** Pavel L Photo and Video/Shutterstock.com; **viibg** PhotoSky/Shutterstock.com; **viii** Mix3r/Shutterstock.com; **ixtr** Aflo Co. Ltd./Alamy Stock Photo; **ixtri** © iStock.com/PragasitLalao; **ixtl** Soloviova Liudmyla/Shutterstock.com; **ixc** Sergio33/Shutterstock.com; **ixbl** Cultura Images/Eugenio Marongiu/Media Bakery; **xt** Serhii Hrebeniuk/Shutterstock.com; **xb** Elenathewise/iStock/Getty Images Plus/Getty Images; **xbi** Mikhail Japaridze/TASS/Getty Images; **xict** Ian Cuming/Ikon Images/Getty Images Plus/Getty Images; **xil** GIPhotoStock/Science Source; **xic** NotionPic/Shutterstock.com; **xicb** Jose Luis Pelaez Inc/DigitalVision/Getty Images; **xir** "Elektrodyna misk-högtalare" by Svjo/Wikimedia Commons/CC BY-SA 3.0

CHAPTER 1

2 Matthew Horwood/Alamy Stock Photo; **4** Adobe Stock/Patrizio Martorana; **5bg** Alexander Limbach/Shutterstock.com; **5** (earth, leaf) shaimaadesigns/Shutterstock.com; **5** (light bulb, rocket, atom model) derter/Shutterstock.com; **5** (beaker) DRogatnev/Shutterstock.com; **6, 7bg** New Line/Shutterstock.com; **6** (broccoli) Kert/Shutterstock.com; **6, 7** (sun, moon) 21kompot/Shutterstock.com; **6** (snowflake) Kichigin/Shutterstock.com; **6** (microscope) DRogatnev/Shutterstock.com; **6** (dominos) Africa Studio/Shutterstock.com; **6** (nautilus shell) aaltair/Shutterstock.com; **7** (sunflower) Ian 2010/Shutterstock.com; **7** (scientist) Andrew Brookes/Cultura Exclusive/Getty Images; **7** (earth) LuckyVector/Shutterstock.com; **7** (robot) MONOPOLY919/Shutterstock.com; **8l** Bryan Chan/Los Angeles Times/Getty Images; **8r** Richard T. Nowitz/Science Source; **9, 22** © iStock.com/shapecharge; **10t** manusapon kasosod/Shutterstock.com; **10b** © iStock.com/stocknroll; **11** Paper Boat Creative/DigitalVision/Getty Images; **11ti** asharkyu/Shutterstock.com; **11ci** Marina Sun/Shutterstock.com; **11bi** Gorodenkoff/Shutterstock.com; **12** both, **21c** ValentinaKru/Shutterstock.com; **13** Matej Kastelic/Shutterstock.com; **14, 21b** Barrett & MacKay/All Canada Photos/Getty Images; **15** (ruler) Neveshkin Nikolay/Shutterstock.com; **15** (kilogram) donatas1205/Shutterstock.com; **15** (stopwatch) Stepan Bormotov/Shutterstock.com; **15** (ampere meter) Stefan Rotter/Shutterstock.com; **15** (thermometer) Tomas Ragina/Shutterstock.com; **15** (coal) Trum Ronnarong/Shutterstock.com; **15** (candle) Aksenova Natalya/Shutterstock.com; **16** (dam) Yao yilong/Imaginechina/AP Images; **16** (radio) Yuliyan Velchev/Shutterstock.com; **16** (jet) Digital Storm/Shutterstock.com; **16** (candy bar) burnel1

/Shutterstock.com; **16** (cup) Kimberly Hall/Shutterstock.com; **16** (hair) Steve Gschmeissner/Science Photo Library/Getty Images; **16** (sign) photocritical/Shutterstock.com; **17t** hwanchul/Shutterstock.com; **17i** Andrey_Kuzmin/Shutterstock.com; **17bl** © iStock.com/MicroStockHub; **17br** iofoto/Shutterstock.com; **18** sciencephotos/Alamy Stock Photo; **20** © iStock.com/StefaNikolic; **21t** GoncharukMaks/Shutterstock.com; **23** Syda Productions/Shutterstock.com

CHAPTER 2

24 Anant Kasetsinsombut/Shutterstock.com; **24i** Robert Clark/National Geographic/Getty Images; **26, 45t** Soloviova Liudmyla/Shutterstock.com; **27** MatiasDelCarmine/Shutterstock.com; **28t** Trevor Clifford Photography/Science Source; **28cl** Christopher Meade/Shutterstock.com; **28cr** Matias DelCarmine/Shutterstock.com; **28b** Dmitry Lobanov/Shutterstock.com; **29** yitewang/Shutterstock.com; **30** Red monkey/Shutterstock.com; **31** prill/123RF; **32** Prasit Rodphan/Shutterstock.com; **33** (chlorine gas) Ian Miles-Flashpoint Pictures/Alamy Stock Photo; **33** (liquid mercury) Alexey V Smirnov/Shutterstock.com; **33** (diamond) SPbPhoto/Shutterstock.com; **33** (sugar cubes) Rob Stark/Shutterstock.com; **33** (table salt) etorres/Shutterstock.com; **34t** Agustin Vai/Shutterstock.com; **34bl** Turtle Rock Scientific/Science Source; **34bc, 45c** Sergio33/Shutterstock.com; **34br** Africa Studio/Shutterstock.com; **35** University of Rochester; **36t** Sergiy Kuzmin/Shutterstock.com; **36b, 45b** Cultura Images/Eugenio Marongiu/Media Bakery; **37lt** Sebastian Janicki/Shutterstock.com; **37rt** Cromagnon/Shutterstock.com; **37rc** DONOT6_STUDIO/Shutterstock.com; **37lb** Turtle Rock Scientific/Science Source; **37rb** Z miri/Shutterstock.com; **38** Peter Macdiarmid/Getty Images News/Getty Images; **38ti** CERN; **38bi** SPL/Science Source; **39** Stefano Montesi/Corbis News/Getty Images; **40** Flegere/Shutterstock.com; **40ri** Phil Degginger/Alamy Stock Photo; **40li** manfredxy/Shutterstock.com; **41** © iStock.com/YongEe; **41i** Turtle Rock Scientific/Science Source; **42–43bg** HAKKI ARSLAN/Shutterstock.com; **44** Aflo Co. Ltd./Alamy Stock Photo; **44i** © iStock.com/PragasitLalao; **46** MSSA/Shutterstock.com

CHAPTER 3

48 CERN; **50t** ORNL/Science Source; **50b** NLM/Science Source; **51** Pictorial Press Ltd/Alamy Stock Photo; **51bg** Public Domain; **52t** World History Archive/Alamy Stock Photo; **52b** © iStock.com/abzee; **53t** Public Domain; **53b** RGB Ventures/SuperStock/Alamy Stock Photo; **54** Sueddeutsche Zeitung Photo/Alamy Stock Photo; **54bg** Marina Sun/Shutterstock.com; **55l** magnetix/Shutterstock.com; **55r** EugenP/Shutterstock.com; **56l, 57t, 64** Leigh Prather/Shutterstock.com; **56r** tele52/Shutterstock.com; **57b** martan/Shutterstock.com; **58t** Carolyn Franks/Shutterstock.com; **58bl** hans.slegers/Shutterstock.com; **58br** VVZann/iStock/Thinkstock; **59** Vadiar/Shutterstock.com; **60–61bg** Darwin Brandis/Shutterstock.com; **63** TaraPatta/Shutterstock.com;

219t fotog/Getty Images; **203**bg d3sign/Getty Images; **203**r (whipped cream) Nattika/Shutterstock.com; **203**r (mayonnaise) stockcreations/Shutterstock.com; **203**r (paint) jannoon028/Shutterstock.com; **203**r (marshmallow) © iStock.com/subjug; **203**r (opal) Alexander Hoffmann /Shutterstock.com; **203**r (vase) Simon Curtis/Alamy Stock Photo; **204**t Andrey_Kuzmin/Shutterstock.com; **204**b © iStock.com/chictype; **205**tl Aleksandra Gigowska /Shutterstock.com; **205**tc SPL/Science Source; **205**tr F. JIMENEZ MECA/Shutterstock.com; **205**bl (coke) DenisMArt /Shutterstock.com; **205**bl (alcohol) © iStock.com/jfmdesign; **205**bl (gatorade) © iStock.com/EasyBuy4u; **205**br Olena Yakobchuk/Shutterstock.com; **206** Foodcollection RF/Getty Images; **208**t Martyn F. Chillmaid/Science Source; **208**b David GABIS/Alamy Stock Photo; **209**t Mariyana M/Shutterstock .com; **209**b Garsya/Shutterstock.com; **210–11** T.W. van Urk /Shutterstock.com; **210**i Turtle Rock Scientific/Science Source; **211**ti gritsalak karalak/Shutterstock.com; **211**bi udaix /Shutterstock.com; **212**, **219**b Turtle Rock Scientific/Science Source; **213** design56/Shutterstock.com; **214** walterericsy /Shutterstock.com; **215** Richard J Green/Science Source /Getty Images; **216** Adie Bush/Cultura RF/Getty Images; **217**t annick vanderschelden photography/Moment/Getty Images; **217**b (fudge) chris kolaczan/Shutterstock.com; **217**b (caramel) MaraZe/Shutterstock.com; **217**b (nougat) vandycan /Shutterstock.com; **217**b (toffee) © iStock.com/cislander; **218**t MaraZe/Shutterstock.com; **218**b imageBROKER /Alamy Stock Photo; **220**l Turtle Rock Scientific/Science Source; **220**cl Venturelli Luca/Shutterstock.com; **220**cr Turtle Rock Scientific/Science Source; **220**r Joe Belanger /Shutterstock.com; **221** Serhii Hrebeniuk/Shutterstock.com

CHAPTER 10

222 Gerry Ellis/Minden Pictures/Getty Images; **224–25**bg kubais/Shutterstock.com; **224** Dario Lo Presti/Shutterstock .com; **225**t, **226**b grmarc/Shutterstock.com; **225**c PHOTO RESEARCHERS, INC./Science Source/Getty Images; **225**b, **239**t traveliving/Shutterstock.com; **226–27**bg Fotaro1965 /Shutterstock.com; **226**t Unuchko Veronika/Shutterstock .com; **227**t Turtle Rock Scientific/Science Source; **227**b Anucha Tiemsom/Shutterstock.com; **230–31**, **239**c elenabsl /Shutterstock.com; **232** Olya Detry/Shutterstock.com; **233**t Turtle Rock Scientific/Science Source; **233**ti NMeM/Royal Photographic Society/SSPL/Getty Images; **233**cl GIPhoto Stock/Science Source; **233**cr Africa Studio/Shutterstock.com; **233**bl Long Bao/Shutterstock.com; **233**br Haluk Köhserli /iStock/Getty Images Plus/Getty Images; **234** MaraZe/Shutter stock.com; **235**l Lijuan Guo/Shutterstock.com; **235**r Sakarin Sawasdinaka/Shutterstock.com; **237** szefei/Shutterstock .com; **238** OBprod/Shutterstock.com; **239**b StefaniaArca /Shutterstock.com; **240**l TREVOR CLIFFORD PHOTOGRAPHY /Science Source; **240**r MARTYN F. CHILLMAID/Science Source

CHAPTER 11

244 kuri2000/iStock Getty Images Plus/Getty Images; **246–47**bg GoodStudio/Shutterstock.com; **246** Science Source /Getty Images; **247**t "Justus Sustermans - Portrait of Galileo Galilei, 1636"/Wikimedia Commons/Public Domain; **247**c Hulton Archive/Getty Images; **247**bl Ian Dagnall /Alamy Stock Photo; **247**br CERN/Science Source; **248–49**,

265t Nadya_Art/Shutterstock.com; **250–51** Alex Oakenman /Shutterstock.com; **252** DRogatnev/Shutterstock.com; **253**, **265**c coldsnowstorm/E+/Getty Images; **254–55** Xinhua /Alamy Stock Photo; **256** alazur/Shutterstock.com; **258** Hulton Deutsch/Corbis Historical/Getty Images; **259** all Macrovector/Shutterstock.com; **260** © iStock.com /mihtiander; **260**ti dnd_project/Shutterstock.com; **260**bi rzstudio/Shutterstock.com; **261** Joe Raedle/Getty Images; **263**, **265**b Rick Neves/Shutterstock.com; **266** Eric Gevaert /Shutterstock.com

CHAPTER 12

268 Benoist/Shutterstock.com; **270–71** 7Horses/Shutter stock.com; **271**i BlueRingMedia/Shutterstock.com; **272**i GingerArt/Shutterstock.com; **272–73**, **291**t Squared pixels/E+/Getty Images; **276**, **277** Wayne0216/Shutterstock .com; **277**li studioloco/Shutterstock.com; **277**ri wave breakmedia/Shutterstock.com; **277**c, ri ESB Professional /Shutterstock.com; **278–79**bg ckimo/Shutterstock.com; **278** miniwide/Shutterstock.com; **280**t, **281**t, **291**c Alena Che /Shutterstock.com; **280–81**bg Evgenia L/Shutterstock .com; **280**b, **282**i, **283**i NotionPic/Shutterstock.com; **281**b Good_Stock/Shutterstock.com; **282**, **283** D1min/Shutter stock.com; **282–83**bg PremiumArt/Shutterstock.com; **284** Graiki/Moment/Getty Images; **286**, **291**b Kekyalyaynen/Shut terstock.com; **287** IR Stone/Shutterstock.com; **289** gualtiero boffi/Shutterstock.com; **290** Darryl Leniuk/The Image Bank /Getty Images; **293** Buena Vista Images/The Image Bank /Getty Images

CHAPTER 13

294 NurPhoto/Getty Images; **296** AFP/Getty Images; **297**t, bl, **319**t Mascha Tace/Shutterstock.com; **297**br Good_Stock /Shutterstock.com; **298** Avalon_Studio/iStock/Getty Images Plus/Getty Images; **299** Jaren Kane/500px/Getty Images; **300–301** ProStockStudio/Shutterstock.com; **302**tl Stoked | Thinkstock/Media Bakery; **302**bl Konrad Wothe/Minden Pictures/Getty Images; **302**r ESB Basic/Shutterstock.com; **303** kali9/E+/Getty Images; **305** Kartinkin77/Shutterstock .com; **306** Monty Rakusen/Cultura/Getty Images; **306**i Serg64/Shutterstock.com; **307**t, **319**b Fabrice LEROUGE /ONOKY/Getty Images; **307**ti Pirozhehnik/iStock/Getty Images Plus/Getty Images; **307**c loco75/iStock/Getty Images Plus /Getty Images; **307**b Image Source/Getty Images; **308**t, **308**bl GIPhotoStock/Science Source; **308**bcl Sam72/Shutterstock .com; **308**bcr dei-sin/iStock/Getty Images Plus/Getty Images; **308**br T.Dallas/Shutterstock.com; **309** iMoved Studio /Shutterstock.com; **313**, **320** kali9/E+/Getty Images; **315**i Ilya Bolotov/Shutterstock.com; **314–15** Patryce Bak /The Image Bank/Getty Images; **316–17** AlexSava/iStock /Getty Images Plus/Getty Images; **318**l Ian Hooton /Science Source; **318**r Ted Kinsman/Science Source

CHAPTER 14

322 Danita Delimont/Gallo Images/Getty Images; **322**i REUTERS/Jim Drury; **324–25** Gabe Rogel/Aurora Creative /Getty Images; **326**, **345** Mascha Tace/Shutterstock.com; **327**l OZaiachin/Shutterstock.com; **327**r Vadym Zaitsev /Shutterstock.com; **328**, **344**t YellowPaul/Shutterstock.com;

329 Mykhailo Bokovan/Shutterstock.com; 331 Westend61
/Getty Images; 332–34, 344c Barcroft/Barcroft Media/Getty
Images; 335, 336t CW craftsman/Shutterstock.com; 336b
Jiw Ingka/Shutterstock.com; 337 nattapon supanawan
/Shutterstock.com; 338–39 Daniel Schoenen/Getty Images;
340–41bg city hunter/Shutterstock.com; 340t, 344b Jacob
Maentz/Corbis Documentary/Getty Images; 340c "Adam Beck
Complex"/Ontario Power Generation/Wikimedia Commons
/CC BY 2.0; 340b Puripat Lertpunyaroj/Shutterstock.com;
341t RoschetzkyIstockPhoto/iStock/Getty Images; 341b
Doug McLean/Shutterstock.com; 342–43bg Gary Saxe/Shut
terstock.com; 342t "Elwha Dam"/DancingBear/Wikimedia
Commons/Public Domain; 342c "Elwha Dam with hole"/NPS
/Wikimedia Commons/Public Domain; 342b "Elwha Dam
Finished May2013"/Zandcee/Wikimedia Commons/CC BY-SA
3.0; 343l Anton Prado PHOTO/Shutterstock.com; 343r
© iStock.com/pepifoto

CHAPTER 15

346 TheJim999/Shutterstock.com; 348–49 Andriy Blokhin
/Shutterstock.com; 349t, 368t GIPhotoStock/Science Source;
349bi David R. Frazier Photolibrary, Inc./Alamy Stock Photo;
350 Scala/Art Resource, NY; 351 Fouad A. Saad/Shutterstock
.com; 352 Ekaterina Pokrovsky/Shutterstock.com; 354t
Heymo/Shutterstock.com; 354bl bbeltman/E+/Getty Images;
354br Charles D. Winters/Science Source; 355 bogdan
kosanovic/E+/Getty Images; 356–57, 368c Pavel L Photo and
video/Shutterstock.com; 357ti Martyn F. Chillmaid/Science
Source; 357bi Elmar Krenkel/imageBROKE/Getty Images; 358
Creative-Family/iStock/Getty Images Plus/Getty Images; 360
Andrei Nekrassov/Shutterstock.com; 360–61bg a_Taiga
/iStock/Getty Images Plus/Getty Images; 362, 363t Turtle
Rock Scientific/Science Source; 363bl GIPhotoStock/Science
Source; 363bc BlueRingMedia/Shutterstock.com; 363br no_
limit_pictures/E+/Getty Images; 364, 368b Gary Saxe
/Shutterstock.com; 366–67bg gerenme/E+/Getty Images;
366 Kateryna Kon/Shutterstock.com; 367 NASA/JPL-Caltech

CHAPTER 16

370 JC Patricio/Moment/Getty Images; 372 Lissandra Melo
/Shutterstock.com; 374t, 390t Valentyn Hontovyy/Shutter
stock.com; 374bl Berke/Shutterstock.com; 374br Chutima
Chaochaiya/Shutterstock.com; 375 Cavan Images/Alamy
Stock Photo; 376 sub job/Shutterstock.com; 377 Image
Source/Getty Images; 378 Jonathan Bird/Photolibrary/Getty
Images; 379 Martyn F. Chillmaid/Science Source; 380, 381
LineTale/Shutterstock.com; 382 Doug Allan/Science Source;
383, 390c Take Photo/Shutterstock.com; 384 age fotostock
/Alamy Stock Photo; 387 Yevhenii Dorofieiev/Unitone
Vector/iStock/Getty Images Plus/Getty Images; 388–89,
390b Elenathewise/iStock/Getty Images Plus/Getty Images;
389i Mikhail Japaridze/TASS/Getty Images; 391 "Aircraft
venturi 1" by YSSYguy/Wikimedia Commons/Public Domain

CHAPTER 17

394 Ron Dahlquist/Perspectives/Getty Images Plus/Getty
Images; 394i David Livingston/Getty Images Entertainment
/Getty Images; 396–97 mikroman6/Moment/Getty Images;
397i, 416t Frank Zullo/Science Source; 400–401 NH

/Shutterstock.com; 400i Sunwand24/Shutterstock.com; 401i
kvsan/Shutterstock.com; 402 Andy Roberts/Caiaimage
/Getty Images; 403, 416c Rod Jones/Alamy Stock Photo;
404–5 Kite_rin/Shutterstock.com; 405i Image by Chris
Winsor/Moment/Getty Images; 410–11bg TierneyMJ
/Shutterstock.com; 410t Andrew Mayovskyy/Shutterstock
.com; 410b Beth Van Trees/Shutterstock.com; 411tl Olga
Popova/Shutterstock.com; 411tr, 412cl, 416b GIPhoto
Stock/Science Source; 411bl Frans Lanting/MINT Images
/Science Source; 411br Tefi/Shutterstock.com; 412–13bg
TierneyMJ/Shutterstock.com; 412t Esa Hiltula/Alamy Stock
Photo; 412cr tamara_kulikova/iStock/Getty Images Plus
/Getty Images; 412b dropStock/Shutterstock.com; 413t Ted
Kinsman/Science Source; 413b Nicholas Toh/Shutterstock
.com; 414 AP Photo; 415 VectorShow/Shutterstock.com

CHAPTER 18

418, 437t SVSimagery/Shutterstock.com; 418i "Chuck Yeager
X-1 (color)" by Jack Ridley/Wikimedia Commons/Public
Domain; 420 Syda Productions/Shutterstock.com; 421t
ilbusca/DigitalVision Vectors/Getty Images; 421b Classic
Image/Alamy Stock Photo; 422–23b granata68/Shutterstock
.com; 423t trgrowth/Shutterstock.com; 424, 437b Speed
Kingz/Shutterstock.com; 425l, 438bl Vecton/Shutterstock
.com; 425r stockshoppe/Shutterstock.com; 426–27 Avpics
/Alamy Stock Photo; 427i, 438br Sedova Elena/Shutterstock
.com; 428t commerceandculturestock/Moment/Getty
Images; 428–29b Roman Babakin/Shutterstock.com; 429ti, bi
Brian A Jackson/Shutterstock.com; 430t Jose Luis Pelaez Inc
/DigitalVision/Getty Images; 430b "Elektrodynamisk-
högtalare" by Svjo/Wikimedia Commons/CC BY-SA 3.0; 431
Paul Colley/iStock/Getty Images Plus/Getty Images; 432–33,
438t bearacreative/Shutterstock.com; 432li Golden Sikorka
/Shutterstock.com; 432ri U.S. Navy; 434tl Christian Beirle
González/Moment Select/Getty Images Plus/Getty Images;
434bl Public Domain; 434r Richman Photo/Shutterstock
.com; 435t Sidekick/E+/Getty Images; 435li BorupFoto
/iStock/Getty Images Plus/Getty Images; 435ri "Ultrasonic
pipeline test"/Davidmack/Wikimedia Commons/Public
Domain; 435b Monty Rakusen/Cultura/Getty Images;
439t Horatiu Bota/Shutterstock.com; 439b BLOOMimage
/Getty Images

CHAPTER 19

440 Barcroft Media/Getty Images; 442, 467t Ted Kinsman
/Science Source; 445 Kan Taengnuanjan/EyeEm/Getty
Images; 446–47bg Comaniciu Dan/Shutterstock.com;
446b Dave King/Dorling Kindersley/Science Source; 447b
Sonsedska Yuliia/Shutterstock.com; 448t vanbeets/iStock
/Getty Images Plus/Getty Images; 448b Dave King/Dorling
Kindersley/Getty Images Plus/Getty Images; 449 SCIENCE
PHOTO LIBRARY/Science Source; 450–51 Sean Pavone
/Shutterstock.com; 451i, 454t VectorShow/Shutterstock
.com; 452t, 453b Vasilius/Shutterstock.com; 452b, 467c
Science Source/Science Source; 453t Laborant/Shutterstock
.com; 454–55b 9dream studio/Shutterstock.com; 457 kadm
/iStock/Getty Images Plus/Getty Images; 458t In-Finity
/Shutterstock.com; 458ct sumkinn/Shutterstock.com; 458cb
Scrap4vec/Shutterstock.com; 458b fullvector/Shutterstock
.com; 459t RedlineVector/Shutterstock.com; 459c Alexandr III

/Shutterstock.com; **459**b Webspark/Shutterstock.com; **460–61**bg stockchairatgfx/Shutterstock.com; **460**, **461**, **467**b haryigit/Shutterstock.com; **462–63** Tim Allen/iStock/Getty Images Plus/Getty Images; **464–65**bg Mr Twister/Shutterstock.com; **464**t NOPPHARAT STUDIO 969/Shutterstock.com; **464**b Yentafern/Shutterstock.com; **465**t PitukTV/Shutterstock.com; **465**c ony Freeman/Science Source; **465**b chonticha stocker/Shutterstock.com; **466**t Bettmann/Getty Images; **466**b Nerthuz/Shutterstock.com; **469** "Police issue X26 TASER-white"/Junglecat/Wikimedia Commons /CC BY-SA 3.0

CHAPTER 20

470 OLI SCARFF/AFP/Getty Images; **472** New Africa/Shutterstock.com; **472**i (magnet) New Africa/Shutterstock.com; **472**i (William Gilbert) "William Gilbert 45626i"/Wikimedia Commons/Public Domain; **472**i (Petrus Peregrinus) INTER FOTO/Alamy Stock Photo; **472**i (globe) ixpert/Shutterstock .com; **472**i, **483**t (compass) Alex Staroseltsev/Shutterstock .com; **473**t CORDELIA MOLLOY/Science Photo Library/Getty Images; **473**c pippeeContributor/Shutterstock.com; **473**b VectorMine/Shutterstock.com; **474** hoch2wo/Alamy Stock Photo; **475** Peter Sobolev/Shutterstock.com; **478–79**, **483**c Bernd Mellmann/Alamy Stock Photo; **478**i BanksPhotos /iStock/Getty Images Plus/Getty Images; **479**ti stockphotograf/Shutterstock.com; **479**bi Studio GL/Shutterstock.com; **480** leezsnow/E+/Getty Images; **481**l, **483**b Sugrit Jiranarak /Shutterstock.com; **481**r Wpadington/Shutterstock.com; **482** suradech sribuanoy/Shutterstock.com; **485**l Chronicle/Alamy Stock Photo; **485**r "Tesla3" by Napoleon Sarony/Wikimedia Commons/Public Domain; **485**i (toaster) Makc/Shutter stock.com; **485**i (flashlight) paradesign/Shutterstock.com

CHAPTER 21

486 Kevin Foy/Alamy Stock Photo; **488–89** JS`s favorite things/Moment Open/Getty Images; **489**i, **500**b Vector Mine/Shutterstock.com; **490**t Bettmann/Getty Images; **490**b "Michelson speed of light measurement 1930" by H. H. Dunn/Wikimedia Commons/Public Domain; **492**, **500**t Carlos Fernandez/Moment/Getty Images; **493**l INTERFOTO /Alamy Stock Photo; **493**r GL Archive/Alamy Stock Photo; **494** S.Borisov/Shutterstock.com; **495** Richard Hutchings /Science Source; **496–97** VectorMine/Shutterstock.com; **498**t Designua/Shutterstock.com; **498**b Zapp2Photo /Shutterstock.com; **501** Ikhsan Yohanda/iStock/Getty Images Plus/Getty Images; **501**i Ted Kinsman/Science Source

CHAPTER 22

502 WIN-Initiative/Getty Images Plus/Getty Images; **504**i VectorMine/Shutterstock.com; **504–5** lazyllama/Shutterstock .com; **505**tr, **505**ctr H.waritha/Shutterstock.com; **505**cbr fishmonger/Shutterstock.com; **505**br, **518**t Katia Vittoria Di Gioia/EyeEm/Getty Images; **506** Lane Oatey/Getty Images; **507** Nella/Shutterstock.com; **508**tl Sakurra/Shutterstock .com; **508**tr Alexander Tsiaras/Science Source; **508**bl Andrey_ Popov/Shutterstock.com; **508**br David Tadevosian/Shutter stock.com; **509** Bence Katona-Kovacs/Shutterstock.com; **510** chirajuti/Shutterstock.com; **511**, **518**b New Africa /Shutterstock.com; **512** tompics/iStock/Getty Images Plus

/Getty Images; **513**l Josfor/Getty Images/iStock/Getty Images Plus/Getty Images; **513**r Pressmaster/Shutterstock .com; **514** waldru/Shutterstock.com; **515**t Ian Cuming/Ikon Images/Getty Images Plus/Getty Images; **515**bl GIPhoto Stock/Science Source; **515**br NotionPic/Shutterstock .com; **516** Sergey Merkulov/Shutterstock.com; **517**, **519** andresr/E+/Getty Images; **520** Damsea/Shutterstock .com; **521** pixelfit/E+/Getty Images

APPENDIXES

522 "The sun is an MHD system that is not well understood-2013-04-9 14-29"/NASA/SDO/Wikimedia Commons /Public Domain; **523** Becart/iStock/Getty Images Plus /Getty Images; **524** Peter Dazeley/The Image Bank/Getty Images Plus/Getty Images; **531** all Kinjalben Apoorva Patel /DigitalVision Vectors/Getty Images; **536**l, r Nadya_Art /Shutterstock.com; **537** Albert Russ/Shutterstock.com

PERIODIC TABLE OF THE ELEMENTS

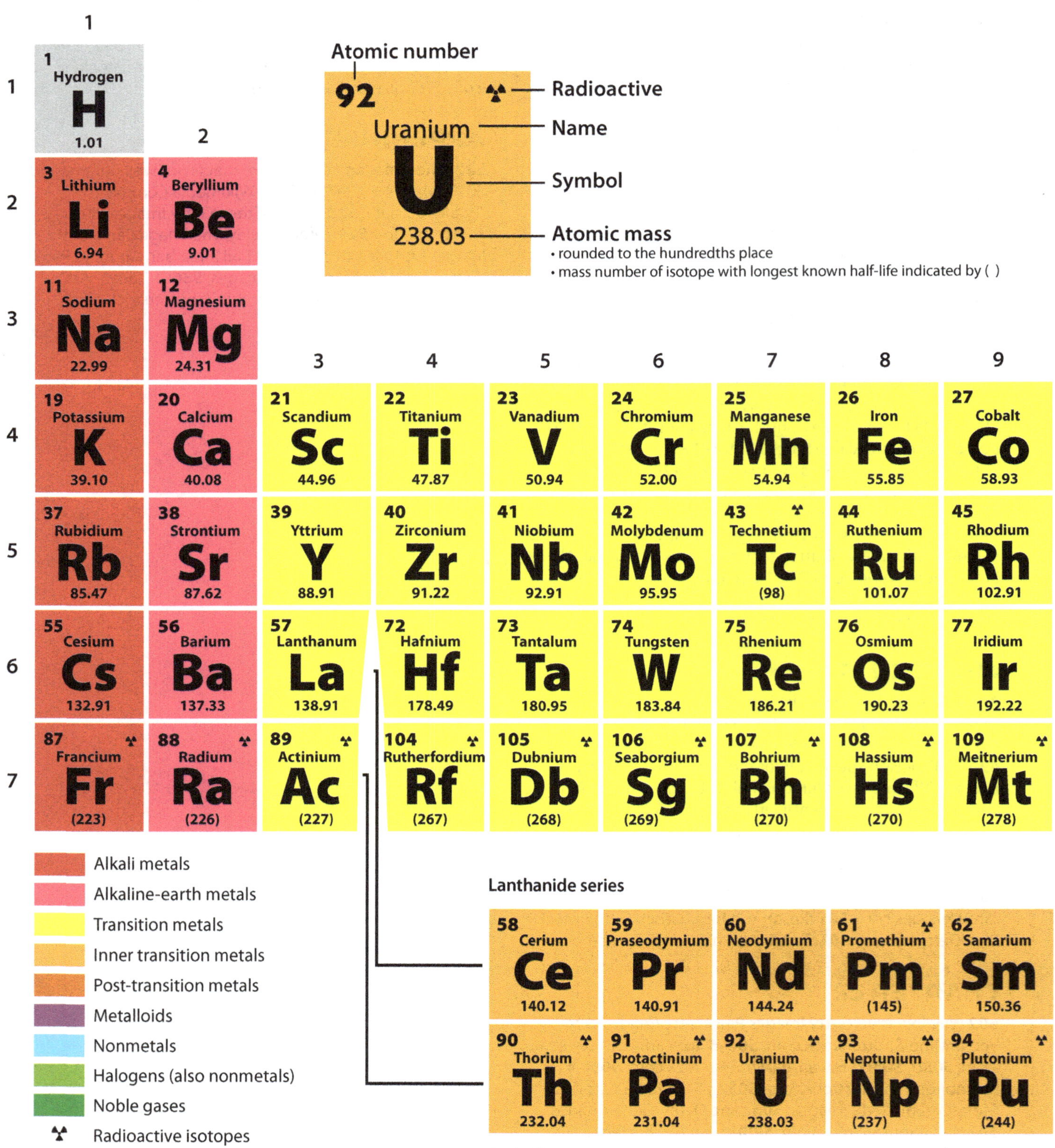

			13	14	15	16	17	18
								2 Helium **He** 4.00
			5 Boron **B** 10.81	**6** Carbon **C** 12.01	**7** Nitrogen **N** 14.01	**8** Oxygen **O** 16.00	**9** Fluorine **F** 19.00	**10** Neon **Ne** 20.18
10	11	12	**13** Aluminum **Al** 26.98	**14** Silicon **Si** 28.09	**15** Phosphorus **P** 30.97	**16** Sulfur **S** 32.06	**17** Chlorine **Cl** 35.45	**18** Argon **Ar** 39.95
28 Nickel **Ni** 58.69	**29** Copper **Cu** 63.55	**30** Zinc **Zn** 65.38	**31** Gallium **Ga** 69.72	**32** Germanium **Ge** 72.63	**33** Arsenic **As** 74.92	**34** Selenium **Se** 78.97	**35** Bromine **Br** 79.90	**36** Krypton **Kr** 83.80
46 Palladium **Pd** 106.42	**47** Silver **Ag** 107.87	**48** Cadmium **Cd** 112.41	**49** Indium **In** 114.82	**50** Tin **Sn** 118.71	**51** Antimony **Sb** 121.76	**52** Tellurium **Te** 127.60	**53** Iodine **I** 126.90	**54** Xenon **Xe** 131.29
78 Platinum **Pt** 195.08	**79** Gold **Au** 196.97	**80** Mercury **Hg** 200.59	**81** Thallium **Tl** 204.38	**82** Lead **Pb** 207.24	**83** Bismuth **Bi** 208.98	**84** Polonium **Po** (209)	**85** Astatine **At** (210)	**86** Radon **Rn** (222)
110 Darmstadtium **Ds** (281)	**111** Roentgenium **Rg** (282)	**112** Copernicium **Cn** (285)	**113** Nihonium **Nh** (286)	**114** Flerovium **Fl** (289)	**115** Moscovium **Mc** (290)	**116** Livermorium **Lv** (293)	**117** Tennessine **Ts** (294)	**118** Oganesson **Og** (294)

63 Europium **Eu** 151.96	**64** Gadolinium **Gd** 157.25	**65** Terbium **Tb** 158.93	**66** Dysprosium **Dy** 162.50	**67** Holmium **Ho** 164.93	**68** Erbium **Er** 167.26	**69** Thulium **Tm** 168.93	**70** Ytterbium **Yb** 173.05	**71** Lutetium **Lu** 174.97
95 Americium **Am** (243)	**96** Curium **Cm** (247)	**97** Berkelium **Bk** (247)	**98** Californium **Cf** (251)	**99** Einsteinium **Es** (252)	**100** Fermium **Fm** (257)	**101** Mendelevium **Md** (258)	**102** Nobelium **No** (259)	**103** Lawrencium **Lr** (266)